SCIENCE 3

Teacher Edition

Fifth Edition

bju press®

Greenville, South Carolina

Note: The fact that materials produced by other publishers may be referred to in this volume does not constitute an endorsement of the content or theological position of materials produced by such publishers. Any references and ancillary materials are listed as an aid to the student or the teacher and in an attempt to maintain the accepted academic standards of the publishing industry.

SCIENCE 3 Teacher Edition
Fifth Edition

Coordinating Writer
Leah R. Thomas, MEd

Writers
Serena Minnick Collins
Nancy Gann Ellis, MEd
Debra Harrold White

Biblical Worldview
Bryan Smith, PhD
Tyler Trometer, MDiv

Academic Oversight
Jeff Heath, EdD
Rachel Santopietro, MEd

Project Editor
Joanna Lynch, MA

Consultant
Milton Ashley, MS

Designers
Emily Heinz
Briseydi Rouse

Page Layout
Jennifer Stuhl

Cover Designer
Emily Heinz

Cover Illustrators
Craig Oesterling
Dana Thompson

Art & Design Facilitator
Jim Frasier

Illustrators
Emily Heinz
Dana Thompson

Project Coordinators
Christopher Daniels
Abby Ray

Permissions
Hannah Labadorf
Ashleigh Schieber
Elizabeth Walker

Photo credits begin on page A135.

The text for this book is set in ASAP by Omnibus-Type, Adobe Minion Pro, Adobe Myriad Pro, Noto Sans by Google, Nunito by Vernon Adams, Simian Display Chimpanzee by House Industries, Wingdings, and Wingdings 2

© 2021 BJU Press
Greenville, South Carolina 29609

Fourth Edition © 2016 BJU Press
First Edition © 1976 BJU Press

Printed in the United States of America

ISBN: 978-1-62856-865-3

15 14 13 12 11 10 9 8 7 6 5 4 3 2

Contents

To the Teacher
Biblical Worldview Shaping in SCIENCE 3 BWS

Can third graders begin to form a biblical view of science? Yes! Using the four biblical worldview themes below, students will begin learning to view all of science biblically.
SCIENCE 3, 5th Edition, answers the questions posed below to help students begin to think about nature and science the way that God intends.

Where did our world come from and how has it changed? (History of Nature)

God created the world in six days. The biblical Flood drastically reshaped the surface of the earth and produced millions of fossils. But modern science believes the earth and the life on it developed without God over millions of years of evolution and erosion.
How can a Christian explain the formation of rocks?
How did so many fossils form if the earth is young?

CH 1	CH 2	CH 3
EV	E, EV	E, E, E, EV, F, F

CH 5	CH 7	CH 8	CH 10
EV	EV	E	F

Why do things work together in nature? (Design in Nature)

God designed things in the world to work together. From the smallest cells in the human body to the vast solar system, God designed forces in nature to work together. Naturalistic science, however, says that nature only appears to be designed.
How can a Christian explain why cells, tissues, organs, and systems work together in the human body?
Is death a natural part of the life cycle?
Are electrical and magnetic forces inevitable physical forces in our universe?

CH 1	CH 2	CH 3
E, F	E	E, EV

CH 4	CH 5	CH 6	CH 7	CH 8	CH 10
E, E	F	E, F, F	E, E, E, E, E, E, F	E, E	E

Why are people important?
(Importance of Humans)

Humans are special because they are made in God's image. That is why we study the human body and why science constantly seeks to better people's lives. Naturalistic science believes that humans are merely highly evolved animals. *How should a Christian explain the shared characteristics of humans and animals?*
What is the basis for ethical decisions that affect humans?
How does a biblical understanding of humans make the study of severe weather important?

CH 2	CH 3
E	E, E

CH 4	CH 5	CH 7	CH 9		CH 10	CH 11
E, E, E, E	E	F	E, E		E	E, E

Why is science important?
(Purpose of Science)

Science is the study of God's world. God created the order in nature. Science is an important tool for obeying God's first command to rule over this world. We learn about nature so that we can take care of it. But much of the scientific community views science as more than just a tool. Science becomes the final authority for wisdom and interpreting the world.
How should a Christian view the use of force and motion for doing work?
Why is it important for Christians to understand the modeling nature of science?

CH 3		CH 6	CH 7	CH 8		CH 11
E		E	E	E		F

Above are the biblical worldview themes that are important for third grade science students to know.
Early in the course the students will recall and explain these themes. However, as these themes are repeated, the students will evaluate ideas within them, formulate a biblical understanding of them, and apply what they have learned about these themes to real-life situations. High levels of internalization are expected wherever the students are required to apply their learning.

KEY

R: Recall biblical teaching/science details

E: Explain biblical teaching/science concepts

EV: Evaluate controversial concepts

F: Formulate a biblical understanding of a controversial concept

A: Apply this biblical understanding to life

■ Unit 1   ■ Unit 2   ■ Unit 3

Building Academic Rigor with SCIENCE 3

1

Desired Learning Outcomes
What do I want students to learn?

- Reading skills in informational scientific texts
- 21st century skills—collaboration, creativity, critical thinking, problem solving, technology literacy
- Facility with scientific tools
- Scientific inquiry skills
- Engineering design process
- Reading and interpreting tables, charts, and graphs
- Biblical worldview development regarding the history of nature, design in nature, the importance of humans, and the purpose of science

2

Teaching and Learning Supports
How will I help students learn?

- Anticipate, evoke, and overcome common student misconceptions
- STEM activities
- STEM career features
- Differentiated instruction
- Visual analysis
- Active learning strategies

3

Evidence of Understanding
How do I evaluate student learning?

- Frequent and varied preassessments and formative assessments
- High-level questioning
- Graphic organizers, tables, charts, graphs
- Rubrics

A Panorama of Academic Rigor

The Learning Environment

Learning happens in a context. Physical and social environments create this context. For students to be stimulated and engaged and to perform at high levels, they need an environment that connects with and shapes their interests and values. Students need a flexible environment that adapts to remedy their deficiencies and capitalize on their proficiencies. They need an interactive environment that intrinsically motivates them by giving them structure, freedom, and choices in learning. They need a safe environment that invites questioning, risk-taking, and honest discussion. Most importantly, students need positive relationships with mentors and peers and one-on-one time with caring, responsible instructors and parents. This kind of environment does not happen by accident—an educational climate like this must be crafted.

The Teacher's Role

The teacher is the person who crafts this learning environment. The teacher creates a learning community with high educational expectations. The teacher ignites interest and passion to help students persevere in challenging educational tasks and inspires them by providing models of learning. The teacher also instructs, supports, critiques, and praises the efforts of the growing learners. He lays a foundation for lifelong learning. Then the teacher helps the students transfer their knowledge, understanding, and skills to life with the purpose of shaping the students' worldview.

The Learning Experience

The textbook opens doors to the world of the learning experience. It captures student interest and significantly contributes to the educational environment. Educational and technological resources enable the teacher to develop knowledge, understanding, and skills in a scaffolded sequence to build a foundation that students can transfer to life. Teachers use educational tools to craft relevant, authentic learning experiences that develop creativity and problem-solving skills in a variety of contexts. These kinds of learning experiences allow students to take responsibility for their learning and discover the joy of serving God with their vocational knowledge and skills.

Instructional Materials

1 Student Edition

The *Science 3 Student Edition* provides grade-appropriate information through text, diagrams, and annotated photographs and illustrations. The *Student Edition* includes a Big Question at the start of each chapter, plus interest boxes and Quick Check questions. It also provides a springboard for each Investigation, Inquiry, Exploration, and STEM activity.

2 Activities

The *Science 3 Activities* provides a variety of pages to aid understanding. Study Guide pages provide a systematic review of concepts. Investigation, Inquiry, Exploration, and STEM activity lessons apply the scientific method, reinforce the basic science process skills, and use the engineering design process to solve real-life problems. Other pages, including graphic organizers, help reinforce and expand on the concepts taught in the lessons.

Answers are provided in the *Science 3 Activities Answer Key*, which is available separately.

3 Teacher Edition

The *Science 3 Teacher Edition* contains 90 lessons. Lessons include additional background information, helps, preparing ahead information, and additional science activities for differentiated instruction.

The *Teacher Edition* also includes useful information about Biblical Worldview Shaping, Academic Rigor, science inquiry skills, the management of activities, and grading. Information about sharing the gospel with children is available in the Teacher Resources in the back of the *Teacher Edition*.

4 Assessments

The *Science 3 Assessments* includes eleven chapter tests. Tests include objective items, the use and labeling of diagrams, and application questions. *Assessments* also includes two quizzes for each chapter as well as a tailored rubric for each Investigation, Inquiry, Exploration, and STEM activity. Test answers can be found in the *Science 3 Assessments Key*, which is available separately.

New to This Edition

If you have used previous editions of BJU Press *Science 3*, you may notice some changes made in the 5th Edition.

The *Science 3* **Student Edition** is divided into three units: Let's Connect with Earth and Space Science, Let's Connect with Life Science, and Let's Connect with Physical Science. Lesson numbers appear on the first page of each new lesson. Each chapter begins with a Big Question to focus attention on a main concept. There are pages to engage the students before each Investigation, Inquiry, Exploration, and STEM activity. Investigation, Inquiry, and Exploration lessons identify the Inquiry Skills (formerly Process Skills) that are emphasized. STEM activities and STEM career pages are designed to build 21st-century skills and to develop interest in science, technology, engineering, and math by using the engineering design process. The Glossary and Index have been expanded to include more than just the boldface vocabulary terms. Some chapters have been added to the *Student Edition*, while others have been revised or removed to better meet the course objectives.

The *Science 3* **Activities** book has additional graphic organizers and activities. STEM activities that follow the engineering design process have been added. Six foundational Inquiry Skills (formerly Process Skills) are emphasized in the Investigation and Exploration lessons to better prepare the students for the upper grades, while the other six Inquiry Skills are integrated into these lessons. The Investigation and Exploration pages include the Materials and Procedure lists with check boxes to aid organization. Many pages include worldview application questions.

The **Teacher Edition** identifies the *Science 3* objectives. The Biblical Worldview Shaping objectives throughout are identified with a **BWS** icon. In addition, an overview of Biblical Worldview Shaping, Academic Rigor, and the teaching cycle have been included in the front of the book. The activities pages for all Investigation, Inquiry, Exploration, and STEM activities have been added to the *Teacher Edition*. A "Differentiated Instruction" folder tab identifies activities for remediation and enrichment. In addition, a "Preparing Ahead" folder tab has been added to identify out-of-the-ordinary materials that will be needed for a future activity. Space has been added at the end of each chapter for notes. Some of the helps that were previously found in the front of the *Teacher Edition* have been moved to the "Getting Started" section in the Teacher Resources in the back.

The *Science 3* **Assessments** includes chapter tests, quizzes, and rubrics. The eleven chapter tests have been revised to align with the updated material on the Study Guide pages in the *Activities* book. Two quizzes are provided for each chapter. Each Investigation, Inquiry, Exploration, and STEM activity has a tailored rubric.

TeacherToolsOnline.com includes many helpful resources. Materials Checklists, Instructional Aids, and Visuals are located at TeacherToolsOnline.com and in the back of this *Teacher Edition*.

SCIENCE 3 Product Objectives

- Analyze patterns and cycles observed in earth and space science, life science, and physical science.

- Analyze the characteristics, structures, functions, reproduction, and ecosystems of living things.

- Analyze the properties of matter and the forces and interactions of motion.

- Apply scientific knowledge through relevant discussions, active participation, and modeling.

- Apply foundational inquiry skills to formulate orderly approaches to real-world problem solving.

- Apply the Bible's teaching regarding the history of nature, design in nature, importance of humans, and the purpose of science.

Teaching Cycle

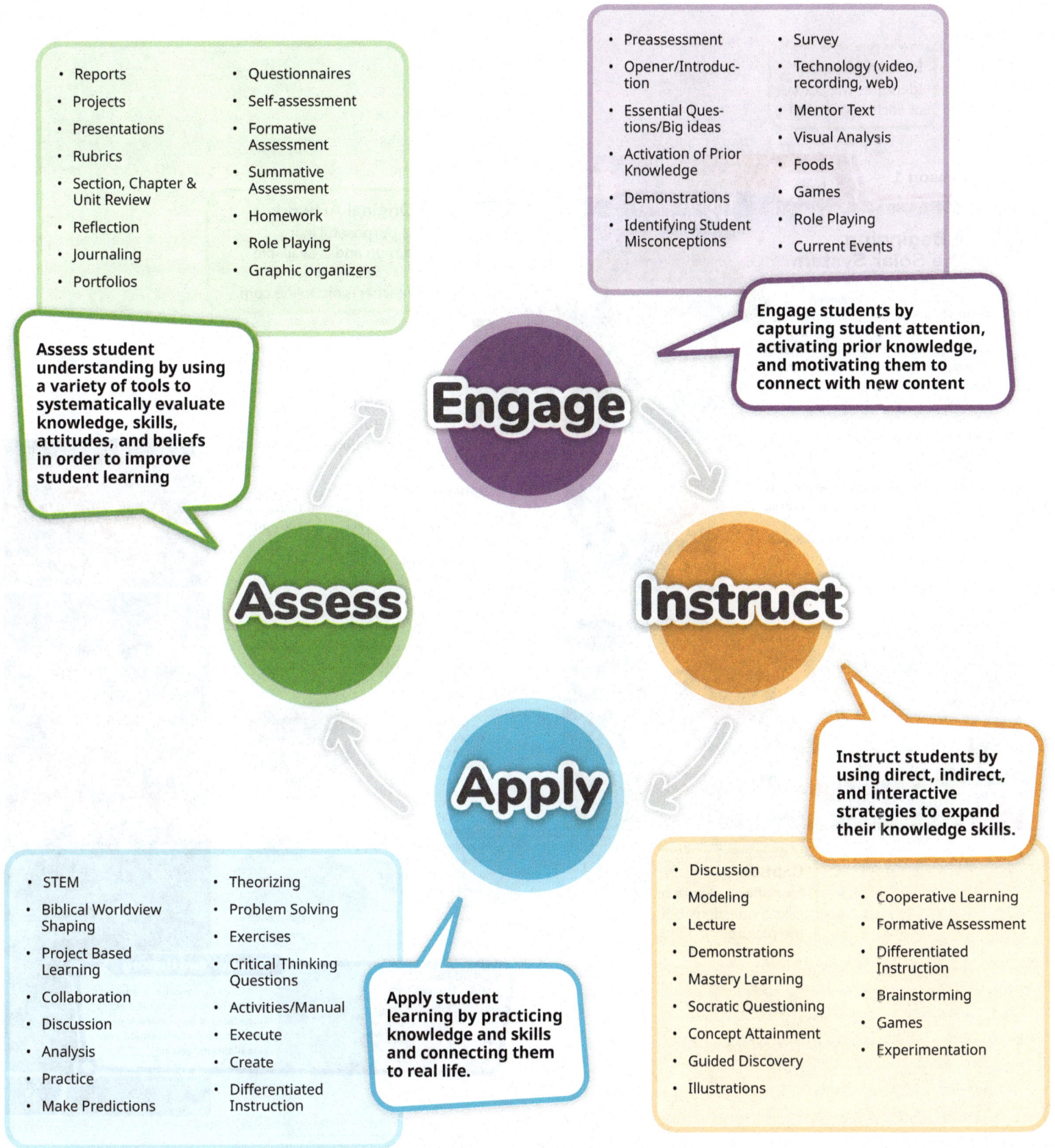

Assess

- Reports
- Projects
- Presentations
- Rubrics
- Section, Chapter & Unit Review
- Reflection
- Journaling
- Portfolios
- Questionnaires
- Self-assessment
- Formative Assessment
- Summative Assessment
- Homework
- Role Playing
- Graphic organizers

Assess student understanding by using a variety of tools to systematically evaluate knowledge, skills, attitudes, and beliefs in order to improve student learning

Engage

- Preassessment
- Opener/Introduction
- Essential Questions/Big ideas
- Activation of Prior Knowledge
- Demonstrations
- Identifying Student Misconceptions
- Survey
- Technology (video, recording, web)
- Mentor Text
- Visual Analysis
- Foods
- Games
- Role Playing
- Current Events

Engage students by capturing student attention, activating prior knowledge, and motivating them to connect with new content

Instruct

Instruct students by using direct, indirect, and interactive strategies to expand their knowledge skills.

- Discussion
- Modeling
- Lecture
- Demonstrations
- Mastery Learning
- Socratic Questioning
- Concept Attainment
- Guided Discovery
- Illustrations
- Cooperative Learning
- Formative Assessment
- Differentiated Instruction
- Brainstorming
- Games
- Experimentation

Apply

Apply student learning by practicing knowledge and skills and connecting them to real life.

- STEM
- Biblical Worldview Shaping
- Project Based Learning
- Collaboration
- Discussion
- Analysis
- Practice
- Make Predictions
- Theorizing
- Problem Solving
- Exercises
- Critical Thinking Questions
- Activities/Manual
- Execute
- Create
- Differentiated Instruction

Student Edition Features

Lesson Number
Identifies the start of each new lesson.

Lesson 1

The Beginning of the Solar System

How did the solar system begin? Not everyone agrees about its beginning. People who look at the solar system can think about it in different ways. The way a person thinks about the world is a **worldview**. It guides what people believe and how they live. Your worldview comes from a big story about the world.

One example of worldview is a biblical worldview. A *biblical worldview* is the belief that the story of the Bible is true. Those with a biblical worldview use the Bible to help them understand the solar system. They believe the Genesis 1 account of the six days of Creation. Nothing was created without God.

Many people believe that Earth and space formed by chance without God. They believe that everything has changed many, many times. And these changes happened slowly over millions of years.

Day 1
day and night

The Six Days of Creation

Day 2
ocean and atmosphere

On which day of Creation did God create the atmosphere?

Day 3
grass, herbs, and trees

Day 4
sun, moon, and stars

Day 5
whales and winged fowl

Day 6
cattle and people

What is a worldview?

4

Original Artwork
Is purposeful in its design and is available for classroom use at TeacherToolsOnline.com.

Vocabulary
Identifies boldface terms that students need to know.

Igneous Rock

Igneous rock forms when melted rock cools. The melted rock comes from deep in the earth. There the rock is a very hot liquid.

A *volcano* is an opening in the earth that allows melted rock to come to the surface. Sometimes the melted rock may erupt, or come out quickly. The melted rock that comes out of a volcano is called *lava*.

Lava cools and hardens. When lava cools very quickly, it forms rocks that are smooth. When it cools slowly, it forms rocks that are rough.

Obsidian is a smooth igneous rock. It is so smooth that it looks like glass.

obsidian

Granite forms when melted rock cools slowly. It is often speckled. It may have shiny crystal spots. It has a rough surface. Some mountains are made of granite.

granite

Erta Ale, Ethiopia

Volcán de Fuego, Guatemala

Science and History

Mount Rushmore is a famous granite mountain in South Dakota. Huge faces of four United States presidents have been carved in the side of this mountain.

63

Caption
Identifies a picture or asks a question about the picture.

Interest Box
Provides extra information about a person or topic.

INVESTIGATION

Lessons 85-86

Magnetism and Matter

Magnetism can pass through matter. It can pull on a metal object even through matter. Magnetic force can move an object. The magnet and the object do not have to touch for the object to move.

In this Investigation, you will predict how the thickness of matter affects magnetism.

Inquiry Skills
- Predict
- Observe
- Infer

Let's inquire into . . .
Magnetism and Matter

Repeat the Investigation with magnets having different magnetism.

In this Inquiry, you will predict how the strength of magnetic force affects the movement of an object through matter.

257

A Way to Reproduce

Fish and most amphibians. What is different about a reptile laying eggs than a fish or an amphibian? Some reptiles dig a hole in sand or soil to bury their eggs. Others make a nest in grasses or hide their eggs in rotting logs. A sea turtle might swim thousands of miles to lay its eggs on the same beach every year.

A reptile can live a long time. Some live as long as most people do. Giant tortoises and alligators can live more than 80 years.

Life Cycle of Reptiles

Most reptiles lay several eggs with a leathery shell at one time. Some lay 20 or 30 eggs in their nests. Others lay more than 100 eggs at a time.

adult

egg

hatchling

young

A *hatchling* is an animal that hatches from an egg. A reptile hatchling breathes with lungs as soon as it is born.

The young reptile looks like its adult parents but is smaller. It does not change form. It does not grow new parts later. It just gets bigger and stronger.

How do most reptiles reproduce?

135

Activities Features

Looking Ahead

Name _____

Write an *X* in the first column if you have never heard the word before.
Write an *X* in the second column if you have heard the word before.
In the last column, write what you think each word means.

Words to Know

Word	Have Not Heard Before	Have Heard Before	What I Think It Means
humus			
bedrock			
weathering			
sedimentary rock			
mineral			
fossil			
adaptation			

Answer the question.

What does the Mohs Scale measure? *the hardness of minerals*
_____

Notebook Connection

Choose two words from the Words to Know chart.
Draw a picture in your notebook showing how you think
words are connected.

SCIENCE 3 *Activities*

Looking Ahead
Provides preassessment exercises to generate a discussion about what the students already know about the chapter subject.

Graphic Organizer
Aids the students' understanding. Helps them to see the organization of facts and details.

Notebook Connection
Encourages note taking, summarizing informational text, researching, and using creative writing and drawing, as well as analyzing and interpreting text, data, and images.

Study Guide
Reviews chapter terms and concepts. "Review" tab identifies the Study Guide pages. Material for the chapter test and quizzes will come from the Study Guides.

Animal Classification 2

Name _____

Complete the organizer.

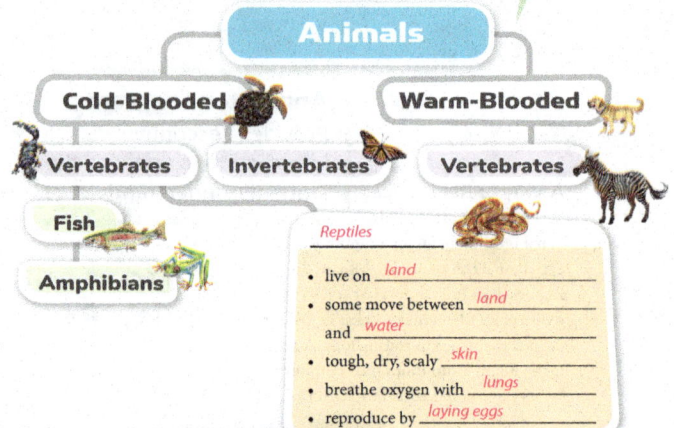

Animals

Cold-Blooded — Warm-Blooded

Vertebrates — Invertebrates — Vertebrates

Fish

Amphibians

Reptiles
- live on *land*
- some move between *land* and *water*
- tough, dry, scaly *skin*
- breathe oxygen with *lungs*
- reproduce by *laying eggs*

Features
- *camouflage* to help it blend in
- does not move much in the cold
- movable upper *jaw*
- sticky *tongue*
- some can lose and regrow tail

Lesson 43 • Pages 133–136   **101**

Review

Study Guide

Name _____

Match each description with the term.

_C_ 1. This rock forms when layers of sediment are pressed together and harden.

_A_ 2. This rock forms when melted rock cools.

_B_ 3. This rock forms when igneous or sedimentary rocks are changed by great heat and pressure.

A igneous rock
B metamorphic rock
C sedimentary rock

Fill in the blanks.

4. Hard pieces of the earth's surface are called *rocks* _____

5. Scientists classify rocks by how they are *formed* _____

6. A solid material in nature that was never alive is called a *mineral* _____

7. Rocks are made of one or more *minerals* _____

8. Three properties that can be used to identify a mineral are *Possib...*
crystal shape, color, streak

Write T if the statement is true. If the statement is false, write the correction for the underlined words.

_T_ 9. An opening in the earth that allows melted ro... surface is called a volcano.

_T_ 10. Each mineral has its own crystal shape.

_T_ 11. Iron is a mineral we can get in the foods we eat.

hardness 12. A mineral is compared to minerals on the Mohs Scale to identify the mineral's color.

SCIENCE 3 *Activities*

Lesson 20 • Pages 61–70
Review of Lessons 19–20   **49**

Science 3

Investigation
Follows the steps of the scientific method (Problem, Hypothesis, Procedure, Observations, and Conclusions) to conduct simple experiments.

Inquiry Skills Box
Identifies the basic science inquiry skills that will be emphasized in the Investigation, Inquiry, or Exploration activity.

STEM
Uses the engineering design process to solve a real-life problem.

INVESTIGATION

Hard or Soft?
Name _____

An **investigation** is a scientific test to solve a problem.

Problem

An investigation begins with a problem. The **problem** comes from an observation you have made. The problem is a question you want answered.

Which minerals can be scratched with each tool?

Hypothesis

A **hypothesis** is a possible answer to the problem. The hypothesis is written as a statement.

Circle all the words that complete your hypotheses.

Apatite can be scratched with a
fingernail, penny, steel nail .

Copper can be scratched with a
fingernail, penny, steel nail .

Gypsum can be scratched with a
fingernail, penny, steel nail .

Quartz can be scratched with a
fingernail, penny, steel nail .

Procedure

The **procedure** is the steps of an investigation.
Follow the steps to do the investigation.

☐ 1. Complete the hypothesis for
☐ 2. Scratch the apatite with your
     a line, or dust on the mineral.
☐ 3. Record the result on the data

SCIENCE 3 Activities

Each investigation will direct attention to two or three **inquiry skills**.

Inquiry Skills
• Predict
• Observe
• Infer

The **materials** are the supplies needed to do the investigation.

Materials

Tools	Minerals
☐ fingernail	☐ apatite
☐ penny	☐ copper
☐ steel nail	☐ gypsum

Materials Box
Identifies the materials needed for Investigation, Inquiry, or Exploration activities. Check boxes aid in organization.

© 2021 BJU Press. Reproduction prohibited.

STEM

Waterproof Roof
Name _____

Ask

What is the problem?
My parents have agreed to let me have a pet bunny if I take care of it. One of the things my bunny will need is a shelter. I will need to make a shelter for my bunny that will keep it dry when it rains.
How can I make a shelter that will keep my pet bunny dry?

Imagine

Think about possible answers to the problem.
_____
_____

Plan

Draw and label your design. List and gather materials.
Design should show a structure with a roof.

	Materials

© BJU Press. Reproduction prohibited.

Lesson 12 • Page 40
STEM **37**

Procedure
Identifies step-by-step directions for conducting an Investigation, Inquiry, or Exploration activity. Check boxes aid in organization.

EXPLORATION

Weather Watcher
Name _____

Inquiry Skills
• Observe
• Measure
• Predict

Materials
☐ thermometer
☐ rain gauge
☐ weathervane

Exploration
Provides hands-on experience as students take a closer look at lesson concepts.

Purpose

I want to observe the weather where I live. I want to compare the weather where I live with the weather in another location.

Procedure

☐ 1. Record two locations for your weather on Charts 1 and 2 under Observations.
☐ 2. Observe the weather where you live for 5 days.
☐ 3. Record the temperature each day in °C.
☐ 4. Record the amount of precipitation each day.
☐ 5. Record the wind direction each day.
☐ 6. Observe the sky conditions each day. Record your observations.
☐ 7. Use a computer or weather app to look up the weather for your second location. Record the temperature, amount of precipitation, wind direction, and sky conditions for each day on Chart 2.
☐ 8. Graph the temperature for each location each day on the bar graph.

SCIENCE 3 Activities

Lesson 10 • Page 34
Exploration **29**

© 2021 BJU Press. Reproduction prohibited.

Teacher Edition Features

Big Question
Begins each chapter and focuses attention on what the chapter will cover.

Materials
Identifies materials that need to be gathered to teach the lesson.

Preparation for Reading
Identifies vocabulary terms and introduces new or more difficult words. Gives the student a purpose for reading.

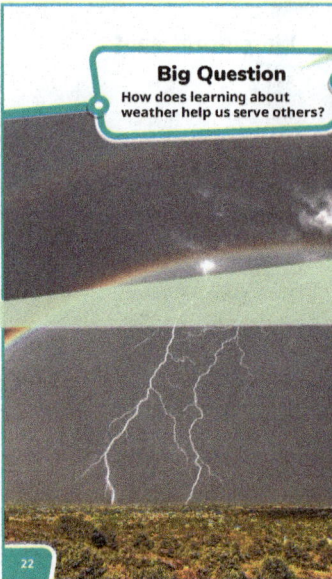

Teacher Resources
Lists the Instructional Aids and Visuals needed to teach the lesson. These are located in the Teacher Resources in the back of this book as well as at TeacherToolsOnline.com.

Preparing Ahead
Alerts the teacher to materials and preparation needed for an upcoming lesson.

Apply
Step of the teaching cycle. Student learning is applied by practicing knowledge and skills and connecting them to real life.

Differentiated Instruction
Provides activities for remediation and enrichment.

Background
Clarifies and enhances the lesson with additional information.

Sample Lesson Page (Lesson 8)

Lesson **8**
Student Edition pages 22–28
Activities pages 21–22

Chapter Objectives
- Develop a biblical explanation for the design of the atmosphere, climate, and weather.
- Infer proper use of meteorological tools to observe and record weather data.
- Identify patterns in the weather.
- Explain how weather and climate vary by seasons and locations.
- Formulate ideas for safety during severe weather.
- Apply the engineering design process to solve a real-life problem.

Lesson Objectives
- Explain how the atmosphere and sun affect our weather. BWS
- Predict what future weather might be, using given weather patterns.
- Infer which meteorological tool should be used in various scenarios.

Materials
- umbrella
- sunglasses
- gloves
- rain gauge
- thermometer
- weather vane
- anemometer

Teacher Resources
- Visuals 2.1–2.2: *Weather Map*; *Weather Patterns*

Vocabulary
- weather
- predict
- meteorologist
- weather satellite
- weather forecast
- map key
- rain gauge
- thermometer
- weathervane
- anemometer

Chapter Introduction
Use completed *Looking Ahead*, *Activities* page 21, to generate a discussion on what the students already know about the weather and climate.

- Direct attention to the chapter title and the picture of rainbow lightning. Provide time for the students to leaf through the chapter, looking at the headings and pictures. Discuss what they think the chapter will be about.
  What does the picture tell you about the weather? Possible answers: It is stormy. There is lightning. There is a rainbow.

Big Question
- Ask a volunteer to read aloud the Big Question.
- Explain that the students will find the answer to the question as they read the chapter.

Big Question
How does learning about weather help us serve others?

22

Background

Chapter Photo
The photo on pages 22–23 is a weather phenomenon called rainbow lightning. This happens most frequently in desert locations where dry air meets humid air in spectacular thunderstorms.

Visit TeacherTools[...]
resources to enhan[...]

Preparing Ahead

Lesson 12 STEM
The Lesson 12 STEM activity will require materials to be gathered ahead of time. Consider asking students to bring in empty cereal [...] that there is enough cardboard for everyone.

Lesson 13 Materials
Look ahead to the materials list in Lesson 13 for required prepara[...] tions should fall clearly within the climate zones that are discussed [...]

Sample Lesson Page (Lesson 17)

Lesson **17**

Preparation for Reading
- Preview and pronounce the vocabulary terms *topsoil*, *subsoil*, and *bedrock*.
- Direct the students to read page 56 silently to find out what the layers of soil are.

Teach for Understanding
- Display the *Layers of Soil* visual. Point out each layer as the page is discussed.
  What layer of soil do you usually see? topsoil
  What grows in topsoil? plants
  What does topsoil contain that helps plants to grow? humus
  What layer of soil is under the topsoil? subsoil
  Does subsoil have any humus? yes, but less than topsoil
  What does subsoil have that topsoil does not have? larger pieces of rock
  What do we call the large, unbroken rock under the subsoil? bedrock
- Direct attention to the "Try It Yourself!" box. You may choose to have the students participate in this activity during recess.

The "Try It Yourself!" activities are designed for the students to do on their own. If you want the activities done at school, you will need to plan ahead to supply the materials needed for some of these activities. You can find the "Try It Yourself!" materials under the Activity heading with the lesson in which the activity appears.

topsoil, subsoil, bedrock

Apply

Activities
Looking Ahead, page 43
This page assesses the students' knowledge prior to the chapter.

Layers of Soil, page 44
This page reinforces the concepts taught in Lesson 17.

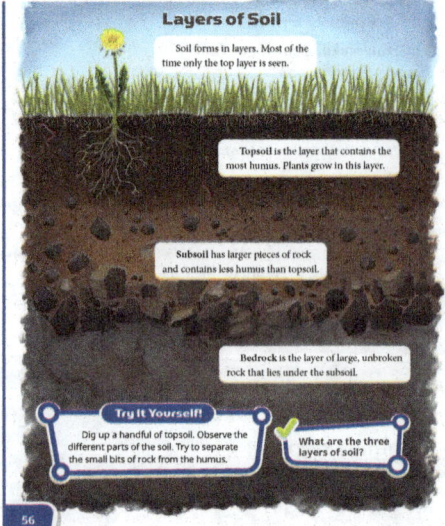

Layers of Soil

Soil forms in layers. Most of the time only the top layer is seen.

Topsoil is the layer that contains the most humus. Plants grow in this layer.

Subsoil has larger pieces of rock and contains less humus than topsoil.

Bedrock is the layer of large, unbroken rock that lies under the subsoil.

Try It Yourself!
Dig up a handful of topsoil. Observe the different parts of the soil. Try to separate the small bits of rock from the humus.

What are the three layers of soil?

56

Activity

Try It Yourself!
Materials: small spade or shovel, handful of soil
Allow students to try this activity on their own.

Differentiated Instruction

Remediation: Dirt Cup Soil Model
Materials (per student): *Dirt Cup* page (Instructional Aid 3.1), 1 clear plastic cup, 266 mL (9 oz); 2 whole vanilla cookies; 236 mL (1 c) of chocolate pudding; spoon; 1 broken vanilla cookie; 1 crushed vanilla cookie
Each student will model the soil layers in a cup. The procedure is included on the Instructional Aid page. The students will be able to eat this "dirt."

Background

Weathering and Erosion
Weathering is different from erosion. Weathering is the process of breaking down the rocks. This can happen mechanically through flowing water, wind, ice, and plants, or it can happen chemically through acid rain, lichens, mosses, and oxidation. Erosion is the movement of the weathered material from one place to another.

Science 3

Vocabulary
Identifies boldface vocabulary terms in the lesson.

Lesson Resource Box
Shows the lesson number and corresponding *Student Edition* and *Activities* pages.

Engage
Step of the teaching cycle. Engage students by capturing their attention, activating prior knowledge, and motivating them with new content.

Quick Check Answer
Provides the answer to the Quick Check question for each text section.

Lesson 2

The star and other stars

A **star** is an object in space that produces its own heat and light energy. It is a ball of gas. At night you can usually see many dots of light in the sky. A few of them are planets. But most of the dots of light are stars.

The star in our solar system is called the sun. The sun is the largest object in our solar system. It looks bigger and brighter than any other star. This is because it is much closer to Earth than any other star. But compared with other stars in the universe, the sun is a medium-size star.

The sun is very important for Earth. It is Earth's main source of energy. The sun gives light and heat energy that are necessary for life on Earth. The sun gives light and heat during the day when the sun is shining on one part of Earth.

✓ What is the main source of energy for Earth?

(i) **Background**

Mass and Gravity of the Sun
The mass of an object determines the gravity that it has. The sun's great mass causes the gravity that holds all the other objects in our solar system in orbit around it.

🜂 **Activity**

Mass of the Sun
Materials: 100 dried beans (or other small objects), clear container
Explain that the sun has the most mass of any object in the solar system. Place the beans in the container and explain that it represents the mass of all the objects in the solar system.
How many of these 100 beans do you think represents the mass of the sun? Accept guesses from volunteers.
Remove 2 beans from the container. Explain that the 98 remaining beans in the jar represents the sun's mass and the 2 represents the mass of all the other objects in the solar system.

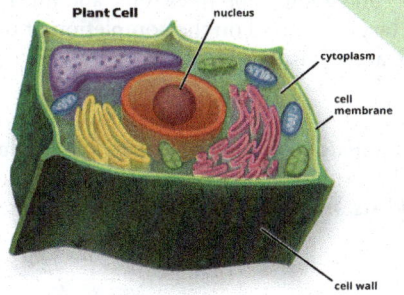

7

Student Edition pages 7–10
Activities pages 3–4
Lesson 2

Objectives
* Recall that the sun is a star.
* Describe the physical characteristics of the sun.
* Identify the earth's main source of energy.
* Explain how constellations are seasonal patterns.
* Recall that a telescope is a magnifying tool used to observe stars other than the sun

Teacher Resources
* Visuals 1.2–1.4: *Star Pattern; Ursa Major; Big Dipper*

Vocabulary
* star
* constellation
* astronomer
* telescope
* data
* astronaut

Engage
* Display the *Star Pattern* visual. Direct the students to try to find a picture in the dots. Allow volunteers to share what patterns they see.
  If you look up at the sky on a clear night, you will see patterns of stars. Some stars seem to make pictures in the sky. Today we will learn about some of these star pictures.

Instruct

Preparation for Reading
* Preview and pronounce the vocabulary term *star*.
* Direct the students to read page 7 silently to find out what the largest object in the solar system is.

Teach for Understanding
* Review what our solar system is and that the sun is at the center of it.
  What is a *star*? an object in space that produces its own heat and light energy
  🜂 What is *energy*? what is needed to cause change or to do work
  What is a star made of? a ball of gas
  What dots of light can you see in the sky? A few dots are planets and most of the dots are stars.
  🜂 Do planets make their own light? No, only stars make their own light.
  What is the name of the star in our solar system? the sun
* Compare the sun to other stars.
  Why does our sun look bigger than the other stars? It is much closer to the earth than any other star is.
  How does the sun compare in size to other stars? It is a medium-size star.
  Why is the sun important to the Earth? It gives light and heat energy that are necessary for life on Earth.
* Direct attention to the picture at the bottom of the page.
  What do you see in the picture? Possible answers: sun, Earth, space
* Explain that the picture is a reflection of the sun in the oceans of Earth at sunrise as seen from space.

✓ the sun

7

Plant Cell
nucleus
cytoplasm
cell membrane
cell wall

Membranes and Walls
A **cell membrane** is a thin covering that holds the cell together. Both plant cells and animal cells have cell membranes.

In addition to a cell membrane, a plant cell also has a cell wall. A **cell wall** is a stiff layer on the outside of the cell membrane. The cell wall gives the plant shape and support. An animal's cell does not need a cell wall. An animal has bones and muscles for shape and support.

✓ What does a plant cell have that an animal cell does not?

93

🜂 **Activity**

Model a Cell
Choose one student to be the nucleus and three students to be the cytoplasm in an animal cell. The cytoplasm should not hold hands but just stand closely around the nucleus. Then choose six more students to be the cell membrane. These students should hold hands to form a boundary. They can move in and out to control the cytoplasm, but they must stay connected.
If there are enough students, choose ten or twelve students to form a cell wall of a plant cell.
The students should hold hands and form a rigid boundary that does not move.
Note: Depending on your class size, you may choose to form two cells or adjust the number of students used for each part of the cell.

---

* Display the *Parts of a Plant Cell* visual.
  What holds the cytoplasm in place? th
  Do all cells have a cell membrane? yes
* Direct attention to the "Plant Cell" dia
  What part of a cell is found only in pl
  wall
  What is a *cell wall*? a stiff layer outside
  brane that gives the plant shape and s
* Complete the "Model a Cell" activity a

✓ cell wall

Apply

Activities
Study Guide, pages 63–64
These pages review the concepts taught i
After completion, direct the students to l
review for Test 4.

Assess

Quiz 4A
The quiz may be given at any time after
lesson.

Gear Icon
Indicates a higher-order question. Requires some analysis, synthesis, or evaluation of the lesson content. Can be supported with prompts, scaffolding, or background as needed to guide the students to the answer.

Instruct
Step of the teaching cycle. Instruct students by using direct, indirect, and interactive strategies to expand and extend their knowledge and skills.

Activity
Provides additional hands-on activities or demonstrations to enhance student understanding of concepts taught.

Assess
Step of the teaching cycle. Student understanding is assessed by using a variety of tools, including quizzes, tests, and rubrics, to systematically evaluate knowledge, skills, attitudes, and beliefs in order to improve student learning.

Assessment and Grading

SCIENCE 3 Assessments provides a variety of tools that teachers may use for assessment and grading. Frequent assessment enables you to adjust your instructional plan to better meet the needs of each student. The chart identifies suggested percentages to use for calculating each student's total grade.

Tests and Quizzes	Written Work	Investigation, Inquiry, Exploration, and STEM Activities	Participation
50%	25%	25%	
34%	33%	33%	
50%	20%	20%	10%
40%	25%	25%	10%

1 Tests and Quizzes

The chapter tests and quizzes in SCIENCE 3 Assessments can serve as the objective part of the evaluation of a student's progress. The most effective assessments are an outgrowth of the teaching process. Accordingly, the tests and quizzes should not replace your individual assessment of a student's understanding and application. The tests and quizzes can be adapted to meet the teaching emphasis and direction as well as the student's maturation level. You may find it necessary to eliminate some items, provide additional items, or do both. Students will find the mastery-level information for the tests and quizzes on the Study Guide pages in SCIENCE 3 Activities.

For True and False sections at this grade level, words are underlined to help the students determine the validity of the statements. In Activities students correct false statements. On tests and quizzes they need only label the statements as "true" or "false."

Most items on a test are valued at one point. However, you may choose to adjust the point values.

2 Written Work

Written work may include daily assignments, Quick Check answers, Reinforcement pages and Study Guide pages in the Activities book, essay and application questions, and journal entries. Scores may be given based on the completion of some assignments and the accuracy of content on others.

3 Participation

You may choose to evaluate each student's participation during lesson discussions. This subjective assessment should be tracked. You could use checklists, rubrics, point values, or letter grades.

Investigation, Inquiry, Exploration, and STEM Activities

A rubric is a useful tool for assessing work that is not objective, such as the Investigation, Inquiry, Exploration, and STEM activities. Specific rubrics for each Investigation, Inquiry, Exploration, and STEM lesson are located in the *Science 3 Assessments*. In a group activity, it is often beneficial to give a grade not only to the group as a whole but also to each student individually. Your scores and comments on a rubric allow the student to see why he or she received a particular grade and what areas need improvement.

STEM Rubric:
Vacation Water Challenge

Name _____

	Excellent (3)	Good (2)	Needs Improvement (1–0)	Points Earned
The Engineering Design Process	Uses all of the steps of the engineering design process to solve the problem of making a system that will water a plant.	Uses some of the steps of the engineering design process to solve the problem of making a system that will water a plant.	Does not use the steps of the engineering design process to solve the problem of making a system that will water a plant.	
Participation	Participates in the STEM activity.	Sometimes participates in the STEM activity.	Does not participate in the STEM activity.	
Attitude	Displays a positive attitude.	Sometimes displays a positive attitude.	Does not display a positive attitude.	
Teamwork	Listens to and encourages others.	Sometimes listens to and encourages others.	Does not listen to or encourage others.	
			Total	

Comments

SCIENCE 3

Rubrics • For use with Lesson 38

Unit 1: Let's Connect with Earth and Space Science
Chapter 1: The Solar System

Lesson	Teacher Edition	Student Edition	Activities	Objectives
1	1–6	1–6	1–2	• Compare and contrast different worldviews on the origins of the solar system. **BWS** • Identify the force that keeps the planets in their revolutions around the sun. • Explain that God designed patterns of movement in the solar system. **BWS**
2	7–10	7–10	3–4	• Recall that the sun is a star. • Describe the physical characteristics of the sun. • Identify the earth's main source of energy. • Explain how constellations are seasonal patterns. • Recall that a telescope is a magnifying tool used to observe stars other than the sun.
3	11–13	11	5–10	**Exploration: Solar Mobile** • Conduct an internet keyword search to identify an interesting fact about each planet. • Classify the planets by physical characteristics. • Sequence the order of the planets in the solar system from the sun outward. • Build a model of the solar system. • Apply the inquiry skills of classify and communicate.
4	14–21	12–17	11–16	• Identify characteristics of the inner planets. • Explain why the earth's design is unique, using Isaiah 45:18. **BWS** • Sequence and describe the phases of the moon. • Differentiate between the waxing and waning of the moon's phases. **Exploration: Moon Mystery** • Apply the inquiry skills of observe and communicate to the moon's phases.
5	22–25	18–21	17–20	• Identify characteristics of outer planets. • Sequence planets in order from the sun outward. • Describe what asteroids are and where they can be found. • Compare and contrast planets and dwarf planets.
6	26	1–21	1–20	**Review** • Recall terms and concepts from Chapter 1.
7	27			**Assessment** • Apply terms and concepts from Chapter 1.

Chapter 2: Weather and Climate

Lesson	Teacher Edition	Student Edition	Activities	Objectives
8	28–34	22–28	21–22	• Explain how the atmosphere and sun affect our weather. **BWS** • Predict what future weather might be, using given weather patterns. • Infer which meteorological tool should be used in various scenarios.
9	35–39	29–33	23–25	• Create models of four types of clouds. • Explain how the water cycle works. • Describe four forms of precipitation.
10	40–44	34	27–32	**Exploration: Weather Watcher** • Observe and record weather data from two locations. • Create a graph to display weather data.
11	45–49	35–39	33–34	• Identify six types of severe weather. • Explain why weather predictions are important. **BWS** • Make a severe weather safety plan. • Give examples of how weather predictions can keep us safe.
12	50–52	40	35–38	**STEM: Waterproof Roof** • Plan and design a roof model that repels water, using the engineering design process. • Construct a roof model that repels water, using the engineering design process. • Evaluate designs to determine which models best repel water. • Redesign the model to make the roof repel more water. • Communicate to others how the design solves the problem.
13	53–57	41–45	39–40	• Identify the three main climate zones. • Identify the characteristics of each climate zone. • List examples of geographic locations in each climate zone.
14	58–63	46–51	41–42	• Identify climate change. • Compare and contrast different views of climate change. **BWS** • Identify possible causes of climate change. • Defend the claim that climate change has limits, using Genesis 8:22. **BWS**
15	64	22–51	21–42	**Review** • Make predictions of future weather based on observed and recorded data. • Recall terms and concepts from Chapter 2.
16	65			**Assessment** • Apply terms and concepts from Chapter 2.

Chapter 3: Soil, Rocks, Minerals, and Fossils

Lesson	Teacher Edition	Student Edition	Activities	Objectives
17	66–70	52–56	43–44	• Explain from Scripture the origin of the earth, including water and dry land. BWS • Identify four components of soil. • Identify the three main layers of soil. • Examine topsoil to identify its various components.
18	71–74	57–60	45–46	• Recall the causes of weathering and erosion. • Describe how landslides are an example of erosion. • Recall that soil is a natural resource. BWS • Explain why it is important to conserve soil. BWS
19	75–78	61–64	47	• Identify the three main types of rocks. • Explain how rocks form and are classified. BWS • Create a way to classify rocks.
20	79–84	65–70	49–50	• Explain how rocks and minerals are related. BWS • Identify some properties of minerals. • Explain how understanding the properties of minerals can be beneficial. BWS • Differentiate between common and precious minerals.
21	85–87	71	51–57	**Investigation: Hard or Soft?** • Formulate a hypothesis to predict the hardness of various minerals. • Record observations. • Draw conclusions from data collected. • Explain why knowing about the hardness of minerals is beneficial. BWS • Rank minerals from softest to hardest, based on the scratch test conducted.
22–23	88–97	72–81	59–60	• Recall what a fossil is. • Compare and contrast different views of how and when most fossils formed. BWS • Explain why there are marine fossils on mountains. BWS • Compare and contrast views of adaptation. BWS • Create a model bone fossil to observe the "fossilization" by the process of mineral replacement. • Formulate a statement explaining how the model bone fossil supports a biblical view of fossil formation. BWS
24	98	52–81	43–60	**Review** • Recall terms and concepts from Chapter 3.
25	99			**Assessment** • Apply terms and concepts from Chapter 3.

Unit 2: Let's Connect with Life Science
Chapter 4: Cells, Tissues, Organs, and Systems

Lesson	Teacher Edition	Student Edition	Activities	Objectives
26	100–106	82–88	61–62	• Explain, using the Bible's teaching, that God knows all about the people He created. **BWS** • Identify the smallest living part of an organism. **BWS** • Identify a tool used to see tiny things. • Label the five main components of a microscope. • Conclude which science tool shows greater detail.
27	107–11	89–93	63–64	• Identify the amoeba as a single-celled organism. • Identify the main parts of an animal cell and a plant cell. • Differentiate between a plant cell and an animal cell.
28	112–14	94	65–66	**Exploration: Edible Cell** • Make a model of an animal cell. • Label the parts of an animal cell. • Communicate information about the animal cell model.
29	115–20	95–100	67–68	• Explain how cells, tissues, organs, and systems are related to each other. **BWS** • Identify four organs of the human body. • Explain why it is important to know how the body works. **BWS** • Identify the largest organ of the body. • Explain why the epidermis is important to the body. **BWS**
30	121–23	101	69–73	**Exploration: Patterns on My Skin** • Classify fingerprint patterns. • Record and graph types of fingerprint patterns.
31	124–27	102–5	75–76	• Identify the parts of the dermis. • Explain two ways that the body is cooled. • Explain why caring for our skin is important. **BWS**
32	128	82–105	61–76	**Review** • Recall terms and concepts from Chapter 4.
33	129			**Assessment** • Apply terms and concepts from Chapter 4.

Chapter 5: Plants

Lesson	Teacher Edition	Student Edition	Activities	Objectives
34	130–34	106–10	77	• Identify three conditions a seed needs to germinate. • Recall the function of each plant part. • Sequence the life cycle of a flowering plant.
35	135–40	111–16	79–81	• Name three things plants need for photosynthesis. • Name two products plants produce during photosynthesis. • Identify where photosynthesis takes place in the plant. • Sequence the steps of photosynthesis. • Explain, using biblical teaching, why photosynthesis occurs. **BWS** • Explain why God gave people plants. **BWS**
36	141–43	117	83–87	**Investigation: A Place to Grow** • Predict the effect of sunlight on plant growth. • Observe and measure the height of plants. • Record and graph data on a bar graph. • Draw a conclusion about the effects of sunlight on plant growth.
37	144–46	118–20	89–90	• Explain why plants have different traits. • Compare and contrast different views about plant adaptation. **BWS**
38	147–49	121	91–92	**STEM: Vacation Water Challenge** • Plan and design a way to water a plant, using the engineering design process. • Create a model to water a plant. • Test and compare models to improve the original design. • Communicate to others how the design solves the problem.
39	150	106–21	77–92	**Review** • Recall terms and concepts from Chapter 5.
40	151			**Assessment** • Apply terms and concepts from Chapter 5.

Chapter 6: Cold-Blooded Animals

Lesson	Teacher Edition	Student Edition	Activities	Objectives
41	152–56	122–26	93–95* (page 95 is used in Lessons 41–42)	• Identify how scientists classify animals. • Differentiate between invertebrates and vertebrates. • Identify the characteristics of warm-blooded and cold-blooded animals. • Differentiate between warm-blooded and cold-blooded animals. • Identify what an animal behavior scientist does.
42	157–62	127–32	95–99	• Identify the characteristics of fish and amphibians. • Identify and describe the structures and features that benefit fish and amphibians in survival and growth. BWS • Construct an explanation, using Scripture, stating why animals reproduce after their own kind. BWS • Identify the ways fish and amphibians reproduce. • Sequence the stages of frog metamorphosis.
43	163–69	133–37	101–6	• Identify the characteristics, structures, and features of reptiles. • Describe how God has provided the structures and features that benefit reptiles for survival and growth. BWS • Sequence the pattern of the reptile life cycle. • Explain, using Scripture, why there will one day be no predators and no prey. BWS **Investigation: Leaping Lizards** • Infer in an Investigation how a cold-blooded animal depends on the temperature of its habitat for survival and growth. • Explain why people should care for reptiles. BWS
44	170–76	138–44	107–10	• Identify the characteristics of insects and spiders. • Describe the structures and features that benefit insects and spiders in survival and growth. BWS • Identify the ways that insects and spiders reproduce. • Sequence the stages of ladybug metamorphosis.
45	177–81	145	111–14	**Investigation: Changes in Dark Places** • Create a hypothesis about the stages in the life cycle of a mealworm. • Construct a living space for mealworms. • Observe and record the changes in a mealworm's life cycle. • Draw conclusions from the data collected.
46	182	122–45	93–114	**Review** • Recall terms and concepts from Chapter 6.
47	183			**Assessment** • Apply terms and concepts from Chapter 6.

Chapter 7: Warm-Blooded Animals

Lesson	Teacher Edition	Student Edition	Activities	Objectives
48	184–92	146–52	115–24	• Recall the differences between invertebrates and vertebrates and between warm-blooded and cold-blooded animals. • Identify the characteristics of birds. • Identify and describe the structures and features that benefit birds in survival and growth. **BWS** • Formulate a biblical explanation of why birds reproduce after their own kind. **BWS** • Sequence the stages of a bird's life cycle. **BWS** **Exploration: Feathery Friends** • Record observations of birds at a bird feeder.
49	193–97	153–57	125–30	• Identify the characteristics of mammals. • Identify and describe the structures and features that benefit mammals for survival and growth. **BWS** • Sequence the stages of a mammal's life cycle. **BWS** • Compare and contrast the views about egg-laying mammals. **BWS** • Compare humans and mammals. **BWS**
50	198–201	158–61	131	• Explain why understanding the different kinds of animals is important for caring for them. **BWS** • Identify the life cycle pattern that all cold-blooded and warm-blooded animals have. • Compare and contrast life cycles of cold-blooded and warm-blooded animals. • Identify similar and different patterns of inherited traits shared between offspring and their parents or among siblings. • Identify a variation of an inherited trait.
51	202–6	162–66	133–37	• Identify animal instincts. • Explain where an animal's instincts come from. **BWS** • Identify learned behaviors of some animals. • Differentiate between inherited traits, learned behaviors, and instincts.
52	207–9	167	139–40	**Exploration: Blubber, Feathers, and Fur** • Identify the purpose of blubber, feathers, and fur. • Observe how blubber, feathers, and fur help keep animals warm. • Record observations. • Draw conclusions from data collected.
53	210	146–67	115–40	**Review** • Recall terms and concepts from Chapter 7.
54	211			**Assessment** • Apply terms and concepts from Chapter 7.

Chapter 8: Ecosystems

Lesson	Teacher Edition	Student Edition	Activities	Objectives
55	212–18	168–74	141–44	• Identify four characteristics of organisms. • Identify what makes up an ecosystem. • Explain the relationship between ecosystem, habitat, community, and population. • Explain how resources meet the needs of organisms. • Identify why there is competition between animals.
56	219–21	175	145–46	**STEM: Habitat Help!** • Plan and design a solution for providing food or water for an animal group during a drought, using the engineering design process. • Create a model of the food system or water system. • Test and compare models to improve the original design. • Communicate to others how the model solves the problem.
57	222–27	176–81	147–48	• Explain how producers get their energy. • Classify organisms as producers, consumers, or decomposers. • Describe the three types of consumers. • Explain, using biblical teaching, why there were no carnivores before the Fall. **BWS** • Analyze the advantages of animals living together in groups.
58	228–31	182–85	149–50	• Recall what a food chain is. • Sequence on a food chain the transfer of energy from one organism to another. **BWS** • Identify what a food web is. • Describe the relationship between a food chain and a food web. **BWS** • Interpret a food web. • Explain what happens when one part of a food web changes.
59	232–36	186–90	151–54	• Name some causes of change in an ecosystem. • Explain a way people should exercise dominion when making changes to ecosystems. **BWS** • Give examples of characteristics that help organisms survive in a changed ecosystem. • Analyze why some organisms survive well and some do not in a changed ecosystem. • Identify how an ecosystem maintains balance.
60	237–39	191	155–56	**Exploration: Ecosystem Tag** • Identify the roles of predator and prey in a food web. • Infer changes in an ecosystem. • Draw conclusions from data collected to make inferences about a population after its predator dies off in an ecosystem.
61	240	168–91	141–56	**Review** • Recall terms and concepts from Chapter 8.
62	241			**Assessment** • Apply terms and concepts from Chapter 8.

Unit 3: Let's Connect with Physical Science
Chapter 9: Matter and Sound

Lesson	Teacher Edition	Student Edition	Activities	Objectives
63	242–50	192–200	157–60	• Identify physical properties of matter. • Explain the difference between mass and volume. • Identify what an atom is. • Compare and contrast the properties of solids, liquids, and gasses.
64	251–55	201	161–66	**Exploration: Which Kind of Matter?** • Compare properties of different kinds of matter. • Measure mass and volume of different kinds of matter. • Classify matter based on properties.
65	256–59	202–5	167	• Describe how heating and cooling can cause matter to change states. • Infer why there is moisture inside a window on a cold day. • Describe the three states of water.
66	260–63	206–9	169–70	• Identify what physical and chemical changes are. • Compare and contrast physical and chemical changes. • Give one example of a physical change and one example of a chemical change. • Explain how people can use changes in matter to serve God and other people. BWS
67	264–68	210–14	171–72	• Recall what sound is. • Identify causes of sound vibrations. • Explain how sound travels. • Describe three ways sound waves interact with matter.
68	269–74	215–20	173	• Describe the characteristics of pitch and volume. • Explain how the physical properties of a musical instrument affect pitch and volume. • Create musical instruments to demonstrate pitch and volume. • Identify ways that sound is used to communicate. BWS
69	275–78	221	175–77	**Investigation: Musical Bottles** • Create a hypothesis to predict how air affects pitch. • Observe and record data. • Draw conclusions about how the amount of air in a bottle affects the pitch.
70	279–81	221	179–81	**Inquiry: Musical Bottles** • Create a hypothesis to predict how the size of a bottle will affect the pitch. • Develop the steps of the procedure in the Inquiry. • Observe and record data. • Draw conclusions about how the size of a bottle will affect the pitch.
71	282	192–221	157–81	**Review** • Recall terms and concepts from Chapter 9.
72	283			**Assessment** • Apply terms and concepts from Chapter 9.

Chapter 10: Forces and Motion

Lesson	Teacher Edition	Student Edition	Activities	Objectives
73	284–89	222–27	183–86	• Recall what force, friction, gravity, weight, and magnetism are. • Formulate a biblical explanation for how we know that God created force. BWS • Identify contact and noncontact forces. • Compare and contrast contact and noncontact forces.
74	290–93	228–31	187–90	• Describe direction, speed, distance, and position as they relate to motion. BWS • Identify balanced and unbalanced forces. • Analyze whether the forces acting on an object are balanced or unbalanced. • Explain why it is important to know about balanced and unbalanced forces. BWS
75	294–96	232	191–92	**Investigation: Balanced and Unbalanced Forces** • Predict how balanced and unbalanced forces will affect the motion of a ball. • Observe and record data in a chart. • Infer how balanced and unbalanced forces affect the motion of a ball. • Apply to the game of bowling the concepts learned about balanced or unbalanced forces.
76	297–300	233–36	193–94	• Write equations for balanced or unbalanced forces on an object. • Identify observed patterns of motion and unseen patterns of motion. • Predict how an object will move next, based on observed patterns of motion. • Identify patterns of motion that God made at Creation. BWS • Explain why some motion cannot be predicted.
77	301–4	237–40	195–98	• Identify what work is. • Identify the living things or nonliving things that can do work. • Determine what is doing the work in a scenario. • Explain, using examples, how forces help in doing work better.
78	305–7	241	199–200	**STEM: A-Maze-ing Marble Game** • Design a game path, using the engineering design process, that gives a marble two different patterns of motion. • Create a path with two patterns of motion for a marble. • Test and compare models to improve the original design. • Communicate how the redesign solves the problem.
79	308	222–41	183–200	**Review** • Recall terms and concepts from Chapter 10.
80	309			**Assessment** • Apply terms and concepts from Chapter 10.

Chapter 11: Electricity and Magnetism

Lesson	Teacher Edition	Student Edition	Activities	Objectives
81	310–15	242–47	201–2	• Identify what electricity is. • Relate atoms to electricity. • Identify when an object is positively or negatively charged or when the object is neutral. • Explain what static electricity is. • Describe what happens when like charges and opposite charges are brought near each other. • Analyze the effects that a charged object has on other objects.
82	316–19	248–51	203–4	• Identify what current electricity is. • Differentiate between static electricity and current electricity. • Compare and contrast conductors and insulators. • Differentiate between a closed simple electric circuit and an open simple electric circuit. • Explain, using biblical teaching, why it is beneficial to understand conductors and current electricity. **BWS**
83	320–22	252	205–6	**Exploration: Conductors Needed** • Predict whether objects are conductors or insulators. • Build a circuit tester. • Identify which objects complete the circuit and which do not. • Infer which materials are conductors and which are insulators.
84	323–26	253–56	207–8	• Recall what a magnet is. • Explain why magnets can act at a distance. • Explain how the invisible magnetic field can be "visible." • Identify the areas of a magnet with the strongest magnetic force. • Analyze the magnetic forces of attract and repel. • Explain how magnets are beneficial. **BWS**
85	327–28	257	209–10	**Investigation: Magnetism and Matter** • Predict the strength of magnetic force. • Observe and record data. • Draw conclusions about the strength of magnetic force and the amount of matter that the magnetic force passes through.
86	329–31	257	211–13	**Inquiry: Magnetism and Matter** • Formulate a hypothesis to predict the magnetic force of several different magnets. • Develop the steps of the procedure in the Inquiry. • Observe and record data. • Organize the data collected onto a graph. • Draw conclusions about the strength of each magnet's magnetic force using the data collected.
87	332–34	258–60	215	• Identify what an electromagnet is. • Explain how the magnetic force of an electromagnet can be made stronger or weaker. • Differentiate between a bar magnet and an electromagnet. • Support the statement that understanding electromagnets has helped people solve real-life problems. **BWS**

Chapter 11: Electricity and Magnetism

Lesson	Teacher Edition	Student Edition	Activities	Objectives
88	335–37	261	217–18	**STEM: Lost and Found** • Design a magnetic or electromagnetic tool, using the engineering design process, that will pick up metal objects from a tight space. • Create a magnetic or an electromagnetic tool. • Test and compare models to improve the original design. • Communicate how the redesign solves the problem.
89	338	242–61	201–18	**Review** • Recall terms and concepts from Chapter 11.
90	339			**Assessment** • Apply terms and concepts from Chapter 11.

Book Introduction

- Allow time for the students to explore the different parts of the book before looking at it together.
- Direct the students to the Contents page.
- Direct attention to the word written in gray above each circle. Guide the students in reading the word *unit*.
- Explain that there are three units in the book.

  Look at the words *Let's Connect with* written in gray at the top of the page. What are you going to explore in the first unit? earth and space science

  What is the unit title? Let's Connect with Earth and Space Science

  Each unit is divided into chapters. Look down to see what the first chapter is about. the solar system

  How many chapters are there in the first unit? three

  What are some of the special features found in Chapter 1? STEM Career: Astronaut; Exploration: Solar Mobile; Exploration: Moon Mystery

- Direct the students to look at the title of the first chapter, *The Solar System*. Then ask them to find the page number where they can find the first chapter.

  Look at the last page of the Contents.

  What are the last two sections in the book? the Glossary and the Index

- Explain that a glossary is like a dictionary. You can find each science vocabulary word and what it means in the glossary.

  Look at the word *Glossary*. Find the page number where you can find the Glossary.

  What page does the Glossary begin on? 263

- Explain that an index tells the page numbers where information about different words, subjects, or people can be found.

  What page can you find the Index on? 277

- Guide the students in finding the Glossary and Index. *Note:* Time will be allotted in a later lesson to help the students use the Glossary.

UNIT 1

ⓘ Background

Unit Photo
This photo is mountain scenery in Tibet.

⚠ Helps

BWS Icon
The BWS icon in the objectives stands for "Biblical Worldview Shaping." Objectives labeled with this icon will shape the students' thinking with a biblical worldview.

Visuals
If you do not have access to multimedia, make for each student or group of students a copy of the visuals that are not in their books.

Open-Ended Questions
To help a third grader relate what he is learning to his own experiences, open-ended questions are included. Answers should not be limited to those given in the *Science 3 Teacher Edition*. If an answer seems unusual, question the student as to why he gave that particular answer. This will help distinguish divergent thinking from silliness. A student should never be made to think his answer is silly when he intended to be serious.

Let's connect with Earth and Space Science

⚠ Helps

Reading Assignments
The reading assignments for each lesson are divided into two- to three-page segments. You may choose to adjust the length of assignments based on the reading ability of the students.

Silent Reading
Teaching the life skill of reading silently for information and pleasure requires an emphasis on silent reading.

Science Groups
Science groups are a good way to encourage cooperative learning. Using groups also helps limit needed supplies while giving each student an opportunity to participate. Groups of two or three students are best for most activities. You may need to assign tasks to the group members, such as getting the supplies, measuring, and recording information.

Unit Introduction

Turn to page 1 in your book. Read the title of the unit. Let's connect with Earth and Space Science

What does the word *connect* mean? Possible answers: to join together, to fasten together

🔧 What do you think the word *connect* means in the unit title? Elicit that *connect* means to see how earth and space science is meaningful to your life.

> The word *elicit* is a pedagogical technique in which you do not give the students the information directly. You ask a sequence of questions to help the students come to the desired conclusion.

What do you see in the picture? Possible answers: mountains, clouds, the moon, blue sky

🔧 How does each part of the picture connect with earth and space science? Elicit that each part of the picture is a part of earth and space science.

> The gear icon (🔧) indicates a higher-order question. These questions are based on information gathered from the text, but they require some analysis, synthesis, or evaluation of the text. Supply prompts or share Background as needed to guide the students to the answer.

Turn to page 2 in your book. What does it say in big letters in the box? Big Question

- Explain that there will be a big question in each chapter. The answer to the question will be in the chapter.

Look at page 3. Find the words that are in a darker black in the paragraphs than the other words. What are these words? universe, atmosphere, solar system

You will find other darker black, or bold, words in the chapters. These words are vocabulary words. We will learn what these words mean.

Find the word that has slanted letters. This is called italic. What is the italic word? energy

- Explain that both the bold words and the italic words can be found in the Glossary.

- Direct attention to page 4.

At the bottom of the page find the box with a green check mark. In this box you will find a Quick Check question. These questions will help you know whether you understand what you have read.

- Guide the students in finding page 10. *Note:* The acronym *STEM* will be explained when discussing the page.

🔧 The big words at the top of the page say "STEM Career." Do you know what a career is? Elicit from the students that a career is a job.

On these pages you will learn about some of the jobs that scientists have.

- Guide the students in finding page 11.

That big word at the top of the page is *Exploration*. An Exploration is a science activity that you will get to do.

Chapter Objectives

The first lesson of each chapter gives a list of chapter objectives. Each chapter objective is part of one of the *SCIENCE 3* Product Objectives (see page x). Each of the lesson objectives is part of one of the chapter objectives.

- Analyze God's design of His creation.
- Analyze the characteristics of the solar system.
- Apply inquiry skills of research, communicate, classify, and observe.

Lesson Objectives

- Compare and contrast different worldviews on the origins of the solar system. **BWS**
- Identify the force that keeps the planets in their revolutions around the sun.
- Explain that God designed patterns of movement in the solar system. **BWS**

Teacher Resources

For your convenience, the Teacher Resources appear in the back of this *Teacher Edition* and are also available on TeacherToolsOnline.com.

- Visual 1.1: *Solar System*
- Instructional Aid 1.1: *Bookmarks*, copy on cardstock

Vocabulary

- universe
- atmosphere
- solar system
- worldview
- planet
- pattern
- axis

Engage

- Direct attention to the chapter title and the picture of the Milky Way. Provide time for the students to look through the chapter, looking at the headings and pictures. Discuss what they think the chapter will be about.
- Ask volunteers to describe the biggest things they can think of as well as things that are far away.

  Today we will learn about some things God made that are very large and very far away.

Each chapter begins with a Looking Ahead page in the *SCIENCE 3 Activities* book. This page is designed to generate a discussion on what the students already know about the chapter subject. You may choose to have the students complete the Looking Ahead page now. The page may be used at the end of the chapter to see what the students have learned.

Big Question

- Ask a volunteer to read aloud the Big Question.
- Explain that the students will find the answer to the question as they read the chapter.

Big Question
What patterns of movement are in the solar system?

ⓘ Background

Chapter Photo

The photo on pages 2–3 is showing the Great Ocean Road in the Milky Way at night.

⚠ Helps

Science Notebook

Encourage each student to keep a science notebook for science-related pages. A three-ring binder or other easily accessible notebook would work best. The students may keep *Activities* pages such as Investigations, Explorations, STEM activities, graphic organizers, and Study Guides in their notebooks. They may also keep other science projects and handouts there.

A "Notebook Connection" will be used periodically throughout the *Activities* book. Its purpose is to make a deliberate connection to the student's science notebook and to encourage note taking, creative writing, drawing, summarizing, analyzing and interpreting text and images, and conducting research.

Keeping a notebook will help the students develop organizational skills. The notebook can also be used in each student's portfolio to demonstrate his progress and ability to think scientifically. Plan to check the content, organization, and neatness of the notebook at least once per unit.

Visit TeacherToolsOnline.com for resources to enhance the lessons.

Chapter 1

The Solar System

Look at the sky on a clear night and you will see thousands of stars. The Bible tells us that, "In the beginning God created the heaven and the earth" (Genesis 1:1). God created the universe. The **universe** is space and all the matter and energy in it. *Energy* is what is needed to cause change or to do work. Energy is needed on the earth. Earth is our home. It is a tiny piece in the universe. You see parts of the universe every time you look at the sky.

The layer of gases surrounding the earth is called the **atmosphere**. The heaven, or outer space, is everything beyond the earth's atmosphere. The sun and the objects that revolve around the sun are called the **solar system**.

3

Preparing Ahead

The section Preparing Ahead will give you a heads-up on materials and/or preparations needed for a future lesson.

Lesson 3 Materials
Look ahead to the materials list in Lesson 3 for the Exploration activity.

Lessons 56, 69, and 70 Materials
A total of three empty 2 liter soda bottles will be needed, per student or group, for the STEM, Investigation, and Inquiry lessons.
One 500 mL empty soda bottle will be needed, per student or group, for Lesson 70.

Background

Songs of David
Many verses in Psalms describe the glory and greatness of God's creation. As a shepherd, David would have had great opportunity to observe the stars and enjoy God's creation. He wrote songs about the God who created such wonderful things. Some of the songs that David wrote are part of the Bible.

Instruct

Preparation for Reading
- Preview and pronounce the vocabulary terms *universe*, *atmosphere*, *solar system*, and the word *energy*.

A vocabulary term is a bold word for which the students learn the definition. Write the terms for display and pronounce them with the students. Vocabulary words can be found in the Glossary. Other important words are in italic. These words can also be found in the Glossary but are not vocabulary terms. Some lessons include words that are not vocabulary terms but may be difficult for the students to decode and pronounce. Assist the students with any words that may be difficult.

- Direct attention to the Glossary. Explain that the definitions to all the bold vocabulary terms, as well as the important words that are italicized, can be found in the Glossary. Point out that the words are listed in alphabetical order.
- Ask the students to collaborate with a partner. Tell them to look up one of the terms above and be ready to share a definition.
- Give different science partners an opportunity to share the definitions that they found.
- Direct the students to read page 3 silently to find out why we see parts of the universe every time we look at the sky.

Reading for information is an important skill for the students to learn. The Preparation for Reading section "find out" (or "find where") helps motivate reading and focuses the students' attention. The students should look for the answer as they read silently. The "find out" will be answered as the information is discussed.

Teach for Understanding
- Read aloud Genesis 1:1 and discuss the truth that God created the earth and everything else in the universe.

What is the *universe*? space and all matter and energy in it

What is *energy*? what is needed to cause change or to do work

We live on Earth. Earth's atmosphere is all around us. What is the *atmosphere*? the layer of gases surrounding the earth

What is beyond the earth's atmosphere? the heaven or outer space

Can we see any of outer space? yes

Where can we see parts of it? in the sky

What term describes the sun and the objects that revolve around the sun? the solar system

Preparation for Reading

- Preview and pronounce the vocabulary terms *world-view*, *planet*, *pattern*, *axis*, and the words *biblical world-view*, *force*, *revolution*, and *rotation*.
- Direct attention to the heading on page 4.

  What will you be reading about? how the solar system began
- Direct the students to read pages 4–6 silently to find out which force God created to hold all things together.

Teach for Understanding

- Direct attention to the pictures, labels, and captions.

  How many days of Creation were there? six

  On which day of Creation did God create the atmosphere? Day 2
- Review the days of Creation as you want.

  What is the way a person thinks about the world called? a worldview

  Where does a person's worldview come from? a big story that they believe about the world

  How does a biblical worldview affect a person's life? It affects what a person believes and how he lives.

  What do those with a biblical worldview use to understand the solar system? the Bible

  What do they believe about the beginning of the solar system? They believe the Genesis 1 account of the six days of Creation.
- Explain that there are people who do not believe the Bible. They have a different worldview.

  What do many people who do not believe the Bible believe about the solar system? They believe that Earth and space formed by chance without God; everything changed slowly many, many times over millions of years.

✓ the way a person thinks about the world

The Beginning of the Solar System

How did the solar system begin? Not everyone agrees about its beginning. People who look at the solar system can think about it in different ways. The way a person thinks about the world is a **worldview**. It guides what people believe and how they live. Your worldview comes from a big story about the world.

One example of worldview is a biblical worldview. A *biblical worldview* is the belief that the story of the Bible is true. Those with a biblical worldview use the Bible to help them understand the solar system. They believe the Genesis 1 account of the six days of Creation. Nothing was created without God.

Many people believe that Earth and space formed by chance without God. They believe that everything has changed many, many times. And these changes happened slowly over millions of years.

On which day of Creation did God create the atmosphere?

Day 3   grass, herbs, and trees

Day 4   sun, moon, and stars

Day 5   whales and winged fowl

What is a worldview?

Day 6   cattle and people

4

ⓘ Background

Worldview Terms

The following vocabulary may be helpful when discussing aspects of a biblical worldview.

Creation is the six-day event described in Genesis 1 when God created all things very good.

Dominion is the job God gave to people to manage the world for the benefit of others and the glory of God. Terms used in the student textbook in conjunction with this concept include *care for*, *manage*, and *rule over*.

The **Curse** is the punishment described in Genesis 3 that God placed on all His creation because of Adam's sin.

The **Fall** is the event described in Genesis 3 when the first man sinned and brought death into the world.

The **Flood** is the event described in Genesis 6–9 when God caused the entire earth to be covered by water.

Redemption is God's work in restoring His creation back to Himself, accomplished through Christ's act of rescuing and freeing from sin.

Both worldviews look differently at the way objects move. Those with a biblical worldview believe that God created the movement of the objects around the sun. Those with a different worldview believe that this movement came about by chance. Your worldview should come from the Bible. The Bible is true and anything that disagrees with the Bible is false.

Gravity

One part of God's design, or plan, in the solar system is the force that He created. A *force* is a push or a pull. This force holds all things together. This force also keeps the solar system in order. This force is called *gravity*. The sun's gravity is stronger than the other objects in the solar system. A **planet** is a large ball of gas or rock that revolves around the sun. The planets revolve, or move, around the sun because of gravity. The sun is not strong enough to pull the planets into the sun. But the sun's gravity is strong enough to make the planets revolve around the sun.

How many planets revolve around the sun?

5

Do people with different worldviews look in the same way at how objects move? no

How do those with a biblical worldview look at the movement of the objects around the sun? They believe that God created the movement.

What do those with a different worldview believe about the movement of the objects around the sun? that it came about by chance

Where should your worldview come from? from the Bible

What do you know about the Bible? It is true.

- Display the *Solar System* visual. Refer to it as the solar system is discussed.

What is a *force*? a push or a pull

Which force did God create that holds all things together? gravity

Which force keeps the planets in order around the sun? gravity

Which object's gravity is stronger than other objects in the solar system? the sun's

What is a *planet*? a large ball of gas or rock that revolves around the sun

What does *revolve* mean? move around

Why do the planets revolve around the sun? because of gravity

Why does the sun not pull the planets into it? The sun's gravity is not strong enough.

- Direct attention to the caption.

Count the number of planets revolving around the sun. How many planets are there? eight

⚠ Helps

In this chapter, most of the time *Earth* is capitalized since it is referring to a proper name for our planet. However, in other places it is generally referred to as "the earth."

ⓘ Background

Solar System

Our solar system consists of many objects revolving around the sun. Most of the planets have at least one moon revolving around them. The solar system also includes asteroids, meteoroids, comets, and dwarf planets.

Number of Planets

For decades scientists identified nine planets in the solar system. However, the last planet, Pluto, did not have some of the characteristics of the other planets. In 2006 scientists refined the definition of a planet, and Pluto was reclassified as a dwarf planet. The solar system now has eight official planets.

Sun as a Fixed Point

All objects in space are in motion. The sun is moving, orbiting with all its planets around the center of the Milky Way galaxy. However, for this grade and this discussion of the solar system, the sun is treated as a fixed point used to determine the motion of the objects around it.

How did God design parts of the solar system to move? He designed them to move in patterns.

What is a *pattern*? a design or natural order that happens over and over again

What are two examples of patterns in the solar system? a revolution and a rotation

What is a *revolution*? the motion of one object around another object

What is a *rotation*? when a planet turns on its axis one time

What is an *axis*? an imaginary line that runs through the center of a planet from the North Pole to the South Pole

- Direct attention to the picture.

Where is the North Pole? at the most northern point on Earth

Where is the South Pole? at the most southern point on Earth

- Ask a volunteer to read the caption.

What is the pattern shown in the picture called? a rotation

What does every planet rotate on? an axis

What does every planet have because it rotates on its axis? days

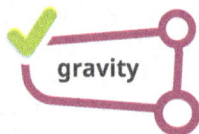

gravity

Apply

Activities

Looking Ahead, page 1

This page assesses the students' knowledge prior to the chapter.

The Solar System and Worldview, page 2

This page reinforces the concepts taught in Lesson 1. *Note:* See "Reading the *Activities* Footer" in Helps before assigning the page.

Patterns in the Solar System

God designed the parts of the solar system to move in patterns. A **pattern** is a design or natural order that happens over and over again. One example of pattern is a revolution. A *revolution* is the motion of one object around another object. Earth and the other planets revolve around the sun. One complete revolution of any planet around the sun is called a year.

Another pattern makes the sun appear to rise in the east. Then the sun appears to move across the sky and set in the west. This pattern is called a rotation, or one spin. A *rotation* is when a planet turns on its axis one time. An **axis** is an imaginary line that runs through the center of a planet from the *North Pole* to the *South Pole*. One rotation of Earth is called a day. Every planet rotates on an axis and has days.

What is this pattern of Earth called?

North Pole

South Pole

What is the force that pulls the objects in the solar system toward the sun?

⚠ **Helps**

Reading the *Activities* Footer

Direct attention to the footer on *Activities* page 2. Point out the page numbers and explain that they are the pages in the student textbook that are used to complete the exercise.

To help the students, give a set of prepared bookmarks (Instructional Aid 1.1) to each student. Direct attention again to the *Activities* footer. Explain that they should mark the beginning, or first, page number in the student textbook with the green bookmark and the end, or last, page number with the red bookmark. This will help the students keep track of the page ranges in the student textbook as they complete their *Activities* assignments.

The Sun and Other Stars

A **star** is an object in space that produces its own heat and light energy. It is a ball of gas. At night you can usually see many dots of light in the sky. A few of them are planets. But most of the dots of light are stars.

The star in our solar system is called the sun. The sun is the largest object in our solar system. It looks bigger and brighter than any other star. This is because it is much closer to Earth than any other star. But compared with other stars in the universe, the sun is a medium-size star.

The sun is very important for Earth. It is Earth's main source of energy. The sun gives light and heat energy that are necessary for life on Earth. The sun gives light and heat during the day when the sun is shining on one part of Earth.

✓ **What is the main source of energy for Earth?**

7

① Background

Mass and Gravity of the Sun

The mass of an object determines the gravity that it has. The sun's great mass determines the amount of gravity that holds all the other objects in our solar system in orbit around it.

⚗ Activity

Mass of the Sun

Materials: 100 dried beans (or other small objects), clear container

Explain that the sun has the most mass of any object in the solar system. Place the beans in the container and explain that this represents the mass of all the objects in the solar system.

How many of these 100 beans do you think represents the mass of the sun? Accept guesses from volunteers.

Remove 2 beans from the container. Explain that the 98 remaining beans in the jar represents the sun's mass and the 2 represents the mass of all the other objects in the solar system.

Objectives

- Recall that the sun is a star.
- Describe the physical characteristics of the sun.
- Identify the earth's main source of energy.
- Explain how constellations are seasonal patterns.
- Recall that a telescope is a magnifying tool used to observe stars other than the sun.

Teacher Resources

- Visuals 1.2–1.4: *Star Pattern*; *Ursa Major*; *Big Dipper*

Vocabulary

- star
- constellation
- astronomer
- telescope
- data
- astronaut

Engage

- Display the *Star Pattern* visual. Direct the students to try to find a picture in the dots. Allow volunteers to share what patterns they see.

  If you look up at the sky on a clear night, you will see patterns of stars. Some stars seem to make pictures in the sky. Today we will learn about some of these star pictures.

Instruct

Preparation for Reading

- Preview and pronounce the vocabulary term *star*.
- Direct the students to read page 7 silently to find out what the largest object in the solar system is.

Teach for Understanding

- Review what our solar system is and that the sun is at the center of it.

  What is a *star*? an object in space that produces its own heat and light energy

  What is *energy*? what is needed to cause change or to do work

  What is a star made of? a ball of gas

  What dots of light can you see in the sky? A few dots are planets and most of the dots are stars.

  Do planets make their own light? No, only stars make their own light.

  What is the name of the star in our solar system? the sun

- Compare the sun with other stars.

  Why does our sun look bigger than the other stars? It is much closer to Earth than any other star is.

  How does the sun compare in size to other stars? It is a medium-size star.

  Why is the sun important to Earth? It gives light and heat energy that are necessary for life on Earth.

- Direct attention to the picture at the bottom of the page.

  What do you see in the picture? Possible answers: sun, Earth, space

- Explain that the picture is a reflection of the sun in the oceans of Earth at sunrise as seen from space.

✓ the sun

7

Preparation for Reading

- Preview and pronounce the vocabulary terms *constellation*, *astronomer*, *telescope*, and *data*.
- Preview and pronounce the name *Lyman Spitzer, Jr.*
- Direct attention to the heading on page 8.
  What will you be reading about? how to observe the stars and planets
- Direct the students to read pages 8–9 silently to find out which tool can help us study objects in space.

Teach for Understanding

What is a *constellation*? a group of stars that form a pattern or picture in one area of the sky

How is a constellation similar to a dot-to-dot picture? The stars are connected with imaginary lines.

What shapes do the imaginary lines form? the shapes of animals, people, or objects

- Display the *Star Pattern* and *Ursa Major* visuals. Direct the students to compare the dots on both of the visuals.
  Can you find Ursa Major in the dot pattern?
  The constellation Ursa Major means "Great Bear."

- Display the *Big Dipper* visual.
  What big constellation do you see? Ursa Major

- Direct a volunteer to trace the smaller pattern in yellow with their finger.
  What smaller pattern is found in the Ursa Major constellation? the Big Dipper
  What does the Big Dipper look like? a large spoon with a long handle

- Explain that although the Big Dipper is a pattern of stars, it is too small to be considered a constellation.

What constellations have you heard of? Possible answers: Orion, the hunter; Cassiopeia; Aries, the ram; Leo, the lion; Cancer, the crab; Draco, the dragon; Gemini, the twins; Pleiades, the seven sisters

Why can you see some constellations in the sky in the winter but not in the spring? because the earth changes location as it revolves

What is the pattern of Earth's location changing as it revolves called? a seasonal pattern

How long have people been studying the stars? since ancient times

- You may want to share the Background information "Constellations in the Bible" at this time.

Who are *astronomers*? scientists who study the stars, planets, and moons

What did the astronomer Galileo see that changed what people knew about the stars? He used a telescope to look at the night sky and found more stars.

What is a *telescope*? a tool that helps us see objects that are far away

- Explain that Galileo did not invent the telescope but was the first to use it to study things in the sky. The use of this tool has helped many people study the sky since then.

Observing Stars and Planets

All through history people have looked at the stars and observed patterns. Many of these patterns are constellations. A **constellation** is a group of stars that form a pattern or picture in one area of the sky. A constellation is similar to a dot-to-dot picture. A dot-to-dot picture connects dots with lines. The stars in a constellation are connected with imaginary lines. The stars and imaginary lines form the shape of an animal, person, or other object. You may have seen the Big Dipper. It looks like a large spoon with a long handle. Its pattern is part of a larger constellation called Ursa Major. The name Ursa Major means "Great Bear."

Our view of outer space changes as the earth revolves around the sun. Some constellations can be observed in the sky in the winter but not in the spring. This is a seasonal pattern. Seasonal patterns happen because the earth changes location as it revolves. This is why some constellations cannot always be seen.

People have been studying the stars since ancient times. Scientists who study the stars, planets, and moons are called **astronomers**. In 1610 an astronomer named Galileo used a telescope for the first time to look at the night sky. He saw stars that no one had seen before. A **telescope** is a tool used to see objects that are far away.

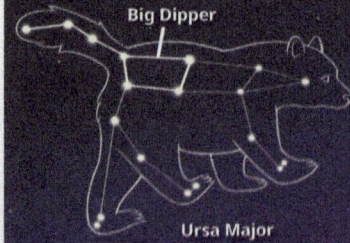

Big Dipper

Ursa Major

telescope

8

⚠ Helps

Identifying Constellations

Constellations are discussed in more detail in later grades. At this level, the students are not required to know the names of all the constellations. However, you may want them to name and recognize one or two of the ones commonly seen in your area.

ⓘ Background

Constellations in the Bible

Several constellations and star clusters are mentioned in Scripture. Examples are Orion and Pleiades (Job 9:9; 38:31) and the crooked serpent, Draco (Job 26:13).

Viewing a Star or Planet?

One way to tell the difference between planets and stars at night is by whether the light you see twinkles. Stars twinkle because of their distance from Earth and how Earth's atmosphere refracts their light. Because planets are nearer to Earth than the stars are and the light does not refract as much, planets appear as solid balls of light and do not twinkle.

Telescopes are still used today. We can use them to study objects in space. Most telescopes are long tubes with curved pieces of glass and mirrors inside. The glass and mirrors make faraway objects appear closer. Astronomers use telescopes to learn many things about the planets and stars. They do not look at the sun, though, because it is too bright. Computers record data from telescopes. **Data** is information or facts collected by making observations. The data helps astronomers learn about the universe.

Meet the Scientist

Lyman Spitzer, Jr.

In 1946 Lyman Spitzer Jr. (1914–1997) suggested putting a telescope in space. He thought a telescope above Earth's atmosphere would get better images. Spitzer worked on the first space telescope for more than 15 years. It is called the Hubble Space Telescope. In 1990 the Hubble Space Telescope was sent into space. Spitzer stayed active in astronomy until his death. In 2003 the new Spitzer Space Telescope was launched into space in his honor.

Hubble Space Telescope

Why can some constellations not be observed in the sky during all seasons?

9

What are most telescopes made of? long tubes with curved pieces of glass and mirrors inside

How do astronomers use telescopes? They use them to observe planets and stars.

What is *data*? information or facts collected by making observations

What records data from telescopes? computers

What does the data help astronomers do? learn more about the universe

- Discuss "Meet the Scientist."

Who suggested putting a telescope in space? Lyman Spitzer, Jr.

Why did he want to put a telescope in space? He thought that a telescope above the earth's atmosphere would get better images or pictures.

Why would a telescope in space take better pictures than one on the earth? Possible answers: Scientists on the earth are looking through the atmosphere. Scientists using a telescope on the earth cannot see through clouds. They must use a telescope on a clear night.

What telescope did Spitzer work on for more than 15 years? the Hubble Space Telescope

Which telescope was launched in 2003 in Lyman Spitzer, Jr.'s honor? the Spitzer Space Telescope

Why did scientists name the telescope after Lyman Spitzer, Jr.? Possible answer: because of his work on telescopes, and all that he did to improve our understanding of space

- Direct attention to the picture of the Hubble Space Telescope.

What do you think the telescope looks like?

because the earth changes location as it revolves; because of seasonal patterns

Background

Astronomer: Galileo Galilei

Galileo Galilei (1564–1642) was a well-known Italian astronomer. He did not invent the telescope, but he improved it and used it to study space. He was the first to look at our moon and discover that it was not smooth but rough and mountainous. He also studied Saturn, the phases of Venus, and sunspots. He discovered four of Jupiter's moons and observed Neptune, which he thought was a star. At the time, Earth was believed to be the center of the solar system. Galileo's observations supported the belief that the solar system revolves around the sun. He was condemned as a heretic and lived under house arrest from 1633 until his death.

Activity

Design Constellations

Materials: star stickers, per student; dark blue construction paper; crayons; writing paper

Provide each student with some star stickers. Direct them to arrange the stickers on the construction paper in any pattern they choose and then draw a line with a crayon to connect the stars to form a picture. Direct them to name their patterns and write two or three sentences describing their newly found constellations.

Preparation for Reading

- Preview and pronounce the vocabulary term *astronaut*.
- Direct the students to read page 10 silently to find out what an astronaut does.

Teach for Understanding

- Display the acronym *STEM*.
  Science
  Technology
  Engineering
  Math
- Explain that each letter stands for a different word. Read aloud each word and pronounce it together.
  What does the *S* in *STEM* stand for? science
  What is science? the study of God's world
  How do we study God's world? by using our five senses
  What are the five senses? sight, touch, taste, hearing, and smell
- Explain that STEM skills are used by people every day to solve or find the answers to problems.
  What is an *astronaut*? a person who is trained to travel and work in space
  What does the word *observe* mean? to use your senses to find out something
  How does an astronaut observe the world? with his senses
  What have astronauts done to help us know more about our planet and space? They take pictures of Earth.
  What can the pictures show us? how Earth has changed
  Where have several astronauts traveled? Possible answers: beyond Earth's atmosphere, to the moon
  What have astronauts collected? moon rocks and moon dust
- Read Genesis 1:28 aloud.

> Some Bible translations use the word *dominion* in Genesis 1:28. If the translation your students are using does, be sure to explain that the word *dominion* means to rule over, to care for, and to manage.

What command is an astronaut obeying? to have dominion over, or care for, the earth
Why is an astronaut's job very important? Elicit that an astronaut is following God's command to care for Earth; people have developed space travel to study Earth and the moon, as well as other heavenly bodies.
What do you think astronauts have done with the moon rocks and moon dust? Possible answers: studied them, displayed them
Why do you think an astronaut has to wear a spacesuit? to be able to breathe in space
Is there any air in outer space? No, it does not have an atmosphere that gives us air to breathe.
- Direct attention to the picture of the astronaut on the moon.
  What is the astronaut John Young doing? He is saluting the American flag.

Astronaut

What would it be like to travel to the top of Earth's atmosphere and beyond? Astronauts really do that. An **astronaut** is a person who is trained to travel and work in space. Many astronauts are scientists who do experiments in space.

We know more about our planet and space because of astronauts. We can see only a tiny part of Earth from the ground. Astronauts can see large parts of Earth from space. Astronauts observe and take pictures of Earth. An astronaut may take a picture of Earth before and after a bad storm. The pictures will show how that part of Earth changed.

Several astronauts have traveled beyond Earth's atmosphere. Some have traveled to the moon and walked on its surface. Some astronauts have collected moon rocks and moon dust.

astronaut spacesuit

Astronaut John W. Young on the moon during Apollo 16

10

Why is he doing that? to show respect or to honor the American flag
Why is the astronaut showing respect to the American flag? Possible answers: because it is his country's flag; because of what the flag stands for; because America did a great thing by landing men on the moon
- Direct attention to the picture of the astronaut spacesuit.
  Would you like to wear the spacesuit?
  Would you like to be an astronaut?
- You may want to share the Background information about Apollo 16 on page 11.

Apply

Activities

Study Guide, pages 3–4

These pages review the concepts taught in Lessons 1–2. After completion, direct the students to keep the pages for review for Test 1.

Assess

Quiz 1A

The quiz may be given at any time after completion of this lesson.

EXPLORATION

Solar Mobile

Inquiry Skills
- Classify
- Communicate

Cameras have been used to take pictures of the planets from space. These pictures were taken by manned and unmanned spacecraft. These pictures show what the planets look like close up. The planets are very far away from each other. Scientists have used the pictures of the planets to find out how the planets revolve around the sun. Scientists have made models to show what the planets look like in their orbits, or paths, around the sun.

In this Exploration, you will make a model of the planets in the solar system. You will put the planets in the same order that they revolve around the sun.

11

Objectives
- Conduct an internet keyword search to identify an interesting fact about each planet.
- Classify the planets by physical characteristics.
- Sequence the order of the planets in the solar system from the sun outward.
- Build a model of the solar system.
- Apply the inquiry skills of classify and communicate.

Materials
- See *Activities* page 5. (See Helps on *Teacher Edition* page 12.)
- Instructional Aid 1.2: *Solar Mobile Crossbars*, per student, copied on cardstock

Engage
- Introduce the Exploration by discussing the purpose of making models.
- What is one way you can find out more about things that are too big or too far away for you to see all at once? Answers will vary, but elicit that one way is by using a model.

  Why are you using a model to show the order of the planets in the solar system? I cannot see all the planets at one time.
- Explain that models do not have to show everything in order to help us learn, but we do need to know what the model does not show us so that we do not misunderstand.

Instruct

Preparation for Reading
- Direct the students to remove pages 5–10 from their *Activities* books. Direct the students to read page 11 and *Activities* pages 5–6 silently before beginning.

Teach for Understanding
What has been used to take pictures of the planets from space? cameras

What took the pictures from space? manned and unmanned spacecraft

What is a manned spacecraft? a spacecraft with a person or persons inside of it

What is an unmanned spacecraft? a spacecraft that operates itself without anyone being inside

What do the pictures from the spacecraft show us? what planets look like close up

Are the planets close together? No, they are far away from each other.

What have scientists created to show us what the planets look like in their orbits around the sun? models of the planets

What will you make a model of in this Exploration? the solar system

ⓘ Background

Space Exploration
Space exploration began when people from the Soviet Union launched Sputnik 1 into space. This was the first manmade object launched into space. The year 2019 marked the 50th year since man walked on the moon.

Apollo 16
On April 21, 1972, astronauts John W. Young and Charles M. Duke, Jr. landed and walked on the moon. This was the fifth time in history that astronauts had walked on the moon. Young and Duke used a space vehicle called a rover to travel nearly 27 km (17 mi) on the surface of the moon. They collected over 90 kg (200 lbs) of rock and soil. They also did experiments and took many pictures while they were on the moon.

James Webb Telescope
A telescope that is 4 times bigger than the Hubble Space Telescope was set to launch in 2021.

- Direct attention to the **Inquiry Skills**.

  What inquiry skills will you be using? classify and communicate

  What will you classify? the planets by their physical properties

  What are some physical properties? color, size, texture

  What will you communicate? a fact about your favorite planet

- Direct attention to the **Purpose**.

  What is the purpose of this Exploration? to model the solar system

Apply

- Direct attention to the **Procedure** list on *Activities* page 5. Instruct the students to follow and to check off Steps 1–2 as each is completed.
- Guide the students as they cut out each piece and punch holes as marked. *Note:* It may be easier to cut out each piece from the back of the pages.
- For Steps 3–5 arrange the students into groups of three or four, each group with a given planet or planets to research. Set up internet searches using the names of the planets as the keywords. You may also choose to provide research books. The students could also use their text-book to find facts.
- As the groups find their fact, write the fact for display. The other groups can use these facts on the back of their planets.
- Share any other guidelines you have established.
- Give help as needed as the students tie strings to the pieces for Step 6.
- Distribute the *Solar Mobile Crossbars* (Instructional Aid 1.2) for Step 7. Guide the students as they cut out and punch holes in the crossbars.

Solar Mobile

Name _____

Purpose

I want to make a model of the solar system.

Inquiry Skills

- Classify
- Communicate

Materials

- Solar Mobile planets
- Solar Mobile Crossbars
- scissors
- hole punch
- 14 pieces of string, 30 cm each

Procedure

1. Cut out the pictures of the sun and planets from pages 7 and 9. Punch a hole in each picture.

2. Find the pictures of the planets that your group has been given. Put the others aside.

3. Do an online keyword search using the name of the planets your group has been given. Identify one interesting fact about each planet.

4. Write one interesting fact on the back of each of the planets.

5. Tie 1 string to each hole in the planets and 2 strings to the hole in the sun.

6. Cut out the Solar Mobile Crossbars pieces that your teacher gives you. Cut a slit on the solid line at the center of each. Punch a hole at each circle on both pieces.

7. Slide the slits of both pieces together so that the bars form an *X*.

8. Tie 1 string to each of the 4 holes labeled *C* at the top center of the crossbars. Tie th other ends of the strings together to form a loop. The loop will be used to hang the mobile.

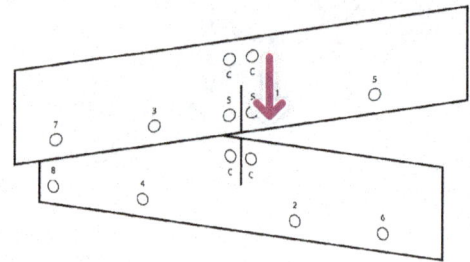

SCIENCE 3 · Activities

⚠ Helps

Solar System Mobile

Cut the string into lengths of 30 cm. You may choose for the students to practice measuring skills and cut their own string.

You may choose to provide additional resources for the students to find facts not included in the lessons.

If you choose to print the mobile sun and planet pieces on cardstock, copy pages 7–10 front to back from the *Science 3 Activities Answer Key*.

9. Tie each string on the sun to a hole marked *S* on the crossbars.

10. Arrange the planets in the order they orbit the sun. Tie the first planet to hole 1 of the crossbars. Tie the other planets to the number place that they are located from the sun.

11. Hold or hang your mobile by the center loop.

12. Share your mobile and tell a fact about your favorite planet.

Conclusions

1. What does your mobile model? *Possible answers: the planets in the solar system; the order of the planets*

2. Describe some characteristics of the solar system that you can observe from the model. *Possible answers would include physical characteristics of size and color.*

3. Describe some characteristics of the solar system that are not shown by the model. *Possible answers: the size of the planets; the distance between the planets; the surface of the planets*

Lesson 3 • Page 11
Exploration

- Direct the students to lay the eight planet pieces on their desks.
  Look at the pieces and name properties that the planets can be classified by. Students should name colors, sizes, or other physical properties of the planets.
- Allow time for the students to sort the pieces according to classification as each is named.
- Give help as needed as the students complete the steps in the Procedure. *Note:* Check each student's order of the planets before he begins tying the planets on the crossbars to save time in having to redo any tying.
- Discuss the **Conclusions**.

Activities
Exploration, pages 5–10
In this Exploration the students will make a model of the solar system.

Assess

Rubric
Use the prepared rubric or design a rubric to include your chosen criteria.

A rubric is a useful tool for assessing work that is not objective, such as the Exploration, Investigation, and STEM activities. Specific rubrics for each Exploration, Investigation, and STEM activity are located in *SCIENCE 3 Assessments*.

Objectives

- Identify characteristics of the inner planets.
- Explain why the earth's design is unique, using Isaiah 45:18. **BWS**
- Sequence and describe the phases of the moon.
- Differentiate between the waxing and waning of the moon's phases.
- Apply the inquiry skills of observe and communicate to the moon's phases.

Materials

- video of the solar system to scale
- large black circle and large white circle, 1 of each (both the same size)
- 1/2 a large white circle, to fit over the large black circle

Teacher Resources

- Visuals 1.5–1.7: *Inner Planets*; *Planet Order*; *Moon Mystery*

Vocabulary

- characteristic
- phases
- waxing
- waning

Engage

- Play the video of the solar system.

- Are the objects in the solar system the same size? No, some planets are larger than other planets; some planets are smaller than other planets; the sun is larger than all of the planets.

- What features did you notice about the planets? Possible answers: They have different sizes, different colors, some have moons, and some have rings.

  In this lesson you will learn about the first four planets in the solar system.

Instruct

Preparation for Reading

- Preview and pronounce the vocabulary term *characteristic*.
- Direct attention to the headings on pages 12–13.
  What planets will you read about first? Mercury and Venus
- Direct the students to read pages 12–13 silently to find out which planet looks like a very bright star.

Teach for Understanding

- Direct attention to the row of planets along the bottom of the pages.

- How are the planets shown along the bottom of the pages? Elicit that this shows the order of the planets from the sun.

- Point out that as a planet is discussed, it will be outlined in yellow at the bottom of the page.
- Point out that the different sizes do not show a comparison of the actual sizes of the planets. The row of planets is just a model.
  What is another name for the inner planets? the rocky planets

- Why do you think scientists call them "rocky planets"? because they are made up of rock

Lesson 4

Inner Planets and Moons

There are eight planets in our solar system. The first four planets, closest to the sun, are called the inner planets. The inner planets are also called rocky planets. These four planets share similar characteristics. A **characteristic** is a special feature. Inner planets are mostly made up of rock. Mercury, Venus, and Mars are more like Earth than the last four planets in the solar system. Only the inner planets have solid surfaces with land.

Mercury: Closest to the Sun

Mercury is the closest planet to the sun. Its orbit around the sun is the shortest. This means that Mercury takes less time than any other planet to revolve around the sun.

Though Mercury revolves quickly, it rotates slowly. Its slow rotation causes great temperature changes. The side that faces the sun is very hot. The side that faces away from the sun is very cold. The cold temperatures fall far below freezing.

Mercury is the smallest planet. It does not have a moon. Mercury is only a little larger than Earth's moon. Sometimes, Mercury can be seen in the sky after sunset.

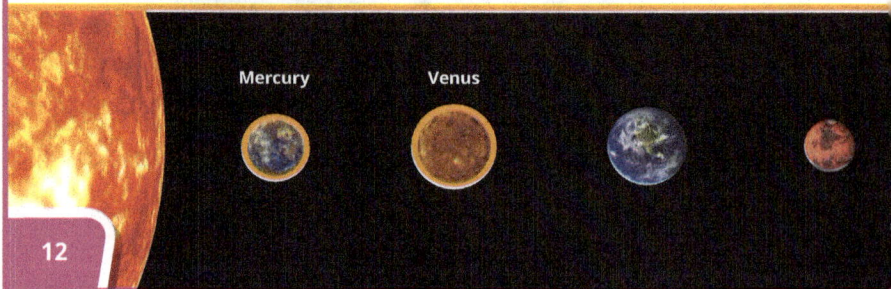

Mercury

Mercury   Venus

12

What is a *characteristic*? a special feature

What makes the inner planets different from the other planets? Only inner planets have solid surfaces with land.

- Why do you think Mercury, Venus, Earth, and Mars are called "inner planets"? Possible answers: They are in the inner part of the solar system; they are the closest planets to the sun, which is at the center of the solar system.

- Direct the students to refer to their solar mobiles and the row of planets modeled along the bottom of the pages as needed during the discussion.
- Display the *Inner Planets* visual. Guide the students as they complete the organizer on *Activities* page 11 as each planet is discussed.

A graphic organizer helps students make pictures that show relationships and connections between facts, information, and terms.

- Discuss characteristics of Mercury.
  What is the size of Mercury compared to the other planets? It is the smallest.
  Where is Mercury located in the solar system? It is the closest planet to the sun.
  Why does Mercury take less time to revolve around the sun than other planets? Possible answers: because it is the closest planet to the sun; its path around the sun is the shortest.

Venus: Earth's Twin

Venus is the second planet from the sun. The atmosphere of Venus makes it the hottest planet. The planet's thick layer of clouds makes it impossible to cool.

Venus is sometimes called Earth's twin. It is just a little smaller than Earth. Scientists have taken pictures of the surface of Venus using space probes. *Space probes* are unmanned spacecraft. The pictures show landforms like those found on Earth. Venus has mountains, valleys, plains, and volcanoes. Unlike Earth, no water can be found on Venus. Venus rotates in the opposite direction from Earth. Venus does not have a moon.

The clouds on Venus reflect the sun's light. This makes Venus look like a very bright star in the sky. Venus is sometimes called Earth's morning star or evening star. Venus is best seen just before sunrise and just after sunset.

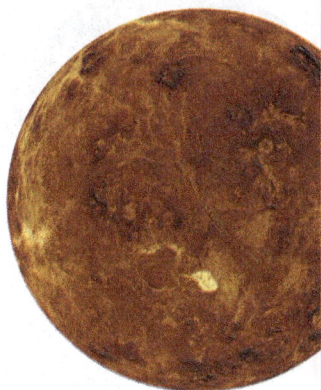

Venus

Fantastic Facts

On both Mercury and Venus, a day is longer than a year. This happens because the planets rotate slower than they revolve.

✓ **What landforms do Venus and Earth have in common?**

13

Why does Mercury have great temperature changes? It rotates slowly.

Why would the side facing away from the sun be so cold? Possible answer: The sun cannot warm it.

> Teaching for page 13 starts here.

What is the second planet from the sun? Venus

Why is Venus the hottest planet in the solar system? because it has a thick atmosphere

Which planet is Venus compared to? Earth

What is Venus sometimes called? Earth's twin

What are twins? Possible answers: Twins are two off-spring,"or babies," born at the same time. Twins often look like each other; they have similar characteristics.

What have scientists used to take pictures of the surface of Venus? space probes

What are *space probes*? unmanned spacecraft

What characteristic does Venus have that makes it similar to Earth? It has landforms such as mountains, valleys, plains, and volcanoes.

What are some differences between Venus and Earth? Venus does not have water; Venus rotates in the opposite direction from the earth; it does not have a moon.

If Venus rotates in the opposite direction from the earth, where does the sun rise and where does it set? The sun rises in the west and sets in the east.

What are other names that Venus is called? "morning star" because it looks like a star before sunrise and "evening star" because it looks like a star after sunset

What causes Venus to look like a star? The clouds reflect the sun's light.

- Direct attention to "Fantastic Facts."

Do Mercury and Venus rotate quickly or slowly? They rotate slowly.

✓ **mountains, valleys, plains, volcanoes**

ⓘ Background

Planet Names

Names for the planets have varied from country and culture since ancient civilization. Except for Earth, the common names we now use for the planets come from Greek and Roman mythology. Several of the planets were named by the International Astronomical Union (IAU). This group was founded in 1919 and includes distinguished astronomers from many different countries.

Preparation for Reading

- Preview and pronounce the vocabulary terms *phases*, *waxing*, *waning*, and the words *Isaiah*, *inhabited*, and *quarter*.

  Read the heading. What planet will you read about next? Earth

- Direct the students to read pages 14–15 silently to find out how Earth's moon appears to change shape.

Teach for Understanding

- Discuss the characteristics of Earth.

  What number is Earth in order from the sun? third

  What two characteristics about Earth allow life? It is the only planet with an atmosphere and liquid water that allow life.

  What is in the atmosphere that meets a need of people? air or oxygen

  Why is water important to life? Water is one of the needs of living things; no living thing can live without water.

  What makes Earth special? No other planet in our solar system was created for people.

- Read Isaiah 45:18 aloud.

  Why did God form Earth? to be inhabited

  What does the word *inhabited* mean? It means to be filled or occupied. Elicit that God created the earth to be inhabited by people.

  What three important characteristics of Earth give Earth its special design and make it possible for it to be inhabited? its atmosphere and liquid water, being at the perfect distance from the sun to provide the earth with temperatures warm and cool enough for living things, and having a perfect speed of rotation to keep the temperatures from being too hot or too cold

  How long does it take for Earth to rotate one time? 24 hours

  How long does it take Earth to revolve around the sun? one year (365¼ days)

- Share the Background information about a leap year, as desired.

  How many moons does Earth have? one

  How long does it take the moon to revolve around Earth? about 29 days

  What does the moon reflect? the sun's light

  Why is only one side of the moon lit at a time? because the side facing the moon acts like a mirror in reflecting the sun's light

  What is the pattern called that makes the moon move around Earth? a revolution

  Why does the earth's moon appear to change shape? because our view of the lit side of the moon changes as the moon revolves around the earth

  What are the different shapes of the lit part of the moon called? phases

Earth

moon

Earth: Our Home

Earth is the third planet from the sun. It is the only planet with both an atmosphere and liquid water that make life possible. No other planet in our solar system was created for people. The Bible tells us in Isaiah 45:18 that God formed Earth to be inhabited, or occupied. He perfectly designed Earth for people.

God placed Earth at the right distance from the sun. If it were closer to the sun, it would be too hot for people to live. If it were farther away, it would be too cold.

Earth rotates on its axis once every 24 hours, so one day on Earth is 24 hours. The speed of rotation helps keep the temperatures from being too hot or too cold. Earth revolves around the sun every 365¼ days. We call one orbit, or one complete trip around the sun, a year.

Earth's Moon

Earth has one moon. It takes about 29 days for the moon to revolve around Earth. The moon does not make its own light. It reflects the sun's light. Only one side is lit at a time. As the moon revolves around Earth, we usually cannot see all of the lit area from Earth. So the moon appears to change shape. The different shapes of the lit part of the moon are called the **phases** of the moon. If you watch the moon for about one month, you can see all of the phases.

Earth

14

ℹ Background

Leap Year

The calendar year does not have 365¼ days. There are 365 calendar days each year. Every four years there is an extra day in our calendar year. This extra day is February 29th.

Adding together four years of 365¼ days is the same as adding three years of 365 days and one year of 366 days.

365 ¼	365
365 ¼	365
365 ¼	365
+ 365 ¼	+ 366
1,461	1,461

When the side of the moon facing the earth is completely dark, the phase is called a *new moon*. After a new moon, a small lit part of the moon appears. The lit part that we can see begins **waxing**, or growing larger. The size of the moon appears to grow bigger each night. The lit moon waxes until half of the lit side of the moon can be seen. This phase is called the *first quarter*. The lit moon waxes each night until you can see a complete circle lit. This phase is called a *full moon*.

After a full moon, the phases appear in the opposite order. The lit part of the moon begins **waning**, or becoming smaller. The full moon wanes until half of the moon appears to be lit. This phase is called the *third quarter*. The moon continues to wane until it is a new moon again. It takes about 29 days for the moon to go from one new moon phase to the next new moon phase.

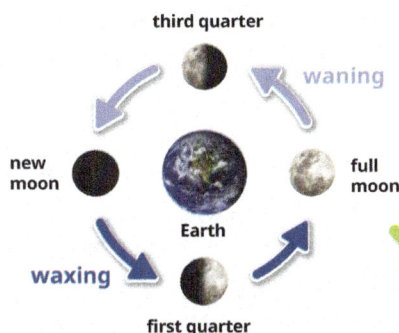

Which characteristics make the earth special?

- Display the large black circle.
  What is the phase of the moon when the moon is completely dark? new moon
- Place the white half circle over the right half of the black circle.
  Which phase of the moon comes after the new moon? first quarter
  How much of the moon is lit? half of the moon
- Explain that the moon does not go from all dark to half lit. The shape of the moon changes a little each night because the moon moves around the earth. As the moon moves, we usually cannot see all of the area that is reflecting the sun's light.
  What is it called when the size of the moon appears to grow larger? waxing
- Explain that when the moon waxes, the lit side of the moon is on the right.
- Display the white circle. Explain that the first quarter moon continues to wax.
  Which phase of the moon does it wax into after the first quarter? full moon
  Which phase of the moon is a completely lit circle? full moon
- Explain that the moon does not go from all light to half dark. The light begins to go away a little bit at a time until the moon is half dark.
  What is it called when the size of the moon appears to grow smaller? waning
- Explain that when the moon wanes, the lit side of the moon is on the left.
  Which phase of the moon comes after the full moon and is when half of the moon is lit? third quarter
- Place the white half circle over the left half of the black circle.
  The third quarter moon continues to wane. Which phase of the moon does it wane into after the third quarter? new moon
  How many days does it take the moon to go from one new moon to the next new moon? about 29 days
- Direct attention to the pictures of the phases of the moon.
  The moon is waxing between what moon phases? the new moon to the first quarter moon to the full moon
  The moon is waning between what moon phases? the full moon to the third quarter moon to the new moon
- Practice the phases by asking the students to name the moon phase as you illustrate the phases with the circles and the half circle.

an atmosphere, liquid water, being the perfect distance from the sun to provide temperatures for living things, and a perfect speed of rotation to keep the temperatures from being too hot or cold

👥 Differentiated Instruction

Note: The models use different scales and should not be combined.

Enrichment: Relative Sizes of the Planets Model
Materials: assorted beads, marbles, and balls as listed
Show the relative sizes of the planets with the items listed. Compare the finished model with the picture on *Student Edition* page 5 and discuss differences.

Mercury: small bead	Mars: small marble	Uranus: softball
Venus: large marble	Jupiter: basketball	Neptune: softball
Earth: large marble	Saturn: soccer ball	

Enrichment: Relative Distances of the Planets Model
Materials: 3 m string, 9 labels, meter stick
Make a label for the sun and each planet. Attach the string to the wall and the sun label at the left end of the string. As each planet is discussed, measure its distance from the sun and attach its label on the string. *Note:* Each measurement is from the sun, not from the previous planet. Compare the finished model with the picture on *Student Edition* page 5 and discuss differences.

Mercury: 4 cm	Mars: 14 cm	Uranus: 180 cm
Venus: 7 cm	Jupiter: 48 cm	Neptune: 230 cm
Earth: 9 cm	Saturn: 89 cm	

Preparation for Reading

- Preview and pronounce the words *Olympus Mons*, *Phobos*, and *Deimos*.
- Direct attention to the heading.

  What planet are you going to read about? Mars
- Direct the students to read page 16 silently to find out what Mars is sometimes called.

Teach for Understanding

- Discuss the characteristics of Mars.

  What number planet from the sun is Mars? fourth

  How much smaller is Mars than Earth? It is half the size of Earth.

  Have astronauts traveled to Mars? no

  How do scientists know so much about Mars? because space probes have landed on Mars and sent back pictures, video, and other data
- Share the Background information, as desired.

  Why is Mars called the red planet? because of its reddish color

  What is a landform? a shape of the land such as a mountain, canyon, hill, valley, plain, or island

  Which landform does Mars have that scientists believe to be the tallest in the solar system? a mountain

  What is the name of the mountain? Olympus Mons

  How many moons does Mars have? two
- Display the *Planet Order* visual and introduce the mnemonic device for learning planet order. (**M**y **V**ery **E**ducated **M**other **J**ust **S**erved **U**s **N**achos.) Point out that the first letter of each word is the same as the first of each planet in order from the sun.

the red planet

Mars: The Red Planet

Mars is the fourth planet from the sun. Mars is one of the brightest objects in Earth's night sky. Mars is about half the size of Earth.

Mars is the most explored planet other than Earth. No person has been to Mars, but scientists still know much about it. Space probes have landed on Mars and sent back pictures, video, and other data.

The surface of Mars is covered with red dust. Because of its reddish color, it is called the red planet. Scientists believe that Mars has the highest mountain in the solar system. It is called Olympus Mons. Mars also has two tiny moons named Phobos and Deimos.

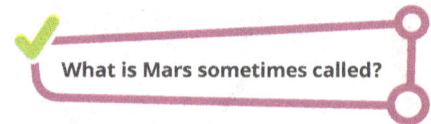

Mars

What is Mars sometimes called?

Mars

16

ⓘ Background

Space Probes and Rovers

A space probe is an unmanned robotic spacecraft that travels in space and sends pictures and other data to scientists on Earth. A space rover is a type of space probe. It is a vehicle with scientific equipment and can move around on the surface of another planet or a moon.

Moon Names

The Greeks used to call Earth's moon Selene or Artemis. The Romans called it Luna. In English it was called Moon. When moons of other planets were later discovered, the word moon was also used to describe them.

EXPLORATION

Moon Mystery

Inquiry Skills
- Observe
- Communicate

Earth's moon is round, but it does not always look round. When the lit part of the moon waxes or wanes, the moon looks like it is changing shape. The shapes that you see are the moon's phases.

In this Exploration, you will observe and record data about the phases of the moon.

Preparation for Reading

- Direct the students to remove pages 13–14 from their *Activities* books. Direct the students to read page 17 and *Activities* pages 13–14 silently before beginning.

Teach for Understanding

Why does Earth's moon not always look round? because the moon looks like it is changing shapes as it waxes and wanes

What are the shapes of the moon called? phases

⚠ **Helps**

Moon Exploration Follow-Up
After a month's time follow up on the Exploration. At that time you will want to discuss the Conclusions with the students.

- Direct attention to the **Inquiry Skills**.

  What inquiry skills will you be using? observe and communicate

  What will you observe? the phases of the moon

  What are the phases of the moon? new moon, first quarter, full moon, third quarter

  What will you communicate? what I learned about the phases by observing the moon

- Direct attention to the **Purpose**.

  What is the purpose of this Exploration? to observe the phases of the moon

- Direct the students to pick one night a week and follow Steps 1–3 each time that they observe the moon for four weeks.

You may want to consult a lunar calendar for your area to help any students that are having difficulty with their observations.

Moon Mystery

Name _____

Dear Parent,

   Your child needs to view the moon phase one time each week for four weeks. If the moon is not visible, your child may try another night. Sometimes it is easier to see the moon phase when your child first gets up in the morning instead of staying up to do so. Each night that your child observes the moon, the date and moon phase should be recorded on the Observation Chart. The Exploration should be returned on _____.
Thank you,

Your Teacher

Purpose

I want to observe the phases of the moon.

Procedure

☐ 1. Record the date of your observation on the Observation Chart.

☐ 2. Observe the moon in a clear night sky once a week.

☐ 3. Shade the circle, if needed, on the chart to create the phase of the moon that you see.

☐ 4. Repeat steps 1–4 for three more weeks during the month.

Inquiry Skills

- Observe
- Communicate

Materials

☐ clear nighttime sky

SCIENCE 3   *Activities*

⚠ Helps

Sky Conditions

The moon may not be visible every night when the student makes his observations. The phases of the moon are visible only when clouds are not covering the view of the moon and when the moon is not new. Direct the students to cross out any calendar days when they try to observe the moon and it is not visible. If the students are unable to observe the moon at night, sometimes the moon is visible for observation early in the morning. Direct the students to observe the moon at least once a week for four separate weeks, and create the phase of the moon for those days on the observation chart.

Observation Chart

Week	Date	Moon Phase
1		◯
2		◯
3		◯
4		◯

onclusions

. How many times did you see a full moon? _____

2. Were you unable to see the moon on some nights?　◯ yes　　◯ no

Why were you unable to see the moon? _Possible answers: because of cloud cover;_

because there was a new moon

3. What did you learn about the phases of the moon by observing the moon? _Sample_

answers: I learned that each phase of the moon takes several days to change from one

to another; the moon waxes for several days before a full moon; the moon wanes for

several days before a new moon.

Lesson 4 • Page 17
Exploration

- Direct attention to the "Observation Chart" on *Activities* page 14.
  How many days each week are you supposed to observe and record the moon phase? 1 day
  How many days will you record on the chart? 4 days
  Do you think the moon phase could be the same for several days? yes
- Demonstrate how to record a moon phase on the *Moon Mystery* visual.
- Write the date you would like the Exploration to be returned for display.
- Direct the students to write the date on the line in the Parent Letter on *Activities* page 13.

Apply

Activities
Inner Planets, page 11
This page was completed during the lesson.

Exploration, pages 13–14
In this Exploration the students will observe the phases of the moon.

Study Guide, pages 15–16
These pages review the concepts taught in Lesson 4. After completion, direct the students to keep the pages for review for Test 1.

Assess

Rubric
Use the prepared rubric or design a rubric to include your chosen criteria.

Objectives
- Identify characteristics of outer planets.
- Sequence planets in order from the sun outward.
- Describe what asteroids are and where they can be found.
- Compare and contrast planets and dwarf planets.

Materials
- English pea
- baseball
- globe

Teacher Resources
- Visuals 1.8; 1.6; 1.1: *Outer Planets*; *Planet Order*; *Solar System*

Engage

Which planets have you learned about? Mercury, Venus, Earth, and Mars

What do we call this group of planets? inner planets or rocky planets

- In this lesson we will learn about four other planets in our solar system. These planets are very different from the inner planets.

Instruct

Preparation for Reading
- Preview and pronounce the words *Jupiter*, *Uranus*, and *Titan*.
- Direct the students to read pages 18–19 silently to find out what Saturn's rings are made of.

Teach for Understanding

What are the surfaces of the outer planets made of? gas

What are the four outer planets sometimes called? gas giants

Why do you think scientists call them gas giants? Possible answer: They have surfaces made of gas and are very large.

- Display the *Outer Planets* visual. Guide the students as they complete the organizer on *Activities* page 17 as each planet is discussed. Additional cells and facts may be added.
- Discuss characteristics of Jupiter.

What number planet is Jupiter from the sun? fifth

How does Jupiter compare in size with the rest of the planets? It is the largest.

- Display the pea and baseball.

Which of these would represent Earth? the pea

Which would represent Jupiter? the baseball

How long does it take Jupiter to rotate? about 10 hours

How long does it take Earth to rotate? 24 hours

How much faster does Jupiter rotate than Earth? about 14 hours faster

What covers Jupiter? colorful clouds of gas

- Direct attention to the close-up picture of Jupiter's Great Red Spot. Explain how the inset shows more detail of what the spot looks like.

What do scientists think Jupiter's giant red spot is? a huge storm

---

Outer Planets

The last four planets from the sun are called the outer planets. The outer planets are Jupiter, Saturn, Uranus, and Neptune. These planets have surfaces that are made of gas. Scientists call these planets gas giants because they are so large. They are also farther apart than the inner planets. The distance between two of the outer planets is greater than the distance from the sun to Mars.

Jupiter: The Largest Planet

Jupiter is the fifth planet from the sun. It is the largest planet in our solar system. If Earth were the size of a pea, Jupiter would be the size of a baseball.

Jupiter also spins faster than any other planet. Jupiter takes only about 10 hours to rotate, so a day on Jupiter lasts for only 10 hours.

Jupiter is covered with colorful clouds of gas. It has a giant red spot that scientists believe is a storm.

Jupiter has 79 known moons. The four largest moons are each bigger than the planet Mercury.

Jupiter

Great Red Spot

18

How many known moons does Jupiter have? 79

How big are the four largest moons? Each is bigger than the planet Mercury.

Saturn: The Ringed Planet

Saturn is the second-largest planet. It is almost as large as Jupiter. The distance from the sun to Jupiter is about the same as the distance from Jupiter to Saturn. Saturn is best known for its rings. Its rings are made of ice and rocks. Each ring travels in its own orbit around the planet. Some pieces of the rings are as small as dust, but others are as big as buildings!

Saturn has 53 named moons. The largest one is called Titan. In old Greek stories, the Titans were giants. The moon Titan is a giant as well. It is larger than the planets Mars and Mercury combined.

Saturn's rings

Saturn

Which planet rotates faster than any other planet?

Jupiter

Saturn

19

- Discuss the characteristics of Saturn.

Which planet is the second largest? Saturn

How could you describe its distance from the sun? Possible answers: Saturn is as far from Jupiter as Jupiter is from the sun; it is almost twice as far from the sun as Jupiter is.

Which position is Saturn from the sun? sixth

What is Saturn best known for? its rings

What are Saturn's rings made of? ice and rocks

Are all the pieces of ice and rock the same size? No, some are small, but others are as big as buildings.

Why can we not see the large pieces of ice and rock in Saturn's rings? Saturn is so far away from the sun that we would need a telescope to see those details.

- Direct attention to the inset picture of Saturn's rings. Explain how it shows more detail of what the rings look like.

How does the number of Saturn's moons compare with the number of Jupiter's moons? Saturn has fewer moons than Jupiter.

How many of Saturn's moons have been named? 53

What is the name of the largest one? Titan

Why was it probably called Titan? It is a giant moon like the Titan giants in the old Greek stories.

How big is Titan? bigger than Mars and Mercury combined

- Explain that Titan is so large that it has its own atmosphere like planets do.

Jupiter

(i) Background

Galilean Moons

The four largest moons of Jupiter are called the Galilean moons after Galileo, who discovered them in 1610. These moons can often be seen with a small telescope or even binoculars.

Saturn's "Handles"

When Galileo observed Saturn's rings, he thought they looked like handles or that Saturn had two large moons on either side. Over several years Galileo observed the rings as different shapes as Saturn was aligned differently with Earth. From Galileo's writings, it appears that he never understood that the "handles" were actually rings.

Planets with Rings

Jupiter, Saturn, Uranus, and Neptune each have rings, but Saturn is best known for its colorful and numerous rings.

Preparation for Reading

- Preview and pronounce the vocabulary terms *equator* and *asteroids*.
- Preview and pronounce the words *Uranus*, *Triton*, *dwarf*, *Ceres*, *Pluto*, and *Eris*.
- Direct the students to read pages 20–21 silently to find out where most asteroids in the solar system are found.

Teach for Understanding

- Discuss characteristics of Uranus.

  What number planet from the sun is Uranus? seventh

  What gives Uranus its pale blue-green color? the gases in its atmosphere

  What is special about the way Uranus rotates? It spins on its side so that the sun shines mostly on its poles.

- Use the discussion about Uranus to review terms such as equator, axis, rotates, and poles.

- Display the globe. Point out its slight tilt. Explain that most planets tilt slightly as Earth does.

  Where do you think Earth and most of the other planets get the most sun? the areas around their equators

- Tip the globe to the side to show how Uranus is tilted.

  What parts of Uranus get the most sunlight? the poles

  How many rings does Uranus have? at least 13

  How many moons does Uranus have? 27

- Discuss the characteristics of Neptune.

  What number planet from the sun is Neptune? eighth

  Why do you think Neptune is very cold? It is far away from the sun and the sun's warmth cannot reach it.

  What color is Neptune? blue

  How fast does the wind blow on Neptune? more than 1600 km (994 mi) per hour

  Have you ever been in a strong wind?

- Explain that winds in a strong thunderstorm can gust in the 80–104 km (50 to 65 mi) per hour range.

  What is one of the coldest objects in space? Neptune's moon, Triton

  Does Neptune have rings? yes

- Review the order of the planets using the *Planet Order* visual.

Teaching for page 21 begins here.

What are asteroids? small rocky objects in space that revolve around the sun

Do asteroids all look alike? No, they are different sizes and shapes.

Where are most asteroids found? in the area between Mars and Jupiter

What is this area called? an asteroid belt

- Direct attention to the *Solar System* visual. Point out the asteroid belt between Mars and Jupiter. Explain that the asteroid belt forms a ring-shaped area much like a belt forms a ring around someone's body.

Uranus

Neptune

Uranus: The Sideways Planet

Uranus is a pale blue-green color. Its color comes from the gases in its atmosphere. Uranus has at least 13 rings and 27 moons.

Uranus rotates in a way no other planet does. The axis of Uranus runs sideways. Other planets are tilted slightly on an axis, so the sun shines mostly on their equators. The **equator** is the imaginary line around the widest part of a planet. Uranus spins on its side so that the sun shines mostly on its poles.

Neptune: Farthest Planet from the Sun

Neptune is known as a blue planet. It is the farthest planet from the sun. Neptune is a very cold planet covered with gas. The wind on the surface of Neptune sometimes blows over 1600 kilometers (994 miles) per hour.

Neptune has several pale rings and 14 moons. Triton is Neptune's largest moon. It is one of the coldest objects in space.

20

ⓘ Background

Uranus

Uranus is considered to be the first planet discovered in modern times. It was discovered by William Herschel in 1781. He wanted to name the planet after King George III of England, but it was finally decided to continue the pattern of naming the planets after mythological gods. Most information that we have about Uranus is a result of the space probe Voyager II.

Neptune

There is much debate over who actually discovered Neptune. Both Urbain-Jean-Joseph Leverrier of France and John C. Adams of England identified it in 1846. Finally both men were given credit for its discovery. What is not debated is that it was found because of mathematical calculations based on irregularities in Uranus's orbit. Neptune cannot be seen without a telescope. Most information that we have about Neptune is a result of the space probe Voyager II.

Blue Color of Uranus

The methane gases on Uranus and Neptune are what give the planets their bluish colors.

Asteroids

Asteroids are small rocky objects in space that revolve around the sun. They are not round like planets but instead are different shapes and sizes. Some are hundreds of miles across. Others are as tiny as pebbles. Most asteroids are found in the asteroid belt. The asteroid belt is located between Mars and Jupiter.

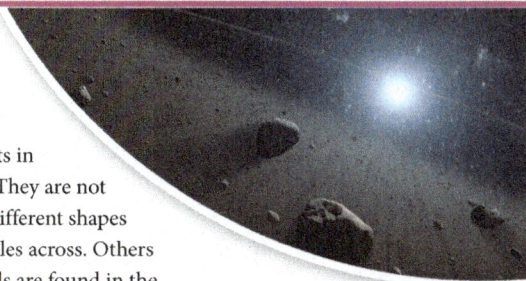

the asteroid belt

Dwarf Planets

Some objects in the solar system are planets and some are dwarf planets. A dwarf planet revolves around the sun. It is the size of a large asteroid but is nearly round like a planet. A dwarf planet is smaller than a planet.

Ceres, Pluto, and Eris are dwarf planets. Ceres is the largest object in the asteroid belt. Pluto and Eris are not in the asteroid belt. The size and shape of Ceres make it a dwarf planet. Pluto was once called a planet, but it was classified as a dwarf planet in 2006. Eris is one of the largest dwarf planets in our solar system.

Planets, asteroids, and dwarf planets show the greatness of God. No matter where you look in the universe, you can be amazed at God's creation.

Pluto

✓ Which planet rotates on its side?

Uranus   Neptune

21

How are dwarf planets different from asteroids? Dwarf planets are the size of a large asteroid but are nearly round like a planet. Asteroids are not round and have many different shapes and sizes.

What dwarf planet is located in the asteroid belt between Mars and Jupiter? Ceres

What are the names of two other dwarf planets? Pluto and Eris

Are Pluto and Eris in the asteroid belt? no

When was Pluto classified as a dwarf planet? in 2006

Which dwarf planet is one of the largest in the solar system? Eris

How are planets and dwarf planets alike? Possible answers: nearly round, orbit the sun

How are planets and dwarf planets different? Possible answers: Planets are larger than dwarf planets; dwarf plants are the size of a large asteroid.

The universe is huge. What does the universe contain? stars, planets, moons, asteroids, dwarf planets

What does the size of the universe show us about God, who created it all? Possible answer: It shows the greatness of God.

- Read aloud Psalm 8:1, 3–4.

What is the psalmist referring to as "heavens"? moon and stars

What does the psalmist say about the heavens? how great and big God's creation is

Even though God made the universe, what did God still think of? God is mindful of, or thinks about, man

How do these verses describe the heavens, moon, and stars? as the work of God's fingers as ordained, or set in motion, by God.

- Explain that only our wonderful God could make such a grand and glorious universe.

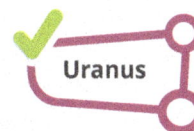

✓ Uranus

Apply

Activities
Outer Planets, page 17
This page was completed during the lesson.

Study Guide, pages 19–20
These pages review the concepts taught in this lesson. After completion, direct the students to keep the pages for review for Test 1.

Assess

Quiz 1B
The quiz may be given at any time after completion of this lesson. Material for this quiz comes from the Study Guides in Lessons 4 and 5.

ⓘ **Background**

Ceres
Ceres was discovered in 1801 and was initially thought to be a planet. Later it was classified as an asteroid. In 2006 it was reclassified as a dwarf planet. It is smaller than Pluto.

Pluto
Pluto was discovered in 1930 and was considered a planet for many years. In 2006 it was reclassified as a dwarf planet. Its orbit overlaps the orbits of Eris and Neptune.

Eris
Eris was discovered in 2005. Its classification led to a universal debate about the definition of a planet. Eventually it was decided that Ceres, Pluto, and Eris would all be classified as dwarf planets. The orbit of Eris is elliptical. At its farthest point it is almost twice the distance from the sun as Pluto is at its farthest point.

Objective
• Recall terms and concepts from Chapter 1.

Apply

Review
• Material for Test 1 will come from the Study Guides on *Activities* pages 3–4, 15–16, and 19–20, as well as from Quizzes 1A and 1B. You may review any or all of the material during the lesson.
• You may review Chapter 1 by playing the Orbit the Sun review game or a game from the Game Bank in the back of this book and on TeacherToolsOnline.com.

🏆 Review Game

Orbit the Sun

Materials: magnetic token, for each team

Group the students into two or more teams, with each team representing a planet. Draw a sun and the same number of orbits as there are teams for display on a magnetic board. On each orbit draw eight stops. Place a token at one of the stops on each orbit as its starting point.

For each correct answer to a review question, the student moves the team's token to the next stop. The winner is the first "planet" to complete its orbit, or the one that completes the most orbits in the given time.

NOTES

NOTES

_____

_____

_____

_____

_____

_____

_____

_____

_____

_____

_____

_____

_____

_____

_____

_____

_____

_____

_____

_____

_____

_____

_____

_____

_____

_____

_____

Objective
- Apply terms and concepts from Chapter 1.

Assess
- Administer Test 1.

Chapter Objectives

- Develop a biblical explanation for the design of the atmosphere, climate, and weather.
- Infer proper use of meteorological tools to observe and record weather data.
- Identify patterns in the weather.
- Explain how weather and climate vary by seasons and locations.
- Formulate ideas for safety during severe weather.
- Apply the engineering design process to solve a real-life problem.

Lesson Objectives

- Explain how the atmosphere and sun affect our weather. **BWS**
- Predict what future weather might be, using given weather patterns.
- Infer which meteorological tool should be used in various scenarios.

Materials

- umbrella
- sunglasses
- gloves
- rain gauge
- thermometer
- weathervane
- anemometer

Teacher Resources

- Visuals 2.1–2.2: *Weather Map*; *Weather Patterns*

Vocabulary

- weather
- predict
- meteorologist
- weather satellite
- weather forecast
- map key
- rain gauge
- thermometer
- weathervane
- anemometer

Chapter Introduction

> Use completed Looking Ahead, *Activities* page 21, to generate a discussion on what the students already know about the weather and climate.

- Direct attention to the chapter title and the picture of rainbow lightning. Provide time for the students to leaf through the chapter, looking at the headings and pictures. Discuss what they think the chapter will be about.
- What does the picture tell you about the weather? Possible answers: It is stormy. There is lightning. There is a rainbow.

Big Question

- Ask a volunteer to read aloud the Big Question.
- Explain that the students will find the answer to the question as they read the chapter.

Big Question
How does learning about weather help us serve others?

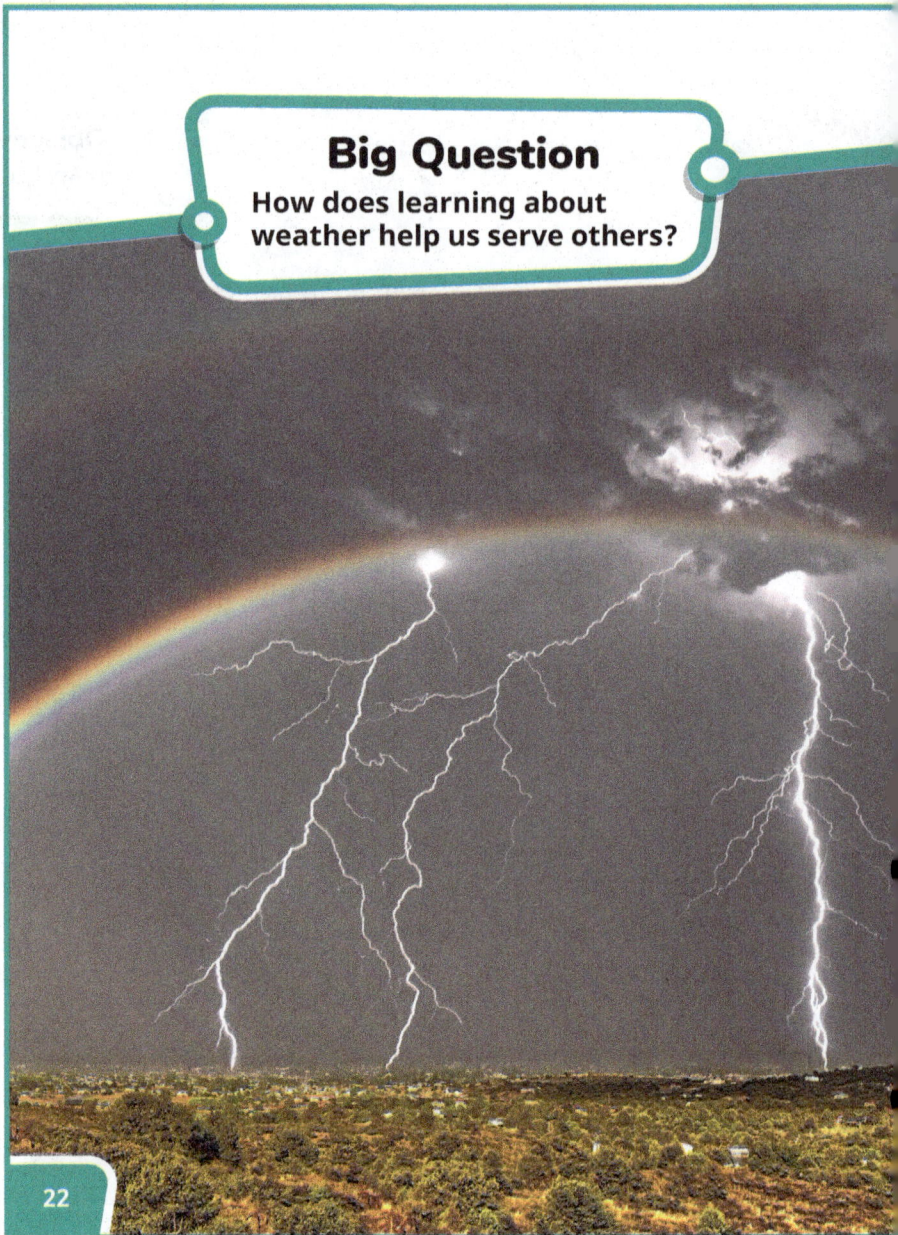

22

ⓘ Background

Chapter Photo

The photo on pages 22–23 is a weather phenomenon called rainbow lightning. This happens most frequently in desert locations where dry air meets humid air in spectacular thunderstorms.

> Visit TeacherToolsOnline.com for resources to enhance the lessons.

↪ Preparing Ahead

Lesson 12 STEM

The Lesson 12 STEM activity will require materials to be gathered and prepared ahead of time. Consider asking students to bring in empty cereal or tissue boxes so that there is enough cardboard for everyone.

Lesson 13 Materials

Look ahead to the materials list in Lesson 13 for required preparation. The locations should fall clearly within the climate zones that are discussed in Lesson 13.

Chapter 2

Weather and Climate

At sunrise, the sky was a beautiful red. Now the clouds are gray and cover the sky. The wind is starting to blow harder. The birds have stopped chirping. What do these things tell you? They tell you that a storm is coming. God has given us many clues that tell us what the weather might be.

Knowing what kind of weather is coming helps you to plan. If a snowstorm is coming, you need warm clothes, boots, and snowsuits. If it is stormy weather, you need to stay indoors. If it is hot and sunny, you can grab a towel and go swimming. Learning about different types of weather is exciting!

23

- Display an umbrella, sunglasses, and gloves.
  When would you wear sunglasses? when the sun is shining
  When would you need an umbrella? when it is raining
  When would you wear gloves? when it is cold outside
  What do you call times of sun and rain and cold? weather
  What kind of weather is your favorite? Why?

Instruct

Preparation for Reading

- Direct the students to read page 23 silently to find out what kinds of clues God has given to tell about the weather.

Teach for Understanding

What are some clues that God has given that tell what the weather might be? Clouds turn gray. The wind is blowing. Birds stop chirping.

- Share other clues from nature about the weather as time allows.
  Why do you check to see what kind of weather is coming? to plan
  What kinds of plans could you make by knowing the weather? warm clothes for snow, stay indoors if a storm is coming, go swimming if it is hot

- Explain that in this chapter the students will learn about changes in the weather and climate.

ⓘ Background

Living Things Respond to Weather

Plants and animals do not predict the weather, but they often notice and respond to weather clues faster than people do. For example, a dog can hear thunder from farther away than humans can and may become restless and fearful before its owner even realizes a storm is coming. Cows often lie down before a thunderstorm comes. They also stay close together if bad weather is coming. Birds tend to roost close to the ground before rain or snow comes. The scarlet pimpernel flowers close their petals when the humidity is high, often right before a rainstorm. Tulips open when the weather is fair and dry.

Preparation for Reading

- Preview and pronounce the vocabulary term *weather*.
- Review the vocabulary terms *atmosphere* and *gravity* from Chapter 1.

  What is *atmosphere*? the layer of gases that surround the earth

  What is *gravity*? a force that pulls all objects to the center of the earth

- Direct the students to read page 24 silently to find out how the sun and atmosphere affect our weather.

Teach for Understanding

What is *weather*? the condition of the air at a certain time and place

How does the sun affect our weather? It heats the air and causes the air to move.

Where does all our weather happen? in the atmosphere

How does the atmosphere help protect the earth? It protects the earth like a blanket. It keeps the temperature on the earth from being too hot or too cold for people, plants, and animals to live.

- Explain that our atmosphere shows that God cares for us. He planned every detail about the earth. None of it happened just by chance.

How do the sun and the atmosphere work together to keep the earth at the right temperature for people, plants, and animals? The sun heats the earth and the atmosphere covers the earth like a blanket to keep the heat in.

- Direct attention to the lighthouse pictures.

  What is the weather like in the top picture? The sky is blue; the sun is shining.

How do we know the weather has changed in the second picture? The waves are high and rough. The sky is gray.

- Direct attention to the picture and caption at the bottom of the page.

What part of the sky can give you clues about the weather? Possible answer: clouds

in the atmosphere

Weather

Weather is the condition of the air at a certain time and place. When you say it is hot or cold outside, you are talking about the weather. Weather changes during the year, and sometimes changes during the day. When you wake up, it might be cool and sunny. After lunch it might be hot and stormy.

The sun and the atmosphere affect our weather. The sun heats the air and causes the air to move. All our weather takes place in the atmosphere. The atmosphere is pulled toward the earth by gravity. Changes in the atmosphere affect the weather. God designed the atmosphere to protect the earth like a blanket. It helps keep the temperatures on the earth from being too hot or too cold. It keeps the earth at the right temperature for people, plants, and animals to live.

How has the weather changed in these pictures?

Where does all weather take place?

People can study the sky to give them clues about the weather.

24

ⓘ Background

Atmosphere

If the earth were the size of an apple, the atmosphere would be thinner than the apple's skin. The atmosphere has several layers that work together to protect the earth and give it a controlled climate.

Studying Weather

Have you ever wondered what the weather will be like on a certain day? How can people know what the weather will be? Only God knows the future, but people can make predictions about what might happen. To **predict** means to make a careful guess at what may happen. For thousands of years people have looked at the sky, the wind, and other signs in nature to predict the weather. Today we know more about how to study and predict the weather.

A **meteorologist** is a scientist who studies weather. He uses tools to gather data about the weather. One important tool is a weather satellite. A **weather satellite** is used to gather information about the weather and climate of the earth. A weather satellite is sent high into the atmosphere. It can report wind speeds in the atmosphere. Weather satellites also measure temperatures and take pictures of the earth's surface.

A meteorologist uses computers to research past weather patterns. Those patterns help him predict what kind of weather is coming. A prediction of coming weather is called a **weather forecast**. These predictions help to keep people safe. We study the weather to obey God by taking care of the earth and serving other people.

25

Preparation for Reading

- Preview and pronounce the vocabulary terms *predict*, *meteorologist*, *weather satellite*, *weather forecast*, and *map key*, as well as the word *symbol*.
- Review the vocabulary term *data* from Chapter 1.
- Direct the students to read pages 25–26 silently to find out what a weather map shows.

Teach for Understanding

Who is the only one who truly knows the future? God

Can people make a guess at what can happen in the future? yes

What do we call a careful guess at what may happen? to predict

- Direct attention to the picture at the bottom of the page.

What do we call a scientist who studies the weather? a meteorologist

What are some things a meteorologist might do? He uses weather tools to gather data about the weather. He uses computers to research past weather patterns and make predictions about future weather. He records data and predictions on weather maps.

What does research mean? to study something, to gather facts about something

What are some kinds of weather data that a meteorologist might record? Possible answers: temperature, wind speed, amount of precipitation, wind direction

- Direct attention to the picture on the right side of the page.

What weather tool in the atmosphere do meteorologists use? a weather satellite

- Discuss Background on weather satellites as time allows.

What information can a meteorologist get from a weather satellite? wind speeds in the atmosphere, temperatures on earth, pictures of the earth's surface

How does knowing the wind speeds in the atmosphere help meteorologists predict the weather? Changes in the atmosphere affect the weather. If the wind speeds change, the weather might change.

What is a prediction of future weather called? a weather forecast

ⓘ Background

Weather Satellites

There are two main types of weather satellites used to collect data about the earth.

One type of weather satellite is the Geostationary Orbiting Environmental Satellite (GOES). The GOES hovers over a single point above the earth and is useful for monitoring the weather. Most weather monitoring and forecasting comes from the GOES satellites.

The United States currently operates two GOES satellites to receive weather information. One is over the East Coast and one is over the West Coast. Meteorologists can overlap the images from the GOES and use them to create weather maps of the entire United States.

Another type of satellite is the Polar Orbiting Environmental Satellite (POES). These satellites travel in a circular orbit from pole to pole. This allows the POES to see the earth twice in a 24-hour period. These satellites provide details and images of storms, volcanoes, wildfires, and other natural events on Earth.

What is a weather map? a map that shows weather data and predictions

What kinds of information might be on a weather map? temperature, rain, snow, or wind

- Display the *Weather Map* visual.

Note: The temperatures given in the *Weather Map* visual, as well as the *Activities* pages, are in degrees Fahrenheit. Most weather apps will allow the user to select Fahrenheit or Celsius. If Celsius is used in your location, make adjustments as needed.

What do the colors on this map show? different temperatures

- Direct attention to the temperature key on the visual.

Which temperatures are warmer—20s or 80s? 80s

What colors show warmer temperatures? yellow, orange, red

What colors show cooler or cold temperatures? green, blue, and purple

How do you know the warm temperatures are yellow, orange, and red? The map key shows what the colors for the temperatures mean.

What do the symbols in the blue and yellow areas of the map tell you about those places? It is probably cold in the blue area because the symbols show snow. It is warmer in the yellow area because those symbols show rain.

- Direct attention to the weather map at the top of the page.

Which areas on the map are warmer? the ones that are yellow, orange, and red

Which areas on the map are colder? the ones that are green and blue

- Direct attention to the weather map at the bottom of the page.

What kind of weather does this map show? sunny, snowy, rainy, stormy, cloudy

- Direct attention to the *Weather Patterns* visual.
- Invite students to observe patterns in the weather forecast. Direct attention to the top chart.

What might temperatures be like on Thursday and Friday? Temperatures will probably be in the 30s or 40s.

- Direct attention to the bottom chart.

What type of weather might happen on Friday and Saturday? It could be sunny or cloudy on Friday. It might rain or snow on Saturday.

- Direct students to the "Try It Yourself!" activity. See the web links on TeachersToolsOnline.com for weather maps, or use your favorite weather app.

a weather satellite

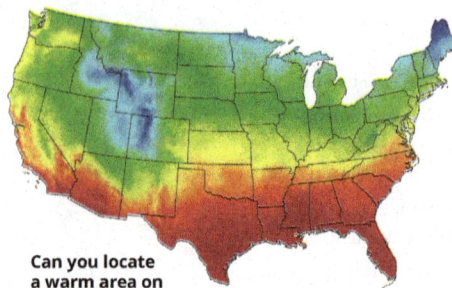

Can you locate a warm area on the map?

Some meteorologists record data and predictions on weather maps. These maps may show the temperatures for cities and towns. Colors are used on a weather map to show temperature. Yellow, orange, and red usually show warm temperatures. Green, blue, and purple usually show cool temperatures.

Symbols are also used on weather maps. Symbols are small pictures that are used to show data. Symbols can show where rain or snow is falling. Symbols can also show where it is windy. Some weather maps have keys. A **map key** tells us what the symbols mean. Meteorologists use weather maps to share weather information with others.

What types of weather does this map show?

Try It Yourself!

Look at a weather map on a computer or on a weather app. Can you tell if it will be warm or cold where you live?

What tool gathers information about the weather and climate from the atmosphere?

26

Activity

Read Weather Maps

Materials: weather maps from different newspapers, or printed from websites. Provide a different weather map for each group.

Place the students in groups and give each group a weather map.

Each group should write down three weather questions that can be answered by reading the map. Then have the groups trade maps and questions. Allow time for each group to answer the questions and discuss their answers.

Try It Yourself!

Materials: weather map on a computer or a weather app

Helps

The "Try It Yourself!" activities are designed for the students to do on their own. You will need to plan ahead to supply the materials needed for some of these activities.

Weather Tools

Rain Gauge

Meteorologists use tools to measure different features of weather. One tool is called a rain gauge. A **rain gauge** measures the amount of rain that falls. Each day meteorologists record the rainfall data. They collect measurements for a long time. The data tells them what the normal rainfall for an area is. They use the data to look for weather patterns. This information can help farmers and landscapers. To do their work, they need to know whether it is going to rain.

How much rain has the rain gauge measured?

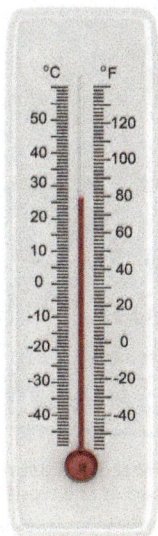

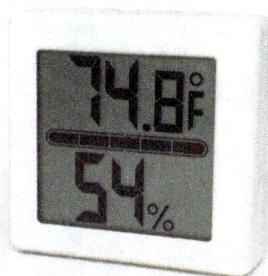

Thermometer

Another feature of weather that we often measure is temperature. Temperature is the measure of how hot or cold something is. When we measure the weather's temperature, we are measuring how hot or cold the air is.

A **thermometer** is a tool used to measure temperature. There are different kinds of thermometers. Some thermometers have degree markings. Many thermometers used for weather are digital. They show the temperature in numbers on a display.

Which picture shows a digital thermometer?

27

ⓘ Background

Measuring Precipitation

Measuring precipitation can be tricky. An inch of snow is not the same as an inch of rain. A meteorologist uses a snow gauge to measure how much snow has fallen. If you do not have a snow gauge, snow can be measured in centimeters (or inches) with a ruler. To compare the amount of snow to rainfall, you would need to measure melted snow in a rain gauge.

Preparation for Reading

- Preview and pronounce the vocabulary terms *rain gauge, thermometer, weathervane,* and *anemometer.*
- Direct the students to read pages 27–28 silently to find out what kinds of tools a meteorologist uses.

Teach for Understanding

What are four tools that a meteorologist uses to collect data about the weather? rain gauge, thermometer, weathervane, and anemometer

Which tool is used to measure rainfall? rain gauge

Why do scientists measure rainfall? Possible answers: to find out what the normal rainfall for an area is; to look for weather patterns

Why do you think it is important for meteorologists to measure and record rainfall data accurately? They need accurate data to compare measurements and weather patterns.

Do you think a meteorologist could accurately measure snow using a rain gauge? No; an inch of snow is not the same as an inch of rain.

- Share Background information "Measuring Precipitation" as time allows.
- Direct attention to the picture at the top of the page.

How much rain is in this rain gauge? 24 mm

What is *temperature?* the measure of how hot or cold something is

What do we use to measure temperature? a thermometer

- Look at the pictures of the thermometers.

How are the thermometers different? One has degree markings, and one shows the temperature on a display.

Which picture shows a digital thermometer? Possible answers: the one on the right; the one that shows the temperature on a display

What are some places that you might find a digital thermometer? on a bank or business sign, in a car, at a doctor's office, at home

- Direct attention to the weathervane on the page.

  What is a *weathervane*? a tool that shows which direction the wind is blowing

  How does a weathervane show the direction of the wind? The arrow on it points in the direction the wind is coming from.

  Name a reason we might want to know the direction of the wind. Possible answers: It helps us know something about the wind. Winds from the north are usually cold and winds from the south are usually warm.

  What tool measures the wind's speed? an anemometer

- Direct attention to the picture of the anemometer. Explain that an anemometer looks like several small cups, or cones, attached to a post. As the wind blows into the cups, they spin around the post. A dial that looks similar to the speedometer in a car shows how fast the wind is blowing. Cooler winds usually blow more swiftly than warmer winds do.

Apply

- Discuss with students that meteorologists are not the only ones who might use weather tools. The following scenarios might help students think of a time when they have or could have used a weather tool.

  Ramon is planning a picnic for his friends. He does not want to have a picnic on a windy day. Which tool would help him decide if today is a good day for a picnic? anemometer or weathervane

  Josie needs to know if it rains in the night so she can decide if she needs to water her plants. Which tool could she use? rain gauge

  Juliette and Krista want to fly kites today. Which tool will tell them if it's a good day for kite flying? anemometer

  Marcus would like to spray paint his desk outside. He does not want the paint to blow in his face while spraying. Which tool can help him decide where to stand while painting the desk? weathervane

  Merrilee is getting ready for school. She doesn't know if she should wear a sweater. Which tool can help her decide? thermometer

  Bradley wants to make a graph showing the amount of rainfall at his house for a month. Which tool should he use? rain gauge

- Invite students to share times when they have used a weather tool.

an anemometer

Apply

Looking Ahead, page 21
This page assesses the students' knowledge prior to the chapter.

Weather Predictions, page 22
This page reinforces the concepts taught in Lesson 8.

Weathervane

One of the oldest weather tools is a weather vane. A **weathervane** shows which direction the wind is blowing. It is also called a wind vane. The arrow on a weathervane points to the direction the wind is coming from. Most weathervanes have letters that stand for the directions. *N* is for north, *S* is for south, *E* is for east, and *W* is for west. When the wind blows from the north, the weathervane points to the *N*.

The direction of the wind can help us know more about the weather. Winds from the north are usually colder. Winds from the south are usually warmer.

Anemometer

A meteorologist also needs to know how hard the wind is blowing. This is a wind's speed. Very strong winds usually mean a storm is coming. An **anemometer** is a weather tool that measures wind speed.

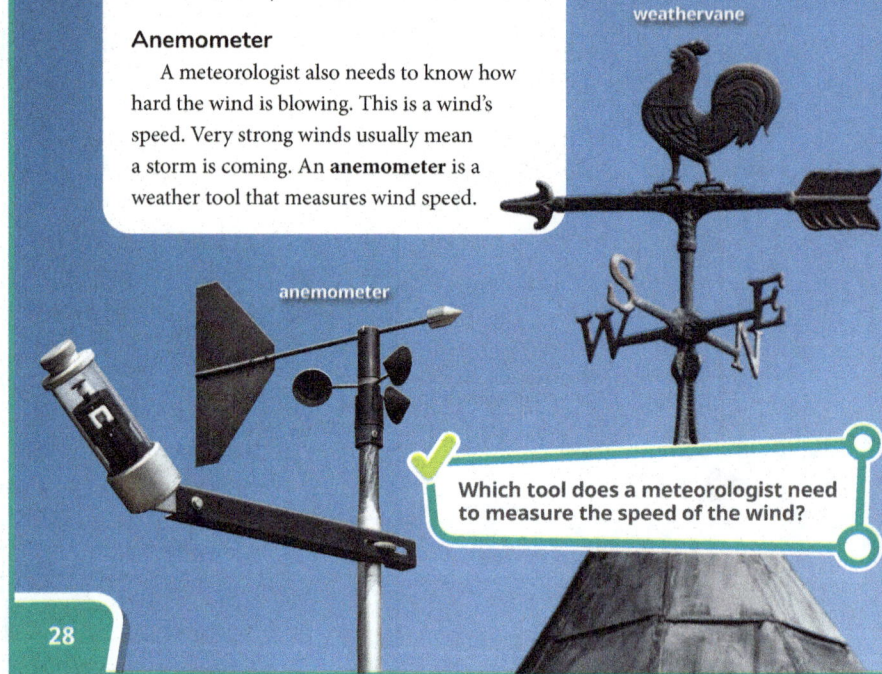

weathervane

anemometer

Which tool does a meteorologist need to measure the speed of the wind?

Activity

Cloud Formations
Materials: blue construction paper; cotton balls, 6 per student; liquid glue

Distribute one sheet of blue construction paper to each student. Fold the sheet of paper in half. Then fold in half again. Open the page up so that there are four sections showing. Label each section with the name of one type of cloud (stratus, fog, cumulus, cirrus). Using cotton balls and glue, make a representation of each type of cloud. Write a one-sentence description of each type of cloud.

Background

Fog
Fog is a very low stratus cloud that lies on the surface of the earth.

Nimbus Clouds
Both stratus and cumulus clouds can bring rain. When the clouds are gray colored and contain rain, a form of the word nimbus is added to the name of the clouds. Nimbostratus clouds often produce light rain. Tall, dark clouds that produce thunderstorms are called cumulonimbus.

Lesson 9

Clouds

A **cloud** is a collection of very tiny water droplets or ice crystals. The droplets and crystals are so small that they can float in the air. Clouds form in the atmosphere. When clouds come between the earth and the sun, the earth gets cooler. Different kinds of clouds give us clues about the weather.

Stratus clouds are a wide, thin layer of clouds. They are low clouds that look like a blanket over the earth. Stratus clouds often cause gray skies. They do not bring heavy rain.

Fog is a cloud that forms close to the ground. Fog forms when warm air blows over cold air near the ground. The water vapor in the warm air cools close to the ground and condenses.

Cumulus clouds are fluffy, cotton-like clouds. They usually mean that the weather is fair. But these clouds can grow large and tall and can form thunderclouds. The dark gray thunderclouds can bring stormy weather.

Cirrus clouds are high, thin, wispy clouds. They often mean that a change in the weather is coming. Some cirrus clouds look like the scales on a fish. A cirrus cloud that appears thin and curly is sometimes called a mare's tail. It looks like the tail of a horse.

stratus clouds

fog

cumulus clouds

cirrus clouds

✓ **What is a cloud?**

29

Lesson **9** begins here.

Objectives
- Create models of four types of clouds.
- Explain how the water cycle works.
- Describe four forms of precipitation.

Materials
- blue construction paper, 1 per student
- liquid glue
- cotton balls, 6 per student

Teacher Resources
- Visuals 2.3–2.4: *Cloud Formations; Water Cycle*

Vocabulary
- cloud
- stratus clouds
- fog
- cumulus clouds
- cirrus clouds
- precipitation
- water cycle
- evaporation
- condensation
- wind

Engage
- Read Christina Rossetti's poem, "Clouds." Read the poem first, before reading the title.
  White sheep, white sheep on a blue hill,
  When the wind stops, you all stand still.
  When the wind blows, you walk away slow.
  White sheep, white sheep, where do you go?
  What do you think are the white sheep in the poem? clouds
  Let's find out what kind of clouds might have reminded the poet of sheep.

Instruct

Preparation for Reading
- Preview and pronounce the vocabulary terms *cloud*, *stratus clouds*, *fog*, *cumulus clouds*, and *cirrus clouds*, as well as the word *crystal*.
- Direct the students to read page 29 silently to find out what cumulus clouds look like.

Teach for Understanding
What is a *cloud*? a collection of very tiny water droplets or ice crystals
- Direct attention to the Glossary. Explain that the definitions to all the bold vocabulary terms, as well as the important words that are italicized, can be found in the Glossary. Point out that the words are listed in alphabetical order.
- Direct students to find the word *crystal* in the Glossary. Choose a volunteer to read the definition.

The word *crystal* is italicized in Chapter 3.

- Direct attention to the photos as each cloud type is discussed.
  What can clouds give us clues about? the weather
  What do stratus clouds look like? a thin blanket; a wide, thin layer of low clouds
  What kind of stratus cloud forms close to the ground? fog
  What do cumulus clouds look like? fluffy and cotton-like
  When cumulus clouds look large, tall, and dark, what kind of weather is probably coming? stormy weather
  What are high, thin, wispy clouds called? cirrus clouds
- Display the *Cloud Formations* visual.
- Complete "Cloud Formations" activity at this time.

✓ a collection of very tiny water droplets or ice crystals

Instruct

Preparation for Reading

- Preview and pronounce the vocabulary terms *precipitation*, *water cycle*, *evaporation*, and *condensation*.
- Direct the students to read pages 30–31 silently to find out what evaporation is.

Teach for Understanding

What is *precipitation*? water that falls from the sky to the ground

What is a cycle? Answers will vary, but elicit that a cycle is something that occurs over and over.

- Write the word *bicycle* for display. Underline *cycle* in the word. Tell the students to think of a spinning bicycle wheel to help them remember what a cycle is.

What is the *water cycle*? the movement of water from the earth to the sky and back to the earth again

- Display the *Water Cycle* visual.

What word describes when water goes into the atmosphere and becomes a gas? evaporation

- Ask a volunteer to point to the part of the water cycle that shows evaporation.

What causes the water on the earth's surface to evaporate? the sun

What gas forms when water evaporates? water vapor

Look at the diagram again. What part of the diagram represents evaporation? the wavy lines going up

As water vapor rises, it cools. What happens to water vapor as it cools? It changes to a liquid.

Precipitation

One part of weather that is easy to observe and measure is precipitation. **Precipitation** is water that falls from the sky to the ground. Precipitation is part of the water cycle. The **water cycle** moves water from the earth to the sky. The water then moves back to the earth again.

The Water Cycle

For the water cycle to occur, water must get into the air. Remember that water is a type of matter that easily changes states. **Evaporation** is the part of the water cycle when water changes from a liquid to a gas. The gas that forms is called water vapor.

Warm water vapor rises in the atmosphere. Air temperature is cooler above the earth's surface. The cooler air cools the water vapor.

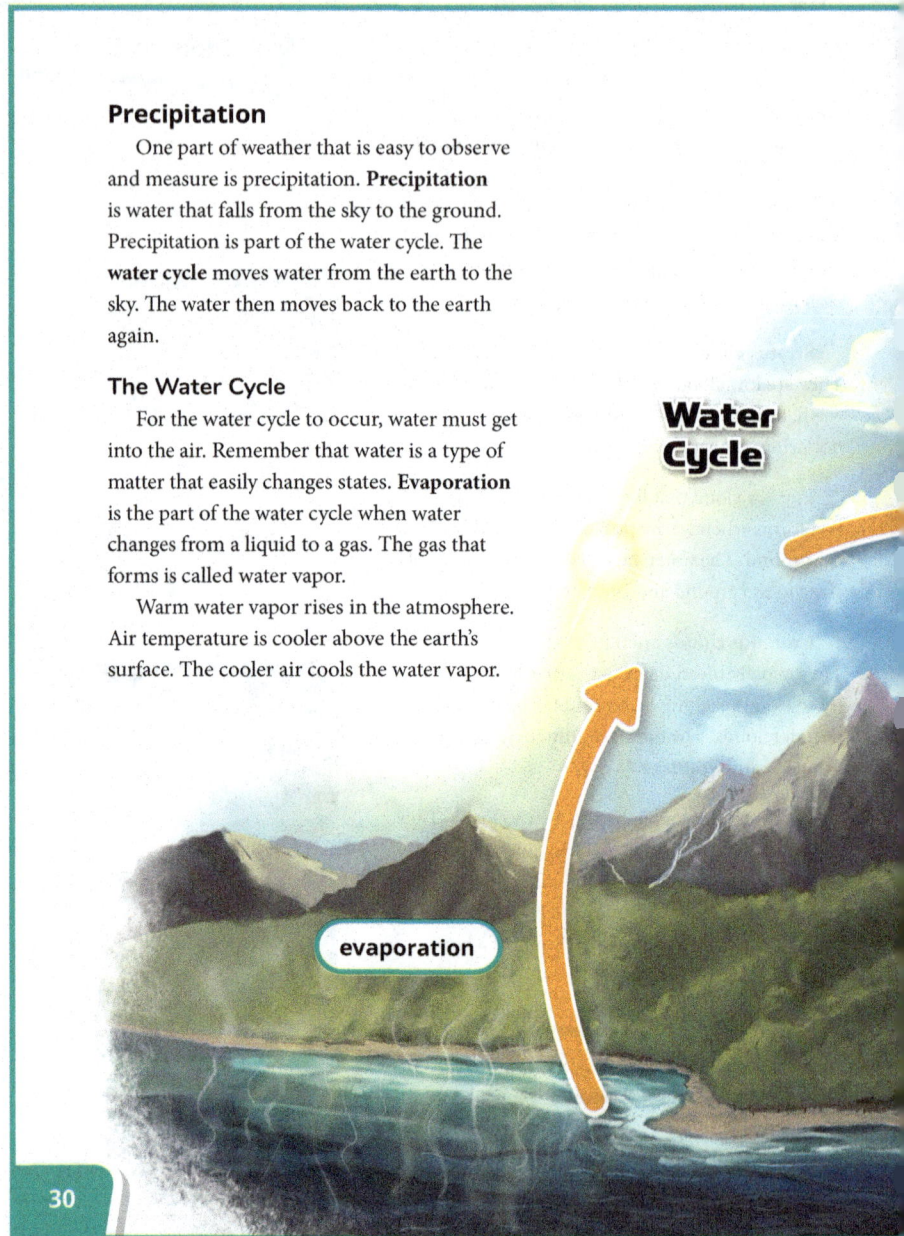

Water Cycle

evaporation

30

ⓘ Background

Dew
When the air cools at night, water vapor in the air condenses on objects near the ground. This forms dew.

Groundwater
Groundwater (precipitation that has soaked into the ground) is another part of the water cycle. It provides fresh water for many living things.

Water Cycle
The water cycle is also called the hydrological cycle.

Condensation is the part of the water cycle when cooled water vapor changes to a liquid. The condensed water vapor forms clouds. The clouds grow as more and more water vapor condenses.

Water droplets form in the clouds. The droplets of water vapor attach to each other to form raindrops. When the drops become heavy, they fall to the ground as precipitation.

As water goes through the water cycle, it is used again and again. God placed exactly the right amount of water on the earth for people, animals, and plants to use.

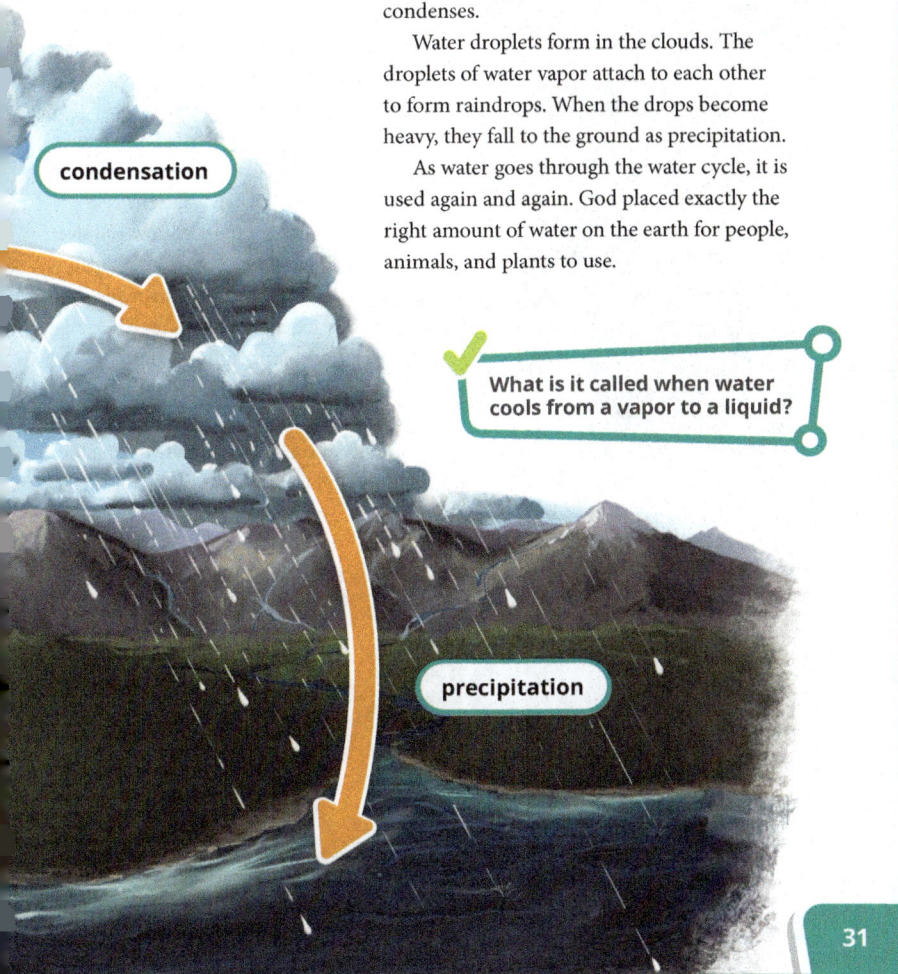

condensation

precipitation

✓ What is it called when water cools from a vapor to a liquid?

31

What is *condensation*? the changing of water vapor to a liquid

- Ask a volunteer to point to the part of the cycle that shows condensation.

What forms clouds? the condensed water vapor

What happens when the droplets of water vapor attach to each other? Raindrops form.

What happens when raindrops become too heavy? They fall to the ground as precipitation.

- Ask a volunteer to point to the part of the cycle that shows precipitation.

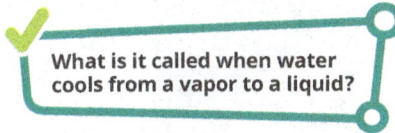

✓ condensation

🞤 Differentiated Instruction

Water Cycle

Remediation: Draw a picture of the water cycle. Be sure to use arrows to show the path of the water through the cycle.

Enrichment: Write a story as if you are a drop of water traveling through the water cyle. Add illustrations to show your journey through the water cycle.

Preparation for Reading

- Preview and pronounce the vocabulary term *wind*.
- Direct the students to read pages 32–33 silently to find out what causes wind.

Teach for Understanding

Which kind of precipitation is a liquid? rain

If it is raining, what do we know about the air temperature? It is above freezing.

How do you know that? If the air temperature were below freezing, the precipitation would not be liquid.

- Direct attention to the picture at the top of the page.

How is rain helping people in this picture? Possible answers: watering the crops; helping plants grow

- Direct attention to the picture in the middle of the page.

How is this boy enjoying the rain? splashing in puddles

What does rain provide? water for all living things

- Read Matthew 5:45 aloud.

Who is responsible for the rain? God

Who does God make the rain to fall on? the just and the unjust, or the righteous and the unrighteous

- Explain that God sends the rain even on people who do not deserve this grace.

What happens to precipitation when parts of the air are below freezing? The precipitation may fall as a form of ice.

Which kinds of precipitation fall as a form of ice? snow, sleet, and hail

- Direct attention to the pictures at the bottom of the page. Choose a volunteer to read the captions aloud.

Have you ever seen sleet or hail?

- Share some of the Background information about hail.

What kind of weather can hail occur in? thunderstorms

Do thunderstorms usually occur in cold weather or in warm or hot weather? in warm or hot weather

How does that make hail different from snow and sleet? Hail occurs when the temperature is warm or hot, but snow and sleet occur only when the temperature is cold.

Rain, Sleet, Snow, and Hail

Rain is a liquid. It can fall only if the air temperature is above freezing. Rain falls on the ground and in the ocean and in other bodies of water. It provides water for all living things. Rain is a natural process that God created. The Bible tells us in Matthew 5:45 that God sends the rain even on people who do not deserve this grace.

When parts of the air are below freezing, precipitation may fall as a form of ice. Snow, sleet, and hail form when water freezes before it reaches the ground. Snow and sleet occur in cold weather. Snow is ice, shaped into tiny crystals. Snowflakes are made up of many tiny ice crystals stuck together. Sleet starts as water that freezes into tiny ice pellets. Sleet usually falls in the winter. Hail is also made of ice pellets. Hail forms in thunderstorm clouds, usually during the summer. Hail is larger than sleet.

Sleet is usually made of small ice pellets.

Hail is also made of ice but can be quite large.

32

ⓘ Background

Supercooled Water

Though water can exist in liquid form at air temperatures below freezing, it is better at this grade to reinforce the idea that rain changes to ice at the freezing point.

Hail

Most frozen precipitation occurs during cold weather, but hail usually falls during warm weather. In some thunderstorms, the upward-moving warmer air pushes some raindrops higher into the tall clouds. There the raindrops freeze into tiny pellets that fall back to the lower clouds. The upward-moving air pushes the pellets back up. Moisture in the clouds adds a layer of ice to the pellets each time they are blown upward. Eventually the pellets become too heavy and fall to the earth as hail. Hailstones vary in size from being as small as a pea to larger than a grapefruit.

Wind

Wind is another characteristic of weather. Though we cannot see the wind, we can see what it does. A gentle breeze causes tree branches to sway back and forth. A strong wind bends trees and may break branches.

Wind is moving air. Changes in temperature cause air to move. The sun makes the earth's surface warm. This causes air closer to the ground to be warmer than air that is higher. Warmer air rises. Cooler air falls. As warmer air rises, cooler air moves toward the earth to replace it. This movement of air causes wind.

Which trees shown are moving in a strong wind?

Key
● warm air
● cool air

Causes of Wind

✓ **What causes air to move?**

33

What is *wind*? moving air

- Direct attention to the pictures of the two trees. Choose a volunteer to answer the caption.

How can you tell that the tree in the top picture is moving in a strong wind? Possible answers: The branches are all blowing to one side; the branches on the bottom tree are hanging down.

- Direct attention to the diagram at the bottom of the page.

What causes wind? changes in temperature

What do the red and blue arrows show? The red arrow shows warm air, and the blue arrow shows cool air.

How does a change in temperature cause wind? Air close to the surface of the earth is warmer than air that is higher. Warm air rises, and the cooler air moves toward the earth to replace it. This movement of air causes wind.

- Discuss Background information "Trade Winds" as time allows.

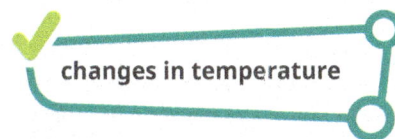

✓ **changes in temperature**

Apply

Activities

Study Guide, pages 23–25

These pages review the concepts taught in Lessons 8–9. After completion, direct the students to keep them for review for Test 2.

Assess

Quiz 2A

The quiz may be given at any time after completion of this lesson.

ⓘ Background

Trade Winds

Some large wind patterns cause wind to blow all the time in certain directions. One large wind pattern causes the trade winds. These winds always blow from Europe to the Americas. Trade winds were used by sailing ships as they came to the New World.

Large wind patterns, such as the trade winds, are a result of the uneven heating of the earth's surface by the sun. The tilt of the earth causes different parts of the earth to be heated more than others. The difference in temperature between warmer places (with rising air) and cooler places (with sinking air) causes the global wind patterns.

Objectives
- Observe and record weather data from two locations.
- Create a graph to display weather data.

Materials
- See *Activities* page 27.

Teacher Resources
- Visual 2.5: *Tongue Twister*

Engage

- Display the *Tongue Twister* visual. Encourage students to try saying the tongue twister.

    Whether the weather be fine,
    Or whether the weather be not,
    Whether the weather be cold,
    Or whether the weather be hot,
    We'll weather the weather,
    Whatever the weather,
    Whether we like it or not!

    Who controls the weather? God

- Explain that God knows exactly the kind of weather that is needed for every place on the earth. For this activity we will be observing the weather that God gives us.

Instruct

Preparation for Reading

- Direct the students to remove pages 29–32 from their *Activities* books. Direct them to read page 34 and *Activities* 29–32 silently before beginning.
- Explain that it is helpful to look over all the pages before starting an Exploration. Point out that it will help them to see what they are going to do and when they will be writing information down.
- Direct attention to the art at the bottom of page 34.
- What weather tool is the girl holding? an anemometer
- What can you observe about the weather in the picture? Possible answers: It is windy or breezy. It is cool. It is cloudy.
- What season do you think it is in the picture? It is fall, or autumn.
- What clues suggest that it is fall? The leaves on the trees are red, orange, and yellow. The leaves are falling off the trees.

Teach for Understanding

- Direct the students to remove pages 27–28 from their *Activities* books.
- Direct the students to place the pages in their science notebooks. These pages will be used as reference throughout the course.
- Read and discuss the Measure Up: Temperature on *Activites* page 27.
- Reinforcement on *Activities* page 28 may be completed as a class, or assigned as time allows.

Lesson 10

EXPLORATION

Inquiry Skills
- Observe
- Measure
- Predict

Weather Watcher

Meteorologists use many tools to study and predict the weather. You can use tools to study the weather, too!

In this Exploration, you will observe and measure the weather at your home or school. You will record your weather data. You will also use a computer or weather app to observe the weather in a location far away. Maybe you will study the weather where your grandparents live. Or maybe you will study the weather where a missionary lives. You will be a weather watcher.

What can you observe about the weather in this picture?

34

Weather Watcher

Name _____

Inquiry Skills
- Observe
- Measure
- Predict

Materials
- ☐ thermometer
- ☐ rain gauge
- ☐ weathervane

Purpose

I want to observe the weather where I live. I want to compare the weather where I live with the weather in another location.

Procedure

- ☐ 1. Record two locations for your weather on Charts 1 and 2 under Observations.
- ☐ 2. Observe the weather where you live for 5 days.
- ☐ 3. Record the temperature each day in °C.
- ☐ 4. Record the amount of precipitation each day.
- ☐ 5. Record the wind direction each day.
- ☐ 6. Observe the sky conditions each day. Record your observations.
- ☐ 7. Use a computer or weather app to look up the weather for your second location. Record the temperature, amount of precipitation, wind direction, and sky conditions for each day on Chart 2.
- ☐ 8. Graph the temperature for each location each day on the bar graph.

SCIENCE 3 *Activities*

Lesson 10 • Page 34
Exploration **29**

- Direct attention to the **Inquiry Skills**.
- Remind the students that scientists use science inquiry skills to gather and use facts.

  Which inquiry skills will you be using in this Exploration? Observe, Measure, and Predict

  What does it mean to *observe*? to use your senses to find out something

  What does it mean to *measure*? to find the size, number, or amount of something

  What does it mean to *predict*? to make a careful guess at what may happen after observing; using what you know to say what may happen

- Choose a student to read the **Purpose** aloud.

Apply

- Invite students to share a location that they would like to use for their second location. Possible locations could include somewhere they would like to travel, a place where a missionary lives, or a place where friends or family live.
- Instruct the students to follow and check off **Procedure** Steps 1–8 as each is completed. Point out that Steps 2–8 will be repeated daily for 4 additional days.

- Direct attention to the "Observations" chart. Remind the students of the importance of accurately recording their **Observations**.
- As needed, help the students record accurate weather data.

Observations

Chart 1

Location:_____

Day	Temperature	Amount of Precipitation	Wind Direction	Sky Observations	
				Types of Clouds	Precipitation
Day 1	___°C	___ inches	from the _____		
Day 2	___°C	___ inches	from the _____		
Day 3	___°C	___ inches	from the _____		
Day 4	___°C	___ inches	from the _____		
Day 5	___°C	___ inches	from the _____		

Chart 2

Location:_____

Day	Temperature	Amount of Precipitation	Wind Direction	Sky Observations	
				Types of Clouds	Precipitation
Day 1	___°C	___ inches	from the _____		
Day 2	___°C	___ inches	from the _____		
Day 3	___°C	___ inches	from the _____		
Day 4	___°C	___ inches	from the _____		
Day 5	___°C	___ inches	from the _____		

EXPLORATION

Weather Watcher

Name _____

Use the data from your observations to complete the bar graph.

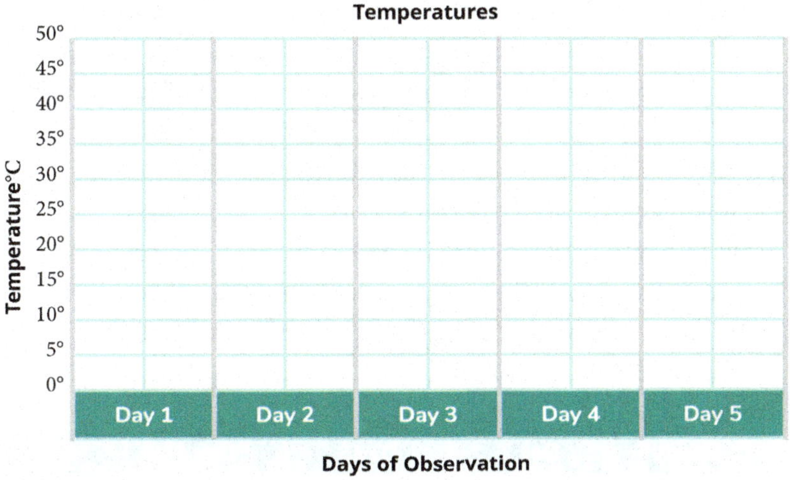

Temperatures

Days of Observation

Graph Key

☐ _____
(location 1)

☐ _____
(location 2)

- Guide students in accurately completing their graph each day.

Some locations may experience temperatures below 0°C. Adjust the temperatures on the graph as needed for your location.

- Direct the students to draw **Conclusions** using their observations.
- Select several students to share their weather predictions based on their observations.
- Guide a discussion of the students' conclusions.

Activities

Measure Up: Temperature, pages 27–28

This page was discussed during the lesson.

Exploration, pages 29–32

Students will observe, record, and graph weather data from 2 locations.

Assess

Rubric

Use the prepared rubric or design a rubric to include your chosen criteria.

Conclusions

1. What weather patterns did you observe in each of your locations? *Accept reasonable answers based on observations made in given locations.*

2. Which tools did you use to make your measurements and observations? *Accept thermometer, rain gauge, weathervane.*

3. Write one interesting fact about your weather observations. *Accept reasonable answers that could include observations about temperature, wind direction, precipitation, and sky conditions.*

4. Which location had the warmest temperature? _____

5. Which location had the coolest temperature? _____

6. Make a prediction about what you expect the temperature to be for the next five days for one of your locations. *Students should make predictions based on their observations.*

Location: _____

Day 1	Day 2	Day 3	Day 4	Day 5
___°C	___°C	___°C	___°C	___°C

32 Lesson 10 • Page 34
Exploration

Lesson 11

Severe Weather

Sometimes a meteorologist predicts severe weather. Severe weather includes droughts, floods, strong thunderstorms, tornadoes, hurricanes, and blizzards. These kinds of weather can cause damage to homes and land. They can even be dangerous to people. Predicting severe weather is very important. When meteorologists predict severe weather, people have time to plan. People are able to keep themselves and their families safe.

Drought

A **drought** happens when there is a long period of very little precipitation. A drought can affect people, plants, and animals. When there is not enough water, crops and plants can die. Some animals must move to another location to find water. People cannot water their plants or wash their cars. People must be careful to use as little water as possible. A drought ends when enough precipitation falls to bring the water levels back to normal.

Flood

A **flood** is water that flows over onto dry land. Floods can cause a lot of damage. Sometimes floods cause water to go into people's homes and cars. Very fast moving floodwater can carry large objects with it. Usually floods are a result of heavy rains during thunderstorms or hurricanes.

Plants cannot grow well in a drought.

35

Lesson **11** begins here.

Objectives
- Identify six types of severe weather.
- Explain why weather predictions are important. **BWS**
- Make a severe weather safety plan.
- Give examples of how weather predictions can keep us safe.

Materials
- audio of weather alert
- video of a derecho
- map of United States
- canned food
- bottled water
- flashlight
- batteries
- weather or battery-powered radio
- first-aid kit
- blanket
- plastic box or other container for emergency supplies
- other items that are not necessary for an emergency kit (candy, toys, soft drinks)

Teacher Resources
- Visual 2.6: *Severe Weather Safety*

Vocabulary
- drought
- flood
- thunderstorm
- tornado
- hurricane
- blizzard
- civil engineer

Engage
- Play a severe weather alert sound from a weather radio or a sound clip. Ask students if they have heard the sound before.
- Explain that the weather alert is to help people stay safe in severe weather. Sometimes the message gives information about weather that may happen soon. Sometimes the message alerts people to weather that is happening right then so that they can be safe.
- Provide time for students to share times when they have experienced severe weather.

Instruct

Preparation for Reading
- Preview and pronounce the vocabulary terms *drought*, *flood*, *thunderstorm*, and *tornado*.
- Direct the students to read pages 35–36 silently to find out what kind of severe weather happens most often.

Teach for Understanding
- Explain that severe weather means weather that is very strong and could cause harm to people, animals, or buildings.

  What kinds of severe weather might a meteorologist predict? thunderstorms, tornadoes, hurricanes, blizzards, droughts, floods
- Direct attention to the picture at the top of the page. Choose a volunteer to read the caption.

  What is it called when an area does not receive as much precipitation as it needs? a drought

  What are some things that could happen because of a drought? Possible answers: Plants may die. Animals may move away because they do not have enough water. Cities and towns may limit how much water people may use.

  When does a drought end? when enough precipitation falls to bring the water levels back to normal
- Direct attention to the picture at the bottom of the page.

  Do you see any problems in this picture? Water is in the houses and buildings. Water is covering the cars. There is water where it does not belong.

  What is water that flows over onto dry land called? a flood

What are some possible causes of flooding? heavy rains, hurricanes, melting snow and ice

What should you do if you see an area that is flooded? Stay away. Move to a safer location.

- Direct attention to the picture at the top of the page. Choose a volunteer to read the caption.

What kind of severe weather occurs most often? thunderstorms

What kind of weather happens during a thunderstorm? heavy rain, strong winds, lightning, thunder, and sometimes hail

What causes thunder? lightning

- Share Background information about lightning as time allows.

What kind of storm can produce a tornado? a thunderstorm

What is a *tornado*? a funnel-shaped cloud of swirling winds that reaches the ground

- Direct attention to the picture at the bottom of the page. Choose a volunteer to read the caption.

What is a tornado over the water called? a waterspout

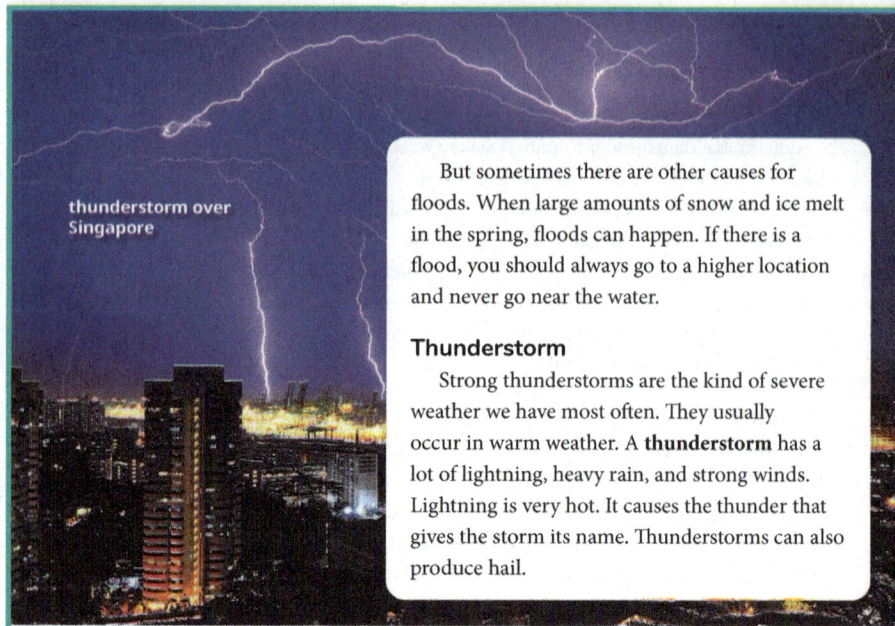

a drought

thunderstorm over Singapore

But sometimes there are other causes for floods. When large amounts of snow and ice melt in the spring, floods can happen. If there is a flood, you should always go to a higher location and never go near the water.

Thunderstorm

Strong thunderstorms are the kind of severe weather we have most often. They usually occur in warm weather. A **thunderstorm** has a lot of lightning, heavy rain, and strong winds. Lightning is very hot. It causes the thunder that gives the storm its name. Thunderstorms can also produce hail.

Tornado

A thunderstorm can produce another kind of severe weather called a tornado. A **tornado** is a funnel-shaped cloud of swirling winds that reaches the ground. Usually a tornado is only a few hundred meters (yards) wide. Its winds are very strong. Tornadoes can cause a lot of damage. They are strong enough to knock down trees. Sometimes they break apart buildings. A tornado over the water is called a waterspout.

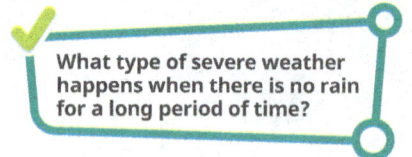

A tornado can travel for many kilometers (miles) before it stops.

What type of severe weather happens when there is no rain for a long period of time?

36

ⓘ Background

Lightning

Lightning opens a small hole in the air called a channel. Then more air rushes in to fill the hole. The air makes the sound that we know as thunder.

Does lightning strike the same place twice? Yes! Lightning often strikes the same place repeatedly, especially if it is the tallest location around. This is why many tall buildings have lightning rods to direct the lightning to the ground instead of damaging the building. The Empire State Building in New York City is struck by lightning an average of 23 times each year. During one storm, it was struck eight times in 24 minutes!

Fantastic Facts

A shelf cloud is sometimes seen at the front of a derecho.

A *derecho* is a fast moving windstorm. The word *derecho* means "straight." Derechos usually move in a straight line. Derechos happen during thunderstorms. Their winds can cause a lot of damage. To be called a derecho, the storm must travel over 400 km (250 mi). In June 2012, a derecho from eastern Iowa traveled more than 1126 km (700 mi)—all the way to the coast of North Carolina! Most derechos happen in the summer months when temperatures are hot.

Hurricane

Another kind of severe weather is a hurricane. A **hurricane** has strong, swirling winds and produces heavy rain. A hurricane can be hundreds of kilometers (miles) wide. It covers a much larger area than a tornado does. Hurricanes form over warm ocean waters. They mostly affect places near the ocean.

The center of a hurricane is called the eye. In the eye of the storm, the winds are calm. But very soon, the rest of the storm comes. Hurricanes are very strong storms.

Blizzard

Blizzards are snowstorms with strong winds. During a blizzard the wind blows the snow very hard. It can be hard to see in a blizzard, and it is easy to get lost. The blowing snow and freezing temperatures make it unsafe to be outside.

The swirling winds of a hurricane move in a circle around the eye.

37

Preparation for Reading

- Preview and pronounce the vocabulary terms *hurricane* and *blizzard*, and the words *derecho* and *emergency*.
- Direct the students to read pages 37–38 silently to find out what kind of severe weather happens in freezing temperatures.

Teach for Understanding

What kind of weather does a hurricane produce? strong, swirling winds with heavy rain

What places do hurricanes usually affect? areas near the ocean Why? Hurricanes happen over warm ocean waters.

- Point out that although hurricanes begin over warm ocean waters, the storms can move over land. Hurricanes can bring heavy rains and winds far from the ocean.

Both tornadoes and hurricanes have strong, swirling winds. How are they different? Hurricanes cover a much larger area and form over warm ocean waters. Tornadoes are funnel-shaped clouds that touch the ground. Tornadoes do not need water to form.

- Direct attention to the picture on the right of the page. What is the calm center of a hurricane called? the eye

Why do you think it is called the eye? Possible answers: It is the center of the storm. It is shaped like a human eye. It is the point that the storm rotates around.

- Direct attention to the picture at the bottom of the page. What is a *blizzard*? a snowstorm with strong winds

Why are blizzards dangerous? It can be hard to see, so it is easy to get lost. Freezing temperatures make it unsafe to be outside.

- Play the *Derecho Radar Loop* or *Satellite* video (TeacherToolsOnline.com).
- Direct attention to the "Fantastic Facts" box. Display a map of the United States. Invite students to trace the path of the derecho from Iowa to the coast of North Carolina.
- Discuss the kinds of weather that occur during a derecho.

Why are people sometimes fearful of these types of weather? Possible answers: The weather may damage homes and land. People and animals may be hurt or killed.

- Direct attention to the picture at the top of the page.
- **What is on the phone?** a weather warning, a severe weather alert, a severe thunderstorm watch

  How do meteorologists warn people that severe weather is coming? Warnings are seen on TV and weather apps and are heard on weather radio.

  Why are weather predictions for severe weather important? It gives people time to plan to be safe.

  Why is it important for people to have time to plan to be safe? Because people are made in God's image. Keeping people safe during severe weather is part of caring for the people that God made.

- Display the *Severe Weather Safety* visual.
- Discuss the types of severe weather and how weather predictions can keep us safe.
- **If a severe thunderstorm or a tornado is predicted, what is the best way to stay safe?** Go indoors. Stay away from doors and windows. Go downstairs. Take cover.
- **What would the safest place be if the house or building you are in does not have a downstairs?** in a bathroom or other room that does not have windows
- **How can a weather prediction of a hurricane help people to stay safe?** Possible answers: People will know to go away from the path of the hurricane. People will be able to get to safety. People will be able to get supplies they might need in case they will not have electricity or water.
- **How can a prediction of a winter storm or blizzard help people to stay safe?** Possible answers: People will be able to get supplies. People will know to stay indoors and not be caught out in the cold and the storm.

Apply

- Direct attention to the "Health and Safety" box.
- Display the various items that could go into an emergency kit and the items that would not be part of an emergency kit. Explain that the materials for an emergency kit should be things that can be stored for a long period of time. Also explain that those things should not require the use of electricity and water.
- Guide students as they choose materials that should be included in an emergency kit. Ask why each item would (or would not) be important for an emergency kit. For example:
- **Why might the flashlight be needed?** Possible answers might include that if electricity were not available, then lights and lamps would not work. A flashlight would help you see in the dark.
- **How would you prepare for a severe thunderstorm?** Possible answers: Go to a safe place indoors. Have an emergency kit with supplies.

a hurricane

Weather Warnings

Meteorologists have developed warnings to tell people that severe weather is close by. Weather warnings are shown on TV and also on weather apps. You could hear them on a weather radio.

A severe thunderstorm warning means that a thunderstorm with strong winds and possibly hail is happening. A tornado warning means that a strong storm with a tornado has formed. A winter storm warning means that a storm with heavy snow or ice is happening.

It is important for meteorologists to predict severe weather. Predicting severe weather before it happens can give people time to plan. You should listen to the warnings that meteorologists give so you can stay safe.

Health and Safety

Before severe weather happens, you need to know how to be safe. Talk with your family about what to do if severe weather is coming. Your plan should include where to go. If your area has a type of severe weather often, your town may have shelters to keep you safe.

Your plan should include making an emergency kit. Include supplies like bottled water, blankets, and canned food. You should also have flashlights, a battery-powered radio, and extra batteries. Include a cell phone charger and emergency phone numbers. Also include medicine for anyone who needs it. Remember supplies for your pets.

Which type of severe weather that forms over the ocean has strong, swirling winds and produces heavy rain?

ⓘ Background

Storm Warnings

The National Weather Service is an organization in the United States that tracks the weather. Sometimes they issue a weather watch. This means that the conditions are right for a type of severe weather but the weather is not currently happening. Once the type of weather is observed, a weather warning is announced. These warnings are announced on special weather radios as well as on radio and television stations.

STEM Career

Civil Engineer

A **civil engineer** is an engineer who has studied design and construction. Civil engineers plan the construction of bridges and buildings. They study to learn about different building materials. Some kinds of stone and metal are stronger than others. Civil engineers work hard to make buildings that will be safe during an earthquake. They make bridges that will last in a hurricane.

Civil engineers also spend a lot of time studying water. They make plans for water systems to clean the water that we drink. They make systems to clean dirty water. They design dams strong enough to hold back water during a flood.

A civil engineer goes to college to learn how to do his job. Civil engineers need to know a lot about math and science to do their job well. We need civil engineers to help us stay safe. If you like to plan and build things, you could become a civil engineer!

39

Preparation for Reading

- Preview and pronounce the vocabulary term *civil engineer*.
- Direct the students to read page 39 silently to find out what a civil engineer does.

Teach for Understanding

What is a *civil engineer*? an engineer who has studied design and construction

What does a civil engineer plan? ways to build bridges and buildings safely

What else do civil engineers plan? systems to clean drinking water

What can civil engineers build to help with flooding? dams to hold back the water

What do civil engineers need to study to do their jobs well? math and science

Assess

Study Guide, page 33

This page reviews the concepts taught in Lesson 11. After completion, direct the students to keep it for review for Test 2.

Enrichment, page 34

This page extends the concepts taught in Lesson 11.

Objectives

- Plan and design a roof model that repels water, using the engineering design process.
- Construct a roof model that repels water, using the engineering design process.
- Evaluate designs to determine which models best repel water.
- Redesign the model to make the roof repel more water.
- Communicate to others how the design solves the problem.

Materials

- spray bottle with water
- newspapers
- 15 cm cardboard square, four per student
- several rolls of tape
- wooden craft sticks of various sizes
- plastic straws
- aluminum foil
- wax paper
- fabric scraps
- yarn or string
- roll of paper towels
- scissors

> Not all students will use each type of building material, but provide enough materials for students to choose from.

Teacher Resources

- Visual 2.7: *STEM: The Engineering Design Process*

Engage

- Invite the students to tell about a problem they needed to find the answer to.
- Direct attention to the picture.

  Have you ever owned a pet?

  What are some things that your pet needed? Possible answers: food, water, shelter

Instruct

Preparation for Reading

- Preview and pronounce the word *model*.
- Direct the students to look up *model* in the Glossary. Ask a volunteer to read the definition aloud.
- Direct the students to remove *Activities* pages 35–38. Direct the students to read page 40 and *Activities* pages 35–38 silently before beginning.

Teach for Understanding

What is the boy holding in the picture? a bunny

- Think about the needs of a bunny. One of those needs is a dry shelter.

  What are the needs of a bunny? Possible answers: food, water, shelter, air

  What are some ways to provide a dry shelter for this pet bunny? Possible answers: a cage for inside a house; a bunny house with a roof for outside

---

Lesson 12

STEM

Waterproof Roof

Your parents have agreed to let you have a pet bunny. Your bunny needs a warm, dry place to sleep outside. In this STEM activity, you will design and build a model of an outdoor shelter for your bunny. The model should have a roof that keeps the inside dry. You will need to make sure that your bunny does not get wet if it rains. You will use the engineering design process to solve the problem.

40

⚠ Helps

STEM Supplies

The supplies listed in the Materials list are suggested items to complete the STEM activity. Supply enough of each item that the students can include several in their designs if needed. Refrain from giving each student a certain number of each item. Rather, put the supplies in a central location for the students to choose what they need after they have completed their designs. This will allow the students to be more creative rather than trying to include each item in their designs.

Waterproof Roof

Name _____

Ask

What is the problem?

My parents have agreed to let me have a pet bunny if I take care of it. One of the things my bunny will need is a shelter. I will need to make a shelter for my bunny that will keep it dry when it rains.

How can I make a shelter that will keep my pet bunny dry?

Imagine

Think about possible answers to the problem.

_____

_____

_____

Plan

Draw and label your design. List and gather materials.

Design should show a structure with a roof.

Design	Materials

Direct attention to *Activities* pages 35–38.

What is your problem in this STEM activity? How can I make a shelter that will keep my pet bunny dry?

What does an engineer do? finds answers to real-life problems

What do we call the plan that is used by an engineer to find the answers to a problem? engineering design process

- Display the *STEM: The Engineering Design Process* visual.

While the engineering design process that is used in *Science 3* has six steps, there are variations that include fewer or more than six steps. Often the steps are illustrated cyclically to indicate that the process or parts of the process may need to be repeated many times to reach a workable solution.

What does the diagram show you? the steps in the engineering design process

- Explain that the engineering design process will be used to solve the problem.

Apply

- For this activity, each student may design his own solution to the problem, or the students may work in science groups to collaborate as they design a solution.

- Distribute four 15 cm cardboard squares to each student or science group. Students should use the tape to fasten the sides of the squares together to make a house shape.

- Instruct the students to *Imagine* a solution to the problem and *Plan* a design.

- Set out enough materials for each student or group to design a solution to the problem. Encourage the students to choose materials that they believe will best solve the problem of keeping the bunny shelter dry.

- Direct the students to *Create* a model of their design.
- Instruct the students to *Test* their model by placing the model on a dry sheet of newspaper. Using the spray bottle with water, the teacher should spray the roof of the bunny shelter five times. Students should lift their bunny shelter off the newspaper to see if the space under the roof is dry. Students should *Improve* their design if water was able to get into the bunny shelter.
- Invite the students to *Share* their designs with a partner or another science group.
- Instruct the students to explain how their design solves the problem.

Activities

STEM: The Engineering Design Process, page 35

This page is to be put in the students' science notebooks.

STEM, pages 37–38

The students will design a way to keep their bunny shelter dry.

Assess

Rubric

Use the prepared rubric or design a rubric to include your chosen criteria.

Create

Follow the plan and make a model.

Test and Improve

What changes need to be made to the design?

Accept any reasonable answer.

_____

_____

Improved Design

Share

Tell others how your design solves the problem.

Climate

Climate is the pattern of weather in an area for a long period of time. It includes the temperature and the amount of precipitation in an area. Climate causes different kinds of weather patterns in an area.

The pattern of warm and cold weather or wet and dry weather is called a **season**. Some places have four seasons. Some places have two seasons. The kinds of seasons in an area are also part of the climate.

Not all places on earth have the same kind of climate. The climate is different because each part of the earth gets a different amount of sunlight. The earth is tilted on its axis. It revolves around the sun. That is why some areas get more sun and some get less sun. The sun's light energy heats the earth.

The equator is the imaginary line around the widest part of the earth. It is the place that receives the most light and heat from the sun. The North Pole is at the top of the earth. The South Pole is at the bottom of the earth. The poles receive the least amount of light and heat from the sun.

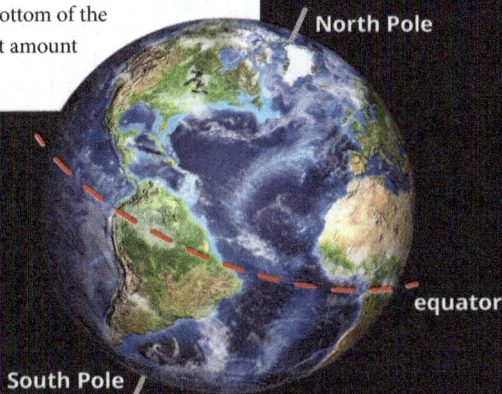

North Pole

equator

South Pole

41

Lesson **13** begins here.

Preparing Ahead

"Where in the World" Activity

This activity will need slips of paper cut out. Each slip of paper should have a country, or city and country, written on it. The locations should fall clearly within the climate zones that are discussed in this lesson.

⚠ Helps

For more information about climate change from a biblical perspective, see Answers in Genesis (www.answersingenesis.org) and Institute for Creation Research (www.icr.org) online.

Objectives
- Identify the three main climate zones.
- Identify the characteristics of each climate zone.
- List examples of geographic locations in each climate zone.

Materials
- world map
- slips of paper with one country (or city, country) written on each
- container or jar for slips of paper

Teacher Resources
- Visual 2.8: *Climate Zones*

Vocabulary
- climate
- season
- climate zone

Engage

- Display the following quote often attributed to Mark Twain:
  "Climate is what you expect. Weather is what you get."
- Invite the students to share what they think the quote means.
- Display a world map. Guide the students in locating Alaska.
- ⚙ What would you expect the weather to be like in Alaska in the winter? cold, snowy
- Guide the students in locating Mexico on the map.
- ⚙ What would you expect the weather to be like in Mexico? hot
- ⚙ Why do you expect it to be cold in Alaska and hot in Mexico? Elicit that the pattern of weather is usually cold in Alaska and hot in Mexico.

Instruct

Preparation for Reading
- Preview and pronounce the vocabulary terms *climate*, *season*, and *climate zone*.
- Direct the students to read pages 41–43 silently to find out what climate is.

Teach for Understanding

What is *climate*? the pattern of weather in an area for a long period of time

What does climate include? temperature and the amount of precipitation

What is the pattern of warm and cold weather or wet and dry weather called? a season

⚙ Why do some places get more energy from the sun than other places? The earth tilts on its axis. The parts that are tilted closer to the sun receive more of the sun's energy than the parts that are tilted away from the sun.

Which area gets the most light and heat from the sun? the equator

Which areas receive the least amount of light and heat from the sun? the North Pole and the South Pole

What is an area of the world that has a certain type of climate called? **a climate zone**

- Direct attention to the diagram of the earth.

What are the three main climate zones on Earth? **polar, temperate, and tropical**

Where are the polar climate zones located? **at the top of the earth near the North Pole and at the bottom of the earth near the South Pole**

What are some places in the polar climate zone? **parts of Alaska and Canada, parts of Denmark, Norway, Finland, Sweden, Iceland, and Russia; also the continent of Antarctica**

Climate Zones

A **climate zone** is an area of the world that has a certain type of climate. A climate zone is based mainly on precipitation and temperature. The three main climate zones on Earth are polar, temperate, and tropical.

polar zone

temperate zone

tropical zone

temperate zone

polar zone

Polar Climate Zones

One polar climate zone is found at the top of the earth, near the North Pole. Parts of Alaska and Canada are in this polar climate zone. Parts of Denmark, Norway, Finland, Sweden, Iceland, and Russia are also in the polar climate zone. The other polar climate zone is at the bottom of the earth, near the South Pole. The continent of Antarctica is in this polar climate zone.

42

ⓘ Background

Climate Zones

The climate zones introduced in this lesson are the broadest classification of climate zones. In addition to the polar, temperate, and tropical zones given in the lesson, there are three more that include arid, mediterranean, and tundra. Sub-climate zones are used to further differentiate between locations within these six climate zones.

Arid zones are hot and dry all year. They include the deserts of North Africa and central Asia, as well as the southwest United States and inland Australia.

The Mediterranean zones have mild winters and hot, dry summers. They include the land around the Mediterranean Sea, southern South America, and southern California.

The tundra zones are very cold all year, although they do support plant life. The northern areas of North America and Asia, the southern coast of Greenland, and mountaintops around the world are considered part of the tundra zones.

The polar zone is almost always cold, even in the summer. It is usually dry and does not get much precipitation. The precipitation that does fall is mostly snow. Very thick ice covers the land. The animals that live in the polar zone usually have light-colored fur to blend in with their surroundings. They have thick coats to help them survive in the cold. Polar bears, penguins, and seals are some animals that live in the polar zone. Very few plants grow in the polar zone.

The polar zones get very little sunlight for most of the year. In the winter the polar zones have months of complete darkness. In the summer they have 24 hours of sunlight. The polar zones have only two seasons. The winter season is very cold and long. The summer season is short and cool.

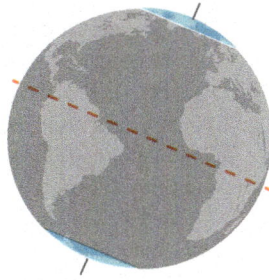

What are the three main climate zones?

What is the temperature like in the polar climate zone? It is always cold.

What kind of precipitation do you think falls in the polar zone? snow and ice

What kinds of animals live in the polar zone? polar bears, penguins, and seals

What seasons are there in the polar zone? summer and winter

Would the polar climate zone be a good place for people to live? No, it is too cold.

polar climate zone, temperate climate zone, and tropical climate zone

43

Preparation for Reading

- Preview and pronounce the words *temperate* and *tropical.*
- Direct the students to read pages 44–45 silently to find out where the temperate climate zones are.

Teach for Understanding

What are some places that are in the temperate climate zone? parts of Asia, parts of South America, parts of Africa, Europe, and parts of North America including the United States

What does *temperate* mean? in the middle

Why do we say that the temperate zone is in the middle? It is in the middle of the polar and tropical zones. The temperatures are in the middle. They are not too hot or too cold.

What kind of precipitation falls in the temperate zone? some rain and some snow each year

- Direct attention to the picture at the bottom of the page. Point out the low lying clouds in the picture.

What do we call clouds that form close to the ground? fog

What makes the temperate zones a good place for growing food? The temperatures are not too hot or too cold, and they get enough rain to grow crops. The temperate zones get a good amount of sunlight and heat each year.

What are the seasons in the temperate climate zone? spring, summer, fall, and winter

Do you think the temperate zone is a good place for people to live? Yes. It is not too hot or too cold. There are four seasons. There is a good amount of sunlight and rain.

Temperate Climate Zones

The temperate climate zones are found between the polar zones and the tropical zone. Parts of Asia, South America, and Africa are in the temperate zone. Europe and the United States are also in the temperate climate zone. The word *temperate* means "in the middle." Temperatures are not too hot or too cold for long periods of time.

Most places in the temperate climate zones receive rain and snow each year. Many plants, animals, and people live in the temperate climate zones. They are a good place to grow food.

The temperate zones get sunlight and heat for most of the year. Most places here have four seasons—spring, summer, fall, and winter. Spring in the temperate climate zones is usually wet. The summers are warm. Fall is usually cool and crisp. The winters are usually cold, and some areas have snowy weather.

What makes the temperate zone good for growing food?

44

Activity

Climate Zones

Divide a sheet of paper into three columns. Write the name of one climate zone at the top of each column. List one fact that you have learned about each climate zone. Give one example of a location in each climate zone.

Differentiated Instruction

Remediation: Climate Zones

Divide a sheet of paper into three columns. Write the name of one climate zone at the top of each column. In each column, write or draw one fact that you have learned about that climate zone.

Enrichment: Climate Zones

Divide a sheet of paper into three columns. Write the name of one geographic location at the top of each column. Identify the climate zone of that location. Give two reasons that support the climate zone you chose.

Tropical Climate Zone

The tropical climate zone is near the equator. Central America, and parts of South America, Africa, and Asia are in the tropical climate zone. Most parts of the tropical climate zone are very hot and have times of very heavy rainfall. Rainforests are part of the tropical climate zone. Large deserts are also part of the tropical climate zone. Many plants, animals, and different kinds of bugs live in the tropical climate zone.

The tropical climate zone gets the most sunlight and heat of anywhere on the earth. That is why most areas are very hot all year. There are two seasons in the tropical climate zone. The wet season is a time of very heavy rainfall. The dry season is a time of little or no rain.

Which climate zone do you live in?

polar zone

tropical zone

temperate zone

temperate zone

polar zone

Fantastic Facts

The coldest temperature ever recorded was at Vostok Station in Antarctica. It was -89.2°C (-128.6°F)! The warmest temperature ever recorded was at Death Valley, California. It was 56.7°C (134°F)!

Why is the tropical climate zone hot almost all year?

45

Where is the tropical climate zone located? near the equator

What are some places that are in the tropical climate zone? Central America, parts of South America, parts of Africa, and parts of Asia

What is the temperature like in the tropical climate zone? very hot most of the time

What kind of precipitation falls in the tropical climate zone? rain

What seasons does the tropical climate zone have? wet season and dry season

- Direct attention to the globe with the climate zones labeled. Help the students find the climate zone where they live.
- Direct attention to the "Fantastic Facts" box. Use the world map to locate Antarctica and California.

Which climate zone is Antarctica in? polar

Which climate zone is California in? temperate

- Explain that Death Valley, California, is a desert.

Why do you think the desert is so hot? Possible answers: It does not get much precipitation. Not many trees grow in the desert for shade.

It gets the most sunlight and heat of anywhere on Earth.

Apply

Activities
Study Guide, pages 39–40
These pages review the concepts taught in Lesson 13. After completion, direct the students to keep the completed pages to review for Test 2.

Assess

Quiz 2B
The quiz may be given at any time after completion of this lesson. Material for this quiz comes from the Study Guides in Lessons 11 and 13.

ⓘ Background

Snow in the Tropical Zone
Most places in the tropical climate zone are hot. But places such as the mountains of Africa or the Himalayas are only tropical at the base of the mountain. These mountains have snow in the higher elevations.

Activity

Where in the World?
Materials: world map, slips of paper with locations written on each, jar or container for slips of paper

Students participate all together, or in groups. One student draws a slip of paper out of the jar and reads the location. Student (or group) locates the city on a world map. Student (or group) identifies which climate zone the city is located in.

Objectives
- Identify climate change.
- Compare and contrast different views of climate change. **BWS**
- Identify possible causes of climate change.
- Defend the claim that climate change has limits, using Genesis 8:22. **BWS**

Teacher Resources
- Visual 2.9: *Map of Alaska*
- Video *Climate Change* from Institute for Creation Research

Vocabulary
- climate change
- Ice Age
- glacier

Engage
- Play the *Climate Change* video (TeacherToolsOnline.com).
- Explain that we cannot know everything about climate change, but we are going to learn some things about climate change.

Instruct

Preparation for Reading
- Preview and pronounce the vocabulary terms *climate change*, *glacier*, and *Ice Age*, and the words *Grinnell*, *Gould*, and *Knik*.

> The pronunciation of the Knik glacier is unusual. In this case, both /k/ sounds are pronounced.

- Direct the students to read pages 46–48 silently to find out when the Ice Age happened.

Teach for Understanding
What is *climate change*? a change in weather patterns over a long time

What was the earth like when God created it? It was beautiful and perfect.

Why did God send the Flood? The sin of the people caused God to send a flood.

How much of the earth was covered by the Flood? the whole earth

What caused the earth's climate to change? the Flood

- Direct attention to the pictures and captions.

How did the Flood change the earth's climate? Possible answers: The earth was beautiful and perfect when God created it. The waters from the Flood covered the whole Earth. Plants and animals died.

Lesson 14

God created a beautiful, perfect world for His glory.

Climate Change
Climates can change in different places on the earth. You have learned that climates change in different seasons. Scientists have studied changes in climate for many years. **Climate change** is a change in weather patterns over many years.

A Biblical View of Climate Change
In Genesis 1:31 the Bible says that God created a perfect world. But the world was changed by sin. The sin of the people caused God to send a flood. The Flood covered the whole earth. All the land animals and people who were not on the ark died. After the Flood ended, the earth's climate changed.

The earth was changed forever by the Flood.

46

Some scientists believe that a lot of rain and snow fell after the Flood. This caused ice to form in many places. Large areas of land were covered in glaciers. A **glacier** is a large sheet of very thick ice. It moves downhill very slowly. Many glaciers still exist today around the world.

Some scientists believe the Bible's account of the Flood. These scientists think that a short Ice Age happened after the Flood. During the **Ice Age** glaciers covered large parts of the earth. The earth's temperature was cooler. Since the Flood, the earth has had patterns of warming and cooling. Now the earth is in a pattern of warming up.

The Knik Glacier is one of many large glaciers in Alaska. It is about 40 km (25 mi) long.

47

What caused ice to form in many places after the Flood? a lot of rain and snow fell after the Flood

What is a *glacier*? a sheet of very thick ice that moves downhill very slowly

How much of the earth was covered in glaciers? large areas of land

Do you think you could see a glacier moving downhill? No, glaciers move so slowly that people cannot see that they are moving.

- Direct attention to the picture and caption.

  What is shown in the picture? a glacier, Knik glacier

- Display the *Map of Alaska* visual.

- Point out the location of the Knik Glacier in the south-central region on a map of Alaska. The Knik Glacier is one of the largest glaciers in south-central Alaska.

- Explain that Alaska has more than 600 named glaciers and about 100,000 total glaciers.

Why do you think there are so many glaciers in Alaska? Possible answers: It is cold in Alaska. There is a lot of snow and ice in Alaska.

In what climate zones do you think you might find a glacier? Possible answers: polar climate zone, temperate climate zone, tropical climate zone

- Explain that glaciers are found in all three climate zones. There are glaciers in the tropical climate zone near the equator. They are found at the tops of very tall mountains where the temperatures are cold.

What do we call the period of time after the Flood when glaciers formed? the Ice Age

Why did the Ice Age happen after the Flood? Rain and snow fell after the Flood. The rain and snow caused ice to form in many places. Large areas of land were covered in ice.

What do many scientists who believe the Bible's account of the Flood think about the earth's temperature since the Flood? There was a short Ice Age after the flood. The earth's temperature was cooler. Since then, the earth has had patterns of warming and cooling. Now the earth is in a pattern of warming up.

ⓘ Background

Knik Glacier

The Knik Glacier is one of the largest glaciers in south central Alaska. It is located about 80 km (50 mi) east of Anchorage, Alaska, near the town of Palmer in the Mat-Su Valley. The glacier averages about 61 m (200 ft) thick. It is about 40 km (25 mi) long and 8 km (5 mi) wide. Its end is at the head of the Knik River and is almost 5 km (3 mi) across. In the spring, the melting ice from the glacier feeds the head of the Knik River.

A scientist who has a different worldview believes that the earth is about how old? **4.5 billion years**

What do they believe about the ice age? **that the earth has had several long ice ages**

Why are some scientists worried about some plants and animals? **They think some plants and animals will die because the earth is getting too warm.**

- Direct attention to the picture and caption.

  What differences do you see in the pictures? **Possible answers: There is more rock in one picture. There is less snow and ice in the other picture. The glacier has gotten smaller.**

- Share Background information about the Grinnell Glacier as time allows.

  Why do you think the Grinnell Glacier has been melting? **Possible answers: The earth has gotten warmer. There has not been as much snow.**

  Can scientists know for sure what the earth's temperatures have been in the past? **No, because we do not have any record of the earth's temperatures from long ago.**

- Explain that scientists make observations and collect data so that they can make educated guesses about the past. Only God knows everything that has happened in the past.

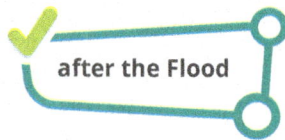

after the Flood

A Different View of Climate Change

Most scientists think that the earth is very old. They believe the earth is 4.5 billion years old. They say that the earth has had several long ice ages. These scientists think that most of the earth was covered with ice millions of years ago. They believe that now the earth is getting warmer. They think that it is warmer now than it has ever been. Some scientists are worried that plants and animals will die. They think the earth is getting too warm.

1938

2019

The Grinnell Glacier from Mt. Gould in Montana has been slowly melting and shrinking. Scientists think it is shrinking because the earth is getting warmer.

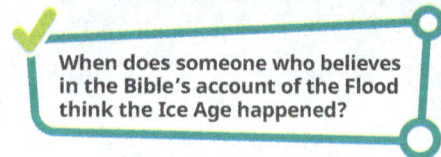

When does someone who believes in the Bible's account of the Flood think the Ice Age happened?

48

ⓘ Background

Grinnell Glacier

Grinnell Glacier is located in Glacier National Park in Montana. It is one of at least 35 named glaciers in Glacier National Park. Grinnell Glacier was named for George Bird Grinnell who was instrumental in the founding of Glacier National Park. In 1850 Grinnell Glacier measured 710 acres. In 1966 the glacier measured 252 acres. By 2005, the glacier measured about 152 acres. While the glacier appears to be receding, it should be noted that growing and receding are a part of the glacial cycle. When temperatures are cold and snow falls, glaciers grow. The growing period is usually from the end of September until June. Then there is usually a three-month period of melting before they grow again.

Possible Causes of Climate Change
Natural Causes

We know that the earth's climate has had patterns of warming and cooling. Scientists do not know all the reasons why the climate changes. But they do know some of the reasons that climates can change.

One reason that the climate can become warmer is changes in the sun's energy. The sun goes through patterns. It gives off more energy. Then it gives off less energy. When the sun gives off more energy, it causes the earth to be warmer. When the sun gives off less energy, the earth is cooler.

Volcanoes, dust storms, and forest fires are other reasons that climates can change. Mt. Pinatubo is a volcano that erupted in 1991. It caused the earth's temperature to be cooler for about two years! Large dust storms and forest fires can also cool the earth. This is because dust and ash act like large clouds. They block some of the sun's light and warmth from reaching the earth.

Mt. Pinatubo is located in the Philippines. When it erupted in 1991, smoke and ash could be seen for hundreds of miles.

49

Preparation for Reading
- Preview and pronounce the words *Mt. Pinatubo*, *volcano*, and the *Philippines*.
- Direct the students to read pages 49–50 silently to find some possible reasons for climate change.

Teach for Understanding

Is it possible for scientists to know all the reasons why the climate is changing? no, because science is based on observations

Can we know any of the reasons the climate is changing? Scientists can make guesses from their observations, but they do not know for sure why the climate is changing.

What pattern does the sun go through? giving off more energy and giving off less energy

How does the sun's energy affect the earth's climate? When the sun gives off more energy, the earth is warmer. When it gives off less energy, the earth is cooler.

- Direct attention to the diagrams of the sun. Discuss that the sun goes through cycles, or patterns, of giving off more and less energy.

What are some natural causes of climate change? changes in the sun's energy, volcanoes, dust storms, and forest fires

How did the eruption of Mt. Pinatubo affect the earth's climate? The earth's temperature was cooler for about two years.

Why did Mt. Pinatubo's eruption cool the earth? The smoke and ash from the eruption acted like large clouds and blocked some of the sun's light and warmth from the earth.

- Direct attention to the picture and caption.
- Explain that the eruption of Mt. Pinatubo was so powerful that the smoke and ash lingered in the atmosphere for almost two years, causing cooler temperatures worldwide.
- Share additional Background information as time allows.

(i) Background

Mt. Pinatubo

Mt. Pinatubo was the largest eruption to affect a heavily populated area. It erupted on June 15, 1991, on the island of Luzon in the Philippines. The ash cloud rose more than 40 km (about 25 mi) into the air and was hundreds of kilometers across. Along with the ash from the eruption, sulfur dioxide was released into the atmosphere. The result was a drop of 0.5°C (1.0°F) in the global temperature. So much rock and magma erupted that the summit of Mt. Pinatubo collapsed and formed a large volcanic depression.

Clark Air Base, owned by the United States, is located about 16 km (10 mi) east of Mt. Pinatubo. It was evacuated just days before the eruption on June 15. The powerful eruption covered Clark Air Base in a layer of soot, ash, and debris that was up to 60 cm (2 ft) thick in some places. Deciding the cost of restoration was too great, the United States surrendered the air base to the Philippine government on November 26, 1991.

What do most scientists believe about the earth's temperature in the last 50 years? It has warmed up.

Can we know that the earth is warmer now than it has ever been? No, scientists do not know what temperatures were in the past. They can only make a guess about how warm the earth was before they made observations and kept records.

What are some ways that some scientists believe people have caused the earth to warm up? Gases from factories, cars, and trucks go into the air causing air pollution. Trees have been cut down, so there are not as many trees to use these gases to make their food.

- Explain that some gases released into the air come from burning fuel that powers the factories and vehicles.

- Explain that when the fuel is burned, or used, it gives off a gas. Trees and plants use this gas to make their food. But if too many trees and plants are cut down, more of this gas will stay in the air. This gas could be one reason why the earth is warming up.

> Carbon dioxide and photosynthesis will be discussed in Chapter 5.

Can we know that people have caused the earth to warm up? No, scientists can make guesses about what causes the earth to warm up, but they do not know for sure.

- Direct attention to the pictures. The bottom picture is a paper plant in Wisconsin. Invite the students to share about times when they have seen air pollution.

> ✔ The sun gives off less energy. There is air pollution from burning fuel. Too many trees and plants are cut down.

Human Causes

Most scientists agree that the earth has warmed up in the last 50 years. Some scientists believe that people have caused the warming of the earth. They believe the earth is warmer now than it has been before. Scientists look at clues they find and make a careful guess about the earth's past temperatures. But we cannot know how warm the earth was in the past.

One possible human cause for a warming earth is air pollution. Pollution is anything that makes the air, water, or land dirty. Some things that people use produce gases that go into the air. Some factories let out gases into the air. Cars and trucks let out gases when they burn fuel. Trees and plants use some of these gases to make their food. But when trees and plants are cut down, more gases stay in the air. These gases might make the earth warmer.

> ✔ **What are some possible reasons that the earth is warming up?**

50

What People Can Do

God told people to take care of the earth. We do not know that air pollution is a cause of climate change. But we do know that people should do their best to keep the air clean. People can think of ways to reduce pollution. People can be careful to cut down only as many trees as they need. They can also plant trees to take the place of the ones they cut down. Planting new trees and other plants can help keep the air clean. God gave us this wonderful Earth to use. We need to use it wisely. How can people help to keep the earth clean?

God's Promise

The earth's climate has changed many times since Creation. But God promised Noah after the Flood that He would take care of the earth. In Genesis 8:22, God gave us four promises about the earth. He said that the earth will always have seedtime and harvest. The earth will also have cold and heat. He promised that summer and winter and day and night will not end. All these things will happen as long as the earth remains. We know that God will take care of us!

What promises did God give about the earth?

What seasons are shown in these pictures?

51

Preparation for Reading

- Direct the students to read page 51 silently to find out what promise about the earth God has given to us.

Teach for Understanding

What are some things that people can do to take care of the earth? keep the air clean, reduce pollution, cut down only as many trees as they need, plant new trees

- Direct attention to the pictures of the trees and caption.

- Point out that all four pictures show the same tree in the same location.

What is different about the tree in each picture? Possible answers: The leaves are different colors. The leaves are off the tree in one picture.

Why does the tree go through changes? It changes in different seasons.

What seasons are shown in the pictures? spring, summer, fall, and winter

Has the earth's climate been the same since Creation? No, it has changed many times since Creation.

How do we know that climate change will not destroy the earth? God gave a promise in Genesis 8:22 that as long as the earth remains there will always be seedtime and harvest, cold and heat, summer and winter, and day and night.

- Explain that God designed the earth with patterns of change. Seasons are one of those patterns. We can do our part to take care of the earth that God has given us and trust God's promise that He will take care of the earth.

There will always be seedtime and harvest, cold and heat, summer and winter, and day and night.

Apply

Activities
Study Guide, pages 41–42
These pages review Lesson 14. After completion, direct the students to keep the completed pages to review for Test 2.

Objectives

- Make predictions of future weather based on observed and recorded weather data.
- Recall terms and concepts from Chapter 2.

> Invite the students to share graphs from Lesson 9. Guide the students in making predictions about future weather patterns.

Apply

Review

- Material for Test 2 will come from the Study Guides on *Activities* pages 23–25, 33, 39–40, and 41–42, as well as Quizzes 2A and 2B. You may review any or all of the material during the lesson.
- You may choose to review Chapter 2 by playing the Weather Mystery Word review game or a game from the Game Bank (Teacher Resources).

🏆 Review Game

Weather Mystery Word

Divide the students into two teams. Choose a vocabulary word from the chapter and display one blank for each letter.

Alternate asking questions to the teams. Each time a team member answers a question correctly, he gets to guess a letter. Fill in each correct letter as it is guessed. A team receives one point for correctly guessing the vocabulary word and another point if it is able to give a definition of the word.

Continue with other vocabulary words. The team with the most points when play ends is the winner.

Note: Avoid review questions that ask for the definitions of the vocabulary words used in the game.

NOTES

NOTES

_____

_____

_____

_____

_____

_____

_____

_____

_____

_____

_____

_____

_____

_____

_____

_____

_____

_____

_____

_____

_____

_____

_____

_____

_____

_____

Objective

- Apply terms and concepts from Chapter 2.

Assess

- Administer Test 2.

Chapter Objectives

- Develop an explanation for the origin of soil, rocks, minerals, and fossils, based on biblical teaching.
- Analyze the composition and characteristics of soil, rocks, minerals, and fossils.
- Apply inquiry skills of observe, predict, infer, and communicate to soil, rocks, minerals, and fossils.
- Evaluate different explanations for how most fossils formed and for the role of adaptation in the post-Flood world.

Lesson Objectives

- Explain from Scripture the origin of the earth, including water and dry land. **BWS**
- Identify four components of soil.
- Identify the three main layers of soil.
- Examine topsoil to identify its various components.

Materials

- 3 soil samples, from three different locations

Choose soil samples that look very different from each other from places familiar to your students.

Teacher Resources

- Visual 3.1: *Layers of Soil*

Vocabulary

- soil
- decay
- humus
- topsoil
- subsoil
- bedrock

Chapter Introduction

- Direct attention to the chapter title and the picture of Arches National Park. Provide time for the students to leaf through the chapter, looking at the headings and pictures. Discuss what they think the chapter will be about.

Use completed Looking Ahead, *Activities* page 43, to generate a discussion on what the students already know about soil, rocks, minerals, and fossils.

Note: The Notebook Connection will be used periodically throughout the *Activities* book. Its purpose is to make a deliberate connection to the student's science notebook and to encourage note taking, creative writing, drawing, summarizing, analyzing and interpreting text and images, and conducting research.

Big Question

- Ask a volunteer to read aloud the Big Question.
- Explain that the students will find the answer to the question as they read the chapter.

Big Question
What do rocks, minerals, and fossils tell us about the earth's history?

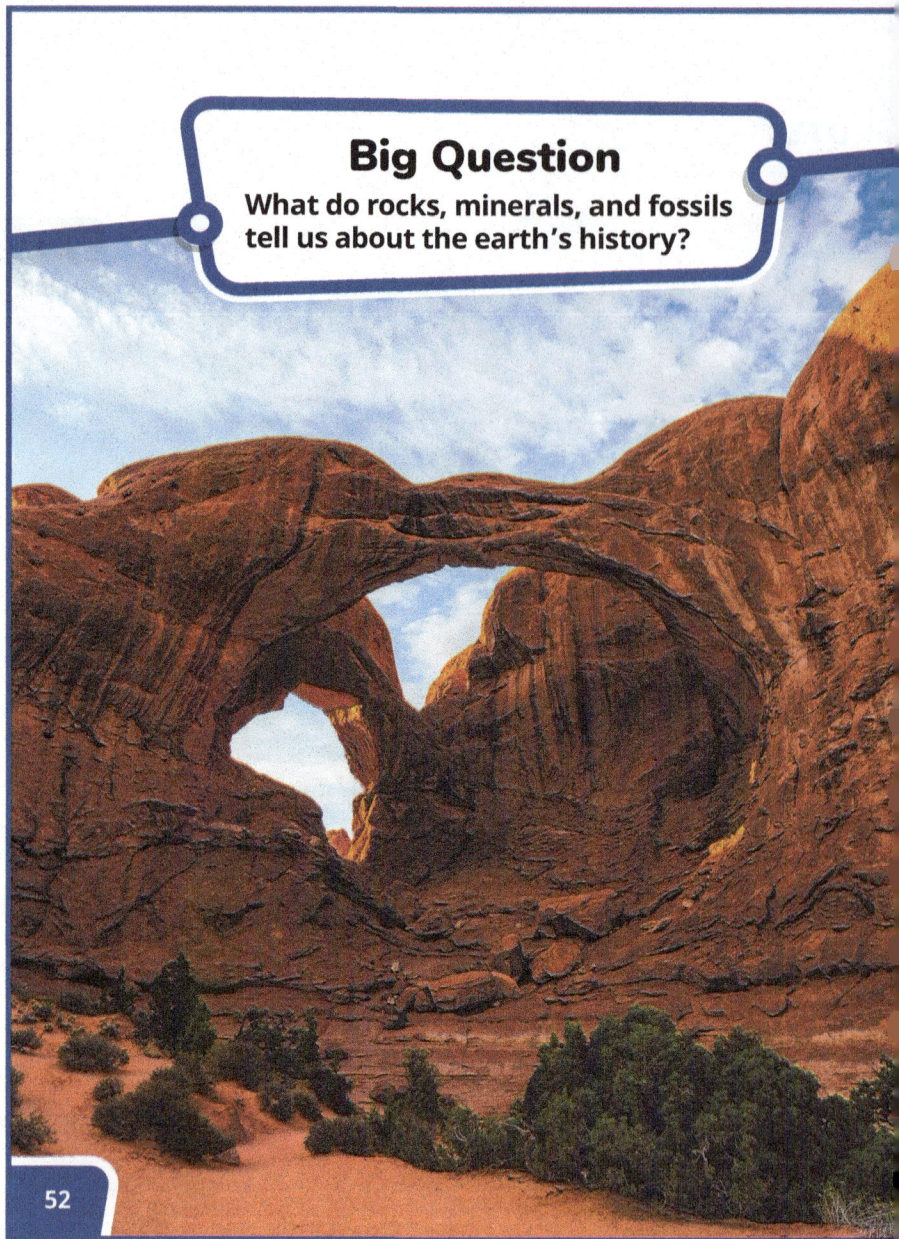

52

Visit TeacherToolsOnline.com for resources to enhance the lessons.

Chapter 3

Soil, Rocks, Minerals, and Fossils

The Bible tells us in Genesis 1 that God created the sun, moon, and stars. He created the earth. God created the atmosphere. He separated the dry land from the seas. The land is a wonderful gift from God. It provides many of the things we need to live.

But the land today is different from what God created. Because of the sin of people, God covered the whole earth with a great flood. The moving water changed mountains, valleys, and rocks. After the Flood, the earth was very different. But trees and other plants grew again in the dirt that settled on the earth's surface.

53

Engage

- Display the soil samples.
  What are these? Possible answers: dirt, soil, ground
  Do the samples of dirt look the same?
- Explain that dirt is not always the same. It can be different depending on the place it came from. But point out that in this lesson they will learn that dirt is made of the same things. *Note:* The soil samples will be used again later in this lesson.

Instruct

Preparation for Reading

- Direct the students to read page 53 silently to find out where the earth, with its water and dry land, came from.

Teach for Understanding

Where did the earth, with its water and dry land, come from? God created the earth and everything in it, including the water and dry land.

Where in the Bible are we told about God's creation? Genesis 1

What other names do we use for dry land? Possible answers: dirt, soil, ground, earth

Why is the land a wonderful gift from God? It provides many of the things we need to live.

Why is the land today different from the land God created? Because of sin, God covered the whole earth with a great flood.

In what ways did the Flood change things on the earth? The moving water from the Flood changed mountains, valleys, and rocks.

After the Flood, what started to grow again on the earth's surface? trees and other plants

ⓘ Background

Chapter Photo
The photo on pages 52–53 is of Arches National Park, located in eastern Utah. The stone arches bear evidence of weathering and erosion, consistent with the Genesis Flood.

Preparing Ahead

Lessons 19–21 Materials
Look ahead to the materials lists in Lessons 19–21 for required preparation.

Lesson 26 Materials
Look ahead to the materials list in Lesson 26 for required preparation.

⚠ Helps

Rock and mineral samples will be used in Lessons 19–21. Rock and mineral collections can be ordered from science catalogs. Rock and mineral shops, as well as some museum and tourist gift shops, are also good sources for rock and mineral samples. Being able to see and handle actual rocks and minerals will greatly enhance the students' comprehension.

Preparation for Reading

- Preview and pronounce the vocabulary terms *soil*, *decay*, and *humus*.
- Direct the students to read pages 54–55 silently to find out why there are different types of soil on the earth.

Teach for Understanding

What is another name for dirt? soil

What is soil? loose rock and bits of decaying plant and animal material

What does it mean to *decay*? to break apart or rot

What are some ways living things depend on soil? Possible answers: Some animals live in soil; most plants need soil to grow; both animals and people need plants as food.

What are four things that make up soil? small bits of broken rock, humus, water, and air

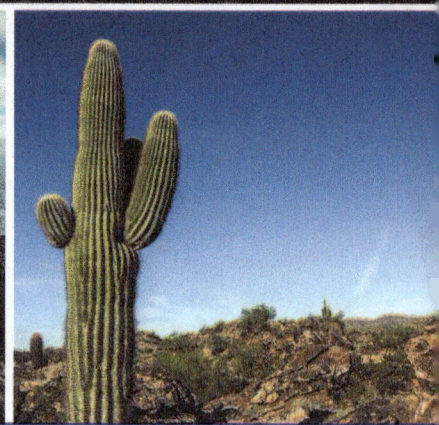

Soil

Another name for dirt is soil. **Soil** is made of loose rock and bits of decaying plant and animal material. When plants and animals **decay**, they break apart, or rot. Many living things depend on the soil. Some animals live in the soil. Most plants need soil to grow. Both animals and people need plants for food. Soil has nutrients in it that help plants to survive and grow. Plants can grow in loose soil or even in the cracks of rocks.

Parts of Soil

If you pick up a handful of soil, you might see tiny pieces of rock in it. You may observe other things as well. Soil is made up of small bits of broken rock, humus, water, and spaces filled with air.

In the fall, leaves drop from the trees. They become brown and dry. After a while they crumble into small pieces. The leaves seem to disappear, but they do not.

54

ⓘ Background

Parts of Soil

Not all soil has the same look and feel. Soil is usually a combination of three different kinds of particles: sand, clay, and silt. Sand particles are the biggest and usually feel gritty. Silt particles are slightly smaller and usually feel soft and silky to touch. Clay particles are the smallest. Clay is smooth when dry but sticky when wet.

Squanto

When the Pilgrims first came to the New World, one of the most important skills they needed was knowing how to grow food. Squanto was a Native American who befriended the Pilgrims. He taught them to grow corn by putting fish in the soil while planting the seeds. The decayed fish added humus to the soil which helped the corn grow better.

Saguaro Cactus

Saguaro cacti grow very slowly. The average height of a 10-year-old saguaro is about 1 to 1.5 inches. A saguaro will produce flowers when it is between 35 and 70 years of age. Between 50 and 70 years of age, the saguaro will begin to sprout its arms. The adult saguaro averages about 40 feet in height, while some very old cacti can be over 70 feet tall.

Over time the remains of the leaves become part of the humus in the soil. **Humus** is the remains of living things that have died and decayed. Plants and animals decay after they die. After a while, they become part of the soil. Humus is the soft dark part of the soil. It has many nutrients that plants need to survive and grow.

Soil contains water and air as well as humus. Water is stored in the soil. There are also tiny spaces of air in the soil. To grow well, plants need both water and air.

Though plants need certain things to survive, not all plants need the same amounts of humus, water, and air. God designed the earth with many types of soil so that different kinds of plants can grow.

✔ **What is soil made of?**

55

What is *humus*? Possible answers: the remains of living things that have died and decayed; the soft dark part of the soil; the part of the soil that has many nutrients that plants need to survive and grow

After dead things decay, what do they become part of? the soil; humus

Besides humus, what else does the soil contain? water, air, broken pieces of rock

Why do you think it is important that water is stored in the soil? Possible answer: Plants need water to survive and grow.

Why do you think it is important that soil has spaces of air? Possible answers: The air spaces give oxygen to plant roots; plants need air to survive and grow.

What is one reason that God designed the earth with many types of soil? Possible answers: Not all plants need the same amounts of humus, water, and air; many kinds of soil allow many different kinds of plants to grow.

- Direct attention to the soil samples again. Discuss their differences in appearance and texture.
- Explain where you found the samples, but do not identify which sample came from which location.
- Guide the students in matching the samples and locations.
- Direct attention to the pictures at the bottom of the pages.

How would you describe the land in each picture? Possible answers from left to right: rocky coast; dry and sandy desert; grassy plains; woodland forest

Which pictures show places that would be the best for growing plants? the pictures on page 55

Why do you think plants would not grow well in the places on page 54? Elicit that few plants are shown growing; there is no soil in the first picture; the second picture is a desert, which means there is little water; most plants need more water to survive and grow than the dry conditions shown in the second picture.

✔ **small bits of broken rock, humus, air, and water**

Preparation for Reading

- Preview and pronounce the vocabulary terms *topsoil*, *subsoil*, and *bedrock*.
- Direct the students to read page 56 silently to find out what the layers of soil are.

Teach for Understanding

- Display the *Layers of Soil* visual. Point out each layer as the page is discussed.

  What layer of soil do you usually see? topsoil

  What grows in topsoil? plants

  What does topsoil contain that helps plants to grow? humus

  What layer of soil is under the topsoil? subsoil

  Does subsoil have any humus? yes, but less than topsoil

  What does subsoil have that topsoil does not have? larger pieces of rock

  What do we call the large, unbroken rock under the subsoil? bedrock

- Direct attention to the "Try It Yourself!" box. You may choose to have the students participate in this activity during recess.

The "Try It Yourself!" activities are designed for the students to do on their own. If you want the activities done at school, you will need to plan ahead to supply the materials needed for some of these activities. You can find the "Try It Yourself!" materials under the Activity heading with the lesson in which the activity appears.

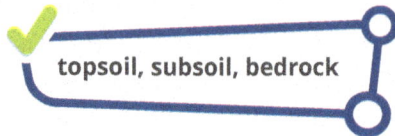

✔ topsoil, subsoil, bedrock

Apply

Activities

Looking Ahead, page 43

This page assesses the students' knowledge prior to the chapter.

Layers of Soil, page 44

This page reinforces the concepts taught in Lesson 17.

Layers of Soil

Soil forms in layers. Most of the time only the top layer is seen.

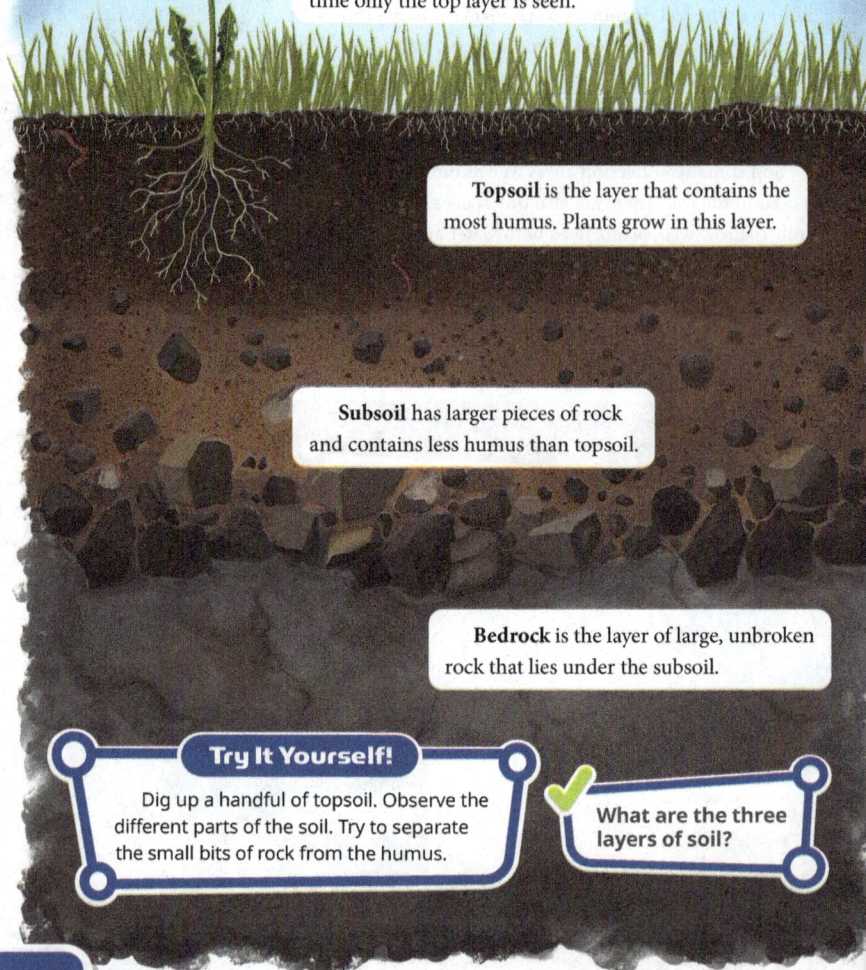

Topsoil is the layer that contains the most humus. Plants grow in this layer.

Subsoil has larger pieces of rock and contains less humus than topsoil.

Bedrock is the layer of large, unbroken rock that lies under the subsoil.

Try It Yourself!

Dig up a handful of topsoil. Observe the different parts of the soil. Try to separate the small bits of rock from the humus.

✔ What are the three layers of soil?

🜍 Activity

Try It Yourself!
Materials: small spade or shovel, handful of soil
Allow students to try this activity on their own.

👥 Differentiated Instruction

Remediation: Dirt Cup Soil Model
Materials (per student): *Dirt Cup* page (Instructional Aid 3.1), 1 clear plastic cup, 256 mL (9 oz); 2 whole vanilla cookies; 227 mL (1 c) of chocolate pudding; spoon; 1 broken vanilla cookie; 1 crushed vanilla cookie

Each student will model the soil layers in a cup. The procedure is included on the instructional aid page. The students will be able to eat this "dirt."

ⓘ Background

Weathering and Erosion
Weathering is different from erosion. Weathering is the process of breaking down the rocks. This can happen mechanically through flowing water, wind, ice, and plants, or it can happen chemically through acid rain, lichens, mosses, and oxidation. Erosion is the movement of the weathered material from one place to another.

Weathering, Erosion, and Conservation

The surface of the earth is always changing. Sometimes changes to the earth's surface happen quickly. Sometimes the changes happen slowly. This is part of God's design.

Huge rocks are broken into smaller rocks. The smaller rocks are broken into even smaller pieces. These tiny pieces of rock become part of the soil. What causes rocks to break apart? Flowing water, wind, ice, and plants can cause rocks to break apart. The breaking down of rocks into smaller pieces is called **weathering**. Water and wind also move rocks and soil. This movement is called **erosion**.

Flowing Water and Wind

Flowing water weathers and erodes rocks. The moving water tumbles the rocks, wearing away any sharp edges. The bits of broken rock are carried away by the water and become part of the soil. Weathering and erosion can help to renew the soil.

57

Lesson **18** begins here.

Objectives
- Recall the causes of weathering and erosion.
- Describe how landslides are an example of erosion.
- Recall that soil is a natural resource. **BWS**
- Explain why it is important to conserve soil. **BWS**

Materials
- smooth rock
- rough rock
- small piece of soap, per student
- sandpaper, per student
- paper towel, per student

Vocabulary
- weathering
- erosion
- landslide
- conserve

Engage

- Display the smooth and rough rocks. Allow the students to feel the texture of each rock.

    How are the rocks alike? Possible answer: Both are hard.

    How are they different? Possible answer: One is smooth and the other is rough.

    How do you think the rocks became different?

- Explain that sometimes a rough rock can become smooth, and sometimes a smooth rock can be broken and become rough. Point out that in this lesson the students will learn about what causes this to happen.

Instruct

Preparation for Reading
- Preview and pronounce the vocabulary terms *weathering*, *erosion*, and *landslide*.
- Direct the students to read pages 57–59 silently to find out how plants weather rocks.

Teach for Understanding

Why is the surface of the earth always changing? It is part of God's design.

What happens to rocks that are broken into tiny pieces? They become part of the soil.

What causes rocks to break apart? flowing water, wind, ice, and plants

What is the breaking down of rocks into smaller pieces called? weathering

What moves rocks that have been weathered? water and wind

What is this movement of rocks and soil called? erosion

How does water weather rocks? The moving water tumbles the rocks, wearing away any sharp edges.

- Direct attention to the picture. Explain that the force of the water itself can also weather rocks.

    How does water erode rocks? Moving water carries away bits of weathered rock.

    How are weathering and erosion helpful? They can help to renew the soil.

    What does it mean to *renew* the soil? to put back into the soil what has been lost from it

How can fast flowing water change the surface of the earth? Quickly falling rain can cause a flood; flood waters can erode rocks and soil.

- Direct attention to the picture of Bryce Canyon at the bottom of the page. Ask a volunteer to read the caption aloud.

  When does flooding happen? when rain falls quickly and the topsoil cannot soak up all the water

  What can heavy rain and flooding on a mountain or hill cause? a landslide

  What is a *landslide*? the rapid downhill movement of a large amount of rocks and soil

  What force causes a landslide? gravity

  What does a landslide erode? large areas of the earth's surface

  Do landslides erode the earth's surface quickly or slowly? quickly

- Direct attention to the before and after pictures of a landslide. Point out the area of land eroded by the landslide.

  How does wind weather rocks? Answers should include some of the following: wind blows sand that rubs the surfaces of rocks; the continual rubbing wears away bits of the rocks; the rocks become rounded and smooth.

  What happens to the weathered bits of rock? They become part of the soil.

  What can wind do to loose topsoil? erode it

  What helps to keep the topsoil in place? plant roots

- Give each student a small piece of soap, sandpaper, a damp cloth, and a paper towel.
- Direct each student to use the sandpaper and damp cloth to gently smooth (round) the edges of the soap. Tell the students to work over the paper towel.

  What do you think the piece of soap represents? a rock

  What do you think the sandpaper represents? sand that wind blows against rocks

  What do you think the damp cloth represents? water that flows over rocks

- Point out that the sandpaper and the cloth both smooth the soap.

  Which one gave you smoother edges? the damp cloth

- Emphasize that real rocks weather very slowly today. It could take years for sand or water to wear away small pieces of rock. Point out that during the Flood of Noah's day, weathering from the force of the flowing flood waters took place very quickly.

Fast flowing water can also change the surface of the earth. Remember that the Genesis Flood covered the whole world. But smaller floods today cover parts of the ground that are usually dry. When rain falls quickly, the topsoil may not be able to soak up all the water. Flood waters can erode rocks and soil. Sometimes heavy rain and flooding can cause a landslide. A **landslide** is the rapid downhill movement of a large amount of rocks and soil. The force of gravity pulls the rocks and soil down a mountain or hill. A landslide erodes large areas of the earth's surface very quickly.

Wind also weathers and erodes rocks. The wind blows sand that rubs the surfaces of the rocks. This continual rubbing eventually wears away bits of the rocks. The rubbing makes the rocks rounded and smooth. The weathered bits of rock become part of the soil. Wind can also erode loose topsoil. But plant roots can help to keep the topsoil in place.

Water from the Genesis Flood, as well as wind since the Flood, have weathered and eroded these canyons.

58

ⓘ Background

Bryce Canyon National Park, Utah
The hoodoos of Bryce Canyon National Park bear evidence of weathering and erosion, consistent with the Genesis Flood.

Ice and Plants

Flowing water and wind can weather rocks by breaking off small bits. Ice and plants can also weather rocks. When water flows into the cracks of a rock and freezes, the water becomes ice. The ice takes up more space than the liquid water did. The ice pushes the rock apart and breaks it into pieces.

Over time, plants can also weather rocks. Even a small plant can break apart a rock. A seed can fall into a small crack in a rock. As a plant grows from the seed, the plant's roots and stem get larger. They push with great force. The growing plant can eventually weather, or break, the rock into pieces. The smaller pieces of rock can be eroded by the movement of flowing water or wind. The bits of rock then become part of the soil.

A growing plant can weather rocks.

Try It Yourself!

Fill a plastic cup half full with water. Place the cup on a flat surface and mark the level of the water. Place the cup in the freezer overnight. Compare the level of the ice with the mark on the cup.

✔ **What are four things that can weather rocks?**

59

In addition to flowing water and wind, what else can weather rocks? ice and plants

How can ice weather rocks? Answers should include some of the following: water flows into the cracks of a rock; the water freezes and becomes ice; the ice takes up more space than the liquid water did; the ice pushes the rock apart and breaks it into pieces.

- Direct attention to the picture and illustrations of the rock broken by ice. Point out that even very large rocks can be broken apart by ice.

How can plants weather rocks? Answers should include some of the following: a seed can fall into a crack in a rock and grow; as the plant grows, the plant's stem and roots get larger; the stem and roots push with great force, breaking the rock.

- Direct attention to the picture of the rock broken by the growing tree roots. Point out that even very tiny seeds can grow into plants with enough force to break a rock.

How can the rock that is weathered by ice and plants be eroded? Flowing water and wind will move the broken pieces of rock.

What happens to the smaller bits of eroded rock? They become part of the soil.

Suppose you find a rock that has been broken apart. Which more likely weathered the rock—wind or ice? ice, because wind smooths rocks, and ice breaks rocks apart

- Direct attention to the "Try It Yourself!" box.

✔ flowing water, wind, ice, plants

⚗ Activity

Try It Yourself!
Materials: clear plastic cup, water, marker, freezer
Allow students to try this activity on their own.

Preparation for Reading

- Preview and pronounce the vocabulary term *conserve*, as well as the words *natural resource*.
- Direct attention to the heading and read it aloud. Explain that *conservation* is a form of the vocabulary term *conserve*.
- Direct the students to read page 60 silently to find out why soil should be used wisely.

Teach for Understanding

Why is soil considered a natural resource? It is something that God made for people to use.

What does it mean to *conserve* a natural resource? to take care of the natural resource; to use it wisely

What are some ways people can conserve soil? plant trees to slow or stop wind erosion of the soil; plant trees and shrubs on a hillside to help prevent landslides caused by water erosion

- Direct attention to the picture of the farm fields surrounded by trees.

Why do you think the farmer planted trees around his fields? to slow or stop wind erosion

- Direct attention to the picture of the trees and shrubs on a hillside.

Why do think there are trees and shrubs planted on the side of the hill? to slow or stop water erosion; to stop landslides

- Share the Background information about the picture as desired.

Why is it important for people to conserve soil? Possible answers: Conserving soil can help to slow or stop erosion; clean soil, or soil that has not been polluted, can be used to grow food for people and animals; some plants grown in soil can be used for medicine; soil can be used to make bricks for homes and walkways; soil can be used to make roads, plates, and cups.

Share the Background information about echinacea as desired.

✔ Soil is a natural resource; clean soil can be used to grow food for people and animals; soil can grow plants used for medicine; it can be used to make bricks for homes and walkways, to make roads, and to make plates and cups.

Apply

Activities
Study Guide, pages 45–46
These pages review the concepts taught in Lessons 17–18. After completion, direct the students to keep them for review for Test 3.

Assess

Quiz 3A
The quiz may be given at any time after completion of this lesson.

Conservation

Soil is a natural resource. A *natural resource* is anything in nature that God made for people to use. It is important for people to conserve this part of God's creation. To **conserve** means to take care of and to use wisely the natural resources.

Planting trees helps to conserve soil. The roots of trees help to hold the soil in place. Trees planted around a farmer's field can help slow or stop wind erosion. Trees and shrubs planted on the side of a hill can help to prevent landslides caused by water erosion. Slowing or stopping erosion is one way to care for the soil.

Soil should be used wisely. Soil that has not been polluted, or has not been made dirty, has many important uses. Soil is used to grow food for people and animals. Some plants grown in soil can also be used for medicine. Some types of soil can be used to make bricks for homes and walkways. Soil can also be used to make roads. Soil can be used to make plates and cups. It is good to care wisely for soil. It is an important natural resource.

Why is it important to conserve soil?

60

ⓘ Background

Landslide Prevention Photo
This photo shows trees planted to control erosion in the Lhasa Valley of Tibet.

Echinacea
The flower in the photo is echinacea. Before the discovery of antibiotics, echinacea was a remedy used by Native Americans living on the Great Plains. It is believed that there is medicinal value in the leaves, flower, and roots. The different parts of the flower were used to fight infections. Today, some people use it to fight colds and flu. It has also been used to treat boils, burns, and other skin conditions.

Rocks

Perhaps you have dug up rocks in a garden or walked on large rocks in a stream. **Rocks** are hard pieces of the earth's surface. Some rocks were created by God. Some formed during the Genesis Flood. Some rocks can even form today. Rocks have many sizes and shapes. Some are small and can easily be picked up. Others are larger and much heavier than you can lift. A rock might be as big as a car or even a tall building. Some rocks are even bigger!

Though rocks come in many shapes and sizes, scientists classify them by how they are formed. Scientists classify rocks into three groups. The groups are sedimentary, igneous, and metamorphic rocks.

Sedimentary Rock

Water and wind can weather and erode pieces of rock. Small bits of weathered rock are called **sediment**. Sediment drops and falls to the bottom of water. The water presses down on the sediment. **Sedimentary rock** forms when layers of sediment are pressed together and harden. Because sedimentary rock is made of sediment, you can often see pieces of other rocks in it.

61

Lesson **19** begins here.

Objectives
- Identify the three main types of rocks.
- Explain how rocks form and are classified. **BWS**
- Create a way to classify rocks.

Materials
- rock samples of limestone, sandstone, obsidian, granite, and marble

Teacher Resources
- Visual 3.2: *Rocks, Rocks, and More Rocks*

Vocabulary
- rock
- sediment
- sedimentary rock
- igneous rock
- metamorphic rock

Engage
- Display the rock samples.
  Do all these rocks look the same? no
- How are they different? Possible answer: They are different sizes, colors, and shapes.

How are they alike? Possible answer: They are all hard. Today we will look at why rocks are different.

Instruct

Preparation for Reading
- Preview and pronounce the vocabulary terms *rock*, *sediment*, and *sedimentary rock*, as well as the words *shale*, *igneous*, and *metamorphic*.
- Direct the students to read pages 61–62 silently to find out how sedimentary rock forms.

Teach for Understanding
What are rocks? hard pieces of the earth's surface

Where have rocks come from? Possible answers: Some rocks were created by God; some formed during the Genesis Flood; some can even form today.

How can rocks be different? Possible answers: They have different sizes and shapes; some rocks are small and easy to pick up, and others are too large and heavy to pick up.

- Poll the class to find out how many students have ever picked up a rock that is bigger than a football.
- Direct attention to the picture at the top of the page. Discuss what it looks like the student is doing.
- Poll the class again to find out how many have picked up or moved a rock as large as the one the student is pushing against and discuss why they have not been able to do this.
- Explain that some very large rocks were weathered and eroded by the Genesis Flood. Other rocks were pushed by the glaciers that formed during the Ice Age after the Flood. When the ice melted, the rocks were left behind. Help the students to understand the great power of the flowing water during the Flood and the power of the glaciers, or moving ice, after the Flood to move such large rocks.

What does *classify* mean? to group things that are alike

How do scientists classify rocks? by how they are formed

How many groups are rocks classified into? three

What are the three groups of rocks? sedimentary rock, igneous rock, and metamorphic rock

What do water and wind often do to rock? They weather and erode rock.

What is *sediment*? bits of weathered rock that drop and fall to the bottom of water

How does sedimentary rock form? Layers of sediment are pressed together and harden.

What presses down on the sediment? water

Because sedimentary rock is made of sediment, what can you often see in it? pieces of other rocks

- Direct attention to the picture at the bottom of the page. Point out the layers of rock. Explain that this is Fire Wave rock formation at Valley of Fire State Park, Nevada.

What are three kinds of sedimentary rock? limestone, sandstone, and shale

What does limestone mainly form from? shells of sea animals

- Display the limestone rock sample. Allow the students to handle the sample. Discuss the texture and color.

Why do you think the limestone is a light color? It is made from seashells, and the material that makes up many seashells is light in color.

How is sandstone made? when layers of sand get pressed together

How does shale form? from layers of mud pressed together

- Display the sandstone and shale rock samples. Allow the students to handle the samples. Discuss their texture and appearance.

- Display the *Rocks, Rocks, and More Rocks* visual. Explain how the organizer is arranged and read. Guide the students as they complete the center of the web and the "sedimentary" portion of the web on *Activities* page 47. Cells with additional information may be added.

- Direct attention to the "Try It Yourself!" box.

hard pieces of the earth's surface

There are many different kinds of sedimentary rock. The kinds come from the different things that are mixed in to make the rock.

Limestone forms mainly from the shells of sea animals. The seashells break apart after the animals die. When pressed together, the bits of shell form limestone.

limestone

Sandstone is a sedimentary rock that is made when layers of sand get pressed together.

sandstone

Shale forms from layers of mud.

shale

What are rocks?

Try It Yourself!

Take a clear jar that has a lid and fill the jar about half full of soil. Add water to within 2.5 cm (1 in.) of the top of the jar. Put the lid on the jar and shake it. Place the jar on a flat surface. Leave the jar untouched overnight. Observe what happens to the soil.

62

Activity

Try It Yourself!
Materials: clear jar with a lid, soil, water
Allow students to try this activity on their own.

Differentiated Instruction

Enrichment: Form "Sedimentary Rock"
Materials: old crayons of various colors, crayon sharpener, wax paper, heavy-duty aluminum foil, 2 small pieces of wood, vise or wood clamps

"Weather" one crayon at a time by sharpening it. Collect the shavings on wax paper. Continue until you have a sizable pile of shavings. Try to keep the shavings in layers. Push the shavings into the center of the wax paper and fold the wax paper around them like an envelope.

Wrap the wax paper envelope in aluminum foil. Place that between the 2 small pieces of wood. With adult supervision, put the pieces of wood inside the vise and tighten it. Keep the vise in place for at least an hour. Loosen the vise. Remove the aluminum foil and wax paper and display the "sedimentary rock."

How does this activity model sedimentary rock? The "weathered" crayon shavings were pressed together. The vise acts like the pressure of the water on the sediment. Sedimentary rock forms from layers of sediment that are pressed together and harden.

Igneous Rock

Igneous rock forms when melted rock cools. The melted rock comes from deep in the earth. There the rock is a very hot liquid.

A *volcano* is an opening in the earth that allows melted rock to come to the surface. Sometimes the melted rock may erupt, or come out quickly. The melted rock that comes out of a volcano is called *lava*.

Lava cools and hardens. When lava cools very quickly, it forms rocks that are smooth. When it cools slowly, it forms rocks that are rough.

Obsidian is a smooth igneous rock. It is so smooth that it looks like glass.

obsidian

Granite forms when melted rock cools slowly. It is often speckled. It may have shiny crystal spots. It has a rough surface. Some mountains are made of granite.

granite

Erta Ale, Ethiopia

Volcán de Fuego, Guatemala

Science and History

Mount Rushmore is a famous granite mountain in South Dakota. Huge faces of four United States presidents have been carved in the side of this mountain.

63

Preparation for Reading

- Preview and pronounce the vocabulary terms *igneous rock* and *metamorphic rock*, as well as the words *volcano*, *lava*, *obsidian*, and *gneiss*.
- Direct the students to read pages 63–64 silently to find out how metamorphic rock forms.

Teach for Understanding

What kind of rock forms when melted rock cools? igneous rock

What state of matter—solid, liquid, or gas—is rock when it is deep inside the earth? liquid

How does the melted rock get to the surface? through a volcano

What is a *volcano*? an opening in the earth that allows melted rock to come to the surface

What is the melted rock that comes out of a volcano called? lava

- Direct attention to the erupting volcano and the flowing lava.

What happens to the lava when it gets to the surface? It cools and hardens.

What do the new rocks look like when the lava cools very quickly? They are smooth.

What igneous rock cools so quickly that it looks like glass? obsidian

- Direct attention to the picture of obsidian. Display the obsidian rock sample. Allow the students to handle the sample. Discuss the texture and color.

What do the rocks look like when the lava cools slowly? They are rough and speckled.

What is an example of an igneous rock that is rough and speckled? granite

- Direct attention to the picture of granite. Display the granite rock sample. Allow the students to handle the sample. Discuss the texture and appearance.
- Discuss "Science and History."

What famous mountain is made of granite? Mount Rushmore

What has been carved in the side of Mount Rushmore? faces of four of the United States presidents (George Washington, Thomas Jefferson, Theodore Roosevelt, and Abraham Lincoln)

- Display the *Rocks, Rocks, and More Rocks* visual. Guide the students as they complete the "igneous" portion of the web on *Activities* page 47. Cells with additional information may be added.

ⓘ Background

Volcanoes

Volcanoes happen around the world, both on land and under water. They happen whenever a crack in the earth's crust allows the lava, or melted rock, and gases to come to the surface. Most active volcanoes are found in an area called the Ring of Fire, located around the edges of the Pacific Ocean.

Differentiated Instruction

Enrichment: Form "Igneous Rock"

Materials: old crayons, empty soup can, water, saucepan, hot plate (or stove), tongs, silicone ice cube tray

Remove all the paper from the old crayons. Place the crayons in the bottom of the can. With adult supervision, fill the saucepan about half full with water and put the can of crayons in the water. Use the tongs to hold the can upright. Water should not get inside the can. Heat the crayons 5–10 minutes until melted. Do not stir.

Carefully remove the can from the water and pour the melted crayons into the ice cube tray. Let cool until hardened. This may take up to an hour, depending on the size of the mold. Remind the students that a liquid changes to a solid when it is cooled enough. When cooled, pop out the new crayons.

How does this activity model igneous rock? The crayons were melted and cooled to form new crayons. Igneous rock forms from melted rock that cools.

What is the third kind of rock? metamorphic

What does *metamorphic* mean? to change

How are metamorphic rocks formed? Igneous and sedimentary rocks are changed by great heat and pressure.

What does limestone change into when heat and pressure are applied to it? marble

- Direct attention to the picture of marble. Allow the students to handle the sample of marble. Discuss their observations. Allow them to compare the sample of marble with the sample of limestone.

What does shale change into? slate

How is slate often used? for walkways and yard decorations, as a roofing material

- Direct attention to the picture of slate. Allow students to share how they have observed slate being used.

What does granite change into? gneiss

How is gneiss often used? as a building material

- Direct attention to the picture of gneiss.

Why do you think gneiss makes a good building material? It is strong.

What happens to rocks? Possible answers: They are always changing. Big rocks break down into smaller rocks. Melted rock cools and changes to solid rock. One kind of rock is changed into another kind by heat and pressure.

How do scientists classify rocks? by how they form

- Using the rock and mineral collection, or photos of various rocks and minerals, guide the students as they think of ways, other than how they form, to classify rocks. Guide the students to think of the physical properties of rocks.

How would you classify rocks based on their physical properties? Possible answers: by color, hardness, texture, specks or no specks

- Display students' suggestions. Allow students time to classify the rocks in the rock and mineral collection using their classification systems.

- Display the *Rocks, Rocks, and More Rocks* visual. Guide the students as they complete the *metamorphic* portion of the web on *Activities* page 47. Cells with additional information may be added.

✓ sedimentary rock, igneous rock, and metamorphic rock

Apply

Activities
Rocks, Rocks, and More Rocks, page 47
This page was completed during the lesson.

Metamorphic Rock

The third kind of rock is metamorphic rock. Metamorphic means "to change." **Metamorphic rock** forms when igneous or sedimentary rocks are changed by great heat and pressure.

Limestone is a soft sedimentary rock. If it is placed under heat and pressure, it becomes marble. Artists often use marble because of its beauty and hardness.

marble

A rock called slate forms from shale. Slate can be split into thin sheets. Landscapers use it for some walkways and yard decorations. It is also used as a material for roofs.

slate

Heat and pressure can change granite into a rock called gneiss. Gneiss is strong, and it is often used as a building material.

gneiss

Rocks are always changing. Big rocks break into smaller rocks. Melted rock changes to solid rock. One kind of rock becomes another.

✓ What are the three groups that scientists classify rocks into?

👥 Differentiated Instruction

Enrichment: Form "Metamorphic Rock"
Materials: wax paper, "sedimentary rock" from the activity on *Teacher Edition* page 76, piece of fabric, iron

Fold wax paper around the "sedimentary rock." Cover with fabric and, with adult supervision, press down firmly with a hot iron. Keep the iron moving and apply pressure. After a few minutes, remove the "metamorphic rock." *Note:* The "rock" should have a stained-glass appearance.

How does this activity model metamorphic rock? The sedimentary rock was changed by heat and pressure. It has a smoother, shinier appearance.

Lesson 20

Minerals

Pirates sometimes had treasures of gold and silver. Their treasure sparkled and shined. Gold and silver are valuable treasures. However, not all treasure sparkles and shines.

The rocks around us may have many valuable treasures in them. All rocks are made of one or more minerals. **Minerals** are solid materials in nature that were never alive. The earth's minerals are a great treasure to us.

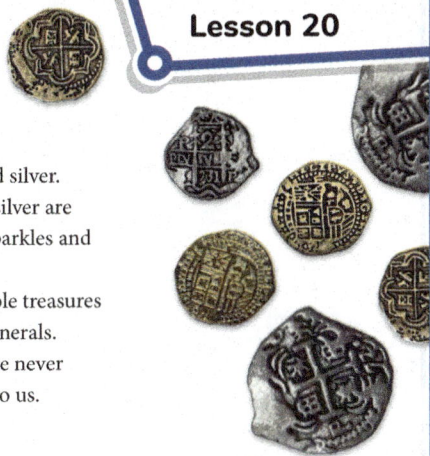

65

Objectives
* Explain how rocks and minerals are related. **BWS**
* Identify some properties of minerals.
* Explain how understanding the properties of minerals can be beneficial. **BWS**
* Differentiate between common and precious minerals.

Materials
* mineral samples

Vocabulary
* mineral

Engage

What is something that is very valuable and important to you?
* Share with the class something that has value and importance to you. Explain why the item is valuable.
* Ask volunteers to name something of value to them and ask them to explain why they consider the item to be valuable and important.

What is another way to describe something that has great value to you? Elicit the word *treasure*.
* Poll the class, by a show of hands, to find out how many students consider salt, pencils, cereal, and toothpaste to be treasures.
* Explain that the students will learn about some valuable and important treasures that God has given us on the earth.

Instruct

Preparation for Reading
* Preview and pronounce the vocabulary term *mineral*, as well as the name *Friedrich Mohs* and the word *physics*.
* Direct the students to read pages 65–67 silently to find out what the Mohs Hardness Scale is used for.

Teach for Understanding
* Direct attention to the picture at the bottom of the page.

What kind of treasure did pirates often have? gold and silver

How are rocks and minerals connected? All rocks are made of one or more minerals.

What are *minerals*? solid materials in nature that were never alive

Where did the minerals come from that make up rocks? Possible answers: God created the minerals; some minerals formed as a result of the Flood.

Why are the earth's minerals a great treasure to us? Possible answers: They are valuable; they are important to our lives; they are useful to our everyday lives.

Why do scientists test how hard or soft a mineral is? It helps to identify the mineral.

What did Friedrich Mohs do with the minerals he was studying? He arranged, or put, them in order from softest to hardest.

What is his arrangement called? the Mohs Hardness Scale

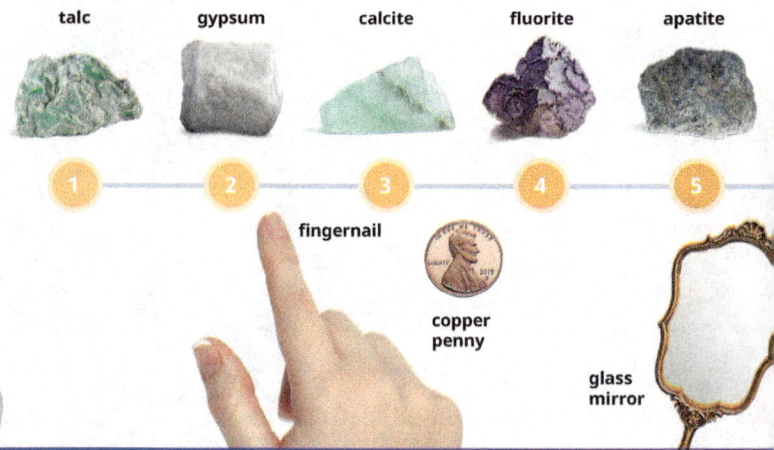

The Mohs Hardness Scale is introduced in third grade as a background for discussing a mineral's property of hardness. It is also used to give the students practice reading graphs. The students will not be assessed over any of the minerals or objects labeled on the scale. They will, however, use concepts from the Mohs scale to conduct the Investigation in Lesson 21.

How can you find out how hard a mineral is? by doing a scratch test

How does a scratch test show which mineral is harder? A harder mineral is able to scratch a softer mineral.

- Direct attention to the "Mohs Hardness Scale" diagram and read the labels together.

How many minerals are numbered on the scale? 10

What mineral is number 1 on the scale? talc

What mineral is number 10 on the scale? diamond

Which is softer—talc or diamond? talc

What number is quartz on the scale? seven

Would quartz scratch a diamond or would a diamond scratch quartz? The diamond would scratch quartz. Explain why. The diamond is harder.

- Direct attention to the pictures underneath the scale. Explain that the pictures show things that can scratch minerals at that level on the scale.

What can scratch any minerals that are softer than calcite? a copper penny

- Explain that the copper penny should be dated 1981 or older to ensure that it is made mostly of copper (95%). Pennies minted after 1981 are made mostly of zinc (97.5%).

Feldspar is number 6 on the scale. What material is harder than feldspar? a steel nail

Find fluorite on the scale. Would glass be able to scratch fluorite? yes How do you know? Possible answer: Glass can scratch any mineral softer than feldspar, and fluorite is softer.

- Point out that all minerals could be placed on this scale. Explain that some are almost as soft as talc, and others are much harder. Scientists compare other minerals to the Mohs Hardness Scale to determine the hardness of the mineral.

Properties of Minerals
Hardness

Minerals are not all alike. Some are hard, and some are soft. Scientists can test how hard or soft a mineral is. This is one way they can identify the mineral. A scientist named Friedrich Mohs studied some minerals and put them in order from softest to hardest. His arrangement formed a scale. The Mohs Hardness Scale ranks minerals from one to ten. Other minerals can be compared to the scale that Mohs made.

A scratch test may be done to find out how hard a mineral is. A harder mineral can scratch a softer one. The hardness of a mineral is shown by what it can or cannot scratch.

Talc (1) is a very soft mineral. It is used to make a soft powder. Almost every other mineral can scratch talc. Diamonds (10) are very hard. They can cut, or scratch, all other minerals.

Mohs Hardness Scale

talc   gypsum   calcite   fluorite   apatite

1   2   3   4   5

fingernail

copper penny

glass mirror

66

Meet The Scientist

Friedrich Mohs (1773–1839) was born in Germany. He studied physics, chemistry, and mathematics. Later he studied mining. He also studied minerals. In 1801 he was hired by a man who had a large mineral collection. The man wanted Mohs to identify the unknown minerals. Mohs created a test to classify the hardness of the minerals. This is called a scratch test. Mohs later became a professor. Today scientists use the Mohs Hardness Scale to help identify minerals.

✔ **What is the purpose of a scratch test?**

- Discuss "Meet the Scientist."

Where was Friedrich Mohs born? Germany

Why did Mohs develop a hardness scale? He wanted to classify, by hardness, the unknown minerals in a large mineral collection.

What do we call the test that is used with the Mohs Hardness Scale? a scratch test

Why is the Mohs scale an important tool today? It is used to help scientists identify minerals.

✔ to find out how hard a mineral is

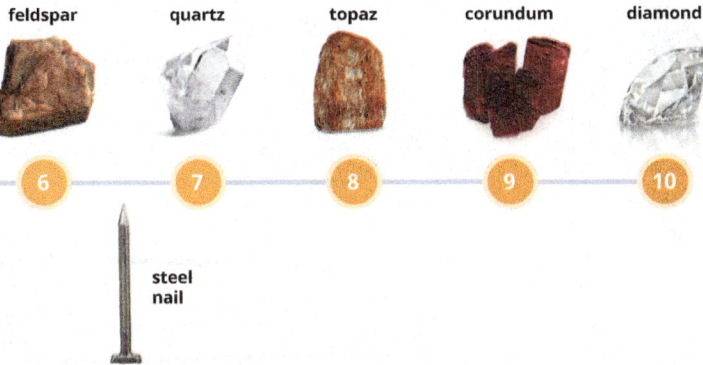

feldspar   quartz   topaz   corundum   diamond

6   7   8   9   10

steel nail

67

Chapter 3: Soil, Rocks, Minerals, and Fossils

Preparation for Reading

- Preview and pronounce the words *crystal* and *pyrite*.
- Direct the students to read page 68 silently to find out three ways to identify minerals in addition to hardness.

Teach for Understanding

How can scientists use the shape of a mineral's crystal to identify the mineral? Each mineral that forms crystals has its own crystal shape.

- Direct attention to the picture of the halite crystals. Point out that halite is also known as rock salt.

What shape are halite crystals? box-shape

Why is it sometimes hard to identify a mineral by color? Some minerals have colors that look alike.

- Direct attention to the pictures of amethyst and fluorite.

How are these minerals different? They have different-shape crystals.

How are these minerals alike? They are similar in color; they are both purple.

If two minerals have colors that look alike, what test can scientists use to identify the minerals? a streak test

How is a streak test done? The mineral is rubbed on a light-colored tile. The color of the mark helps identify which mineral it is.

How do gold and iron pyrite look alike? They are both yellow.

- Direct attention to the caption.

Is the mineral in the picture gold or iron pyrite? iron pyrite

How do you know? It left a greenish-black streak on the tile.

- Share the Background information about iron pyrite, as needed, at this time.

Possible answers: crystal shape, color, streak (hardness)

Halite, or salt, crystals look like little boxes.

Crystal Shape

Most minerals form crystals. A *crystal* is a solid material whose particles are arranged in a special pattern. Each mineral has its own crystal shape. Scientists can use the crystal's shape to help them know what mineral it is.

Color and Streak

Minerals are different colors, but the colors of some minerals can look alike. Scientists may use a *streak test* to help them identify a mineral. They rub the mineral on a light-colored tile. The color of the streak, or mark, on the tile helps them know what mineral it is.

Gold and iron pyrite look alike. They are both yellow, but a streak test can be used to tell them apart. Gold leaves a yellowish streak. Iron pyrite leaves a greenish-black streak.

amethyst

fluorite

Is this mineral gold or iron pyrite?

What are properties that scientists use to help tell minerals apart?

ⓘ Background

Mineral Properties

The *Student Edition* mentions hardness, crystal shape, color, and streak, but luster and cleavage are two other mineral properties that scientists use to identify minerals. The luster of a mineral is the way it reflects the light. Cleavage describes a mineral's ability to break into flat sheets or along smooth, straight lines.

Crystal Shape

Most minerals have crystals, but not all crystals are minerals. Many crystals have straight sides and sharp corners, but some mineral crystals form in rounded shapes.

Iron Pyrite

Iron pyrite is often called "fool's gold" because its color and shape are so similar to gold. Many prospectors and miners have been fooled into thinking they had found gold when in fact they had found iron pyrite, which is worth much less. Iron pyrite is a common mineral and is used in fertilizer, car batteries, jewelry, and machinery as well as other things.

Uses of Minerals

Minerals are used in many ways. You probably have eaten some minerals today. The grains that make up many breakfast cereals contain iron. Iron is a mineral that your body needs to work properly. Calcium is another mineral you need. Milk is a good source of calcium. Learning about the uses of minerals can help you honor God. You honor God when you care for your body by eating the right foods.

Preparation for Reading

- Preview and pronounce the vocabulary term *gem*, as well as the words *precious mineral* and *common mineral*.
- Direct the students to read pages 69–70 silently to find out what makes some minerals precious.

Teach for Understanding

Why is it good to learn about minerals? Possible answers: People can use knowledge of minerals to help other people; we can use our knowledge of minerals to manage the earth as God commanded.

- Make the following T-Chart for display. Direct the students to copy the chart into their science notebooks. As the uses of different minerals are identified, direct the students to write them in the correct column.

Using Knowledge of Minerals	
For Our Bodies	For Tools and Other Things

What are two minerals that you often eat? iron and calcium From what food do you get some of the iron that is needed for your body? Possible answers: grains, breakfast cereal What food is a good source of calcium? milk

- Direct attention to the nutrition labels on the cereal box and the milk container on the page. Explain that it is important for students to eat some of these minerals every day in order for their bodies to stay healthy.

- Display and identify mineral samples during the discussion.

  Why are some minerals considered to be precious minerals? They are hard to find or hard to get.

  Name two precious minerals. Possible answers: gold, silver, diamonds

  What kind of precious minerals are gold and silver? They are metals.

  Why are the crystals of some metals cut and polished? to reflect light

  What are these kinds of minerals called? gems

  What are precious metals and gems often used in? jewelry

- Direct attention to the picture of the gold nugget and the gold and gem jewelry. Explain that some gems are very valuable because they are so hard to find. Point out that the white gems in the jewelry shown are diamonds and the green gem is an emerald.

  Besides jewelry, what is another use for diamonds? They are used in drills and other cutting tools.

  Why can diamonds be used to cut hard things? Possible answers: Diamonds are very hard; they are number 10 on the Mohs Hardness Scale; diamonds can scratch all other minerals.

  What do we call minerals that are easier to find? common minerals

  Name some common minerals. Possible answers: halite, graphite, quartz

  What do we get from the mineral halite? salt

  What is one way graphite is used? It is mixed with clay to make pencil lead.

  What is one of the most common minerals? quartz

  What are some things that quartz is used in? Possible answers: glass, computers, cell phones

Using Knowledge of Minerals		
For Our Bodies	For Tools and Other Things	
calcium	computers	drills
iron	cell phones	glass
salt	pencil lead	jewelry

✓ They are easier to find than precious minerals.

Apply

Activities

Study Guide, pages 49–50

These pages review the concepts taught in Lessons 19–20. After completion, direct the students to keep them for review for Test 3.

Assess

Quiz 3B

The quiz may be given at any time after completion of this lesson.

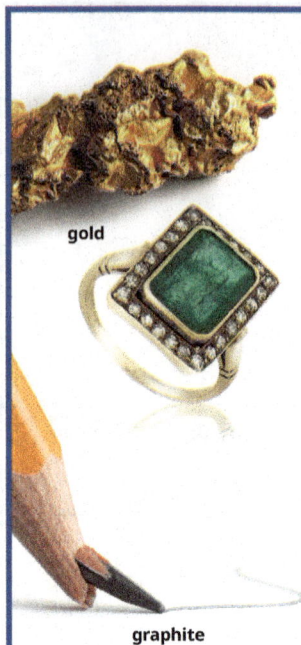

gold

graphite

Some minerals are hard to find or hard to get. Because of this, they are considered to be *precious minerals.* Gold and silver are two precious minerals. They are also metals. The crystals of some precious minerals are cut and polished to reflect light. These minerals are called **gems**. A diamond is a gem. Precious metals and gems are often used in jewelry. Diamonds are used in jewelry. But because diamonds are so hard, they are also used in drills and other tools that cut hard things.

Other minerals are easier to find. We say they are *common minerals.* Halite, graphite, and quartz are some common minerals. We get salt from the mineral halite. The "lead" in our pencils is not really lead at all. It is really a mixture of clay and the mineral graphite. Quartz is probably the most common of all the minerals. Most sand is made of tiny bits of quartz. It is an important mineral used to make glass. Quartz is also used in computers and cell phones. Knowing how to use precious and common minerals helps people to manage the earth as God commanded us to do.

quartz

✓ What makes some minerals common?

⚠ Helps

Lesson 21 Visuals

The *Science Inquiry Skills, Scientific Investigation,* and *Science Safety Tips* visuals (Visuals 3.3–3.5) have been included so you may display them in your room after the lesson has been completed.

Lesson 21 Controlled Investigation

In the previous edition of *Science 3* a controlled investigation was called a fair test.

Lesson 21 Materials

If actual mineral samples are not available, you may choose to use items made from these minerals. For example, pieces of copper pipe can be used for the copper sample and a piece of drywall can be used for gypsum.

Hard or Soft?

Name _____

An **investigation** is a scientific test to solve a problem.

Problem

An investigation begins with a problem. The **problem** comes from an observation you have made. The problem is a question you want answered.

Which minerals can be scratched with each tool?

Hypothesis

A **hypothesis** is a possible answer to the problem. The hypothesis is written as a statement.

Circle all the words that complete your hypotheses.

Apatite can be scratched with a

fingernail, penny, steel nail .

Copper can be scratched with a

fingernail, penny, steel nail .

Gypsum can be scratched with a

fingernail, penny, steel nail .

Quartz can be scratched with a

fingernail, penny, steel nail .

Procedure

The **procedure** is the steps of an investigation.

Follow the steps to do the investigation.

☐ 1. Complete the hypothesis for each mineral.

☐ 2. Scratch the apatite with your fingernail. Check whether your fingernail left a scratch, a line, or dust on the mineral. A scratch will not be able to be rubbed off.

☐ 3. Record the result on the data chart.

Each investigation will direct attention to two or three **inquiry skills**.

Inquiry Skills

- Predict
- Observe
- Infer

The **materials** are the supplies needed to do the investigation.

Materials

Tools
☐ fingernail
☐ penny
☐ steel nail

Minerals
☐ apatite
☐ copper
☐ gypsum
☐ quartz

INVESTIGATION

Lesson 21

Hard or Soft?

Inquiry Skills
- Predict
- Observe
- Infer

Some minerals are soft. They can be scratched easily. Other minerals are hard. They are more difficult to scratch.

In this controlled Investigation, you will test the hardness of several minerals. A **controlled investigation** changes only one thing at a time. All other parts of the investigation are kept the same.

71

Objectives

- Formulate a hypothesis to predict the hardness of various minerals.
- Record observations.
- Draw conclusions from data collected.
- Explain why knowing about the hardness of minerals is beneficial. **BWS**
- Rank minerals from softest to hardest, based on the scratch test conducted.

Materials

- See *Activities* page 55.

Teacher Resources

- Visuals 3.3–3.5: *Science Inquiry Skills; Scientific Investigation; Science Safety Tips*

Vocabulary

- controlled investigation

Engage

- Direct the students to look back at the "Mohs Hardness Scale" diagram on pages 66–67.

  What is the softest mineral on the scale? talc

  What is the hardest? diamond

  Today you will be testing some other minerals to find out how hard they are.

Instruct

Preparation for Reading

- Direct the students to remove pages 55–57 from their *Activities* books. Direct them to read page 71 and *Activities* pages 55–57 silently before beginning.
- Explain that it is helpful to look over all the pages before starting an Investigation. Point out that it will help the students to see what they are going to do and when they will be writing information down.

Teach for Understanding

- Direct the students to remove pages 51–53 from their *Activities* books and to place the pages in their science notebooks. These pages will be used as reference throughout the course.
- Display the *Science Inquiry Skills* visual. Ask volunteers to read each skill aloud. Review and discuss each skill. *Note:* The Inquiry Skills and their descriptions can be found on *Activities* page 51. *Note:* The Science Inquiry Skills are the same as the Science Process Skills in the previous edition.
- Display the *Scientific Investigation* visual.
- Direct attention to *Student Edition* page 21.

  What is a controlled investigation? an investigation that changes only one thing at a time; except for one thing, all parts of the investigation stay the same

- Ask volunteers to read aloud each step on the *Scientific Investigation* visual. Review and discuss the steps.
- Display the *Science Safety Tips* visual. Allow volunteers to read each safety tip. Review and discuss each tip. Explain where the students should keep their science tools and how they should care for them.

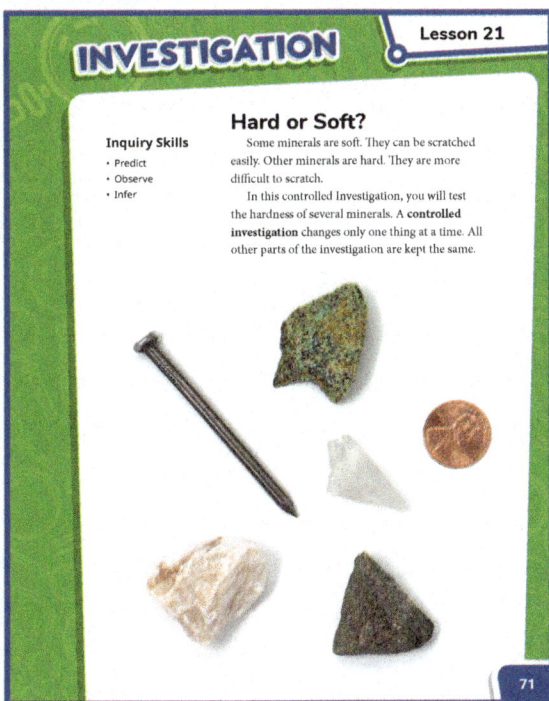

- Direct attention to the **Inquiry Skills**.
- Remind the students that scientists use science inquiry skills to gather and use facts.

  Which inquiry skills will you be using in this investigation? Predict, Observe, and Infer

- Review what it means to *predict*, *observe*, and *infer*.
- Choose a student to read the **Problem** aloud.

  What will you be using to test the hardness of the minerals? a scratch test

  What tools will you use? fingernail, penny, steel nail

  What minerals are you testing? apatite, copper, gypsum, quartz

  If other minerals are substituted for the ones suggested in this investigation, direct the students to change the names of the minerals on their *Activities* pages. You will need to prepare your own answer key.

- Direct the students to complete the **Hypotheses** to answer the problem. Point out that the word "hypotheses" means that there is more than one hypothesis.
- Explain that more than one tool may scratch each mineral. Instruct the students to circle their prediction for each mineral that will be tested.
- Instruct the students to follow and check off **Procedure** Steps 1–7 as each is completed. Point out that Step 8 is directing them to repeat Steps 2–7 for each of the three remaining minerals.

  Teaching for *Activities* page 56 begins here.

- Direct attention to the "Observations" chart. Remind the students of the importance of accurately recording their **Observations**.
- As needed, help the students determine whether the tool scratched the mineral.

  The penny may leave a colored mark on the mineral. This means that the mineral scratched the penny, not that the penny scratched the mineral. Direct the students to rub off the color and then look for any scratches on the mineral.

☐ 4. Scratch the apatite with the penny. Check whether the penny left a scratch on the mineral.

☐ 5. Record the result on the data chart.

☐ 6. Scratch the apatite with the steel nail. Check whether the steel nail left a scratch on the mineral.

☐ 7. Record the result on the data chart.

☐ 8. Repeat procedure steps 2–7 for each mineral.

Observations

Scientists observe what happens, using the five senses. Scientists record data from their **observations** in charts, lists, or graphs.

Results			
Mineral	Fingernail	Penny	Steel Nail
apatite			
copper			
gypsum			
quartz			

56 Lesson 21 • Page 71
Investigation

Hard or Soft?

Name _____

Conclusions

Conclusions are the answers to the test.

1. Which mineral can be scratched using your fingernail? _gypsum_____

2. Explain why the mineral could be scratched using your fingernail. _because gypsum is_
   _softer than my fingernail_____

3. Which minerals can be scratched using the penny or steel nail? _apatite, copper, and_____
   _gypsum_____

4. Explain why the minerals could be scratched using the penny or steel nail. _because_____
   apatite, copper, and gypsum are softer than the penny or steel nail

5. Which mineral is the hardest? _quartz_____

6. Explain how you know this mineral is the hardest. _because quartz could not be_____
   _scratched by any of the tools_____

7. How is knowing about the hardness of minerals useful? _Possible answers: Knowing a_____
   mineral's hardness will help in deciding whether it is hard enough to be used for tools,
   for kitchen or bathroom counter tops, for jewelry, or for pipes.

8. Do an internet keyword search of "Mohs Hardness Scale" to discover whether apatite or
   copper is the harder mineral. Write which mineral is harder. _apatite_____

9. On the chart below, write the order of the minerals you tested from softest to hardest.

softest   _gypsum (2)_____

   _copper (3.5)_____

   _apatite (5)_____

hardest   _quartz (7)_____

- Direct the students to draw **Conclusions** using their observations. The conclusions may be done individually or with their science group.
- Guide the students as they conduct an internet keyword search of "Mohs Hardness Scale."
- Guide a discussion of the students' conclusions.

Apply

Activities
Investigation, pages 55–57
Students will predict the hardness of sample minerals.

Assess

Rubric
Use the prepared rubric or design a rubric to include your chosen criteria.

Objectives

- Recall what a fossil is.
- Compare and contrast different views of how and when most fossils formed. **BWS**
- Explain why there are marine fossils on mountains. **BWS**
- Compare and contrast views of adaptation. **BWS**
- Create a model bone fossil to observe the "fossilization" by the process of mineral replacement.
- Formulate a statement explaining how the model bone fossil supports a biblical view of fossil formation. **BWS**

Materials

- fossil samples, or photos of fossils
- sponge
- scissors
- shallow dish
- beaker
- very warm water
- Epsom salt or table salt
- spoon

Teacher Resources

- Instructional Aid 3.2: *Fossilization and Minerals*, per student
- Visual 3.6: *Ark on Flood Waters*

Vocabulary

- fossil
- environment
- extinct
- adaptation
- taphonomist

Two days have been alloted for this lesson. You may divide the lesson where it works best for your instruction time.

Engage

- Display the *Ark on Flood Waters* visual.
  What is this a picture of? the Flood; Noah's ark during the Flood
- Explain that the students are going to learn how the Flood changed the world and left a record of the things that were alive when Noah was on the earth.

Instruct

Preparation for Reading

- Preview and pronounce the vocabulary term *fossil* and the word *petrified*.
- Direct the students to read pages 72–74 silently to find out why fossils of fish and seashells are found on mountains.

Teach for Understanding

What is a *worldview*? the way a person thinks about the world

What do scientists use to explain their observations? their worldview

Fossils
A Biblical View of Fossils

Do you remember what a worldview is? A worldview is the way a person thinks about and understands the world. Scientists use what they believe to explain their observations. Someone with a biblical worldview believes the Bible is true because it is God's Word. He views fossils in a way that is different from someone who does not believe the Bible.

Fossils Formed How and When?

A **fossil** is the remains or the trace of a living thing that has died. Most fossils form when living things are buried quickly. The softest parts of the living thing usually decay first. The harder parts can sometimes be seen as fossils. A mark made by a living thing can also become a fossil. These marks are called trace fossils. A trace fossil can be a footprint, track, or burrow. Most fossils are found in sedimentary rock.

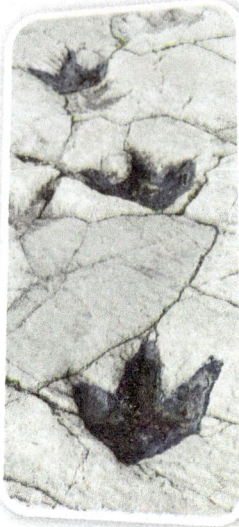

dinosaur tracks

What kind of living thing do you think left behind this trace fossil?

devil's corkscrew

72

What does a person with a biblical worldview believe? The Bible is true.

Why does he believe the Bible is true? because it is God's Word, and God cannot lie

What is a *fossil*? the remains of a living thing that has died and that has been buried quickly

What parts of living things are usually seen as fossils? the harder parts

Why are the soft parts not usually seen as fossils? Possible answers: They decay first; they decay faster than the hard parts.

What kind of fossil is made by a mark of a living thing? a trace fossil

What are some examples of trace fossils? footprints, tracks, burrows

- Direct attention to the pictures and captions.
  What kind of trace fossil did the dinosaur leave behind? tracks, footprints
- Ask a volunteer to read aloud the caption under the Devil's Corkscrew picture.
  What is a burrow? a hole or tunnel that an animal has dug
- Explain that this is a historical photo. Trace fossils of these burrows were first discovered in the late 1800s in America.

Scientists who have a biblical worldview believe most fossils formed during the Flood. The Bible tells us that God judged the earth with a great Flood. The Flood happened about 5000 years ago when Noah was alive. The water from the Flood caused the soil and rocks to wear away quickly. A lot of sediment formed in a short amount of time. The Flood waters moved large amounts of sediment. The sediment buried living things that were not on the ark. The sediment settled into rock layers as the water dried up. A scientist with a biblical worldview believes fossils found in sedimentary rock are support for the worldwide Flood.

73

- Point out the spiral or corkscrew shape of the vertical part of the burrow and how it ends in the horizontal part.
- What do you think the animal used the flat part of the burrow for? Possible answers: sleeping, eating, keeping warm, keeping safe
- What kind of animal do you think made this burrow? Elicit that scientists believe it was dug by a type of beaver.
- What type of rock are most fossils found in? sedimentary
- What type of rock is sedimentary rock? rock that forms when layers of sediment are pressed together and harden

Teaching for page 73 begins here.

A scientist who has a biblical worldview believes most fossils formed when? during the Flood

Why did God send the Flood? to judge the earth; to judge the sin of people on the earth

How long ago did the Flood happen? about 5,000 years ago, when Noah was alive

How did a lot of sediment form in a short amount of time during the Flood? The water from the Flood caused the soil and rocks to wear away; the Flood caused a great amount of weathering and erosion.

What happened to many of the living things that died during the Flood? They were buried in the sediment.

Why are most fossils found in sedimentary rock? The sediment that buried the once living things settled into rock layers as the Flood waters dried up.

When do you think a scientist with a biblical worldview believes the corkscrew, or spiral-shaped, burrow was filled with silt and sand? Possible answer: during the Genesis Flood

- Share the Background information about the Daemonelix, or Devil's Corkscrew, burrow fossil, as needed.
- Direct attention to the rainbow in the picture of Horseshoe Bend in Glen Canyon National Park. Point out that the canyon formed as a result of the Genesis Flood. Remind the students that God promised Noah that He would never again flood the entire earth and that He gave the rainbow as the sign of His promise (Genesis 9:11–17).

ⓘ Background

Burrow Trace Fossil

The Daemonelix, or Devil's Corkscrew, is a spiral-shaped burrow fossil with a horizontal chamber at the bottom. These burrow fossils are filled with silt and sand and have been found in Nebraska and Wyoming. When scientists first observed this type of burrow, they thought the spiral tunnel was made by a freshwater sponge or by the roots of a tree that lived long ago. It was not until paleontologists, or scientists who study fossils, discovered the remains of an animal in a burrow that they identified the kind of animal that they believe dug the tunnel. These burrows are believed to have been dug by a kind of beaver.

Scientists with a biblical worldview believe the burrows were filled with silt and sand during the Genesis Flood around 5,000 years ago. Scientists who believe in evolution believe the tunnels filled with silt and sand over 22 million years ago.

How did the minerals in the Flood waters help some fossils form? **The minerals in the water took the place of the smallest living parts in the remains of the plants and animals.**

⚙ What are minerals? **the solid materials in nature that were never alive**

What happened to the remains of the plants and animals when minerals from the Flood waters replaced their smallest living parts? **The remains turned into rock.**

• Direct attention to the picture and caption at the top of the page.

What is petrified wood? **part of a tree that has been turned into rock; fossilized part of a tree**

• Point out that the word *petrified* means "turned to rock."

What common mineral that you have already learned about is found in petrified wood? **quartz**

Where can fossils of fish and seashells be found today? **on land around the world; on mountains**

• Display the *Ark on Flood Waters* visual.

What is this picture showing? **the Flood; Noah's ark on the Flood waters**

⚙ Can you see any land in the picture? **No, it is under the water.**

• Direct the students to turn to Genesis 7:17–22 in their Bibles. Read the passage aloud while the students follow along.

⚙ How much of the land on Earth was under water during the Flood? **all of it; the entire earth**

⚙ How does the Flood of Noah's day explain why there are fossils of fish and seashells on mountains? **The Bible tells us that the Flood waters covered the entire earth; the highest mountain was covered with about 6.7 m (22 ft) of water; the remains of many fish and seashells were buried quickly by sediment on mountains.**

• Direct attention to the picture and caption at the bottom of the page.

⚙ How do you know these fish died quickly? **One fish is eating another fish.**

• Share the Background information "Fish Eating Fish Fossil" to emphasize how quickly the fish died in order for this fossil to exist.

✓ **during the Genesis Flood**

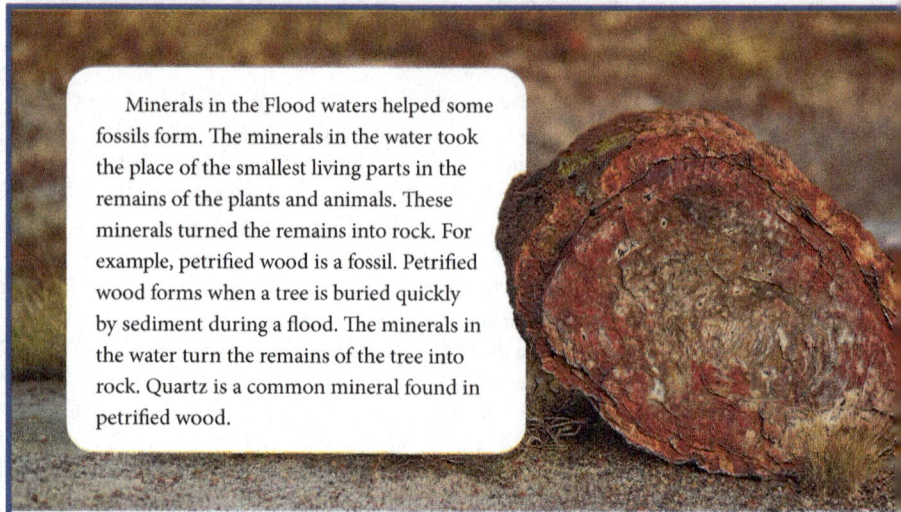

Minerals in the Flood waters helped some fossils form. The minerals in the water took the place of the smallest living parts in the remains of the plants and animals. These minerals turned the remains into rock. For example, petrified wood is a fossil. Petrified wood forms when a tree is buried quickly by sediment during a flood. The minerals in the water turn the remains of the tree into rock. Quartz is a common mineral found in petrified wood.

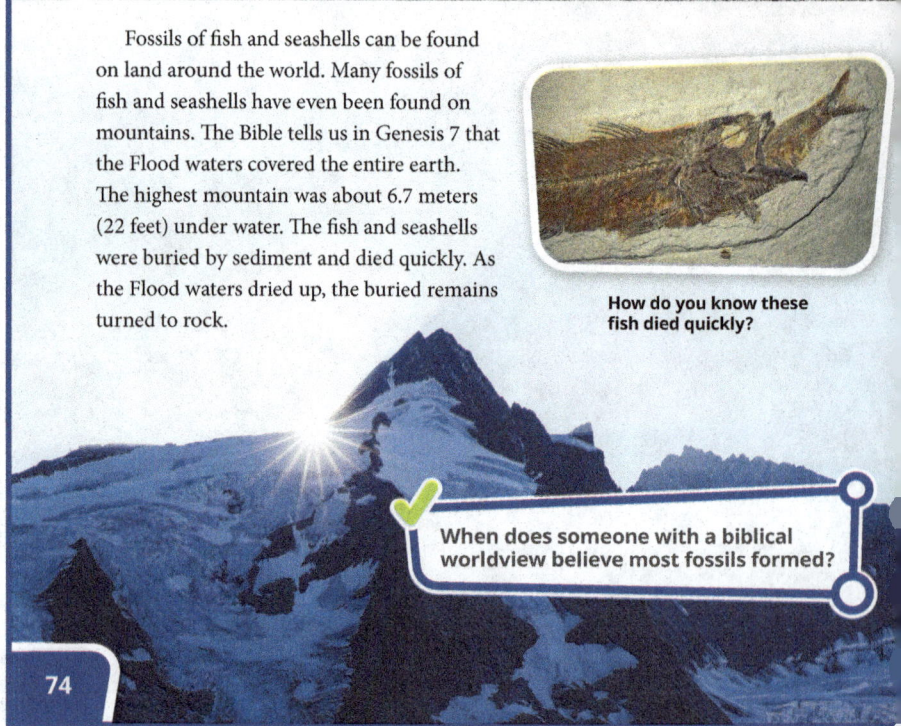

Fossils of fish and seashells can be found on land around the world. Many fossils of fish and seashells have even been found on mountains. The Bible tells us in Genesis 7 that the Flood waters covered the entire earth. The highest mountain was about 6.7 meters (22 feet) under water. The fish and seashells were buried by sediment and died quickly. As the Flood waters dried up, the buried remains turned to rock.

How do you know these fish died quickly?

✓ **When does someone with a biblical worldview believe most fossils formed?**

74

ⓘ Background

Mineral Replacement Fossils or Petrification

Mineral replacement fossils form quickly when minerals take the place of the cellular material in living things. Since the term *cell* will be introduced in Chapter 4, *cell* is described as the "smallest living part of a living thing" in this chapter.

Petrification occurs when the original cellular material in a plant or animal is replaced by minerals. While quartz is a common mineral in petrified wood, other minerals in petrified wood include opal, coal, pyrite, and chromium.

Fish Eating Fish Fossil

The death and burial of living things was very rapid during the great catastrophe of the Flood. The fossil of one fish eating another fish is a demonstration of this fact. The death, burial, and preservation of both fish happened very quickly.

Extinction

After the Flood, the earth changed. Some plants and animals may not have had their needs met. They were not able to survive and grow in the changed environment. The **environment** includes all the living and nonliving things around a plant or animal. As a result, some plants and animals became extinct. **Extinct** means that none of the plants or animals of a certain kind are still alive. Some living things became extinct after the Flood. It is believed that dinosaurs, giant ants, and even some kinds of plants are now extinct. What we know about these plants and animals comes from their fossils. The shapes of the original plants and animals have been preserved as rock.

When do scientists who have a biblical view of fossils, believe most dinosaurs became extinct?

Preparation for Reading

- Preview and pronounce the vocabulary terms *environment*, *extinct*, and *adaptation* and the word *predator*.
- Direct the students to read pages 75–77 silently to find out what adaptation is.

Teach for Understanding

What happened to the earth after the Flood? It changed.

What did the changes to the earth mean for some of the plants and animals that survived the Flood? The environment changed; some plants and animals may not have had their needs met in the new environment.

What happens when the needs of living things are not met? They are not able to survive and grow; they die; they might become extinct.

What does *extinct* mean? None of the plants or animals of a certain kind are still alive.

How do we know all that we do know about extinct plants and animals? Because none of the extinct animals are still alive, everything we know, we have learned from their fossils.

- Direct attention to the picture and caption. Ask a volunteer to read the caption aloud. Elicit that many scientists with a biblical worldview believe most dinosaurs became extinct after the Flood.

Why are some things extinct while other things are not? Many plants and animals that survived after the Flood were able to adapt and did not become extinct.

What is *adaptation*? a characteristic that is passed down, from parents to offspring, and that helps a living thing to survive in its environment

- Explain that an adaptation cannot happen unless the parents pass the characteristic on to their offspring.

According to Genesis 1, what did God create living things to do? make more living things "after their kind"

Can one kind of living thing make another kind of living thing? No, God created each living thing to make more of the same kind of living thing.

What did God create lizards and snakes to do? God created lizards to make more lizards and snakes to make more snakes.

- Direct attention to the pictures and captions of the lizard and snake. Discuss the ways that these animals are different. Explain that the differences help us to see that these animals are different kinds.

Why are plants and animals able to adapt? God created them with the ability to change.

Why is adaptation a good characteristic? Plants and animals that adapt are more likely to survive and not go extinct.

What adaptation has the snowshoe hare passed down from parents to offspring? The color of its fur changes in the winter from brown to white.

How has this adaptation helped the snowshoe hare to survive in an environment that has cold and snowy winters? The rabbit's white fur in the winter makes it harder for predators to see it in the snow. More rabbits with white fur will survive the winter. These rabbits will give birth to bunnies with fur that changes color also.

- Share the Background information at this time.

Adaptation

ocellate lizard

green snake

Why are some things extinct while other things are not? Many plants and animals that survived after the Flood were able to adapt, or change, to the changed world. **Adaptation** is a characteristic, or feature, that helps a living thing to survive in its environment. The characteristic is passed down from parents to offspring. The Bible says that God created all plants and animals. God made them to make more plants and animals "after their kind" (Genesis 1). God made them so that one kind of plant or animal does not make another kind. God created lizards and He created snakes. Lizards are lizards and snakes are snakes.

But God did create living things with the ability to adapt. For example, a snowshoe hare has fur that changes color. In the winter, its fur turns from brown to white. It is harder for a *predator* to see the snowshoe hare in the snow because of its white fur. It would be easier for a predator to see and eat the rabbit in the winter if the rabbit's fur stayed brown all year. This adaptation has helped the snowshoe hare survive in its environment. Over time, more rabbits with fur that could turn white survived the winter. More and more rabbits were born with the adaptation. Soon all the rabbits in the area had fur that turned from brown to white in the winter.

76

(i) Background

Creation: Lizards and Snakes
According to Genesis 1:24–25, God created the land animals, which includes lizards and snakes, on Day Six of Creation. He told these living things to reproduce "after their kind."

Adaptation and the Environment
While the environment can be a factor for or against survival, adaptation can only occur because of God's design of inheritance. An animal cannot adapt unless it has the genetic ability to do so.

Rabbits have adapted. The snowshoe hare does well in a cold habitat where it snows in the winter. The black-tailed jackrabbit is a hare that does well in a desert habitat.

Lizards have adapted. The thorny dragon lizard does well in a dry habitat in Australia. The green anole does well in a habitat where there is plenty of water.

Bears have adapted. The polar bear does well in the cold Arctic. The grizzly bear does well in the woodland forest.

✔ Why did some animals become extinct after the Flood?

77

- Direct attention to the pictures and captions. Ask volunteers to read aloud the captions.
- Discuss the ways in which the animals have adapted. Point out that each of the animals has adapted to a different environment.
- Explain that a change in the animal's environment does not give the animal the ability to pass the characteristic down to its offspring.
- Remind the students that adaptation is a God-given ability and that it is passed down from parents to offspring.

> The term *inherited trait* will be introduced in Chapter 5. Adaptation can only occur through a change in an organism's DNA. The environment does not change an organism's DNA. The ability to adapt was given by God when each living kind was created (Genesis 1).

- Share the Background information as needed.

✔ The earth changed after the Flood and some animals may not have had their needs met.

(i) Background

Snowshoe Hares
Snowshoe hares live in parts of the United States and Canada. The changing fur color, from brown to white in the winter, helps this rabbit to survive.

Thorny Dragon Lizard
The thorny dragon lizard has adapted to the dry Australian Outback. The unique design of its skin helps to bring water to its mouth.

Preparation for Reading

- Preview and pronounce the word *evolution*.
- Direct the students to read pages 78–80 silently to find out what evolution teaches about why there are fish fossils found on mountains.

Teach for Understanding

What is *evolution*? the belief that living things change by chance into new kinds of living things over millions of years

What does evolution teach about *how* fossils formed? Evolution teaches that fossils formed when a living thing was buried quickly in sediment.

What does evolution teach about *when* fossils formed? Evolution teaches that fossils formed at different times over many millions of years.

How does the biblical view of how and when fossils formed compare to the view held by the teachings of evolution? A biblical worldview and the teachings of evolution both hold that fossils formed very quickly when a living thing was buried in sediment. But a biblical worldview believes most fossils formed quickly as a result of the Genesis Flood about 5,000 years ago.

According to evolution, why can fossils of fish and seashells be found on mountains? Evolution teaches that the land where the fish and seashells were buried, at the bottom of rivers, lakes, or oceans, was pushed up and became mountains millions of years after the fossils formed.

- Direct attention to the illustrations. Explain that the diagrams are attempting to explain how mountains formed and why fossils of fish and seashell animals can be found on mountains today.

How is the biblical view of why there are fossils of fish and seashell animals on mountains different from the view held by evolution? The biblical view believes that fish and seashell fossils are found on mountains because all the mountains were under water during the Genesis Flood.

An Evolutionary View of Fossils
Fossils Formed How and When?

Evolution teaches that living things change by chance over millions of years. Living things have changed into new kinds of living things. Evolutionists think this is why there are so many different kinds of living things on earth today. Scientists use evolution to explain fossils, too.

Evolution tries to explain that fossils formed when living things were buried quickly. The living things were buried by sediment and died. It explains that the living things did not decay, or rot. The fossils formed quickly after the living things were buried and died. However, it teaches that these fossils formed at different times in the past. Evolution explains that fossils were layered in sedimentary rock, at different times, over many millions of years.

Why are there fish and seashell fossils found on mountains? Evolution suggests that fish and seashells were buried quickly at the bottom of rivers, lakes, or oceans. It suggests that millions of years later the ocean floor was pushed up. Evolution says that the ocean floor, with the buried fossils, became a mountain.

78

Extinction

Evolution tries to explain why some plants and animals have become extinct. It explains that what we know about these plants and animals comes from their fossils. Evolution teaches that some animals became extinct many millions of years ago. For example, it suggests that dinosaurs became extinct about 65 million years ago.

Adaptation

Why are some things extinct while other things are not? Evolution suggests that some plants and animals have adapted. It teaches that adaptation is a characteristic of a living thing that helps it to survive in its environment. It explains that adaptation can result in one kind of animal evolving, or changing, over millions of years into another kind of animal. For example, evolution tries to explain that snakes evolved from lizards. Evolution says these changes happened little by little, over millions and millions of years.

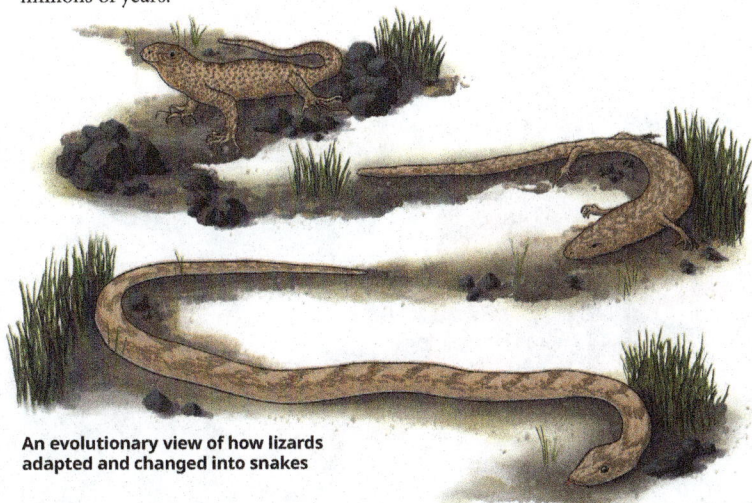

When do scientists who have an evolutionary view of fossils, believe dinosaurs became extinct?

An evolutionary view of how lizards adapted and changed into snakes

79

According to evolution, how do we know what we know about extinct plants and animals? What we know about them comes from their fossils.

When does evolution teach that dinosaurs became extinct? about 65 million years ago

- Direct attention to the dinosaur skull. Explain that this is a skull of an extinct *Triceratops*.

Does evolution need a little time or a lot of time for changes in living things to happen? a lot of time

According to evolution, why are some things extinct while other things are not? Some plants and animals have adapted and some have not.

According to evolution, what can happen to a plant or animal that has adapted to its environment? Over millions of years, it can evolve, or change, into a different kind of plant or animal.

What is one example that evolution uses to explain adaptation? Evolution teaches that lizards adapted, or changed, into snakes.

- Direct attention to the illustration and caption at the bottom of the page. Choose a student to read the caption aloud.

How does someone with a biblical worldview explain the fact that there are lizards and snakes in the world? God created both lizards and snakes on Day Six of the Creation week. There are lizards and snakes today because God created them to make more lizards and snakes "after their kind." Lizards make more lizards and snakes make more snakes.

ⓘ Background

Evolution and Adaptation

Evolution teaches that adaptation is a first step in the change from one living organism to another. Most evolutionists would agree that adaptation can occur without evolution, but that evolution cannot occur without adaptation.

Evolution: Lizards to Snakes

According to standard evolutionary teaching, snakes evolved from lizards. Evolution teaches that snakes are legless reptiles that evolved from four-legged lizards. It teaches that this change from one organism to another may have taken about 150 million years to occur.

🔩 How do you know that evolution is *not* true? Answers should include some of the following: Evolution does not teach that God created all things. Evolution teaches that one kind of animal can adapt to change into another kind of animal. Evolution does not believe that God made lizards and that He also made snakes. Evolution does not teach that God sent a worldwide Flood about 5,000 years ago to judge the people of the earth for their sin. Evolution does not teach that most fossils formed as a result of the Flood. Evolution is not based on the Bible.

How do you know the Bible *is* true? It is God's Word; God cannot lie.

• Direct attention to the pictures at the top of the page.

🔩 Which animal is a snake and which is a lizard? The top picture is of a lizard and the bottom picture is of a snake.

🔩 Why are some lizards different from other lizards, and why are some snakes different from other snakes? because God created lizards and snakes with the ability to adapt

✔️ Evolution believes that adaptation is a characteristic that helps a living thing survive in its environment and can result in one kind of animal changing, over millions of years, into another kind of animal.

Is Evolution True?

A scientist who believes evolution is true does not think that God created all things. He does not think that God made each kind of plant or animal. A scientist who believes evolution is true thinks that one kind of animal can become another kind of animal. He does not think that God created lizards to make only more lizards.

A scientist who thinks evolution is true may not believe that God sent a worldwide Flood about 5000 years ago. He does not think that the Genesis Flood changed the earth. He does not connect fossils with the Flood.

But evolution is not true. Evolution does not explain what really happened in the past because evolution is not based on the Bible. The Bible is true because it is God's Word.

✔️ **What does evolution teach about adaptation?**

80

🐎 Preparing Ahead

Lesson 27 Materials
Look ahead to the materials list in Lesson 27 for required preparation.

STEM Career

Taphonomist

A scientist who studies how fossils are made is a **taphonomist**. A taphonomist tests ways to understand how fossils formed.

Taphonomists have learned that fossils formed very quickly. A fossil must form quickly. If it does not, the living thing will decay. Taphonomists have learned that, under the right conditions, some fossils can form in just days. For some types of fossils to form, water and dissolved minerals are needed.

Millions and millions and millions of fossils have been found all over the world. The Genesis Flood produced the right conditions for fossils to form very quickly.

Try It Yourself!

How do bones become hard like rock? Become a taphonomist to understand one way fossils form.

Cut a sponge into the shape of a bone. Place the sponge in a shallow dish. Add 100 mL of very warm water to a beaker. Add 30 mL of Epsom salt or table salt to the water. Use a spoon to stir the water and salt until the salt disappears in the very warm water. Slowly pour the salt water over the sponge. Place the dish with the sponge in a warm, dry place. Leave the dish uncovered so the water can evaporate. Date and record your observations in your science notebook. When the sponge is dry, break the "fossil" in half. Describe what happened to the sponge.

81

Activity

Try It Yourself!
Materials: *Fossilization and Minerals* page (Instructional Aid 3.2), sponge, scissors, shallow dish, beaker, very warm water, Epsom salt or table salt, spoon
Allow students to try this activity on their own.

Preparation for Reading

- Preview and pronounce the vocabulary term *taphonomist*.
- Direct the students to read page 81 silently to find out what a taphonomist does.

Teach for Understanding

What is a taphonomist? a scientist who studies how fossils are made

What does a taphonomist test? ways to understand how fossils formed

- Remind the students that a *paleontologist* is a scientist who studies fossils, but not *how* the fossils formed.

What have taphonomists learned about fossil formation? Fossils must form quickly or the living thing will decay.

How quickly can some fossils form? in just days

What is needed for some types of fossils to form? water and dissolved minerals

What event created the right conditions for millions and millions of fossils to form all over the world? the Flood

- Direct attention to the "Try It Yourself!" box. This activity may be done individually, as a group, or with the class as a whole. Upon completion of the activity, direct the students to complete the *Fossilization and Minerals* instructional aid. Because this activity will extend beyond Chapter 3, it will not be assessed on the test. Answers for *Fossilization and Minerals*:

1. hard
2. snap; crack
3. salt water
4. salt
5. yes
6. minerals
7. Answers may vary.
8. Answers should include some of the following: The Bible tells us in Genesis that God judged the world because of sin with a great flood. The flood waters covered the entire earth. Land animals and people that were not on the ark died in the Flood. Minerals in the water helped to turn some bones into rock. This happened quickly.

Apply

Activities

Study Guide, pages 59–60

These pages review the concepts taught in Lessons 22–23. After completion, direct the students to keep them for review for Test 3.

Objective

- Recall terms and concepts from Chapter 3.

Apply

Review

- Material for Test 3 will come from the Study Guides on *Activities* pages 45–46, 49–50, and 59–60, as well as Quizzes 3A and 3B. You may review any or all of the material during the lesson.
- You may choose to review Chapter 3 by playing the Rock Types review game or a game from the Game Bank (Teacher Resources).

🏆 Review Game

Rock Types

Materials: a box of cut-out letters or letter tiles from a game

Group the students into two teams. Alternate asking review questions. Each time a team answers correctly, it can choose any letter. The first team to spell out *sedimentary* or *metamorphic* wins.

NOTES

NOTES

_____

_____

_____

_____

_____

_____

_____

_____

_____

_____

_____

_____

_____

_____

_____

_____

_____

_____

_____

_____

_____

_____

_____

_____

_____

_____

_____

Objective

• Apply terms and concepts from Chapter 3.

Assess

• Administer Test 3.

Unit Introduction

- Direct attention to pages 82–83.

  What is the title of Unit 2? Let's Connect with Life Science

  What do you see in the picture? a flower, an insect on a flower

- Explain that the picture is an orchid mantis. You may share the Background information about the orchid mantis.

- Direct the students to look through Chapters 4–8, pages 82–191. Guide them through several pages, stopping to look at headings and pictures.

- What do you think we are going to be learning about in Unit 2? living things, people, plants, animals

UNIT 2

Let's connect with

Life Science

ⓘ Background

Unit Photo

The orchid mantis pictured on pages 82–83 is an unusual insect. It is native to Southeast Asia. The orchid mantis has lobes on its legs that look like flower petals. The female orchid mantis can be 6–7 cm long, but the male is only about 2.5 cm long. The orchid mantis is a popular pet for many people in Southeast Asia.

Chapter Objectives

- Explain the importance of studying parts of the human body.
- Evaluate tools used for magnification.
- Compare and contrast plant and animal cells.
- Apply inquiry skills of observe, infer, classify, and communicate.
- Analyze the relationship between cells, tissues, organs, and systems in the human body.
- Explain the importance of skin to the human body.

Lesson Objectives

- Explain, using the Bible's teaching, that God knows all about the people He created. **BWS**
- Identify the smallest living part of an organism. **BWS**
- Identify a tool used to see tiny things.
- Label the five main components of a microscope.
- Conclude which science tool shows greater detail.

Materials

- hand lens, per group
- microscope
- thinly sliced onion skin, per group
- prepared slides of onion skins

Teacher Resources

- Visual 4.1: *Parts of a Microscope*
- Instructional Aid 4.1: *Look Closer!*, per student or group

Vocabulary

- organism
- cell
- microscope

Chapter Introduction

- Direct attention to the chapter title and the rendering of a cardiovascular system. Provide time for the students to leaf through the chapter, looking at the headings and pictures. Discuss what they think the chapter will be about.

> Use completed Looking Ahead, *Activities* page 61, to generate a discussion on what the students already know about organisms, cells, tissues, organs, and systems.

Big Question

- Ask a volunteer to read aloud the Big Question.
- Explain that the students will find the answer to the question as they read the chapter.

Big Question

Why is it important to study parts of the body?

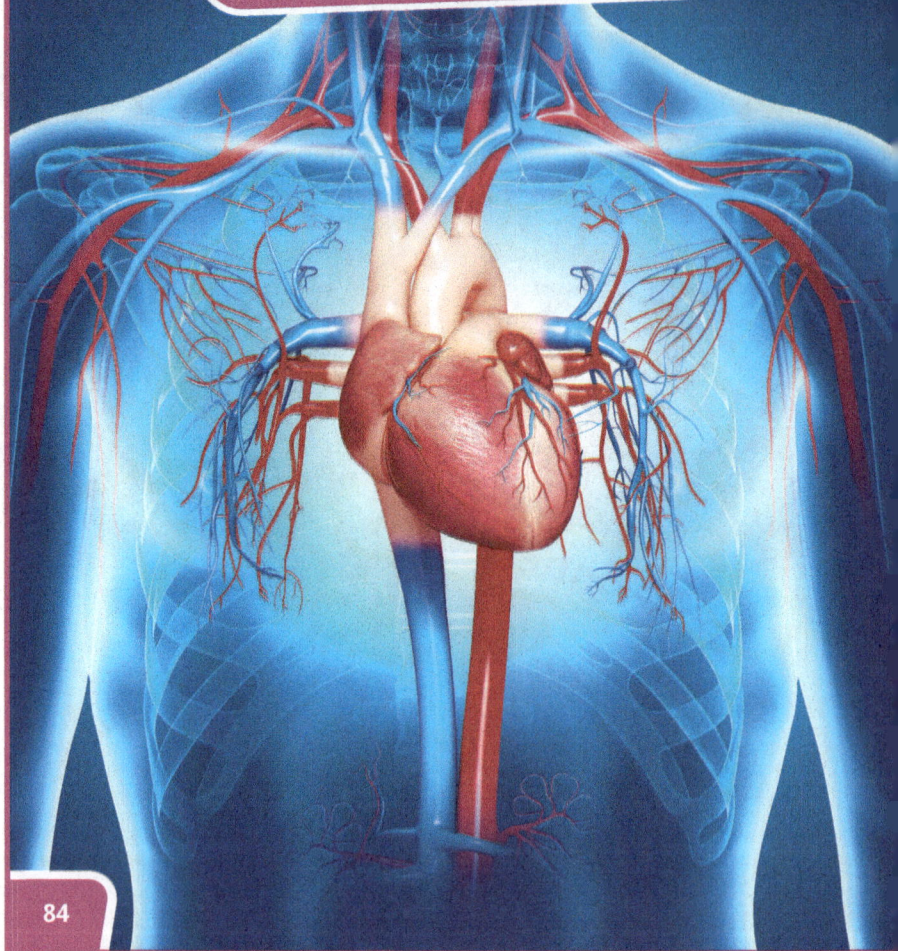

84

ⓘ Background

Chapter Photo

The picture of the cardiovascular system on pages 84–85 shows the flow of blood from the heart to the lungs and body. The cardiovascular system is made up of cells, tissues, and organs all working together to give the body the oxygen and blood that it needs.

> Visit TeacherToolsOnline.com for resources to enhance the lessons.

Chapter 4

Cells, Tissues, Organs, and Systems

For thousands of years, people did not know what plants and animals were made of. Scientists could observe a plant's stems, leaves, and flowers. They could look at an animal's fur and a person's skin. But they could not study the tiny parts that make up each of these. Then a tool was invented that allowed scientists to see the tiny parts of living things.

Scientists needed a tool to see these tiny parts. But we know that God could always see them. God planned how you would look. He planned your eye color and your hair color. He planned the patterns on your skin. Matthew 10:30 says that God even knows the number of hairs on your head. God made every part of you just the way He wanted you to be.

85

- Display a hand lens and a microscope.
- What are these two items? hand lens, or magnifying glass, and microscope
- What are they used for? Elicit that they are used to make small or tiny things look much larger.
- Explain that there are very tiny parts that make up living things that you cannot see with your eyes. Today you will learn about some of those tiny parts of living things. *Note:* The hand lens and microscope will be used again in this lesson.

Instruct

Preparation for Reading

- Direct the students to read page 85 silently to find out who knows about every tiny part of you.

Teach for Understanding

- Direct the students to look around the room and observe the things around them. Choose several students to share what they have observed. Display a list of the objects. *Note:* Keep the list displayed for later use in this lesson.

Lead students to choose both living and nonliving items for their list.

- How can you observe things around you? Possible answers: touching, tasting, smelling, seeing, hearing; through your senses
- Are there things that you cannot see and touch? Yes, some things are too small to see or too far away to see or touch.

Who can see all things, even the things that you cannot see? God

Choose a volunteer to read Matthew 10:30 and Isaiah 43:1 aloud.

What does Matthew 10:30 tell us that God knows? how many hairs are on our head

What does Isaiah 43:1 tell us that God knows? God knows me and calls me by name.

- Choose a volunteer to read Psalm 139:14 aloud. Explain that in this verse the word *fearfully* means causing wonder or amazement. God has made you in a marvelous way!

Preparation for Reading

- Preview and pronounce the vocabulary terms *organism, cell,* and *microscope,* as well as the words *chamber, structures, diseases,* and *Robert Hooke.*
- Direct the students to read pages 86–87 silently to find some ways that microscopes are used today.

Teach for Understanding

- Refer back to the list of objects that the students observed around the classroom.

Which of these are living things?

What makes these *living* things? Possible answers: They breathe. They need food and water.

What are all living things called? organisms

What are the tiny building blocks that organisms are made up of called? cells

Who made all cells? God

- Direct students to look at their arms or hands.

Can you see the cells on your arm? no, because they are too small

What tool is used to see things that are very small? a microscope

Who was the first person to see cells using a microscope? Robert Hooke

What did Robert Hooke observe about the cork plant under the microscope? It was made up of little boxes.

What did he name the little boxes? cells

- Direct attention to the pictures. The boy is using a microscope to look at a piece of red onion skin.

What do you think the onion skin looks like?

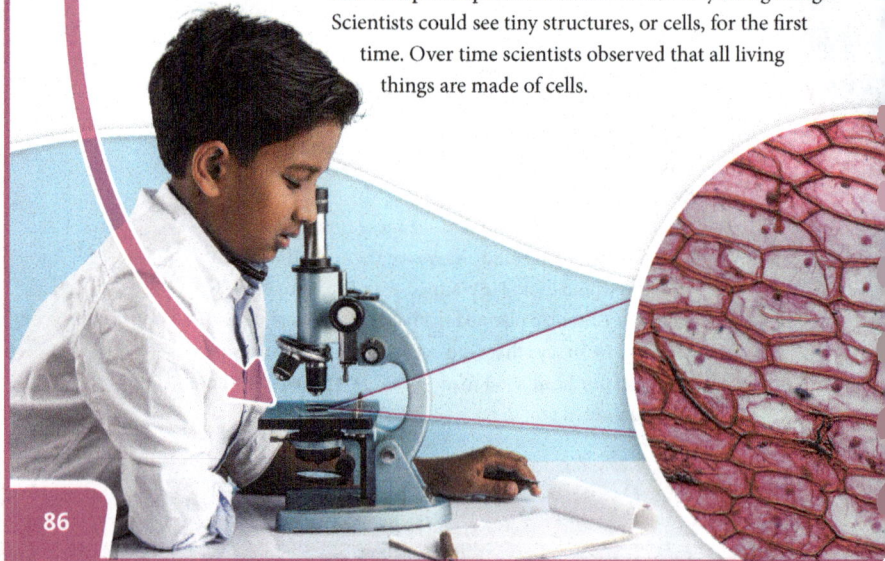

Lesson 26

Cells

There are living things everywhere around us. We call a living thing an **organism**. Organisms are made up of tiny building blocks called cells. A **cell** is the smallest living part of an organism.

Observing Cells

Scientists use many tools to observe the world around us. One of the tools scientists use to observe the tiny parts of an organism is a microscope. A **microscope** is a tool used to see things that are very small. Most cells can be seen only under a microscope.

About 350 years ago, a scientist named Robert Hooke used a microscope to look at a thin piece of a cork plant. He saw that up close it looked like the plant was made up of little boxes. He called these small boxes cells. The word *cell* comes from the Latin word meaning "chamber."

Many other scientists began to use microscopes. The microscopes helped them observe and study living things. Scientists could see tiny structures, or cells, for the first time. Over time scientists observed that all living things are made of cells.

86

ⓘ Background

Microscopes

Hans Janssen and his son Zacharias are credited with inventing the compound microscope. Hans was a lens maker and probably began to make the first compound microscope. Zacharias completed his father's work in the late 1500s. In the 1670s another scientist, Anton van Leeuwenhoek, used a simple microscope to study tiny creatures swimming in water. He called those tiny creatures "animalcules," meaning, "little animals."

Cell Theory

The idea that all organisms are made of cells is considered to be true but is called a theory because, even though it is based on many observations, it cannot be proven.

Robert Hooke

Robert Hooke (1635–1703) was a man of great talent. His ideas and inventions spanned a wide range of science topics. As he experimented, he carefully recorded his observations. In 1665 he published *Micrographia*. The book's finely detailed drawings demonstrate his artistic ability and scientific accuracy.

Meet the Scientist

Robert Hooke

Robert Hooke (1635–1703) drew many pictures of the world around him. He also built the first microscope like we use today. His microscope used a lens to make objects look larger. Hooke wrote a book with drawings of objects that he saw through his microscope. Hooke looked at different plant parts through his microscope. He drew a picture of the little boxes he saw. He called them "cells." His book was used for many years to teach others about cells.

Which kind of microscope do we use today?

People in many jobs use microscopes today. Microscopes help doctors fight diseases. They help scientists know how to make unclean water safe for people to drink.

There are many kinds of microscopes used today. Some microscopes are very large. Others are small. But most microscopes have the same parts and work almost the same way as Robert Hooke's microscope.

What did Robert Hooke call the little boxes he saw with his microscope?

87

- Direct attention to the "Meet the Scientist" box. Choose a volunteer to read the box aloud.
- Do you think Robert Hooke had a camera to take pictures of things he observed? no
  What did he do instead of taking pictures of cells? He drew them in a book.
- Why do you think that it was helpful for him to draw pictures? Elicit that pictures can show what something looks like better than words.
  What are some ways that microscopes are used today? Microscopes help doctors fight diseases. They help scientists know how to make unclean water safe for people to drink.
- Direct attention to the pictures on the page. Choose a volunteer to read and answer the question.

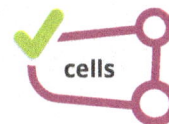

cells

Activity

Words or Pictures

Materials: shoebox or similar container, nail, paper clip, safety pin, thumbtack

Place the small items in the shoebox. Direct a student to secretly choose one item. Tell him to briefly describe it without gesturing or telling what it is used for. Allow the rest of the class to guess the item. If they cannot guess, direct the student to draw it and allow the class to guess again.

Note: You may substitute the items. The purpose of this activity is to show that sometimes a drawing is more effective than words.

Preparation for Reading

- Preview and pronounce the word *eyepiece*.
- Explain that the use of the word *slide* in this case refers to a small glass or plastic rectangle that fits on the microscope.
- Direct the students to read page 88 silently to find out what part of the microscope makes tiny objects look larger.

Teach for Understanding

- Display the *Parts of a Microscope* visual.

  Which part of the microscope is the object put on to observe it? the slide

  Where does the slide go? on the stage

  What shines through the slide and the object? the light

  What part makes the object look bigger? the lens

  What do you look through to see the object? the eyepiece

Apply

- Complete the "Look Closer!" activity at this time.

  This activity can be set up to be viewed and completed throughout the day as students have time.

  What did the onion skin look like using the hand lens?
  What did the onion skin look like using the microscope?

- Choose one or more volunteers to share their drawings of the onion skins.

  Did the hand lens show as many tiny parts as the microscope? No, it did not make the tiny parts large enough to see.

  Which tool is better for making very tiny things look larger? a microscope

- Direct attention to the "Try It Yourself!" box. This activity may be done individually, as a group, or with the class as a whole.

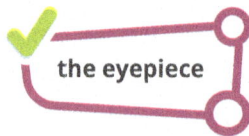

  the eyepiece

Activities
Looking Ahead, page 61

This page assesses the students' knowledge prior to the chapter.

Parts of a Microscope, page 62

This page reinforces the concepts taught in Lesson 26.

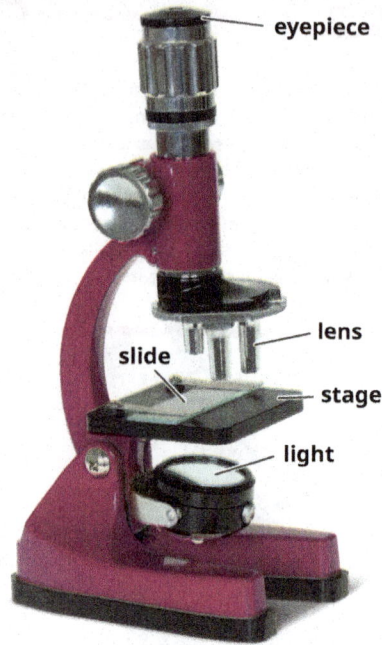

eyepiece
slide
lens
stage
light

One part of a microscope is a slide. A tiny piece of an object is put on the slide to observe it. Then the slide is placed on the stage above the light. The light shines through the slide and through the object. The light also shines into the lens. The lens is curved to make the object look bigger. When you look through the eyepiece, you can see the object through the lens. The lens makes the tiny object look much larger and closer.

Try It Yourself!

A hand lens and a microscope both make small objects look larger. Use a hand lens to observe small objects around you. You could look at leaves, dirt, a flower, a button, or even your shirt.

Now look at some objects using a microscope. Do the objects look different?

Which part of the microscope do you look through?

88

Activity

Look Closer!

Materials: *Look Closer!* page (Instructional Aid 4.1), per student or group; thinly sliced onion skin, per student or group; hand lens, per student or group; microscope; prepared slides with onion skin

Each student will first view the onion skin using the hand lens. The student should draw a representation of the onion skin on the instructional aid page in the box labeled with the hand lens. Then each student should view the onion skin using the microscope. The student should draw a representation of the onion skin in the box labeled with the microscope.

Helps

Try It Yourself!

Materials: hand lens, microscope, various small objects
Allow the students to try this activity on their own.

Lesson 27

Kinds of Cells

Cells are the tiny parts of larger objects. Cells come in many shapes. Some, such as the cork plant observed by Robert Hooke, look like little boxes. Others look like rods, circles, coils, or even blobs of jelly.

Cells also come in many sizes. Some cells are so small that if you lined up 50,000 of them, the row would be only a little more than two centimeters (0.8 inches) long. These cells are so tiny that you can see them only with a microscope. Some cells are much larger. The yolk of an egg is one cell. The largest single cell is the yolk of an ostrich egg. The egg can weigh as much as 1,361 grams (3 pounds)!

Compare the size of the chicken egg yolk and the ostrich egg yolk. Each yolk is still only one cell!

89

Lesson **27** begins here.

Objectives
* Identify the amoeba as a single-celled organism.
* Identify the main parts of an animal cell and a plant cell.
* Differentiate between a plant cell and an animal cell.

Materials
* 3 lb bag of clementines, or other object weighing about 3 lb
* 1 chicken egg
* Kool-Aid® packet, for display
* Kool-Aid drink, per student
* 3 oz paper cup, per student

Prepare enough Kool-Aid for each student in the class.

Teacher Resources
* Instructional Aid 4.2: *Parts of Cells*
* Visuals 4.2–4.3: *Parts of an Animal Cell; Parts of a Plant Cell*

Vocabulary
* food technologist
* nucleus
* cytoplasm
* cell membrane
* cell wall

Engage

* Discuss different jobs the students and their family members have at home.
  Who usually prepares the meals for your family?
  Who takes care of pets?
  Who washes dishes?
  Who takes out the trash?
  Who vacuums or sweeps?
* Point out that family members work together by doing different jobs for each other. Each person has a specific responsibility or way to help.
* Explain that today the students will find out that the parts of a cell also have specific jobs to do.

Instruct

Preparation for Reading
* Preview and pronounce the word *amoeba*.
* Direct the students to read pages 89–90 silently to find out which cells carry messages around the body.

Teach for Understanding
  What are the tiny parts that make up living things? cells

  What shapes can cells be? boxes, rods, circles, coils, blobs
* Direct attention to the pictures of cells on the right side of the page.
  What shapes are these cells? circle, rod or stick, oval
* Display a line that is 2 cm long.
  Some cells are so small that 50,000 of them together would still be only a little longer than this line.
* Direct attention to the pictures at the bottom of the page.
  What is the largest single cell? the yolk of an ostrich egg
* Display the bag of clementines and the egg. Explain that an ostrich egg weighs about as much as the bag of clementines. Pass the bag around to let students feel the weight of three pounds.
* Explain that the ostrich egg yolk and the chicken egg yolk are both single cells, even though they are very different in size.

Do some living things have more than one cell? yes
What is related to the number of cells a living thing has? its size
Would a person have many cells or only a few cells? many

- Point out that although a person is not as large as a redwood tree or an elephant, each person still has an estimated 30 trillion cells.

Do all living things have more than one cell? no
What is an amoeba? a single-celled creature
Can you see an amoeba? only with a microscope
How big is an amoeba? about half the size of a period at the end of a sentence
Where do amoebas live? in water or moist soil

- Explain that an amoeba moves by using a pseudopod. A *pseudopod* is an extension of the amoeba's jellylike body. The pseudopod extends in one direction, and the rest of the body slides to follow the pseudopod.
- You may choose to complete the "Model an Amoeba's Movement" activity at this time.

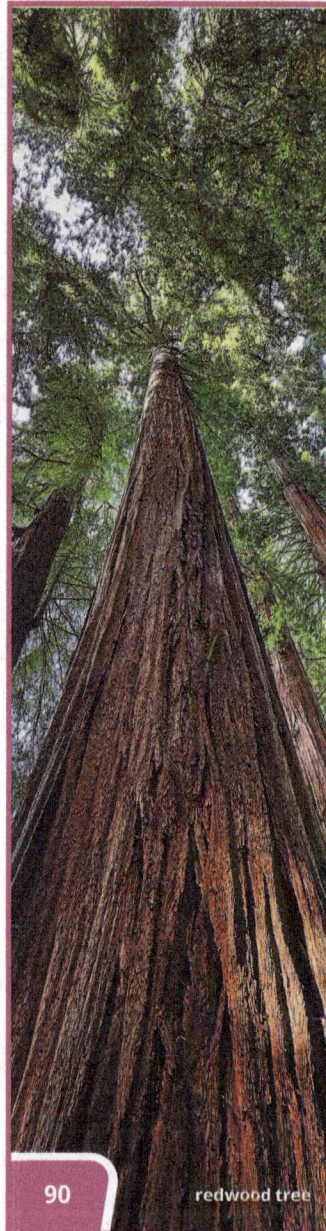

amoeba

Plants and animals are made up of many cells. The size of a living thing is related to the number of cells it has. An elephant and a huge redwood tree have millions of cells. A hummingbird and a flower have fewer cells.

Some living things are made up of only one cell. A microscope can help you see single-cell organisms. An *amoeba* is a single-cell organism. It is about half the size of the period at the end of a sentence. It looks like a blob of jelly and lives in water or moist soil.

There are many different kinds of cells in the human body. Nerve cells carry messages around our body. They tell us that something we touch is hot or cold. They tell us that something feels soft or hard. Brain cells in our body help us think. Muscle cells help us move. All the different kinds of cells in our body work together in the way God created.

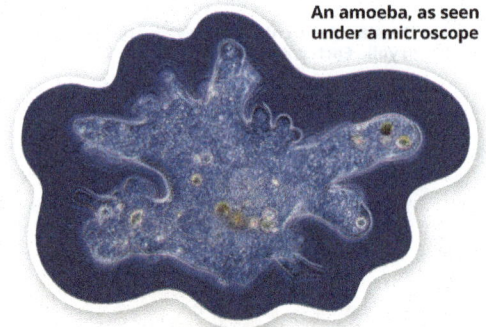

An amoeba, as seen under a microscope

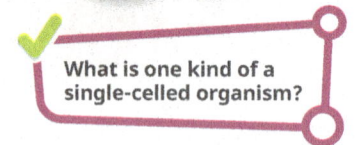

What is one kind of a single-celled organism?

90

redwood tree

Activity

Model an Amoeba's Movement

Materials: several yards of sewing elastic, 1" wide

Choose three students to be an amoeba. Loosely tie the elastic around the three students. Appoint one student to be the pseudopod. Tell him to move forward and stop. The other students should resist a little until the elastic stretches. Then they should follow the other student until the elastic is comfortable again.

STEM Career

Food Technologist

A **food technologist** studies many kinds of foods. He creates new food products. He makes sure foods are safe to eat. To do this, a food technologist uses a microscope to see the tiny cells that food is made of. He needs to make sure that food is safe from anything that could make people sick.

A food technologist also plans recipes. He studies what kinds of foods taste good and creates new types of foods. Sometimes the new foods are not good. Other times they are very good, and many people like them.

One food technologist was Edwin Perkins. He created a drink mix. People could add the mix to water to make a fruit drink. But it was hard to sell the drink mix in glass bottles. So he created a powdered drink mix that could be sold in wax paper envelopes. He worked with different foods and flavors. In 1927 Edwin Perkins began selling Kool-Aid® in six different flavors. He continued to improve the drink mix. Now it is sold in many different flavors!

A food technologist needs a college degree, usually in food science. Food technologists might work in factories or laboratories. If you enjoy trying new kinds of foods, you might enjoy being a food technologist!

91

Preparation for Reading

- Preview and pronounce the vocabulary term *food technologist* and the word *laboratories*.
- Direct the students to read page 91 silently to find out what a food technologist does.

Teach for Understanding

- Distribute the cups and Kool-Aid to each student.

What are you drinking? Kool-Aid

How was Kool-Aid created? Edwin Perkins created a powdered drink mix that people could add to water. He spent many years mixing and testing different kinds of flavors to make Kool-Aid.

What are some things a food technologist does? creates new food products, makes sure foods are safe to eat, plans new recipes

How does a food technologist make sure foods are safe to eat? He uses a microscope to see the tiny cells that food is made of.

What are some places where a food technologist might work? in factories or laboratories

---

ⓘ Background

Kool-Aid

Kool-Aid first began as Fruit-Smack and was made by reducing fruits into a liquid. The fruit liquid was mixed with water to make the fruit drink. Fruit-Smack was sold in glass bottles, which often leaked and broke during shipping. Edwin Perkins continued to experiment with different ways to make the fruit drink better. Perkins finally learned to dehydrate the Fruit-Smack into a powder, mixing it with other flavors and colorings to add variety.

Preparation for Reading

- Preview and pronounce the vocabulary terms *nucleus*, *cytoplasm*, *cell membrane*, and *cell wall*.
- Direct the students to read pages 92–93 silently to find out what job a nucleus does.

Teach for Understanding

- Distribute the *Parts of Cells* instructional aid page.
- Display the *Parts of an Animal Cell* visual.

  What are three main parts of a cell? the nucleus, cytoplasm, and cell membrane

- Guide the students as they complete the *Parts of Cells* page (Instructional Aid 4.3) during the discussion.

  Which part of the cell is the control center? the nucleus

- Explain that the nucleus contains all the directions for what the cell is supposed to do and how it is supposed to work.

  Which part of the cell is a jelly-like liquid? the cytoplasm

  What does the cytoplasm surround? the nucleus

  What is in the cytoplasm? other parts that help the cell live and grow

- Direct attention to the "Animal Cell" diagram.

  In the cytoplasm diagram, there are some things that have different shapes and colors. What do you think those things represent? the other parts that help the cell live and grow

Parts of Cells

Whether cells are big or small, all cells have the same main parts. God designed each part to do its own job for the cell.

Nucleus

The **nucleus** of a cell is its control center. It directs the activities of the cell. It is usually shaped like a sphere.

Cytoplasm

Cytoplasm is the jelly-like liquid that surrounds the nucleus. It includes the parts that help the cell live and grow. It does not include the nucleus. The cytoplasm is made up mostly of water. Without the cytoplasm, the cell would not work properly.

Animal Cell

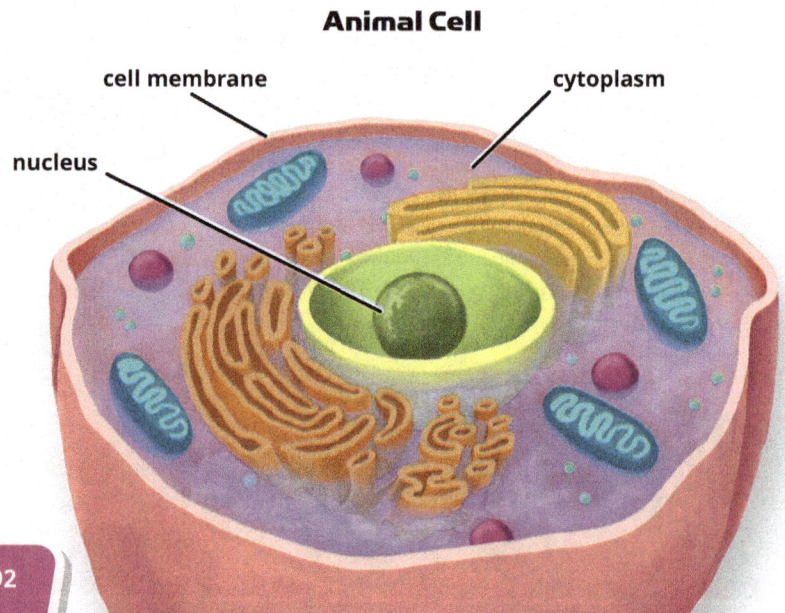

cell membrane

cytoplasm

nucleus

92

ⓘ Background

Cell Parts

Cells have more than three parts. Each part of a cell has a job. The nucleus of a cell contains the DNA, or the coded instructions the cell follows. The DNA is packed into tight bundles called chromosomes. A cell also has mitochondria, which are responsible for breaking down the cell's food and releasing energy. A cell's transportation system is called the endoplasmic reticulum. The ribosomes in a cell make the proteins that the cell needs and carry out the instructions that the DNA gives.

Plant Cell

nucleus

cytoplasm

cell membrane

cell wall

Membranes and Walls

A **cell membrane** is a thin covering that holds the cell together. Both plant cells and animal cells have cell membranes.

In addition to a cell membrane, a plant cell also has a cell wall. A **cell wall** is a stiff layer on the outside of the cell membrane. The cell wall gives the plant shape and support. An animal's cell does not need a cell wall. An animal has bones and muscles for shape and support.

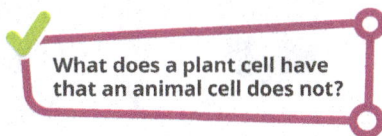

What does a plant cell have that an animal cell does not?

93

- Display the *Parts of a Plant Cell* visual.
  What holds the cytoplasm in place? the cell membrane
  Do all cells have a cell membrane? yes
- Direct attention to the "Plant Cell" diagram.
  What part of a cell is found only in plant cells? the cell wall
  What is a *cell wall*? a stiff layer outside the cell membrane that gives the plant shape and support
- Complete the "Model a Cell" activity at this time.

cell wall

Apply

Activities
Study Guide, pages 63–64
These pages review the concepts taught in Lessons 26–27. After completion, direct the students to keep them for review for Test 4.

Assess

Quiz 4A
The quiz may be given at any time after completion of this lesson.

Activity

Model a Cell
Choose four students to model an animal cell. One student will be the nucleus and three students will be the cytoplasm. The cytoplasm should not hold hands but just stand closely around the nucleus. Then choose six more students to be the cell membrane. These students should hold hands to form a boundary. They can move in and out to control the cytoplasm, but they must stay connected.

If there are enough students, choose ten or twelve students to form a cell wall of a plant cell.

The students should hold hands and form a rigid boundary that does not move.

Note: Depending on your class size, you may choose to form two cells or adjust the number of students used for each part of the cell.

Objectives
- Make a model of an animal cell.
- Label the parts of an animal cell.
- Communicate information about the animal cell model.

Materials
- See *Activities* page 65.

Teacher Resources
- Visual 4.2: *Parts of an Animal Cell*

Engage
- Display the *Parts of an Animal Cell* visual.
  What are the three main parts of an animal cell? the nucleus, the cytoplasm, and the cell membrane
- Choose volunteers to identify the parts of an animal cell on the displayed visual.
- Display the gelatin cup, fruit strips, and gumball. Explain that the students will be using these materials to make a model of an animal cell.
- Point out that one purpose of making a model is to aid in learning about things that are too small to easily see.

Instruct

Preparation for Reading
- Direct the students to remove pages 65–66 from their *Activities* books. Direct them to read page 94 and *Activities* pages 65–66 silently before beginning.
- Explain that it is helpful to look over all the pages before starting an Exploration. Point out that it will help them to see what they are going to do and when they will be writing information down.

Lesson 28

EXPLORATION

Inquiry Skills
- Infer
- Communicate

Edible Cell

Sometimes we use a hand lens to look at things that are small. Sometimes we use a microscope to see things that are very tiny. And sometimes we can use a model to see what a small thing would look like if it were larger. To see a cell, we need to use a microscope. But using a model can help us learn about the parts of a cell.

In this Exploration, you will make a model of an animal cell. You will also identify the parts of an animal cell.

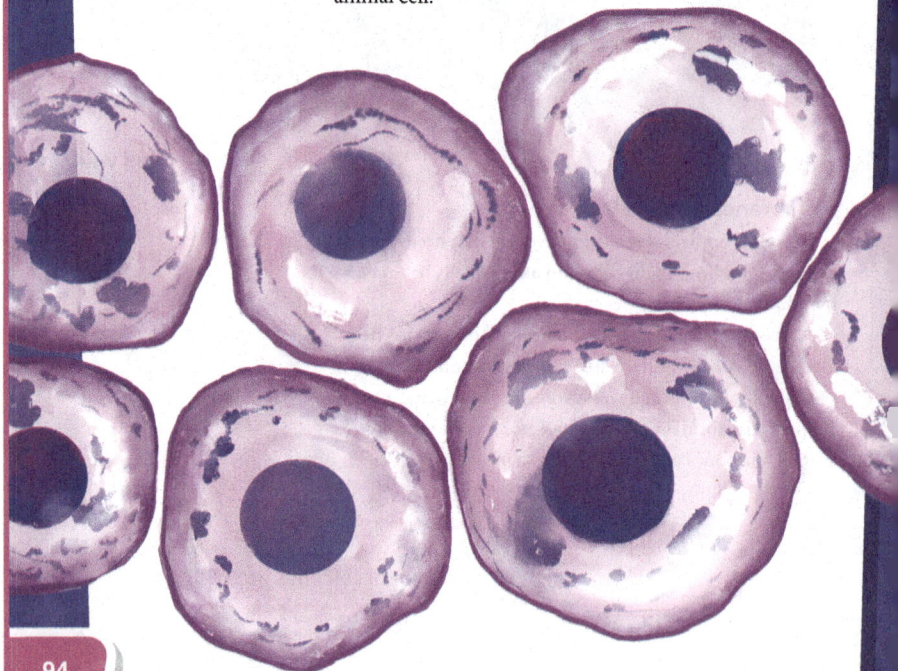

94

⚠ Helps

Edible Cell Exploration

The goal of this activity is to model a cell. You may substitute with other materials or allow the students to provide their own. You may choose for the students to represent other structures within the cell as well.

A light-colored gelatin works best for seeing the structures inside the cell. Instead of using prepared gelatin cups, you may choose to prepare gelatin. One 4 oz package of gelatin makes four models. Pour liquid gelatin several inches deep in a cake pan and cut circles with a cookie cutter after it is set.

Edible Cell

Name _____

Purpose

I want to make a model of an animal cell.

Procedure

☐ 1. Open the gelatin cup. Slide the plastic knife between the gelatin and the container. Push the knife all the way around the container to loosen the gelatin.

☐ 2. Turn the container upside down on the plate. Gently squeeze the gelatin container until the gelatin slides out.

☐ 3. Wrap the fruit strips around the outside of the gelatin. Cover all of the gelatin with the fruit strips.

☐ 4. Use the plastic knife to make a small cut in the top of the gelatin.

☐ 5. Push the gumball into the cut made by the knife.

Inquiry Skills

• Infer
• Communicate

Materials

☐ individual flavored gelatin cup
☐ 2 rolled fruit strips
☐ round gumball
☐ small paper plate
☐ plastic knife

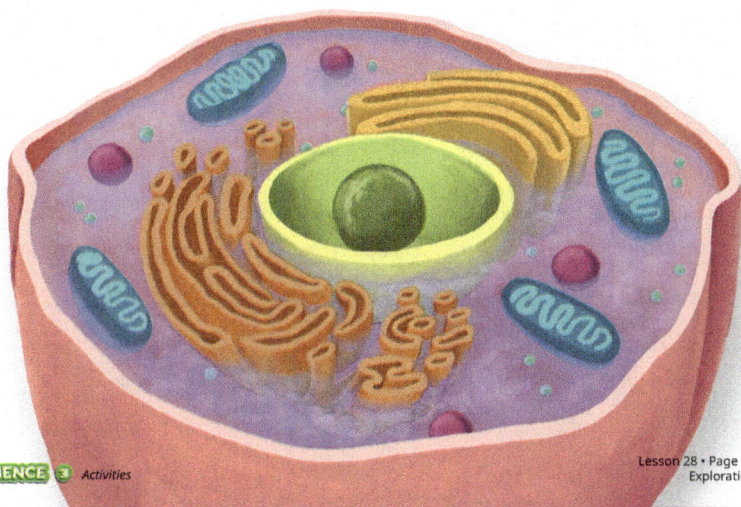

SCIENCE 3 *Activities*

Lesson 28 • Page 94
Exploration **65**

Teach for Understanding

• Direct attention to the **Inquiry Skills**.
• Remind the students that scientists use science inquiry skills to gather and use facts.
  Which inquiry skills will you be using in this investigation? Infer and Communicate
• Review what it means to *infer* and to *communicate*.
• Choose a student to read the **Purpose** aloud.
  What will you be using to make a model of an animal cell? gelatin, fruit strips, and a gumball

If other items are substituted for the ones suggested in this Exploration, direct the students to change the items listed on their *Activities* pages.

Apply

• Direct attention to the **Procedure** list. Instruct the students to follow and check off Steps 1–5 as each is completed.
• Guide the students as they use the plastic knife to loosen the gelatin from the container.
• Give help as needed as the students complete the steps in the Procedure.

- Provide time for students to draw and label their diagrams.
- Discuss the **Conclusions**.
- Remind students that in a real cell the cytoplasm and the cell membrane would completely surround the nucleus.

⚙ How is your model not like a real cell? Answers should include that the model is not alive, it is much larger than a cell, and the model contains different materials than a cell does.

⚙ This model represents an animal cell. If it represented a plant cell, what would need to be added? the cell wall

- Choose volunteers to share their models and diagrams with their group or the class.

Activities

Exploration, pages 65–66

In this Exploration the students will make a model of an animal cell.

Assess

Rubric

Use the prepared rubric or design a rubric to include your chosen criteria.

☐ 6. In the space below, draw and color a diagram of your cell. Label the parts of your cell.

☐ 7. Display your model and diagram. Tell your partner or group about the parts of your cell.

Conclusions

Identify what each part of the model represents.

1. rolled fruit strips: _cell membrane_
2. gumball: _nucleus_
3. gelatin: _cytoplasm_

Write the answer.

4. List the ways that your model is different from a real cell. _Answers should include that the model is much larger than a cell. The model contains different material than a cell does._

5. What could you add to make this a model of a plant cell? _Answers should include the concept of putting another type of material around the outside of the rolled fruit strip to represent the cell wall._

Tissues

A cell has the characteristics of a living thing. It needs food. It grows. It uses energy. It makes more cells.

Sometimes a cell works alone. Single-cell organisms are examples of this. In most organisms, though, groups of cells work together. A **tissue** is a group of cells all doing the same kind of work.

God gave people special kinds of tissues. Muscle tissue and nerve tissue are two kinds of tissues in the human body. Both of them have important jobs.

Muscle tissue allows your body to move. You can see and feel some of your muscles as they work. Other muscles, such as the ones that help you digest your food or cause your heart to beat, cannot be seen. Whether they can be observed or not, all your muscles do very important jobs.

Nerve tissue carries messages between your brain and your body. The messages tell your body what is happening and what to do. They allow you to respond to the things around you. Nerve tissue has many millions of tiny fibers. These fibers "talk" to each other by passing messages back and forth.

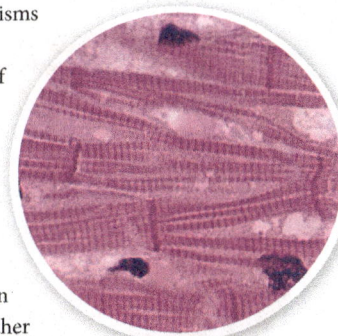

muscle tissue

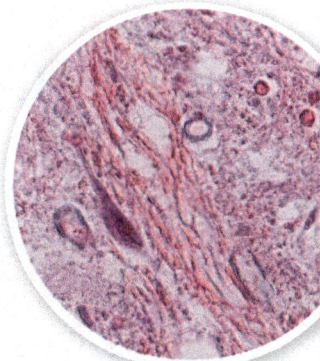

nerve tissue

✔ What are two kinds of tissues found in your body?

95

Lesson **29** begins here.

Objectives
- Explain how cells, tissues, organs, and systems are related to each other. **BWS**
- Identify four organs of the human body.
- Explain why it is important to know how the body works. **BWS**
- Identify the largest organ of the body.
- Explain why the epidermis is important to the body. **BWS**

Materials
- rope for tug of war
- blindfold
- various items of different textures or temperatures
- picture of a knight in armor

Teacher Resources
- Visuals 4.4–4.5: *Upper Body Organs; Layers of the Skin*
- Instructional Aids 4.3–4.4: *Layers of the Skin; Layers of the Skin Key*

Vocabulary
- tissue
- organ
- system
- epidermis
- dermis
- melanin
- fingerprint

Engage
- Choose six volunteers to demonstrate a game of tug of war. Place three volunteers on each team.

  What job do the students on each team have? to pull the other team across the center line

  How is each team doing that? by working together to pull the rope
- Quietly direct two players on one team to stop pulling on the rope.

  What happened when two players stopped pulling? Possible answers: Only one person was pulling; their team did not win.
- Point out that the team members had to work together to do their job. The cells inside the body and inside other living things must work together, too.

Instruct

Preparation for Reading
- Preview and pronounce the vocabulary term *tissue*, as well as the words *muscle tissue* and *nerve tissue*.
- Direct the students to read page 95 silently to find out what kind of tissues carry messages.

Teach for Understanding
What are some characteristics of a cell? It is the smallest living thing; it needs food; it grows; it uses energy; it makes more cells.

Are most cells like single-celled organisms? No, most cells work together in groups.

What is a *tissue*? a group of cells all doing the same kind of work

What are two kinds of tissues in the human body? muscle tissue and nerve tissue

Which kind of tissue allows your body to move? muscle tissue
- Choose several volunteers to demonstrate using muscle tissue. Examples could be blinking their eyes, clapping their hands, stomping their feet, and nodding their head.

  Which kind of tissue carries messages between your body and your brain? nerve tissue

  What is nerve tissue made up of? millions of tiny fibers
- Explain that fibers are thin, threadlike structures that form part of the muscle tissue, nerve tissue, or other tissues in the body.
- Choose a volunteer to wear the blindfold. Direct the volunteer to hold out his hand. Place an item in his hand.

  What message is your nerve tissue sending to your brain about this item? Answers should include something about the texture, temperature, weight, or other characteristic of the item that can be felt.
- Choose other volunteers and other items as time allows.

✔ muscle tissue and nerve tissue

Preparation for Reading

- Preview and pronounce the vocabulary terms *organ* and *system*, and the words *esophagus* and *intestine*.
- Direct the students to read pages 96–97 silently to find out what a group of organs working together is.

Teach for Understanding

What is a group of tissues working together called? an organ

Are organs made up of only one kind of tissue? No, they are made up of several different kinds.

What is one example of an organ that is made up of more than one kind of tissue? the eye

- Display the *Upper Body Organs* visual.

Your eye is an organ. What are some other organs in your body? Possible answers: the lungs, the heart, the stomach, the brain

Organs

An **organ** is a group of tissues working together to do a job. Organs are the major parts of the body. Most organs are made of two or more kinds of tissues. One example of an organ is the eye. Muscle tissue focuses and moves the eye. Nerve tissue sends messages to the brain, telling what is seen. Each tissue in the eye has its own job, but the tissues also work together. When all the tissues do their jobs, then the organ, or the eye, can do its job.

God gave your organs important jobs. Your eyes let you see. Your lungs allow you to breathe. Your heart pumps blood to your body. Your stomach helps you digest, or break down, food. Your brain is the control center for your body. These are just a few of your many organs.

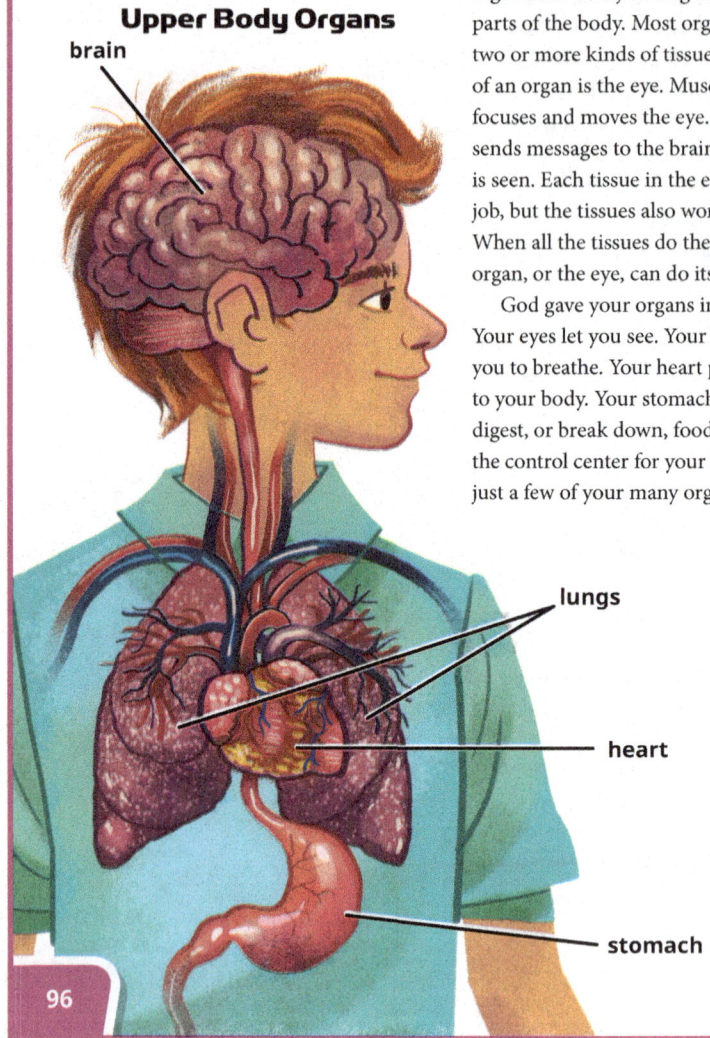

Upper Body Organs

brain

lungs

heart

stomach

96

Systems

God gave us organs to do different things in our bodies. A **system** is a group of organs working together to do a job. One system in your body is the digestive system. The stomach is part of this system, and so are other organs. The tongue, esophagus, liver, small intestine, and large intestine work together with the stomach to digest your food.

Cells make up tissues. Tissues form organs. Organs work together to form systems.

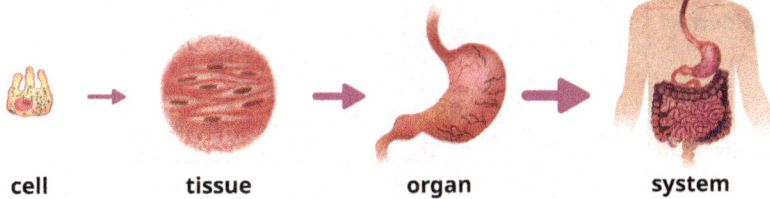

cell → tissue → organ → system

Each part of the body has its own job. But each part also works together with the other parts. When you understand how the parts of your body work, then you will be able to keep your body healthy. Doctors understand how your body works. They give you proper care when your body is not well. Learning how to take care of your body is one way you can glorify God.

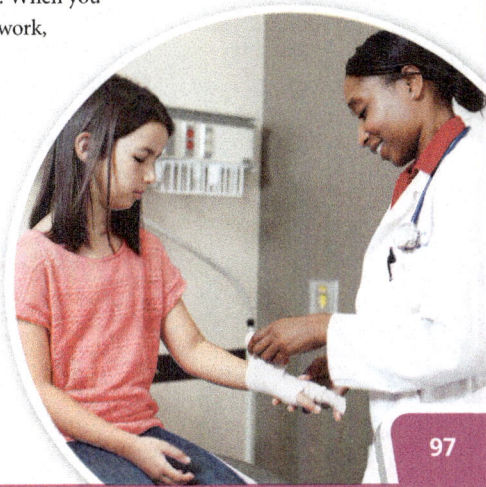

What are three organs in your body?

97

What is a group of organs working together called? a system

What organs are part of the digestive system? stomach, tongue, esophagus, liver, small and large intestines

What is the purpose of the digestive system? to digest food

What are some other systems in your body? Possible answers: the circulatory system, the respiratory system

- Direct attention to the diagram and labels.

In God's design, how are cells related to organs? Cells make up tissues, which make up organs.

In God's design, how are cells related to systems? Cells make up tissues. Tissues make up organs. Organs make up systems.

Did all the cells in your body make up tissues and organs and systems by chance? No, God designed and planned our bodies to work together perfectly.

- Read 1 Corinthians 12:12–21 aloud. Explain the comparison between our physical bodies and the "body" of Christ.

What do these verses compare the people (members) who make up the church to? a body

How many bodies in Christ are there? one

How many people (members) are in the body of Christ? many

Do all the members in the church have the same job? no

Do all the organs in the body have the same job? no

What would happen to the body if your eyes didn't work? You would not be able to see.

What would happen to the body if your legs didn't work? You would not be able to walk.

How should Christians in the church work together? Christians should work together like the organs of the body work together; members should care for each other and work together to do the jobs that God has given them.

- Explain that God designed each part of the body to have its own job.

Why should you learn about how the parts of the body work? So you can keep your body healthy. Doctors learn about parts of the body so they can give proper care when your body is not well.

- Direct attention to the picture.

How is the doctor taking care of the girl? She is putting a cast on her broken bone.

How is taking care of your body one way to glorify God? God created people. People are important to God. I can glorify God by taking care of my body, because it was created by God.

Possible answers: eye, heart, lungs, brain, stomach, tongue, esophagus, small intestine, large intestine

Preparation for Reading

- Preview and pronounce the vocabulary terms *epidermis* and *dermis*.
- Direct the students to read page 98 silently to find out what the top layer of the skin is.

Teach for Understanding

- Direct the students to examine the skin on their knuckles as they bend and straighten their fingers.

  Does the skin stay the same when your fingers move? no

  Describe any changes you see. Possible answers: The skin stretches and becomes smoother when it bends; it wrinkles when it is straightened.

  Do you think your skin is the same size that it was when you were a baby? No, it has grown with the rest of the body.

  When you get a cut does your skin stay open? No, it closes up; it heals itself.

  What is an *organ*? a group of tissues working together

  What are the tissues made of? cells

  What is the largest organ of your body? the skin

- Display the *Layers of the Skin* visual. Choose volunteers to point out the layers as they are discussed.
- Guide the students as they complete the *Layers of the Skin* Instructional Aid page during this lesson and Lesson 31.

  What are the two layers of skin? the epidermis and the dermis

  Which is the top layer of skin? the epidermis

  Which is the lower layer of skin? the dermis

  What prevents the skin from getting thin and wearing out? New skin cells are always growing to replace the old cells.

  Where are new skin cells made? in the lowest part of the epidermis

- Explain that the epidermis tissue is made of many very thin layers of cells. The new cells at the base of the epidermis are column shaped and by the time they flake off, they are thin and flat.

  What happens to the old skin cells? They get pushed to the surface and fall off.

- Direct the students to rub their hands together. Explain that when you do this you are removing dead cells from the surface of your skin.

  Why can you not see the cells that come off? Elicit that most cells cannot be seen without a microscope.

- Show the picture of the knight in armor.

  What is the knight wearing? armor

  What is it used for? protection

- Explain that God made our skin to be a kind of armor. It covers and protects us.
- Discuss with the students the wonder of God's design for their bodies. Emphasize that it is easy to take lightly things that are familiar, such as skin, but the design is a reflection of the greatness of the Creator God.

Skin

Every organ in your body has a job to do. Your body's largest organ is your skin. Though it has many parts, it has just two main layers. The **epidermis** is the top layer of the skin. Under the epidermis is the second layer, called the **dermis**.

Even though you cannot see anything happening, your skin is hard at work. New skin cells are always being made in the lowest part of the epidermis. As the new cells move to the surface, they push older cells ahead of them. After a while, the old, dead skin cells at the surface of the epidermis fall off.

Most of the skin you see is the dead cells waiting to fall off. You do many things that help remove them. Washing your skin or even changing your clothes removes some of the dead cells.

Layers of Skin

epidermis

dermis

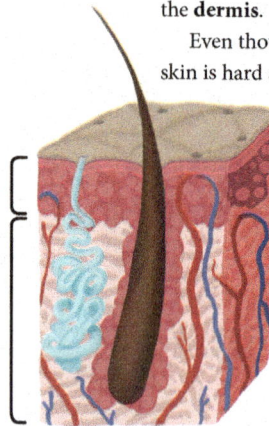

The Epidermis

God made skin as a marvelous protection for your body. Although the epidermis is very thin, it works like a strong shield. It helps keep harmful things out of your body. Dirt and germs cannot pass through the skin. If these things could get through, you would get sick much more often.

✓ What is the largest organ of your body?

98

How does the epidermis protect your body? It keeps harmful things out of your body and keeps in things that your body needs.

What are some harmful things that the epidermis keeps out of your body? dirt and germs

What happens if dirt and germs are not kept out of your body? You would get sick much more often.

✓ the skin

ⓘ **Background**

Skin Layers

Under the epidermis and dermis is a fatty layer called the subcutaneous layer. This layer of fat helps the body stay warm, helps the skin hold on to the tissues underneath it, and acts like a cushion to protect the body against injury from bumps and falls.

New Skin Cells

It takes about a month for new skin cells to work their way to the surface and fall off. The average person loses more than 25,000 skin cells a day.

The epidermis also keeps in things that your body needs. One important thing your body needs is water. Water keeps your body from drying out. It helps keep your body's organs working properly. It also keeps your blood flowing.

The epidermis also contains a substance called melanin. **Melanin** is the coloring in your skin. When you are in the sun, your body makes extra melanin to help protect your skin from damage. The more melanin you have, the darker your skin will be.

Fingerprints

The epidermis has many lines and grooves in it. An interesting part of your epidermis is your fingertips. If you look closely at them, you can see dozens of curved ridges. The prints made by the ridges on your fingertips are called **fingerprints**. Fingerprints are very helpful. They help you grip and hold on to things. Without the ridges on your skin, it would be hard for you to pick up something smooth like a glass.

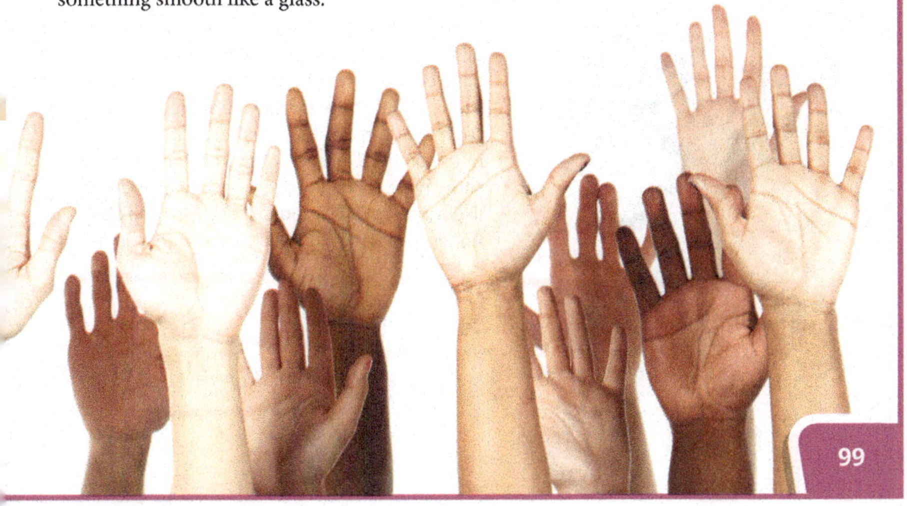

Drinking plenty of water is important for your health.

99

Preparation for Reading
- Preview and pronounce the vocabulary terms *melanin* and *fingerprints*.
- Direct the students to read pages 99–100 silently to find out what makes skin different colors.

Teach for Understanding
What important thing does your skin keep in the body? water
- Direct attention to the picture on the right of the page. Choose a volunteer to read the caption aloud.
  Why is water important to your body? Water keeps your body from drying out. Water keeps your body's organs working properly and keeps your blood flowing.
- Direct attention to the picture at the bottom of the page.
  Why do each of the hands look different? Each person has a different amount of melanin in their skin.
  What is *melanin*? the coloring in skin
  How does melanin affect your skin color? The more melanin you have, the darker your skin is.
  What effect does sunlight have on the melanin in skin? Sun causes the cells in the skin to make more melanin.
- Direct the students to look at the tiny lines on their fingertips. Ask volunteers to describe what they look like.
- Explain that some lines look lighter and others darker. The lighter raised lines are called ridges. They are like little hills. The darker lines are the grooves, or valleys.
  What do we call the prints made by the ridges on your fingers? fingerprints
  When were the ridges on your fingers formed? before you were born
  Do any two people have the same fingerprints? no
  Why can fingerprints help identify you? Every person has his own set of fingerprints.

ⓘ Background

Freckles
Freckles are flat spots of concentrated melanin in the skin. Freckles are thought to be caused partly by genetic inheritance and partly from sun exposure. They occur in people of all skin colors but predominantly in people with fair, or light-colored, skin. The spots are harmless and do not pose a health risk. Identical twins who have freckles may have similar freckle patterns.

Types of Fingerprints
Although everyone's fingerprints are unique, fingerprints can be classified into three broad categories: loops, arches, and whorls. There are many variations within each category. These categories are discussed further during the activity in Lesson 88. A person's fingerprints can be all in the same category or can include two or even all three categories.

⚙ When you gently place your finger on something, do the ridges or the grooves touch first? **the ridges**
What do the ridges on your fingers help you do? **pick things up and hold them**

- Direct attention to the "Technology Connection" box.
What are some ways fingerprints can be used to identify people? **Possible answers: Police use them to identify criminals. Businesses use them to identify who is working for them. Fingerprints can be used like a key for a cell phone or a computer.**

- Direct attention to the "Fantastic Facts" box.

- Discuss with the students God's amazing design in giving even animals stripes and prints that are completely different from those of any other animal.

✓ **prints made by the ridges on your fingertips**

Apply

Activities

Study Guide, pages 67–68

These pages review the concepts taught in Lesson 29. After completion, direct the students to keep them for review for Test 4.

Assess

Quiz 4B

The quiz may be given at any time after completion of this lesson.

Technology Connection

Every person in the world has a different set of fingerprints. Even identical twins each have different fingerprints. God created each individual fingerprint with a different design. There is no one else in the world that has fingerprints just like yours!

Police use fingerprints to help identify criminals. Schools and businesses use fingerprints to identify who works for them. Cell phones and computers use fingerprints as a key. A tiny camera in the cell phone or computer scans a person's fingerprint. If it matches the one stored in the phone or computer, it unlocks the device.

Fantastic Facts

Just as God designed your fingerprints to be different, He also gave animals some unique, or different, features. A tiger's stripes are unique. No two tigers have exactly the same stripes. Dermal ridges, similar to fingerprints, are found on dogs' noses. No two dogs have the same nose prints. A koala has fingerprints just like a person. It is impossible to tell a koala's fingerprints and a human's fingerprints apart!

✓ **What are fingerprints?**

100

ⓘ Background

Fingerprint Wear

The ridges that form the prints may wear away. A person's occupation may cause the ridges on his fingers and thumbs to wear down. The ridges wear away on the hands of people who handle rough materials such as bricks and rocks. Surprisingly, people who handle a lot of paper also often have fading fingerprints. The ridges on the hands of elderly people may become wider, making the prints harder to read.

Patterns on My Skin

Name _____

Purpose

I want to discover what type of fingerprint patterns I have. I want to compare my fingerprints with others' fingerprints.

Procedure

☐ 1. Complete the first two rows of your fingerprint card on page 73.

☐ 2. Find the box for your right thumb.

☐ 3. Gently roll your right thumb across the ink pad at a steady speed. Then press your thumb down lightly in the box for your right thumb, rolling your thumb again.

☐ 4. Repeat with each finger on your right hand. Then use the hand sanitizer and paper towels to clean your hand.

☐ 5. Repeat with your left hand.

☐ 6. Find the box marked *Left Four Fingers*. Ink your left four fingers all at once and press them all into the square at the same time. Clean your fingers.

☐ 7. Repeat with your right hand.

☐ 8. Ink your thumbs and put your thumbprints in the spaces between the fingerprint groups. Clean your thumbs.

☐ 9. Compare your fingerprints with the example patterns. Use the magnifying glass to help you see the details. Decide which pattern best matches each of your prints. Write the type of pattern near each print.

☐ 10. Compare your prints with others' in your group.

☐ 11. Make a tally mark in each column for the type of fingerprint that someone else has.

Inquiry Skills

• Observe
• Classify
• Communicate

Materials

☐ washable-ink pad
☐ hand sanitizer
☐ paper towels
☐ magnifying glass

SCIENCE 3 *Activities*

Lesson 30 • Page 101
Exploration **71**

ℹ Background

Fingerprint Categories

Loops are the most common pattern type, occurring in about 65% of people. Whorls are the next most common, and arches are the least common.

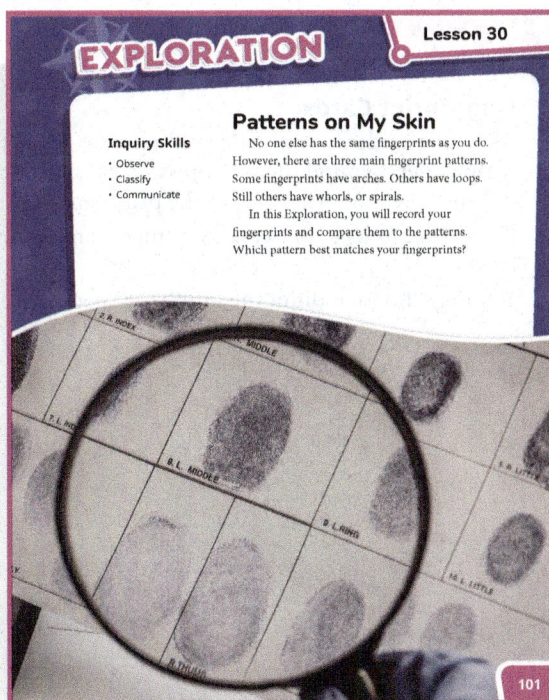

EXPLORATION   Lesson 30

Patterns on My Skin

Inquiry Skills
• Observe
• Classify
• Communicate

No one else has the same fingerprints as you do. However, there are three main fingerprint patterns. Some fingerprints have arches. Others have loops. Still others have whorls, or spirals.

In this Exploration, you will record your fingerprints and compare them to the patterns. Which pattern best matches your fingerprints?

101

Objectives

• Classify fingerprint patterns.
• Record and graph types of fingerprint patterns.

Materials

• See *Activities* page 71.

Teacher Resources

• Visual 4.6: *Fingerprint Patterns*

Engage

• Display the *Fingerprint Patterns* visual.

What differences do you see in these fingerprint patterns? Lines may be loops, arches, or whorls.

• Ask volunteers to point out the unique feature of each print. Try to make a connection to each name and the pattern seen.

Instruct

Preparation for Reading

• Direct the students to remove pages 69–74 from their *Activities* books. Direct them to read page 101 and *Activities* pages 69–74 silently before beginning.

• Explain that it is helpful to look over all the pages before starting an Exploration. Point out that it will help the students to see what they are going to do and when they will be writing information down.

Teach for Understanding

• Direct attention to *Activities* page 69 and to the *Fingerprint Patterns* visual.

What are the three main patterns of fingerprints? loop, arch, and whorl

• You may choose to complete Who Did It, *Activities* page 70, now as a class, or assign it to be done individually or in groups as students finish the Exploration.

• Remind the students that scientists use science inquiry skills to gather and use facts.

Which inquiry skills will you be using in this Exploration? Observe, Classify, and Communicate

• Review what it means to *observe*, *classify*, and *communicate*.

What will you be classifying in this activity? fingerprint patterns

How will you be classifying them? according to the three main patterns: loop, arch, and whorl

• Explain that comparing your prints with those of another person is another way of classifying.

• Choose a student to read the **Purpose** aloud.

Apply

• Direct attention to the **Procedure** list. Instruct the students to follow and to check off Steps 1–5 as each is completed.

• Instruct the students to follow and to check off Steps 6–8 as each is completed.

- Give help as needed as the students complete the steps in the Procedure.

  Which pattern best matches your fingerprints?

  Do your fingertips all have the same patterns, or do you have different patterns on each finger?

  Are your fingerprints similar to anyone else's?

> Each finger's fingerprint can be different, so one person can have multiple fingerprint patterns. For the purpose of classifying and comparing, it would be easiest to assign students to compare a specific finger or thumb, such as the right index finger or left thumb.

- Direct students to make a tally mark in the appropriate column for the type of fingerprint someone else has.
- Guide students to complete the graph of the fingerprint types in their group.
- Guide students in completing the **Conclusions**.
- You may choose to classify all the students according to their most common pattern, or the pattern of the finger or thumb that was used for the comparison. Designate a place in the room for each pattern and direct the students to go to the area that matches their prints. Discuss which pattern is the most common.

Loop	Arch	Whorl

☐ 12. Graph the type of fingerprints in your group below.

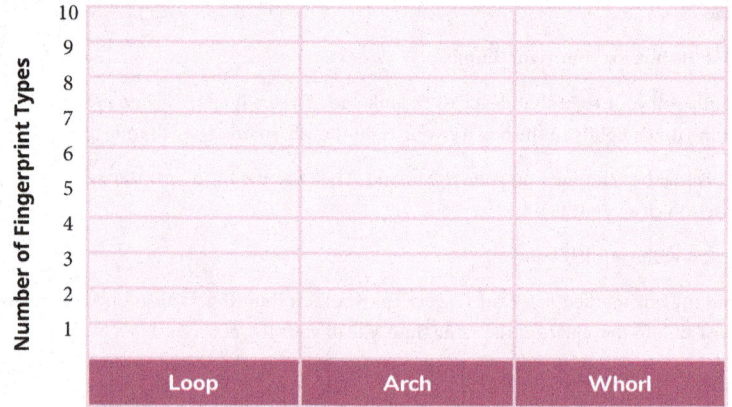

Number of Fingerprint Types

10
9
8
7
6
5
4
3
2
1

Loop	Arch	Whorl

Conclusions

1. Which fingerprint pattern did you see the most? *Answer should match the tally chart above.*

2. Which pattern best matches your fingerprints? _____

3. Are your fingerprints the same as anyone else's? *No*

4. How do you think identifying people by fingerprints is helpful to the police? *Possible answer: Since no one has the same fingerprints, police can tell the difference between people by using fingerprints.*

72 Lesson 30 • Page 101
Exploration

⚠ **Helps**

Fingerprint Cards

Allow the students to practice rolling their inked fingers to make clear prints on scrap paper. Remind them to press down only once and to press lightly. Repeated or heavy pressing will smudge the prints and make them difficult to read.

Direct the students to clean their fingers and thumbs before classifying their prints.

You may choose to limit the number of people they compare prints with.

Fingerprint Card   Name _____

Last Name	First Name	Middle Initial

Signature of Person Fingerprinted	Date of Birth		
	Month	Day	Year

R. Thumb	R. Index	R. Middle	R. Ring	R. Little
L. Thumb	L. Index	L. Middle	L. Ring	L. Little

Left Four Fingers	R. Thumb / L. Thumb	Right Four Fingers

Lesson 30

Activities

Exploration, pages 69–74

In this Exploration the students will compare and classify fingerprint patterns.

Assess

Rubric

Use the prepared rubric or design a rubric to include your chosen criteria.

Objectives
- Identify the parts of the dermis.
- Explain two ways that the body is cooled.
- Explain why caring for our skin is important. **BWS**

Materials
- hand soap

Teacher Resources
- Visual 4.5: *Layers of the Skin*
- Instructional Aids 4.3–4.4: *Layers of the Skin; Layers of the Skin Key*

Vocabulary
- nerves
- blood vessels
- sweat glands
- pores
- hair follicle
- oil glands

Engage

- Direct the students to close their eyes. Tell them to open their eyes when they think 20 seconds have passed. Start timing. Shake your head "no" if anyone opens his eyes before 20 seconds. When 20 seconds have passed, tell the students to open their eyes.
- Explain that it can be hard to tell how long 20 seconds is without using a clock. Singing a song that takes about 20 seconds can help to know how much time has passed.
- Sing "London Bridge" or "Happy Birthday" with the students.
- Point out that there is something else that takes about 20 seconds and is very important. Today we will find out what it is and why we should do it more often!

Instruct

Preparation for Reading
- Preview and pronounce the vocabulary terms *nerves* and *blood vessels*, and the word *automatically*.
- Direct the students to read pages 102–3 silently to find out what important jobs blood vessels have.

Teach for Understanding
- Review that senses are needed to observe matter.
  What sense do we use our skin for? touch
  What can we learn about matter through our sense of touch? Possible answers: how hard or soft something is, how hot or cold something is, how smooth or rough something is
- Display the *Layers of the Skin* visual and point out each feature as it is discussed.
- Guide students in completing the *Layers of the Skin* Instructional Aid page as each part is discussed.
  Is the dermis or the epidermis thicker? the dermis
  Where is the dermis located? under the epidermis
  What does the dermis contain? nerves, blood vessels, sweat glands, and oil glands

---

Lesson 31

The Dermis
The dermis is the second layer of the skin. The dermis is much thicker than the epidermis. It has a different job to do for your body. It contains the nerves, blood vessels, sweat glands, and oil glands.

Nerves
Nerves are a small but important part of the dermis. The nerves allow you to feel things. **Nerves** are the tiny pathways that carry messages between the brain and other parts of the body. Some nerves tell you how things feel. Some tell you when something hurts. Others tell you whether you are hot or cold.

The dermis all over your body has nerves. But some areas have more nerves than others. You use your fingers and hands to touch things. God designed the skin on your fingers and hands to have many nerves. Places such as your back and your earlobes have fewer nerves.

God created nerves in your skin to protect you. They help you prevent injury. For example, without nerves you could not feel whether something is hot. Your nerves cause you to feel pain so that you can avoid getting badly burned.

102

- Discuss what nerves are and where in the skin they are found.
  What are *nerves*? tiny pathways that carry messages between the brain and other parts of the body
  What job do your nerves have? They allow you to feel things.
  How do you know what you are feeling? The nerves send messages to your brain.
  Does all of the dermis have nerves? yes
  Does the dermis in all areas of your body have the same number of nerves? No, some areas have more nerves than others.
  Where are some areas of skin that have more nerves than other areas? the fingers and hands
  Why do you think those areas have more nerves? Our fingers and hands are the main places we use to touch things.
  Where are some places that have fewer nerves? the back and the earlobes
  Why do you think that some places have more nerves than others? Possible answer: More nerves are in the areas we use more often to hold and touch things.
  How can nerves protect you? They give warnings of pain to help you avoid injury.

Blood Vessels

Blood vessels are thin tubes that carry blood around your body. Blood vessels have two very important jobs. They carry oxygen and nutrients to all parts of your body. In the skin, the blood vessels are in the dermis. There are no blood vessels in the epidermis. This is why your skin does not easily bleed when you scratch it.

Your blood vessels also help keep your body at a constant temperature. The blood vessels in the dermis do this by getting wider or narrower. When you get hot, the blood inside your body gets warmer. When this happens, the blood vessels in your skin automatically become wider. The wider vessels bring more blood than usual to your skin. When the blood comes close to the surface of your skin, your body releases the heat. When you get cold, the blood vessels become narrower. This holds most of the blood deeper inside your body to keep the heat in.

Are this boy's blood vessels becoming wider or narrower?

Fantastic Facts

When you get cold, you might notice that your skin is covered with little bumps. Some people call them goose bumps. But they have nothing to do with geese. When you are cold, the nerves send messages to tiny muscles near the hairs. This causes the hairs to stand tall and bumps to form on the skin. You may also experience goose bumps when you are afraid.

✔ **What do nerves allow you to do?**

103

Louis Braille

Louis Braille (1809–52) became blind at the age of three after an accident in his father's workshop. When he was 10, he was able to attend the Royal Institute for Blind Youth. There he heard about a system of dots developed by an army captain named Charles Barbier. Captain Barbier's system was called night writing and was used for nighttime communication on the battlefield. This system was too complicated to be useful for the blind students, but it gave Louis some ideas. He worked on his ideas for a few years and came up with a simple system of patterns of raised dots to represent letters. The raised dots allowed blind people to be able to easily read by touch. His system came to be known as Braille and today is used by blind people all over the world.

- Discuss blood vessels.

  What do blood vessels carry? blood

  What important jobs do your blood vessels have? They carry nutrients and oxygen to all parts of the body, and they help keep your body at a constant temperature.

  How do your blood vessels keep your body at a constant temperature? They get wider when you are hot and get narrower when you are cold.

  Explain how blood vessels help you to feel cooler. The answer should include that when you are hot, blood within your body gets warmer. The wider vessels allow more blood to come near the surface of the skin and release heat to your surroundings.

- Direct attention to the picture. Choose a volunteer to read aloud and answer the caption.

  Are this boy's blood vessels becoming wider or narrower? They are becoming wider so his body can cool off.

  Why do you think a person's face becomes red when he is hot? Possible answer: It becomes red when more blood is at the surface of the skin.

  Explain how blood vessels help you feel warmer. The answer should include that when you are cold, blood vessels become narrower to help hold the heat inside your body.

  Why do you think a person's skin may become paler (lighter) when a person is cold? Possible answer: It becomes paler because less blood is at the surface of the skin.

  Nerves and blood vessels are in the dermis. Is that the only place you can see them? No, they are also in the layer below the dermis.

  Why does the diagram on your handout show the blood vessels and nerves in more places? Possible answer: They connect to all other parts of the body.

  Why do you not bleed if you lightly scratch the palm of your hand? There are no blood vessels in the epidermis.

- Point out God's perfect design for our skin. Explain that God even planned where the blood vessels should be so that we would not bleed too much.

- Direct attention to the "Fantastic Facts" box.

  What part of your skin tells your hair to stand tall, creating goose bumps? the nerves

  Why do you think they are called *goose bumps*? When a goose's (or other bird's) feathers are plucked, a small bump forms at the place where each feather was.

✔ **feel things**

Preparation for Reading

- Preview and pronounce the vocabulary terms *sweat glands*, *pores*, *hair follicle*, and *oil glands*.
- Direct the students to read pages 104–5 silently to find out how to keep their skin healthy.

Teach for Understanding

- Ask volunteers to share times when they have had sweat on their arms or faces.

  Where does sweat come from? sweat glands

  What is the purpose of sweat? to help cool down your body

  What do sweat glands do? make sweat and move it to the surface of the skin

  What part of the skin releases sweat? pores

  How does sweat help your body? When the water in the sweat evaporates, your body cools down.

- Remind the students that water is very important in helping our bodies work properly. Explain that when our bodies lose water through sweat, we need to drink extra water to replace what we lost.

  Why does your hair not hurt when it is cut? It is made of dead cells.

  Why does it hurt when a hair is pulled out? Each hair is in a hair follicle, and there are nerves around the hair follicle.

  What is near the base of each hair? an oil gland

  What do oil glands do? They release oil.

  What are benefits of having oil on your skin? to keep skin soft, to keep moisture inside the skin, and to protect the skin from infection

  How do the oil and hair protect the skin from infection? They trap and hold dirt until it can be washed off.

Sweat Glands

When you run or exercise, your body gets hot. Your skin feels damp. The drops of liquid on the surface of your skin are called sweat. Your sweat is another way that God gave your body to cool down. **Sweat glands** are special parts in the dermis that make sweat. They move the sweat to the skin's surface. The body releases the sweat from tiny openings in the epidermis called **pores**. The sweat is mostly water. It evaporates quickly and cools the surface of your skin.

Oil Glands

Most of your body has hair. Hair is made up of dead cells. It does not hurt to cut hair, but it does hurt to pull out a strand of hair. This is because each strand of hair starts in a **hair follicle** in the dermis. If you pull out a hair, you feel pain from the nerves around the hair follicle.

Near the base of each hair is an oil gland. **Oil glands** release oil, which helps protect your body. Oil keeps your skin soft. It also keeps moisture inside your skin. Without oil, your skin would dry out.

The oil and hair work together to trap dirt on the skin until it can be washed off. This helps protect your skin from infection. It is important to keep your hair and skin clean.

Layers of Skin

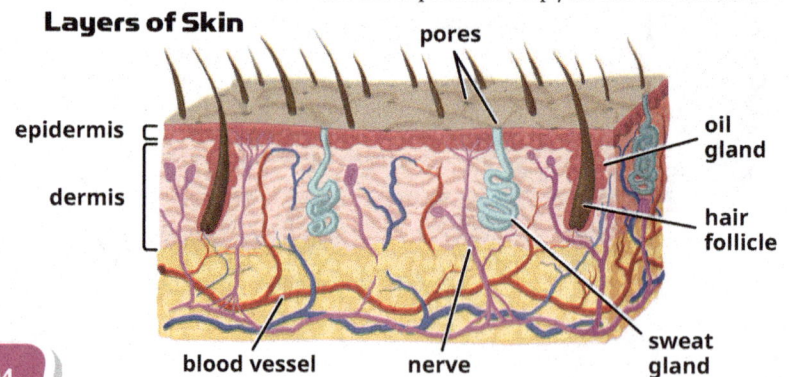

104

Skin Care

You were made in God's image. This means that you are important. You also have a special job from God. You should take care of His world. This includes your body. Your care begins with your skin, the largest organ your body has.

Keeping your skin clean helps you stay healthy. Use soap, water, and a clean washcloth to remove the dirty oil from your skin.

You can also protect your skin by being careful around sharp objects. A cut in your skin requires special care. If you get a cut, clean it with soap and water to remove any dirt. An adult may need to put medicine on it. It is also a good idea to cover a cut with a bandage to keep dirt out.

We often do not think about our skin unless it gets hurt. But it is always working to keep us safe and healthy. God, the Creator, designed each part of the body just right. Psalm 139:14 tells us that we are "fearfully and wonderfully made!"

Health and Safety

Sunlight is good for you, but too much sun can burn your skin. You should be careful to protect your skin from too much sun. Clothing covers some of your skin. Sunscreen can protect your exposed skin. Take care of the skin God has given you!

How does oil protect your body?

105

Activity

Show That Oil Captures Dirt
Materials: newspaper, 2 index cards, vegetable oil, flour, paper towels
Cover the work surface with newspaper. Coat the top of one index card with a thin layer of oil. Place both cards on the newspaper and sprinkle flour on them. Wait several seconds and then pick up the cards and shake them over the newspaper. Observe which card holds the most flour. Choose volunteers to relate the results to how oil in the skin traps dirt.

Hand-Washing Campaign
Discuss with the students that even though they now know the importance of hand washing, many people often forget to do it. Encourage them to begin a campaign to promote proper hand washing. Direct the students to design and make small posters to be hung in restrooms, or write and record service announcements to be shown to others.

Why do you have value? You were made in God's image.

What job has God given you to do? take care of His world

Where should your care for your body begin? taking care of your skin

Explain why caring for your skin is important. We are important to God and are part of His creation. We should take care of what He created.

How can you keep your skin healthy? by keeping it clean

What does washing do? remove dirty oil

- Explain that washing your hands is one of the easiest ways to stay healthy. You cannot avoid all germs, but you can wash away many of them.
- Complete the "Handwashing" activity now.
- Explain that proper handwashing should take about 20 seconds. Use the hand soap and a sink to practice washing hands for about 20 seconds.

When do you think you should wash your hands? Possible answers: before eating, after playing with animals, after using the bathroom, before preparing food, when sick or sneezing, when they are dirty, after playing outside

Why do cuts require special care? They can allow dirt and harmful things to enter your body through the skin.

How should you take care of a cut? Clean it with soap and water, put medicine on it if needed, and cover it with a bandage.

Why is it a good idea to let an adult know about a cut? An adult can determine whether a doctor should check the cut.

- Direct attention to the "Health and Safety" box.

Is sunlight bad for you? no

- Explain that sunlight helps your skin to make vitamin D, which is important to your body's good health. Vitamin D is especially important for strong and healthy bones. It also helps your muscles, heart, lungs, and brain work well and helps your body fight infection.

What can happen if you get too much sun? sunburn

How can you protect your skin from getting too much sun? Possible answers: wear extra clothing to cover your skin, use sunscreen

keeps skin soft, keeps moisture inside skin, traps dirt on the skin until it can be washed off

Apply

Activities
Study Guide, pages 75–76
These pages review the concepts taught in Lesson 31. After completion, direct the students to keep them for review for Test 4.

Objectives

- Recall terms and concepts from Chapter 4.

Apply

Review

- Material for Test 4 will come from the Study Guides on *Activities* pages 63–64, 67–68, and 75–76, as well as Quizzes 4A and 4B. You may review any or all of the material during the lesson.

- You may choose to review Chapter 4 by playing the Cover It Up review game or a game from the Game Bank (Teacher Resources).

🏆 Review Game

Cover It Up

Materials: box of adhesive bandages, chart paper

Group the students into teams. Display one large stick figure on chart paper for each team. For each correct answer to a review question, the student places a bandage on his team's person. The team with the most bandages wins.

Stickers or pieces of tape may be substituted for the bandages.

NOTES

NOTES

_____

_____

_____

_____

_____

_____

_____

_____

_____

_____

_____

_____

_____

_____

_____

_____

_____

_____

_____

_____

_____

_____

_____

_____

_____

_____

_____

_____

_____

Objective

- Apply terms and concepts from Chapter 4.

Assess

- Administer Test 4.

Chapter Objectives
- Analyze what is needed for plants to survive and grow.
- Evaluate the effects of photosynthesis on organisms.
- Compare and contrast inherited traits of plants. **BWS**
- Apply the inquiry skills of predict, observe, measure, infer, and communicate to the needs of a plant.
- Explain the importance of plants. **BWS**

Lesson Objectives
- Identify three conditions a seed needs to germinate.
- Recall the function of each plant part.
- Sequence the life cycle of a flowering plant.

Teacher Resources
- Visuals 5.1–5.4: *Sample Plants*; *Life Cycle of a Plant*; *Plant Parts*; *Parts of a Bean Seed*

Vocabulary
- germinate
- nutrients
- seedling
- reproduce
- pollen
- pollination

Chapter Introduction
- Direct attention to the chapter title and the picture of the clover flower. Provide time for the students to leaf through the chapter, looking at the headings and pictures. Discuss what they think the chapter will be about.

Use completed Looking Ahead, *Activities* page 77, to generate a discussion on what the students already know about soil, rocks, minerals, and fossils.

Big Question
- Ask a volunteer to read aloud the Big Question.
- Explain that the students will find the answer to the question as they read the chapter.

Big Question
Why is it important for plants to reproduce?

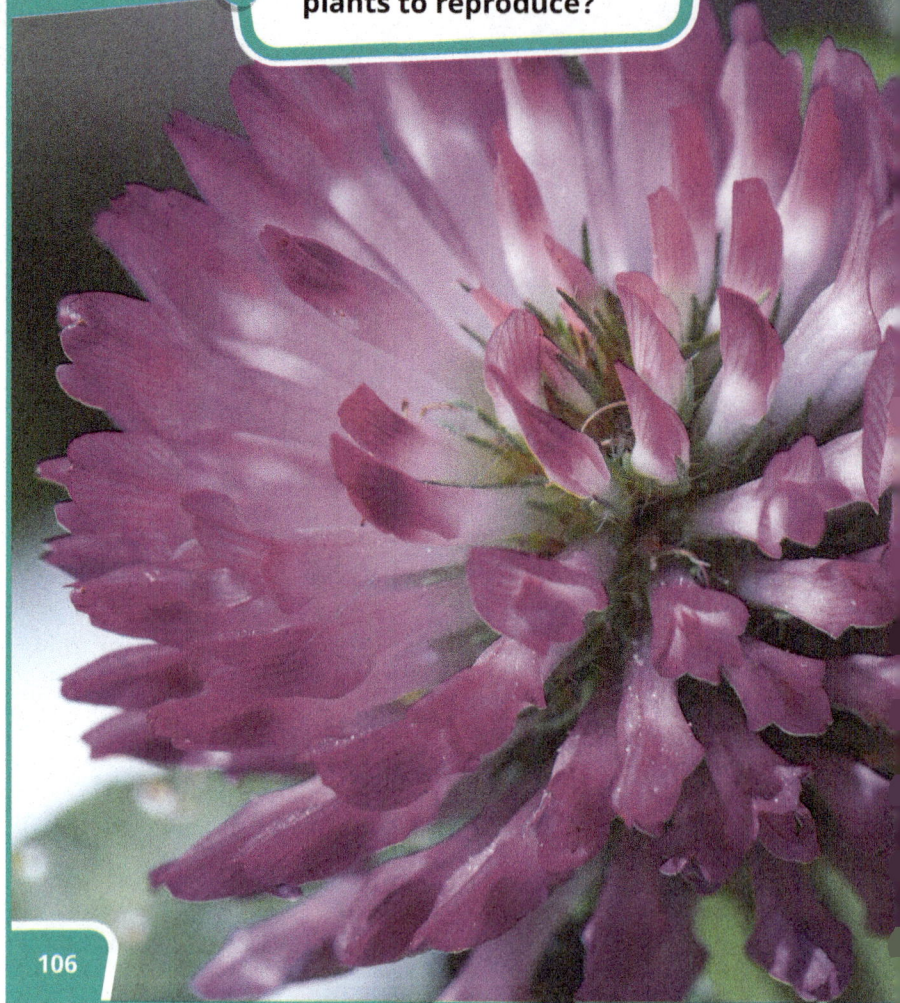

106

Visit TeacherToolsOnline.com for resources to enhance the lessons.

Chapter 5

Plants

If you take a walk in a forest, you might see animals that live there. Birds fly from tree to tree. Squirrels chase each other up and down tree trunks. Spiders spin webs between plants. Rabbits nibble on low bushes. All around these animals are plants. When God created the world, He filled it with many kinds of plants. People and animals use plants in different ways.

People and animals use energy to grow, to move, and to work. Trees and other plants also need energy. A plant is an organism that uses energy to grow and to make food.

Engage

- Display the *Sample Plants* visual. Ask volunteers to identify similarities and differences between the plants shown. Possible answers: the tree is tall; the cactus is short; some plants grow food; some have flowers.
- Point out that plants are more than a green, growing object. Explain that they have some important jobs to do. Today we will find out about an important job that all plants have.

Instruct

Preparation for Reading

- Direct the students to read page 107 silently to find out what people need to grow, move, and work.

Teach for Understanding

- Ask volunteers to describe living things that they have seen outside.

  How do people and animals use energy? to grow, move, and work

- Direct the students to the Glossary. Instruct them to locate the word *energy*. Ask a volunteer to read the definition aloud.

  Do plants need energy? yes

  How do trees and other plants use energy? to grow and make food

---

ⓘ Background

Chapter Photo

The photo on pages 106–7 is a red clover flower blooming. This type of clover is also called cowgrass.

Preparation for Reading

• Preview and pronounce the vocabulary terms *germinate*, *nutrients*, *seedling*, *reproduce*, *pollen*, and *pollination*.

• Direct the students to read pages 108–10 silently to find out what happens when a sunflower nears the end of its life cycle.

Teach for Understanding

• Display the *Life Cycle of a Plant* visual. Refer to the diagram as the "seed" and "growth" stages are discussed.

How do most flowering plants begin life? as a seed

What does a seed do when it *germinates*? It sprouts.

Name three conditions a seed must have to germinate. water, warmth, and air

As the seed germinates, what is the first thing to grow? a root

What direction does the root grow? down

What does the root do for the seed? It takes in water and nutrients from the soil.

What are *nutrients*? things in nature that help plants and animals live and grow

What part of the plant pushes up through the soil and grows above ground? the shoot

Plant Life Cycle

Germination

Some flowering plants begin life as a seed. A seed must have certain conditions met before it can **germinate**, or sprout. The seed must have water, warmth, and air. After these needs have been met, the seed will germinate.

Growth

As the seed germinates, a root begins to grow. As it grows, it pushes down into the soil. Roots take in water and nutrients from the soil. **Nutrients** are things in nature that help plants and animals live and grow. Next the seed sends out a shoot. The shoot is the part of the plant that will grow above the ground. The shoot pushes up through the soil.

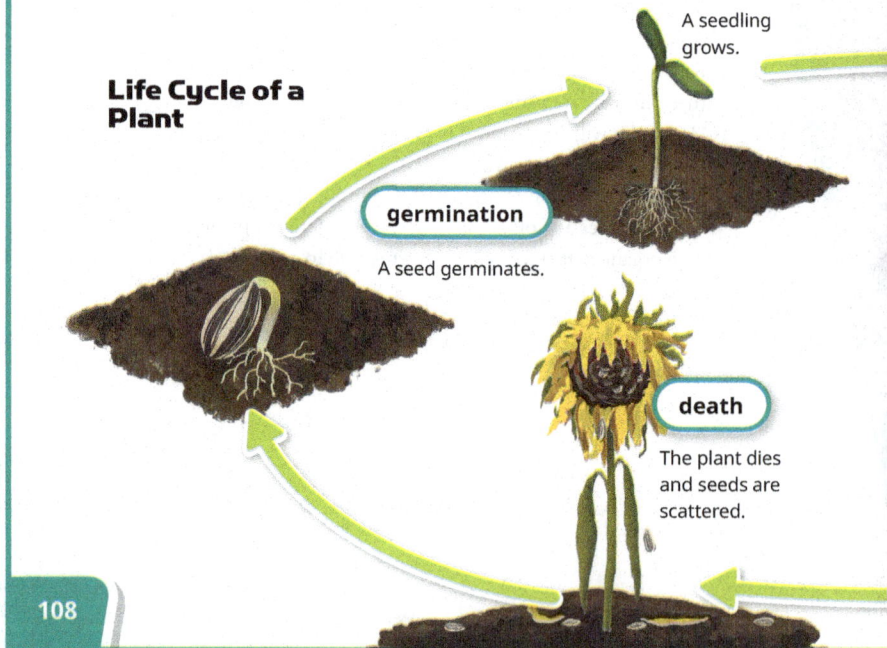

Life Cycle of a Plant

A seedling grows.

germination

A seed germinates.

death

The plant dies and seeds are scattered.

108

👥 Differentiated Instruction

Enrichment: Show How Stems Transport Water

Materials: food coloring, container, water, white carnation or celery stalk

Mix food coloring into a container of water until the water is a dark color. Cut the end off the bottom of the stem. Place the carnation or celery stalk in the colored water and let it stand in the water overnight.

Direct the students to describe changes they see in the color of the flower. Guide the students in drawing the conclusion that the colored water moved through the stem to the flower.

ⓘ Background

Arbor Day

National Arbor Day is observed on the last Friday in April, though some states celebrate earlier or later, depending on the climate. The holiday celebrates the beauty and uses of trees. It was started in Nebraska in 1872 by J. Sterling Morton.

When the new flowering plant first comes out of the soil, its leaves are not ready to make their own food. So the seed coat and stored food stay attached. The tiny plant uses the stored food in the seed coat until its leaves develop. When the leaves can make food for the plant, the remains of the seed coat fall off. The plant is now a **seedling**, or young plant. Soon more leaves begin to grow, and the stem grows thicker and taller. The stem moves water and nutrients to different parts of the plant. The seedling grows until it becomes an adult plant. As an adult plant, it will be able to produce its own seeds.

growth Flowers grow, and the plant becomes an adult.

An insect pollinates the plant.

reproduction

The adult plant produces fruit.

109

What part of the plant is needed for the tiny plant to make its own food? leaves

- Display the *Plant Parts* visual. Ask volunteers to name each part and to recall what each part does for the plant. *Note:* The plant parts and their functions were covered in SCIENCE 1 and SCIENCE 2. This material should be reviewed.

  Plant Parts Key:

  flower   part of many plants, makes seeds

  stem   holds the plant up, moves water and nutrients from the roots to other parts of the plant

  leaf   makes food for the plant

  roots   hold the plan in the ground, take in water and nutrients from the soil

  What does the young plant use for food until its leaves develop? the seed coat

- Direct attention to the *Parts of a Bean Seed* visual. Review each part. *Note:* The parts of a bean seed were covered in SCIENCE 2. This material should be reviewed.

  When do the remains of the seed coat fall off? when the leaves can make food for the plant

  When the seed coat falls off, what is the plant called? a seedling

  As the leaves begin to grow, what happens to the stem of the seedling? it grows thicker and taller

  What does the stem move to different parts of the plant? water and nutrients

  What does a seedling grow to become? an adult plant

  What will the adult plant be able to produce? seeds

ⓘ Background

Stems
In addition to transporting water and food, stems help hold up the leaves so that they get the light needed for photosynthesis.

Roots
Roots have small root hairs that absorb water from the soil.

What did God make each plant with the ability to do? reproduce

What does it mean to *reproduce*? to make more of the same kind of organism, or living thing

What stage in the life cycle of a sunflower does reproduction take place in? adult stage

What does the adult sunflower make in the reproduction part of the life cycle? pollen

What is *pollen*? the dust-like grains of a flower needed for the plant to reproduce

Why are bees and other insects attracted to the pollen in the sunflower? It is sweet smelling.

What is *pollination*? the movement of pollen from one part of the flower to another part

How do insects pollinate a flower? The insects come to a flower looking for nectar. Pollen is picked up by the insect's body, and it falls off the insect as it moves from one part of the flower to another part.

Why is pollination important? It is the only way that a flower can reproduce.

After pollination, what forms inside the flower? seeds

- Direct attention to the "Life Cycle of a Plant" on pages 108–9. Point out that fruit produced by an adult sunflower is the sunflower seed. Explain that not all plant seeds are the fruit, and that not all seeds can be eaten.

⚙ What is the fruit of an apple tree? apples

⚙ Where are the seeds of an apple? in the middle of the fruit

- Explain that while the seeds of a sunflower are safe to eat, the seeds of an apple should not be eaten in large quantities.

When a sunflower nears the end of its life cycle, what happens? The seeds are ready to make new sunflowers. The sunflower dries out, the seeds are scattered and fall to the ground, and the life cycle can begin again.

What do we say happens to a sunflower plant at the end of its life cycle? The plant dies.

- Display the *Life Cycle of a Plant* visual. Direct the students to identify, in order, the life cycle of a flowering plant. seed, growth (seedling), reproduction (adult plant with flowers and fruit), death

✓ Pollination is needed for a flower to reproduce.

Apply

Activities

Looking Ahead, page 77

This page assesses the students' knowledge prior to the chapter.

Reproduction

God made each plant with the ability to reproduce. **Reproduce** means to make more of the same kind of organism. An adult sunflower begins the reproduction part of its life cycle by making pollen. **Pollen** is the dust-like grains of a flower needed for the plant to reproduce. The sweet-smelling pollen in the sunflower invites bees and other insects.

Some insects come to a flower and take its nectar. Nectar is a sweet liquid that some plants produce. While the insects get nectar, their bodies brush against the plant's pollen. Some of the pollen falls off of the insect in different areas of the flower. Pollination is very important for every flower. **Pollination** is the movement of pollen from one part of the flower to another part. It is the only way that a flower can reproduce. Inside the flower, seeds begin to form. The flower holds the seeds.

Death

The sunflower nears the end of its life cycle. The seeds have grown. They are ready to make new sunflowers. The sunflower dries out, and then the seeds are scattered. When a seed falls to the ground, it can begin its life cycle.

✓ Why is pollination important for flowers?

110

ⓘ Background

Nectar

Nectar is the sweet liquid, produced by flowers, that many insects and some birds eat. The sugary liquid gives insects and birds the energy they need to live and grow.

Apple Seeds

Apple seeds have a cyanide-containing compound called *amygdalin*. Eating a large quantity of apple seeds (about 200) could be fatal.

Photosynthesis

When God created the world, He filled it with many kinds of plants. Plants use energy to grow and to make food. Food is important because it contains nutrients. Most plants get nutrients from the soil. People and animals get nutrients from the food that they eat. All living things need nutrients for energy.

God created plants with a special process, or way, to make the food they need. The process that plants use to make food is called **photosynthesis**. For photosynthesis to happen, a plant needs to take in sunlight, carbon dioxide, and water. **Carbon dioxide** is a gas in the air that a plant needs for photosynthesis. Plants then use these things to produce food and oxygen.

Basic Photosynthesis

Key
- light energy
- water
- carbon dioxide
- oxygen

111

Lesson **35** begins here.

Objectives
- Name three things plants need for photosynthesis.
- Name two products plants produce during photosynthesis.
- Identify where photosynthesis takes place in the plant.
- Sequence the steps of photosynthesis.
- Explain, using biblical teaching, why photosynthesis occurs. BWS
- Explain why God gave people plants. BWS

Teacher Resources
- Visuals 5.5–5.6: *Basic Photosynthesis; Photosynthesis Up Close*

Vocabulary
- photosynthesis
- carbon dioxide
- chlorophyll
- horticulturist

Engage

Think of how a car factory works. What are some of the materials needed to make a car? Possible answers: steel, glass, tires, carpet
- Point out that those materials are delivered and taken into the factory. In the factory the materials are sometimes changed, combined, and painted.

When the factory finishes using these supplies, what does it produce? cars

- Explain that the individual materials are put together to make something new. Point out that plants are similar to factories. They take in certain materials and produce, or make, things from those materials.

Instruct

Preparation for Reading
- Preview and pronounce the vocabulary terms *photosynthesis*, *carbon dioxide*, and *chlorophyll* and the word *oxygen*.
- Direct attention to the "Basic Photosynthesis" diagram. Point out that the labels of this diagram are in a key.
- Direct the students to read pages 111–12 silently to find out why there are different types of soil on the earth.

Teach for Understanding
Why do plants need energy? to grow and to make food
Where do most plants get the nutrients that they need? from the soil
Where do people and animals get nutrients from? from the food they eat
Why do all living things need nutrients? for energy
What name is given to the process that plants use to make food? photosynthesis
What is a process? steps in a certain order
- Discuss that a process is a way or set of actions to do something. Refer to the car factory in the lesson introduction and explain that the workers follow processes to make the cars.
- Discuss some processes familiar to students such as making a bed, feeding a pet, lining up in class, or handing in papers.
- Explain that the word *photosynthesis* is made of two word parts: *photo*, meaning "light," and *synthesis*, meaning "put together."
What words do you know that include the word *photo*? Possible answers: photograph, photocopy
- Display the *Basic Photosynthesis* visual. Discuss how to read the colored arrows using the key. Refer to the diagram as each part of the process is discussed.
What three things are needed for photosynthesis to happen? sunlight, carbon dioxide, and water
What is *carbon dioxide*? a gas in the air that plants need for photosynthesis
What do plants produce as a result of photosynthesis? food and oxygen
Why is photosynthesis important to plants? Plants make their own food through photosynthesis.
Why is the process of photosynthesis important to people and animals? Possible answers: Through the process of photosynthesis, plants make oxygen that is needed by people and animals; people and animals eat plants; plants provide people and animals with nutrients; the nutrients in plants provide people and animals with energy.

Where does photosynthesis take place? **in the leaves of plants**

What gives plants the energy needed for photosynthesis? **sunlight**

What color are the leaves of most plants? **green**

Where do plants get their green coloring from? **from chlorophyll**

- Direct attention to the photo and caption of the chlorophyll in plant cells. Point out the cell walls and the green chlorophyll.

What does chlorophyll do? **It makes the plant green; it helps the plant's leaves use energy from sunlight to produce food.**

What gas in the air do plants need for photosynthesis? **carbon dioxide**

How does a plant get carbon dioxide? **through tiny holes or openings (stomata) that open and close on the underside of plant leaves**

What are the tiny holes called? **stomata**

- Direct attention to the picture of the stomata. Explain that the picture is magnified and that plants can have 100 to 1000 stomata per square millimeter.

How does a plant get the water it needs? **Its roots absorb water.**

What part of the plant carries water from the roots to the leaves? **the stem, small tubes in the stem**

- Direct attention to the *Basic Photosynthesis* visual.

Why is the arrow representing sunlight pointing toward the top of the leaf? **because sunlight shines down on the leaf**

Why are the arrows representing carbon dioxide pointing toward the bottom of the leaf? **The openings (stomata) that allow carbon dioxide in are on the underside of the leaf.**

Why is the arrow representing water in the soil? **It represents the water in the soil that is absorbed by the roots.**

sunlight, carbon dioxide, water

What Plants Need

Photosynthesis takes place in the leaves of a plant. Leaves help the plant absorb, or take in, sunlight. The sunlight is the energy the plant needs for photosynthesis to happen.

Most plants have green leaves. The green coloring of a plant comes from **chlorophyll**. The chlorophyll helps the plant's leaves use energy from sunlight to produce food.

A plant also needs carbon dioxide for photosynthesis. Carbon dioxide is a gas that is part of the air. To take in the air that you need, you breathe with lungs. But a plant does not have lungs to breathe in air. God gave the plant tiny openings on the underside of its leaves. They open and close to allow carbon dioxide into the plant.

The third thing a plant needs for photosynthesis is water. The plant's roots take in water from the soil. Then small tubes carry the water through the stems to the leaves. There the water can help the plant make food.

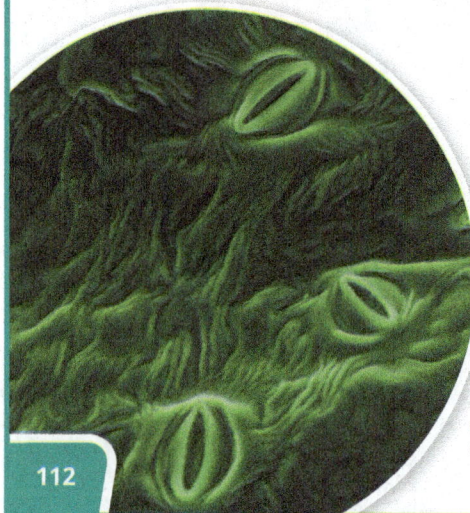

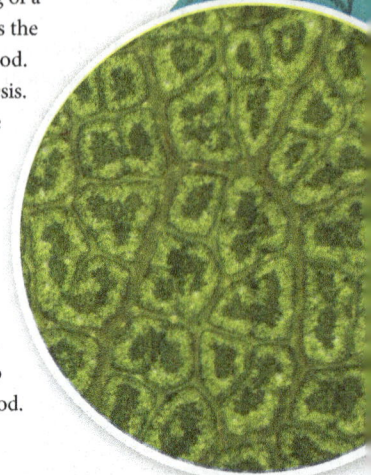

Tiny dots of chlorophyll are seen in this magnified view of a leaf.

What three things does a plant need for photosynthesis to happen?

Openings on the underside of a leaf are called stomata.

112

ⓘ Background

Chloroplasts

Chlorophyll is located in tiny organelles in green plant cells called chloroplasts. Photosynthesis for the entire plant takes place in the tiny chloroplasts located in the plant cells.

Stomata

The stomata (sing. *stoma*) are openings between the guard cells. The stomata allow gas and water vapor to pass in and out of the leaf. The guard cells, one on either side of a stoma, open and close to control the loss of excess water vapor.

Fantastic Facts

In the fall you may say that the leaves are changing color, but they are actually just showing less green. Some plants, such as maple trees, stop making chlorophyll when temperatures begin to get cold. Without chlorophyll the leaves begin to lose their green color. Other colors start to show. The leaves are then red, orange, or yellow.

What a Plant Produces

Once a plant has sunlight, carbon dioxide, and water, it continues the process of photosynthesis to produce food and oxygen. The chlorophyll in a plant uses energy from sunlight to produce food.

The food that a plant makes is a type of sugar. Tiny tubes carry the sugar to all parts of the plant. The sugar becomes food for the plant and gives it energy to grow. Plants make food for their own energy. They use this energy to grow. But sometimes a plant makes more food than it needs. That extra food is stored in the plant to be used later.

Key

- **light energy** (yellow)
- **water** (blue)
- **carbon dioxide** (green)
- **oxygen** (gray)
- **sugar** (purple)

113

Preparation for Reading

- Direct the students to read pages 113–15 silently to find out what one of the main uses of plants is.

Teach for Understanding

- Discuss "Fantastic Facts."

  What causes the leaves of some trees to show colors in the fall? In the fall, some trees stop making chlorophyll. Without the green of chlorophyll, other colors in the leaves show.

- Display the *Photosynthesis Up Close* visual. Refer to the visual during the discussion.

  What does the plant make using sunlight, carbon dioxide, and water? food and oxygen

  What food does the plant make? a type of sugar

  What carries the sugar to all the parts of the plant? tiny tubes

  What does the sugar provide for the plant? food; energy

  What does the plant use the energy for? to grow

  What does the plant do with the extra food it makes? stores it

What gas does the plant produce and release during photosynthesis? oxygen

Why is oxygen important to animals and humans? They need to breathe oxygen to survive.

What part of the plant allows oxygen to go out of the plant? the tiny holes (stomata) on the underside of the leaves

- Direct attention to the close-up of the stomata on page 112.

Which gas is taken in by the leaf? carbon dioxide

Which gas is released by the leaf? oxygen

- Point out that carbon dioxide is taken in and oxygen is released through the same stomata.

What is the boy in the picture on page 114 doing? breathing out

What gas do people and animals breathe out? carbon dioxide

Does the breathing out of carbon dioxide by animals and people help or hurt plants? helps, because plants need carbon dioxide for photosynthesis

Why does photosynthesis happen? Possible answers: God created plants with this process to maintain life on Earth; God made everything in nature to work together.

How did God create plants, animals, and people to work together? Plants use photosynthesis to make food for themselves. The stored energy in plants becomes food for animals and people. Photosynthesis also provides oxygen that animals and people need to survive. Plants use the carbon dioxide that animals and people breathe out to make food during photosynthesis.

- Discuss "Health and Safety."

Name four plants that can be used for medicine. foxglove, aloe, witch hazel, ginger

What can the medicine that we get from foxglove be used to treat? heart problems

What can aloe be used for? to treat minor burns and scrapes

What can you treat with witch hazel? small skin cuts

What is ginger used for? to treat an upset stomach

- Point out that the picture in the "Health and Safety" box is ginger root and the other picture is aloe.

While foxglove has medicinal value, it is also toxic. Children should not touch or eat the foxglove plant. Ingesting the plant can be fatal.

Every plant that goes through photosynthesis produces oxygen. Like carbon dioxide, oxygen is a gas that is part of the air. A plant takes in carbon dioxide and releases, or lets out, oxygen. Carbon dioxide enters a plant through tiny holes in the leaves. The same tiny holes release oxygen into the air. The plant's survival and growth serve animals and people by providing oxygen. People and animals breathe in oxygen and breathe out carbon dioxide. This is part of God's plan for His creation. He designed living things to work together.

Health and Safety

Some medicines are made from plants. A common heart medicine comes from a plant called foxglove. Aloe is a plant that helps to treat minor burns or scrapes. People sometimes treat small skin cuts with liquid from the witch hazel plant. Some plants, such as ginger, can calm upset stomachs.

114

ⓘ Background

Foxglove

Foxglove is no longer used as a heart medicine by itself because the plant is poisonous when too much is used. However, it is an ingredient in a group of pharmaceutically prepared heart medicines called digitalin.

Meet the Scientist

George Washington Carver (1864–1943) was interested in plants from an early age. Carver wanted to learn. He worked hard to get schooling wherever he could.

In the 1890s insects were destroying cotton, the South's main crop. Carver suggested planting peanuts instead of cotton. He became famous for finding hundreds of uses for the peanut plant. In all his work, he gave glory to God as Creator of all things.

Uses of Plants

God gave people the job of managing plants. He gave us plants to use and even enjoy. People use and enjoy plants in different ways in God's world. We can eat many kinds of plants. We use plants to make things. We can see the beautiful colors of the plants around us. We can enjoy the scent of flowers. Plants are a wonderful gift from God.

One of the main uses of plants is food. God created plants to make their own food. He also created them to store food and nutrients for other living things to use. The Bible tells us in Genesis 1:29-30 that God gave people and animals plants for food. When people and animals eat plants, they receive nutrients for energy.

What two things does a plant produce during photosynthesis?

115

- Discuss "Meet the Scientist."
  When was George Washington Carver born? in 1864
  Why did Carver suggest planting peanuts instead of cotton? Cotton was being destroyed by insects.
  What did Carver become famous for? finding hundreds of uses for the peanut plant
  What was Carver's attitude toward God? He gave God glory as the Creator of all things.
- Discuss uses of plants.
  Why did God give people plants? to take care of them, to use and enjoy them
  What can people make out of plants? Possible answers: paper, furniture, houses, clothing
  How can people enjoy plants? Possible answers: looking at their beautiful colors, enjoying the scent of flowers, enjoying the way they taste
- Direct the students to Genesis 1:29–30 in their Bibles. Instruct them to follow along as you read the verses aloud.
  According to Genesis 1:29–30, what is the main purpose that God gave people and animals plants? for food
  What do people and animals receive from plants? nutrients
  What do the nutrients give people and animals? energy
  What are some plants that you like to eat? Possible answers: apples, oranges, grapes, corn, wheat (bread), peanut butter

sugar and oxygen

① Background

Plants for Fabrics

Cotton is not the only plant from which fabric is made. Cotton is used more often because it costs less to harvest cotton fibers than other plant fibers.

Linen is another fabric that is made from a plant. Linen is made from the flax plant. The stems are soaked in water until the fibers can be separated. The fibers are separated and spun into thread.

Fibers from the hemp plant are often used to make ropes and canvas, which is used for sails and tents.

ᑫ Differentiated Instruction

Remediation: Identify Objects Made from Plants

Materials: wooden spoon, straw basket, paper bag, cotton dishtowel, rope, toothpicks, metal spoon, penny, rock, seashell, other items as desired

Display the items and allow the students to decide which items were made from plants.

Preparation for Reading

- Preview and pronounce the vocabulary term *horticulturist* and the word *ornamental*.
- Direct the students to read page 116 silently to find out what a horticulturist is.

Teach for Understanding

- Direct attention to the pictures.

  What is a *horticulturist*? a person who grows and takes care of plants

  What kinds of plants might a horticulturist grow and take care of? fruits, vegetables, ornamental plants

  What does the word *ornamental* sound like? ornament

  Where might you hang ornaments at Christmas? on a Christmas tree

  Why are ornaments hung on a Christmas tree? to decorate it

- Point out that the term *ornamental* means *decorative*.

  What does a horticulturist need to know about plants? Answers should include some of the following: how best to take care of them; the needs of plants; which temperatures are good for growth; what season is best for planting; which soil is best; how much water is needed; the best ways to prune, or cut, a plant.

  Why does a horticulturist make sure plants are safe from insects? Insects can harm plants.

  Where do horticulturists work? gardens, schools, labs, greenhouses

  What is a greenhouse? a building made just for growing plants

- Point out that a greenhouse is often made of glass and has a constant temperature. The environment of a greenhouse allows plants to grow well indoors.

Apply

Activities

Study Guide, pages 79–81

These pages review the concepts taught in Lessons 34–35. After completion, direct the students to keep them for review for Test 5.

Assess

Quiz 5A

The quiz may be given at any time after completion of this lesson.

STEM Career

Horticulturist

A **horticulturist** is a person who grows plants and cares for plants. He might grow vegetables and fruits. He might also grow ornamental, or decorative, plants. Horticulturists know a lot about plants.

Horticulturists can recognize many kinds of plants. They study and find out how best to take care of plants. They know about the needs of plants. They know which temperatures are good for plant growth. They know which time of year or season is best for planting certain plants. They know what kind of soil is best for each plant. They also know the amount of water that certain plants need.

Horticulturists know about the best ways to prune, or cut, certain plants so that they can grow again. They mix liquids to help plants grow or to stop plants from growing. They help other plants by killing weeds.

Horticulturists can tell whether a plant is healthy or sick. They know the kinds of insects that can harm plants. They know where certain insects can be found on plants. They also make sure plants are safe from insects and other pests.

Horticulturists work in many places. Some work in gardens, schools, and labs. Others might work in greenhouses, or buildings made just for plants.

116

INVESTIGATION

A Place to Grow

Name _____

Problem

How does light affect plant growth?

Hypothesis

The plant in the light will grow **shorter, taller** than the plant in the dark.

Procedure

☐ 1. Label one cup *Light* and the other cup *Dark* with a marker and masking tape.

☐ 2. Put the same amount of potting soil in each cup.

☐ 3. Plant 2 of the same kind of seed in each cup. Cover the seeds with soil.

☐ 4. Carefully water the bean seeds in each cup with the same amount of water.

☐ 5. Place the Light cup where it will get sunlight.

☐ 6. Place the Dark cup in a place where it will get no light.

☐ 7. Water the plants as needed to keep the soil moist.

☐ 8. Observe the cups. Record on the data chart the date when you observe a plant in each cup germinate.

☐ 9. Measure each plant every day for 7 days. Carefully place the end of the ruler on top of the soil. Measure the height of the plant. Be careful that you do not pull the plant out as you measure it.

☐ 10. Observe the color and the leaves of each plant.

☐ 11. Record your measurements and observations in the data chart under Observations.

☐ 12. Write the dates of your observations on the bar graph.

Inquiry Skills

- Predict
- Observe
- Measure

Materials

☐ 2 plastic cups, 266 mL (9 oz)
☐ marker
☐ masking tape
☐ potting soil
☐ 4 bean seeds
☐ water
☐ centimeter ruler

SCIENCE 3 *Activities*

⚠ **Helps**

Seed Choice

Bean seeds will work best since they have a short germination period. Check the seed packets to find ones with a germination that best fits your schedule.

INVESTIGATION Lesson 36

Inquiry Skills
- Predict
- Observe
- Measure

A Place to Grow

Plants make food to supply their own energy. They make this food through the process of photosynthesis. They use energy from the food to grow. Would it be possible for plants to grow without light?

In this Investigation, you will compare two plants. You will find out whether the amount of light they receive affects their growth.

117

Objectives

- Predict the effect of sunlight on plant growth.
- Observe and measure the height of plants.
- Record and graph data on a bar graph.
- Draw a conclusion about the effects of sunlight on plant growth.

Materials

- houseplant
- See *Activities* page 85.

Engage

- Display the houseplant.
- What is needed for this plant to grow? Possible answers: water, light, carbon dioxide
- Do you think this plant will continue to grow if you put it in a dark closet? Accept any answer.
- Explain that in this Investigation the students will conduct an experiment to find out what happens to a plant that does not receive light.

Instruct

Preparation for Reading

- Direct the students to remove pages 83–87 from their *Activities* books.
- Direct the students to read page 117 and *Activities* pages 83–87 silently before beginning. *Note:* Reading should include the Measure Up: Length on *Activities* page 83. Students should place this page in their notebooks.

Teach for Understanding

- Direct attention to and discuss the **Inquiry Skills** of Predict, Observe, and Measure. If necessary, refer to the Science Inquiry Skills on *Activities* page 51. This page should be in the students' notebooks.
- Review the **Materials** list.
  What measuring tool are you using in this activity? centimeter ruler
  What will you be measuring? the height of the plants
- Direct attention to the Measure Up: Length page.
- What is the length of the pen? 14 cm
- What is the length of the pencil? about 15 cm
- Discuss the importance of measuring carefully to make accurate comparisons. As needed, provide time for the students to practice their measuring skills with a centimeter ruler.

Apply

- Choose a student to read the **Problem** aloud.
- Direct the students to complete the **Hypothesis** to answer the problem. Instruct the students to circle the answer.
- Help the students understand that they are testing the amount of light. They need to make sure all the other conditions for both cups are the same to have a controlled investigation.

- Instruct the students to follow and check each step of the **Procedure** as it is completed.
- Direct the students to continue to follow the steps of the Procedure and to record the data in the **Observations**.
- Emphasize the importance of carefully measuring the height of each plant.
- Plan to have the students check the cups at the same time each day. Encourage them to look for details about the color and appearance of each plant as well as the number of leaves it has. Observations should be recorded on the "Observations" chart.
- Explain that it is important to use the same amount of water for each plant. Water the plants as needed throughout the activity, but caution students against over watering.

☐ 13. Fill in the Graph Key with a light color for the Light cup measurements and a dark color for the Dark cup measurements.

☐ 14. Graph your results.

Observations

Day	Light Cup		Dark Cup	
	Height	Observations	Height	Observations
1	___ cm		___ cm	
2	___ cm		___ cm	
3	___ cm		___ cm	
4	___ cm		___ cm	
5	___ cm		___ cm	
6	___ cm		___ cm	
7	___ cm		___ cm	

INVESTIGATION

A Place to Grow

Name _____

Plant Growth

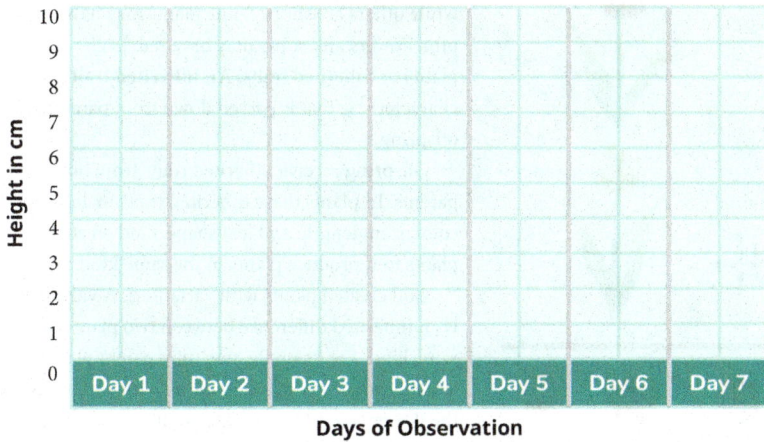

Height in cm

| 10 | 9 | 8 | 7 | 6 | 5 | 4 | 3 | 2 | 1 | 0 |

Day 1　Day 2　Day 3　Day 4　Day 5　Day 6　Day 7

Days of Observation

Graph Key

▢ **Light cup**　　▢ **Dark cup**

Conclusions

1. Which plant grew taller?　○ the plant in the Dark cup　● the plant in the Light cup

2. Did your results support your hypothesis?　○ yes　○ no

3. In what other way are the plants different? _Possible answer: The plant in the dark has yellow leaves while the plant in the sunlight has green leaves._

4. What caused the difference? _Possible answer: The plants were unable to make food in the dark. Plants need light to make food and grow._

5. How do your results affect where you would plant a garden? _Possible answer: I would plant a garden in an area that gets sunlight to help my plants make food and grow._

SCIENCE ③ *Activities*

• Guide the students as they graph the height of the plants in centimeters.

• Discuss how graphs provide a visual way to view the differences. You may choose to create a class graph showing the combined results.

• Direct the students to draw **Conclusions** using their observations. The conclusions may be done individually or with their science group.

• Guide a discussion of the students' conclusions.

Activities

Measure Up: Length, page 83

This page was discussed in the lesson and should be kept in the students' science notebooks.

Investigation, pages 85–87

Students will predict how the amount of light a plant receives affects its growth.

Assess

Rubric

Use the prepared rubric or design a rubric to include your chosen criteria.

---

① **Background**

Darkness and Growth

For most seeds, the plant in the dark will die once the stored food in the cotyledon is used up. Darkness causes the cells in plant stems to grow rapidly, but it inhibits the growth and development of leaves. Light causes the stem to straighten and the growth to slow. The leaves develop, and the plant begins to make its own food. Without light, the stems are pale and weak. Leaves will be small and pale or nonexistent.

Objectives

- Explain why plants have different traits.
- Compare and contrast different views about plant adaptation. **BWS**

Materials

- 2 cut flowers, same kind but different colors
- world map

Vocabulary

- inherited trait
- variation

Engage

- Display the cut flowers.

  How are these flowers alike? Possible answer: Both have a stem, leaves, and petals.

  How are these flowers different? Possible answer: different stem length, number of leaves, color of petals

- Explain that today the students will learn about why the same kinds of plants can look different.

Instruct

Preparation for Reading

- Preview and pronounce the vocabulary terms *inherited trait* and *variation*.
- Direct the students to read page 118 silently to find out why there are variations in plants.

Teach for Understanding

Why do all plants not look alike? Plants inherit traits from their parent plants.

What is an *inherited trait*? a characteristic that is passed down from parents to offspring

What are some inherited traits that offspring plants can receive from their parent plants? Possible answers: color, stem length, leaf shape

How were plants created by God to reproduce? God created plants to reproduce plants of the same kind.

- Display the cut flowers.

  What inherited traits can you observe by looking at the cut flowers? Possible answers: color, stem length, leaf shape

  Note: If the flowers have a scent, ask volunteers to compare the scents of both flowers.

  What did God create plants with? variations

  What is a *variation*? an inherited difference between two plants of the same kind

- Direct attention to the diagram of the tulips.

  Why are the plants all the same kind? Each plant received inherited traits from their parents to become tulips.

- Direct attention to the offspring.

  What inherited traits did the offspring receive? Possible answers: stem length, leaf shape, color

  What variation is the orange offspring showing? color variation

---

Lesson 37

parent

parent

+

offspring

Plant Traits

Not all plants look the same. Some are tall while others are short. Some plants may have pink flowers and some may have red. Each plant has inherited traits. An **inherited trait** is a characteristic that is passed down from parents to offspring.

Offspring receive inherited traits from both parents. In plants these inherited traits include color, stem length, and leaf shape. God created plants to reproduce plants of the same kind.

God created plants with variations. A **variation** is an inherited difference between two plants of the same kind. For example, two tulip parent plants can reproduce and have offspring with variations of the flower color. One of the parents may have red flowers. The other parent may have yellow flowers. Their offspring could include tulips with red flowers, yellow flowers, or orange flowers.

Meet the Scientist

Gregor Johann Mendel (1822–1884) was an Austrian scientist and a monk. He studied pea plants. He discovered that plants inherit traits from their parent plants.

✓ **What is an inherited trait?**

Why are the tulips different colors? Each parent plant received an inherited trait for color from its parents. Each parent, the red tulip and the yellow tulip, passed on traits to their offspring. The orange tulip inherited a color variation from the parent plants.

- Discuss "Meet the Scientist."

  Who was Gregor Mendel? Mendel was an Austrian monk and scientist.

- Point out that Austria is a country in Europe near Germany. Display a world map and ask a volunteer to locate Austria on the map.
- Explain that Mendel was a monk in the Roman Catholic church.

  When did Gregor Mendel live? 1822–1884

  What did Mendel study? pea plants

  What did he discover? that plants have inherited traits

✓ **a characteristic that is passed down from parents to offspring**

Adaptation

Remember that an adaptation is a characteristic that helps a plant to survive in its environment. The characteristic is passed down from parents to offspring. There is a biblical view and there is an evolutionary view of adaptation. Both views believe that adaptation occurs. Variations are adaptations that sometimes produce useful traits. Parent plants pass on variations to their offspring. When a plant adapts, its offspring receives the inherited traits. These inherited traits can help plants survive in their environment.

A Biblical View of Adaptation

Scientists with a biblical worldview believe that God created plants. He created them with the ability to adapt. Parent plants do not change their traits. But they do pass on variations to their offspring. When a plant kind adapts, the offspring receive the inherited traits. These inherited traits help them to survive.

Root length is an example of a variation. Plants of the same kind may have long roots or short roots. Plants can adapt to their environments. They do this by passing down traits that help their offspring survive. In a dry environment, seedlings with short roots would not survive well. The seedlings with long roots would have a better chance of surviving. This is because the seedlings with long roots can reach water that is farther away in the soil. The seedlings with short roots would not be able to absorb the water they need. They would then die. The plants with long roots would make more plants with long roots. Soon all plants in the environment would have long roots.

119

Preparation for Reading

- Direct the students to read pages 119–20 silently to find out what an evolutionist believes about where apples came from.

Teach for Understanding

What is an adaptation? a characteristic that helps a plant to survive in its environment

What are two views of adaptation called? a biblical view and an evolutionary view

How do plants adapt? Parent plants pass on variations to their offspring; the offspring inherit variations.

What do scientists with a biblical worldview believe about the way God created plants? God created plants with the ability to adapt.

Can plants change their traits? No, but they can pass on variations to their offspring.

- Direct attention to the picture.

What kind of environment is pictured? dry environment

What adaptation has helped the cactus to survive in its dry environment? It has inherited long roots to absorb the water it needs to survive and grow.

- Point out that while a cactus has long roots, they are shallow and spread out around the plant. So little rain falls in the desert that it does not normally soak deeply into the ground. The shallow roots allow the cactus to absorb the water it needs from the little rain that falls in its dry environment.

Preparing Ahead

Lesson 45 Materials

Look ahead to the Materials list in Lesson 45 for the Investigation activity. You will need to purchase mealworms and supplies for the living space.

How is an evolutionary view of adaptation similar to a biblical view of adaptation? Scientists who believe evolution believe that an adaptation helps a plant to survive in its environment. They believe that variations in plants are passed down from parents to offspring.

How is an evolutionary view of adaptation different from a biblical view of adaptation? An evolutionary view of adaptation believes that, after millions of years, one kind of plant can evolve into another kind of plant. A biblical view believes God created each kind of plant and that one kind does not evolve, or change, into another kind of plant.

- Direct attention to the diagram at the bottom of the page.

According to evolution, where did apple trees come from? They evolved from a rose plant. The rose plant made many small changes over millions and millions of years. A scientist who believes in evolution thinks the rose adapted into the apple tree.

Who made each kind of plant, including roses and apples? God

How did God make each plant? Possible answers: God made plants with the ability to adapt; one kind of plant does not change into another kind of plant.

✓ Adaptation is a characteristic of a living thing that helps the living thing to survive in its environment; the characteristic is passed down from parents to offspring.

Apply

Activities
Study Guide, pages 89–90
These pages review the concepts taught in Lesson 37. After completion, direct the students to keep them for review for Test 5.

Assess

Quiz 5B
The quiz may be given at any time after completion of this lesson.

An Evolutionary View of Adaptation

Scientists who believe evolution describe adaptation in a similar way as those who have a biblical view. However, scientists who believe evolution is true also say that, after millions of years, adaptation can make one plant kind evolve, or change, into another plant kind. For example, evolution tries to explain that apple trees evolved from a rose plant by making small changes over millions and millions of years. A scientist who believes evolution is true does not think that God created roses to make only more roses. Remember that God made each kind of plant. Each plant has the ability to adapt. But one kind of plant does not change into another kind of plant.

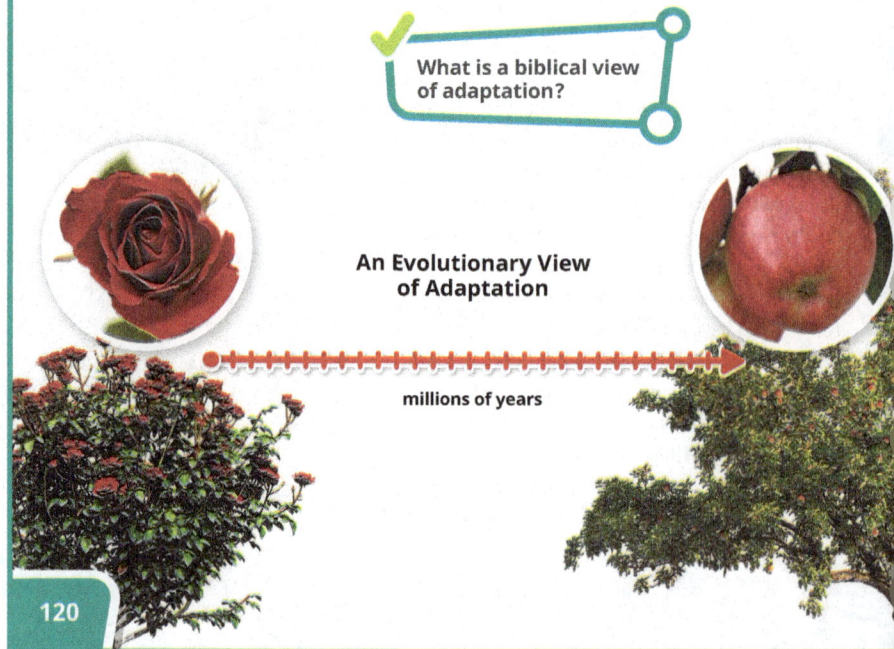

✓ **What is a biblical view of adaptation?**

An Evolutionary View of Adaptation

millions of years

120

STEM

Lesson 38

Vacation Water Challenge

On the last day before summer break, you receive a plant from your teacher. You later find out that you and your family are going on vacation for four weeks. You are excited to go on vacation. But your big brother tells you that your plant will probably die. No one will be home to water it. You do not know what to do. Your plant must have water or it will die. What can you do to keep your plant alive?

In this STEM activity, you will design a way to water your plant. You will create a system that will water your plant while you are on vacation.

121

Lesson **38** begins here.

Objectives

- Plan and design a way to water a plant, using the engineering design process.
- Create a model to water a plant.
- Test and compare models to improve the original design.
- Communicate to others how the design solves the problem.

Materials

- potted house plant or a container of potting soil to mimic environment of a potted house plant
- water bottles
- plastic bags
- plastic wrap
- cotton string
- nylon rope
- rubber bands
- water
- ice cubes
- paper towels

The Materials list includes items that may not work as well to solve the problem as other materials. This will allow the students the opportunity to test their designs and to find the best solution. You may add items of your choosing to the Materials list.

Teacher Resources

- Visual 2.7: *STEM: The Engineering Design Process*

Engage

- Poll the class to find out how many students have ever gone on vacation.
- Allow the students to share ways that they prepared for their vacation. Answers may include: packed a suitcase; found someone to give food and water to a pet; boarded a pet at the vet's office; watered their plants.
- Explain that today the students will learn more about ways to provide water for an indoor plant for a long period of time.

Instruct

Preparation for Reading

- Direct the students to remove *Activities* pages 91–92. Direct the students to read page 121 and *Activities* pages 91–92 silently before beginning.

Teach for Understanding

- Direct attention to the picture.
- What clue in the picture tells you that the family is going on vacation? There are suitcases packed and waiting to be put in the car.

  How long is the family going to be gone on vacation? 4 weeks

  What do the plants in the picture need to survive and grow for four weeks? Answers should include: air, water, light, and space.

  What will happen to a plant that does not get water for four weeks, or one month? Possible answers: It may die; it will not survive and grow; it will wilt.
- Display the *STEM: The Engineering Design Process* visual or direct the students to the copy in their science notebooks.
- Explain that in this STEM activity the students will design a way to water an indoor plant while they are away on a long vacation.

Apply

- Direct attention to *Activities* pages 91–92.
- Ask a volunteer to identify the *Problem*.
- For this activity, each student may design his own solution to the problem, or the students may work in science groups to collaborate as they design a solution.
- Instruct the students to *Imagine* a solution to the problem and *Plan* a design.

> Set out enough materials for each student or group to design a solution to the problem. Encourage the students to choose materials that they believe will best solve the problem of getting enough water to a plant to keep the plant alive for four weeks.

- Direct attention to the materials you have set out. Explain that the students are to list the materials they will use and they are to gather the materials from those provided.

STEM

Vacation Water Challenge   Name _____

Ask

What is the problem?
   I will be going on vacation for four weeks. No one will be home to water my plant. The plant must have water, or it will die. I will need to create a system that will water my plant while I am on vacation.

What kind of system can I design to water my plant?

Imagine

Think about possible answers to the problem.
   Possible designs could include a water bottle and a rope or cotton string, and a resealable

   plastic bag and thread. The design must be able to water the plant in small, steady amount

_____

Plan

Draw and label your design. List and gather materials.

Accept any reasonable design.

Design	Materials

Create

Follow the plan and make a model.

Test and Improve

What changes need to be made to the design?

Accept any reasonable answer.

_____

_____

Improved Design

Share

Tell others how your design solves the problem.

Lesson 38 • Page 92
STEM

- Direct the students to *Create* a model of their design.
- Instruct the students to *Test* their model using the potted house plant or the container of potting soil.
- Discuss what would happen if the designed system allows the water to flow out too quickly or too slowly.
- After testing their designs, direct the students to compare models and *Improve* their design.
- Invite the students to *Share* their designs with a partner or another science group. Instruct the students to explain how their design solves the problem.

Activities

STEM, pages 91–92

The students will design a way to water an indoor plant.

Assess

Rubric

Use the prepared rubric or design a rubric to include your chosen criteria.

Objective

- Recall terms and concepts from Chapter 5.

Apply

Review

- Material for Test 5 will come from the Study Guides on *Activities* pages 79–81 and 89–90, as well as Quizzes 5A and 5B. You may review any or all of the material during the lesson.
- You may choose to review Chapter 5 by playing the Photosynthesis review game or a game from the Game Bank (Teacher Resources).

🏆 **Review Game**

Photosynthesis

Materials: *Needs Cards* and *Food Cards*, on cardstock (Instructional Aids 5.1–5.2)

Shuffle together the sunlight, carbon dioxide, and water cards and place these needs cards face down in a pile. Place the food cards face up in a separate pile. Group the students into teams.

For each correct answer, the team member draws a needs card. Once the team collects a set of the three needs cards (sunlight, carbon dioxide, and water), they should return the set to the bottom of the pile in exchange for one food card. The team with the most food cards at the end wins.

NOTES

NOTES

_____

_____

_____

_____

_____

_____

_____

_____

_____

_____

_____

_____

_____

_____

_____

_____

_____

_____

_____

_____

_____

_____

_____

_____

_____

_____

_____

_____

_____

_____

Objective

- Apply terms and concepts from Chapter 5.

Assess

- Administer Test 5.

Chapter Objectives

- Analyze the characteristics that scientists use to classify animals.
- Describe the characteristics that cold-blooded animals use for survival and growth.
- Explain how cold-blooded animals have unique and different life cycles yet all have birth, growth, reproduction, and death in common.
- Explain, using biblical truth, the steps of the life cycle of cold-blooded animals.
- Explain, using biblical truth, why it is important to care for cold-blooded animals.
- Apply the science inquiry skills of infer, observe, and predict to the characteristics of cold-blooded animals.

Lesson Objectives

- Identify how scientists classify animals.
- Differentiate between invertebrates and vertebrates.
- Identify the characteristics of warm-blooded and cold-blooded animals.
- Differentiate between warm-blooded and cold-blooded animals.
- Identify what an animal behavior scientist does.

Materials

- assortment of pencil erasers

Teacher Resources

- Visual 6.1: *Animal Classification 1*

Vocabulary

- vertebrate
- invertebrate
- warm-blooded
- cold-blooded
- animal behavior scientist

Chapter Introduction

> Use completed Looking Ahead, *Activities* page 93, to generate a discussion on what the students already know about cold-blooded animals. You may want to direct the students to complete the "Notebook Connection" at the bottom of the page at this time.

- Direct the students to turn to the Contents page in their *Student Edition* and locate Unit 2.
- Direct the students to locate the page for Chapter 6 and turn to the beginning of the chapter.
- Instruct the students to look at the picture on the opening pages of the chapter.
  What do you see in the picture? An alligator sunning itself.
  Why do you think the alligator is in the sun? Accept any reasonable answer. It is trying to get warm.

Big Question

What are the characteristics of cold-blooded animals?

122

(i) Background

Chapter Photo

The photo on pages 122–23 is an alligator sunning itself on driftwood in Lake Martin in Alabama.

> Visit TeacherToolsOnline.com for resources to enhance the lessons.

Chapter 6

Cold-Blooded Animals

God made our world and everything that is in it (Genesis 1). He made the land and the water. He made many kinds of organisms. All plants and animals are organisms God created. Last of all, God made people. He named the first man Adam. The Bible tells us in Genesis 1:31 that "God saw every thing that he had made, and, behold, it was very good."

God brought all the animals to Adam. He gave Adam the task of naming the animals. So Adam gave names to the cattle, the birds in the air, and all the other animals (Genesis 2:19–20). Would you like the task of naming all the animals? Have you ever named a pet?

123

Big Question

- Direct attention to the Big Question. Ask a volunteer to read aloud the question.
- Ask the students to look at the chapter to find out what it is about.
  What do you think this chapter is about? As you give your answer, tell what made you think that. Possible answers: amphibians, reptiles, spiders, insects, characteristics of animals

Engage

- Display the assortment of erasers.
- What does it mean *to classify*? to put into groups
- Direct the students to collaborate with their science partner. Ask them to think of different ways the erasers can be classified. color, size, shape
- Allow the students to share ways to classify the erasers and to try several classifications with the erasers.
  Today you will study about ways to classify animals.

Instruct

Preparation for Reading

- Direct the students to read page 123 silently to find out what God said about His creation.

Teach for Understanding

What is another name for plants and animals? organisms

What did God say after He created everything? "It was very good."

Who did God bring the animals to? Adam

- Read Genesis 2:19–20 aloud and discuss Adam's job.
- Why do you think animals need names? Possible answers: to tell them apart, to describe them to other people

Would you like the task of naming all the animals? Have you ever named a pet?

⚠ Helps

Class Pet

Consider having a class pet. Interaction with an animal provides many teaching opportunities. Direct the students to observe the animal's characteristics and behaviors. You may want to have a cold-blooded animal such as a frog, toad, turtle, or lizard as a pet during Chapter 6. Then have a warm-blooded animal such as a hamster, gerbil, mouse, or guinea pig during Chapter 7.

↪ Preparing Ahead

Lesson 48 Materials

Look ahead to the materials list in Lesson 48 for the Exploration activity. You will want to hang the bird feeder now so the birds will find the feeder and begin coming to it by Lesson 48.

Lesson 52 Materials

Look ahead to the materials list in Lesson 52 for the Exploration activity. You will need to purchase fur, feathers, and petroleum jelly (thick shortening, or butter).

Preparation for Reading

- Preview and pronounce the vocabulary terms *vertebrates*, *invertebrates*, *warm-blooded*, and *cold-blooded* and the words *classify* and *gecko*.
- Direct attention to the big heading on page 124.

  What will you be reading about? how to classify animals
- Direct the students to read pages 124–25 silently to find out how scientists classify animals.

Teach for Understanding

How have some kinds of animals received the names they are called today? from scientists

- Refer to the activity with erasers from the lesson introduction.

  Think about the erasers. What did you do with the erasers? classified or grouped them

  Scientists do the same thing with animals. What do scientists look at when classifying animals? the animal's special features or characteristics
- Display the erasers to practice using the term "characteristics." Ask the students to name characteristics of the erasers.

  What are the two main groups of animals? animals with backbones and animals without backbones

  What is a *vertebrate*? an animal with a backbone
- Discuss that the students are also vertebrates. Direct them to lean forward and feel the "bumps" down the middle of their backs. Explain that each bump is called a *vertebra* and that together they form what is called the backbone.

  How are vertebrates different from invertebrates? Vertebrates have backbones, and invertebrates do not.
- Practice using the term *vertebrate* by asking students to name examples of vertebrates using the following sentence structure: "A _____ is a vertebrate because it has a backbone."
- Ask volunteers to describe characteristics of worms. Allow them to share experiences of handling worms.

  Does a worm have a backbone? No, it is an invertebrate.

  Are all invertebrates the same size? No, they are many sizes, from large to small.

  A crab is not squishy like a worm. What makes it stiff? its shell
- Explain that some invertebrates have shells or another kind of outer covering that gives them shape.
- Practice using the term *invertebrate* by asking students to name examples of invertebrates using the following sentence structure: "A _____ is an invertebrate because it does not have a backbone."
- Direct attention to the picture and the caption.

  Which animals are vertebrate and which are invertebrate? vertebrate: shark, fish; invertebrate: jellyfish, octopus, crab, starfish *Note:* If a student mentions corals, corals are classified as animals instead of plants. They do not make their own food and have an exoskeleton instead of a backbone. They are invertebrates.

Which animals are vertebrate and which are invertebrate?

Classifying Animals

We do not know what names Adam gave to all the animals. Scientists give names to animals too. They look at the different characteristics, or special features, of animals. Scientists *classify*, or group, animals with the same characteristics together. As scientists classify animals, they give the groups and animals names.

Vertebrate

Scientists classify all animals into two main groups by their backbones. Animals with backbones are called **vertebrates**. These animals can be as big as a cow or as small as a frog. They might walk, fly, or swim. There are thousands of animals in this group. Animals without backbones are called **invertebrates**. Worms, crabs, and crickets are a few of the invertebrates. There are more animals that are invertebrates than those that are vertebrates.

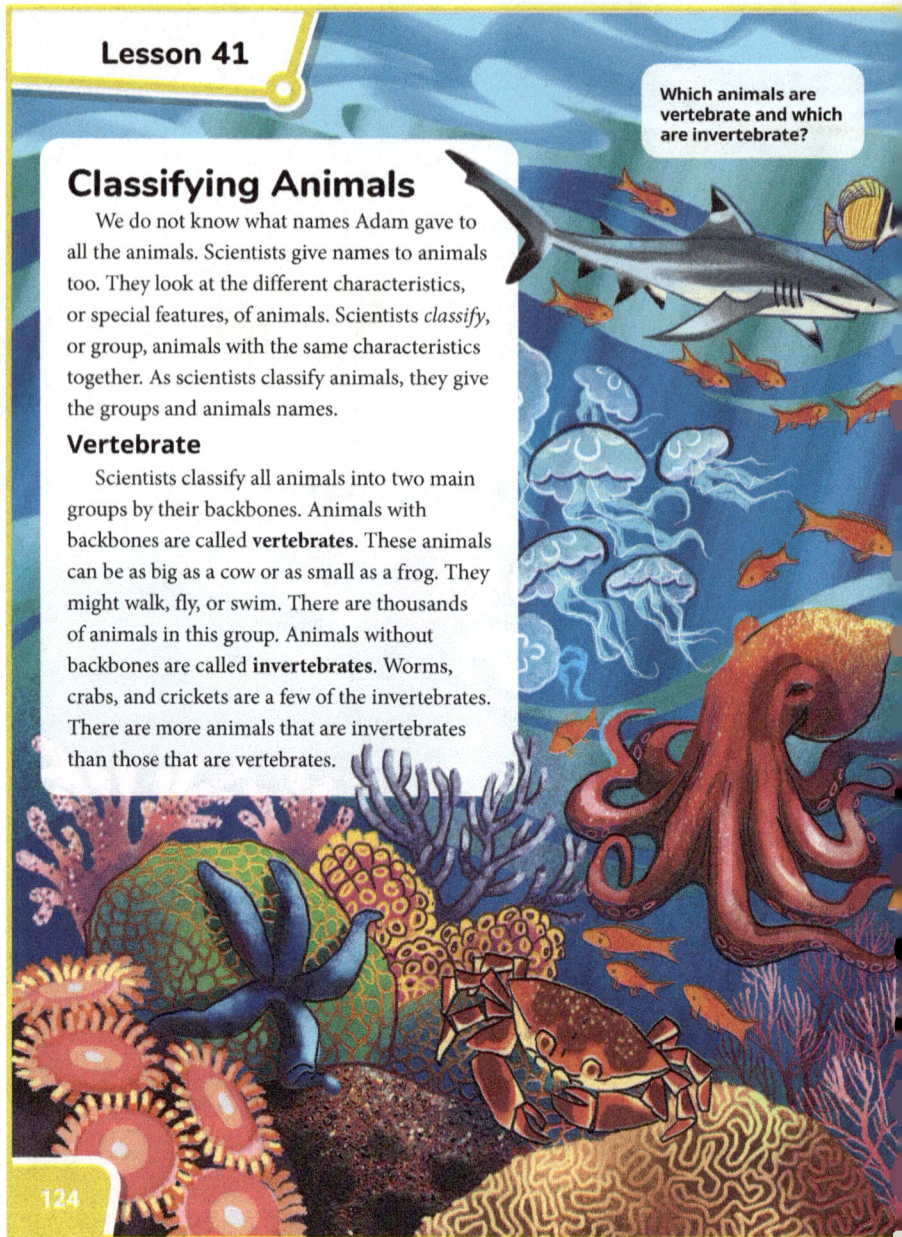

124

ⓘ Background

Warm-Blooded Animals

Warm-blooded animals are called endotherms. An endotherm regulates its body temperature internally. Birds and mammals are warm-blooded. Some mammals, such as whales and dolphins, are not covered in fur. They may have only a few hairs. These animal groups are discussed in Chapter 7.

Cold-Blooded Animals

Cold-blooded animals are called ectotherms. An ectotherm cannot regulate its body temperature internally, so it is dependent on its surroundings. Fish, amphibians, reptiles, insects, and spiders are cold-blooded. All cold-blooded animals are ectotherms.

Temperature

Scientists also classify animals by how their bodies stay warm. Some animals are warm-blooded. Others are cold-blooded. An animal that has about the same body temperature all the time is **warm-blooded**. Animals that have feathers or fur are warm-blooded. Cats, birds, bears, and seals are warm-blooded animals. A cat may sit in the sun for a while. Later it might take a nap in the shade. No matter where a cat is, its body temperature stays about the same.

Lizards, turtles, and fish are cold-blooded animals. A **cold-blooded** animal has a body temperature that changes depending on its surroundings. When a lizard sits in the sun, its body gets warmer. The heat from the sun warms it, and its body temperature rises. Cold-blooded animals move around more when they are warm. But if they get too warm, they look for places to cool off. If a lizard is too hot, it might crawl into the shade under a plant or lie in some cool mud.

How does the temperature of the gecko change?

✓ **How are animals classified by temperature?**

How does the temperature of the cows change?

125

⚠ Helps

Animal Classification Charts

The *Activities* animal classification charts are hierarchy charts. A hierarchy chart shows the relationships between several concepts. For example, a company has an organizational chart that shows the relationships of employees and managers. Hierarchy charts are especially useful in science when studying classifications. When a hierarchy chart is created, think of the different categories that a larger category includes.

ⓘ Background

Heat Sources

In a natural setting, cold-blooded animals use sunlight or things warmed by sunlight, such as the ground or rocks, to adjust their body temperature. In an artificial setting, light and warmth may come from another source.

⍰ Misconception

Blood Temperature

The terms *warm-blooded* and *cold-blooded* describe whether an animal's body temperature stays constant or varies depending on the environment. It does not mean that a cold-blooded animal has cold blood or that a warm-blooded animal has hot blood. In a very warm environment, a cold-blooded animal's blood temperature may be higher than that of a warm-blooded animal.

Chapter 6: Cold-Blooded Animals

What do the words *warm-blooded* and *cold-blooded* tell you about animals? how their bodies stay warm

What does it mean when we say that an animal is warm-blooded? The animal has about the same body temperature all the time.

What characteristic of an animal's appearance can tell you that it is warm-blooded? if it has fur or feathers

What are some examples of warm-blooded animals? Possible answers: cats, birds, bears, seals, deer, cows

- Explain that there are exceptions to classifications. Not all animals easily fit the groupings that people have made. For example, whales and dolphins are not covered in fur, but their other characteristics classify them as mammals.

What does it mean when we say that an animal is cold-blooded? The animal's body temperature changes depending on its surroundings.

What are *surroundings*? Possible answers: the things around an animal; conditions such as the amount of sunlight

What characteristic of an animal's appearance can tell you that it is cold-blooded? if it has scales or smooth skin

Name some cold-blooded animals. Possible answers: lizards, turtles, frogs, salamanders

- Discuss ways that other cold-blooded animals can get warmer and cooler.

- Direct attention to the picture on pages 122–23.

What kind of skin does the alligator have? scales

Is the alligator warm-blooded or cold-blooded? cold-blooded

Why is the alligator in the sun? to warm its body

- Direct attention to the pictures of the gold dust gecko.

What do you notice about the gecko's surroundings? One is in the sun and one is in the shade.

How does the temperature of the gecko change? Answer should include that the gecko has smooth skin, so its body temperature changes depending on its surroundings. It gets warmer from the sun's heat and becomes cooler in the shade.

- Direct attention to the pictures of the cows.

What do you notice about the Angus cows' surroundings? One is in the sun and the other is in the shade.

- Direct attention to the caption.

How does the temperature of the Angus cow change? Answer should include that the cow has fur or hair, so it is warm-blooded and its body temperature stays about the same all the time.

- To review the classification of animals, display the *Animal Classification 1* visual. Explain how the organizer is arranged and read. Direct the students to complete the "Cold-Blooded," "Warm-Blooded," "Vertebrate," and "Invertebrate" cells on *Activities* page 95. Give help as needed. Direct the students to keep their organizers in their science notebooks for use in the next lesson.

✓ **as cold-blooded or warm-blooded; by how their bodies stay warm**

155

Preparation for Reading

- Preview and pronounce the vocabulary term *animal behavior scientist* and the word *predator*.
- Direct the students to read page 126 silently to find out what an animal behavior scientist does.

Teach for Understanding

What does an animal behavior scientist do? studies how animals behave, or act

- Explain that this kind of behavior does not mean how the animal does tricks or whether it is being naughty or nice.
- Instruct the students to use the Glossary to find a definition for behavior.

What does *behavior* mean? how an animal reacts to its environment

- Direct the students to clap their hands loudly several times. Explain that the students reacted to the sound of the clapping hands the same way an animal reacts to things that are in its surrounding or its environment.

What are some things an animal behavior scientist wants to know or study? Possible answers: how animals find food, how animals care for their offspring, how animals avoid predators, what causes certain behaviors, what those behaviors mean to the animal's health

What is an *offspring*? a baby animal

What is a *predator*? an animal that hunts and eats other animals or prey

Why would a scientist want to know about an animal's health? to prevent diseases, to care for the animal

Where are some places where animal behavior scientists work? zoos, aquariums, museums

Would you enjoy being an animal behavior scientist?

- Discuss the pictures of animal behavior scientists.

What do you think the scientist is studying with the cows? Possible answers: How does the food affect the cows? Are the cows healthy eating this food? How is the environment affecting the cows?

Where do you think the scientist is with the seals? in the cold, in the arctic region

What might the scientists be studying about the seals? Possible answers: how they are living, what they are eating, how warm they are

Apply

Activities

Looking Ahead, page 93

This page assesses the students' knowledge prior to the chapter. *Note:* You may want the students to do the "Notebook Connection" at the bottom of the page.

Animal Classification 1, page 95

This organizer page helps the students to classify animals. It was started as part of this lesson and will be used again in the next lesson.

Animal Behavior Scientist

An **animal behavior scientist** studies how animals behave, or act. He may want to know how animals find food. He may study how animals choose their mates. He may study how animals care for their offspring. He may study how animals avoid *predators*.

Animal behavior scientists want to know what causes certain behaviors. They want to know what those behaviors mean to the animal's health. Is the animal getting the right kind of food? Is the animal's water clean? Is the animal safe and comfortable? Learning about animal behavior helps scientists prevent diseases. Scientists can understand how to care for an animal.

Animal behavior scientists work in many different places. They can work with wild animals or tame animals. Some work in zoos, aquariums, or museums.

126

Differentiated Instruction

Remediation: Classify Animals
Materials: *Animal Cards*, on cardstock (Instructional Aids 6.1–6.3)
Direct the students to categorize the animal cards as *vertebrates* or *invertebrates*. This activity can be made to be self-checking by writing *vertebrate* or *invertebrate* on the back of the cards. *Note:* The animal cards can be used again in this chapter and in Chapter 7.

Background

Animal Behavior Scientist: Temple Grandin
Temple Grandin was born on August 29, 1947, in Boston, Massachusetts. She was diagnosed with autism as a child. She studied psychology and animal science. Grandin has written books and consulted on the humane treatment of animals. She believes it is worth the extra expense to decrease the amount of suffering that animals used for food go through. She has improved and invented several devices that help in moving cattle through the stockyards with less trauma to the cattle.

Cold-Blooded Animals

There are thousands of cold-blooded animals. Some are vertebrates, or have a backbone. Some are not. God commanded people to care for the earth. Taking care of cold-blooded animals is part of God's command.

Cold-blooded animals can be classified into smaller groups. The five groups of cold-blooded animals are fish, amphibians, reptiles, insects, and spiders.

Fish
Characteristics

What makes a fish a fish? A **fish** is a cold-blooded vertebrate that lives in water and has gills, fins, and scales.

Fish Characteristics

Fish are cold-blooded vertebrates.

Fish live in the water.

Fish breathe oxygen with gills.

Most fish have fins and scales.

Fish have slime covering their bodies.

Fish reproduce by laying eggs or giving birth to live offspring.

vertebrate
fin
gills
mutton snapper
scales

How is the angelfish reproducing?

queen angelfish

How is the lemon shark reproducing?

lemon shark

127

⚠ Helps

Camouflage Materials
The pictures of camouflaged animals may be from books or printed individually. You may choose to use a video of camouflaged animals instead.
Tape or glue the white construction paper to the dark sheet of construction paper.

ⓘ Background

Oxygen
Although fish breathe by taking in water, they are not getting oxygen from the water compound H_2O. Instead they breathe oxygen gas dissolved in the water. It is much harder to get oxygen from water than from air since air has much more oxygen in it than water does. However, because fish are cold-blooded and have a lower metabolism, they do not need as much oxygen as warm-blooded animals do.

Objectives
- Identify the characteristics of fish and amphibians.
- Identify and describe the structures and features that benefit fish and amphibians in survival and growth. **BWS**
- Construct an explanation, using Scripture, stating why animals reproduce after their own kind. **BWS**
- Identify the ways fish and amphibians reproduce.
- Sequence the stages of frog metamorphosis.

Materials
- pictures of fish from different habitats, for display
- picture of an animal camouflaged in its habitat, per group (or video of camouflaged animals)
- white sheet of construction paper (See Helps.)
- dark brown or black sheet of construction paper

Teacher Resources
- Visuals 6.1–6.3: *Animal Classification 1*; *Life Cycles of Fish*; *Stages of Amphibian Metamorphosis*

Vocabulary
- fish
- camouflage
- prey
- amphibian
- predator
- metamorphosis

Engage
- Display the pictures of fish.
  What are these animals? fish
- Ask how the students knew this and record their responses for reference during the lesson.

Instruct
Preparation for Reading
- Preview and pronounce the vocabulary terms *fish*, *camouflage*, and *prey* and the word *adaptation*.
- Direct attention to the big heading on page 127.
  What will you be reading about? cold-blooded animals
- Direct the students to read pages 127–29 silently to find out how fish reproduce.

Teach for Understanding
Are cold-blooded animals vertebrates or invertebrates? Some are vertebrates and some are invertebrates.
- Read Genesis 1:28 aloud.
  What command did God give you about animals? to have dominion, or rule over, them
- ❀ How do we use *dominion*? by managing, or ruling over, the world for the benefit of others and the glory of God.
  What makes a fish a fish? It is a cold-blooded vertebrate that lives in water and has gills, fins, and scales.
- As you review the characteristics of fish, refer to the responses that were recorded in the Engage activity.

The characteristics of fish were studied in *Science 2* except for reproduction. Use the chart, pictures, and captions to review the characteristics of fish as needed. You may want to review by using the *Animal Classification 1* visual and the "Fish" cell on *Activities* page 95. The reproduction characteristic will be discussed in this lesson.

What is *adaptation*? a characteristic that helps an animal survive in its environment

What does *survive* mean? Possible answer: to live

What is similar in the biblical and evolutionary views of adaptation? A biblical view and evolutionary view both explain adaptation the same way. It is a change being passed down from parents to offspring.

What is the difference in the views of adaptation? A biblical worldview explains that adaptation does not change the kind of animal. God created animals to adapt. Evolution claims that an animal can adapt and evolve into another kind of animal.

- Discuss the features for survival and growth.

  What is *camouflage*? a color or pattern that allows animals to not be seen easily in their surroundings

- Distribute a picture of a camouflaged animal to each group. *Note:* You may choose to show a video of the camouflage feature at this time.

- Direct one student in each group to hold the picture up from a distance of about 15 feet.

  Do you see something hiding in the picture?

  Can you find the animal?

  Why is it hard to find the animal? The animal blends in with its surroundings or is camouflaged.

- Explain that some animals use camouflage to hide from their enemies. Others use it to go unnoticed as they search for food themselves.

- Look at the picture on the previous page.

  Do any of the fish appear to be using camouflage for protection? no

  What is a *prey*? an animal that is caught and eaten by another animal

- Display the two-toned construction paper with the dark side on top. Direct the students to imagine that the construction paper is a predator fish.

- Hold your fist above the construction paper to illustrate the prey.

  What does the prey above the imaginary predator fish see? The prey sees the dark ocean floor.

  What advantage is that for the imaginary predator fish? The prey will not see the imaginary predator fish, so it will be easier for the predator fish to catch the prey.

- Hold your fist below the construction paper.

  What does the prey below the imaginary predator fish see? The prey sees the sun or moon shining.

  What advantage is that for the imaginary predator fish? The prey will not see the imaginary predator fish, so it will be easier for the predator fish to catch the prey.

- Direct attention to the flounder and the caption.

  How is the flounder using color to survive? It is using camouflage to blend in.

- Direct attention to the tuna and the caption. Explain that both fish are tuna.

  How is the tuna using color to survive and grow? The bottom tuna is blending in with the ocean floor so that prey has a difficult time seeing it. The top tuna is blending with the ocean's surface.

Features to Survive and Grow

Remember what adaptation is. *Adaptation* is a characteristic that helps an animal survive in its environment. The adaptation is passed down from parents to offspring. This does not change the kind of animal. God created animals with the ability to adapt.

Evolution explains adaptation the same way. However, it also makes another claim. Evolution claims that an animal can adapt until it evolves into another kind of animal. This change happens slowly over millions of years.

One feature God has given many fish is **camouflage**. They have a color or pattern that allows them to not be seen easily in their surroundings. Clownfish are brightly colored like the coral reefs where they live. Flounder can change colors to help them blend in with the ocean floor.

Colors also help some fish catch prey. A **prey** is an animal that is caught and eaten by another animal. Cod and tuna have a dark color on top. Underneath they have a light color. Any prey looking down on the fish will think they are seeing the dark ocean floor. Any prey looking up will see the ocean surface with the sun or moon shining.

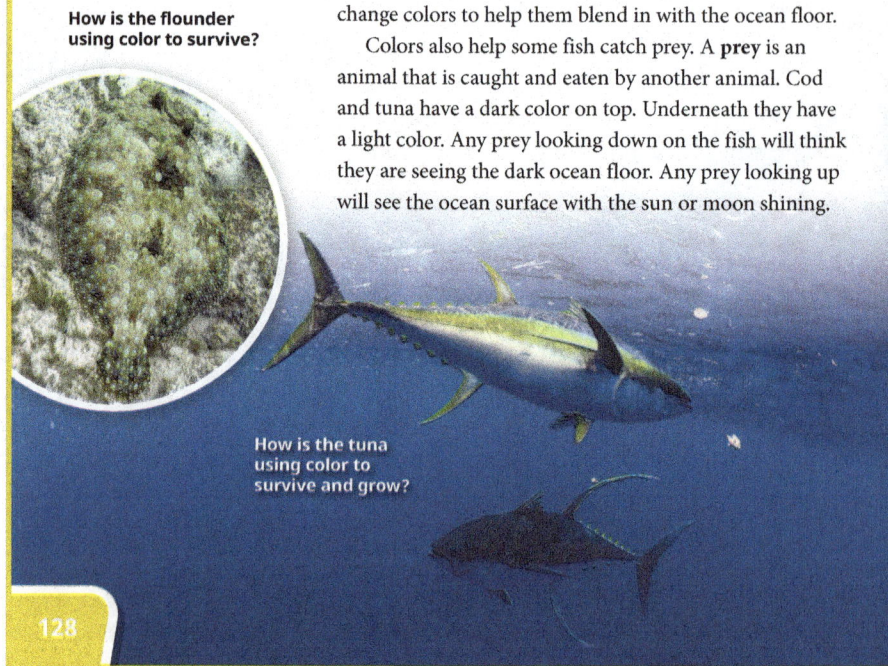

How is the flounder using color to survive?

How is the tuna using color to survive and grow?

128

Background

Scales

Some fish, such as catfish, do not have scales. Other fish, such as sharks, have scales that are so tiny they are hardly noticeable.

Gills and Lungs

Gills and lungs are similar. They both have small blood vessels that take in oxygen and release carbon dioxide. Gills are used by animals that get their oxygen from the water. Lungs are used by humans and by animals that get their oxygen from the air.

Fish and the Sea of Galilee

Luke 5:1–11 records a miracle with fish. The fishermen had not caught anything all night. Jesus told them to cast their nets into the water one more time. This time they caught so many fish that their boats began to sink! The men may have caught sardines, barbels, and musht (a type of tilapia) in the Sea of Galilee. These fish are still caught there today.

Some fish produce light in their bodies. Parts of the fish light up in the dark. Some use this light to attract prey. Some attract other fish with the light.

Many fish have spines, or sharp-pointed parts, as a feature. These spines have venom in them. The spines of stingrays and catfish help the fish survive.

Ways to Reproduce

All organisms reproduce. God said each animal will reproduce after its own kind (Genesis 1:21, 24). A fish reproduces more fish. A fish does not reproduce a turtle.

Most fish reproduce by laying eggs in the water. The eggs are like soft balls of jelly that stick together. A fish can lay thousands of eggs at one time. They lay many eggs so that at least some of the offspring survive.

Other fish reproduce by giving birth to live offspring. The offspring start growing from eggs inside the mother's body. When the unborn fish are the right size, they are born.

A young fish looks much like an adult fish. It does not change its form the way a caterpillar changes to a butterfly. A young fish already looks like an adult fish. It does not grow new parts later. It just gets bigger and stronger.

angler fish

Can you find the venomous spine on the stingray?

salmon

guppy

✓ **What way does color help fish survive?**

129

How do some fish use light? They use light in their bodies to attract prey.

What do stingrays and catfish use to survive? spines or sharp-pointed parts with venom in them

- Direct attention to the first two pictures and their captions.

🌸 What is the angler fish using to survive? light to attract other fish

Can you find the venomous spine on the stingray?

🌸 Where is the venomous spine? on its tail

- Discuss ways fish reproduce.

🌸 What does it mean to *reproduce*? to make more of the same kind of animal

- Read Genesis 1:21, 24 aloud.

When God created animals, how did He create them to reproduce? to reproduce more of the same kind of animals

- Direct the students to listen as you read the verses again aloud to find out what words in the verses tell them that animals reproduce the same animals.

What words tell you that animals reproduce the same kind of animal? Possible answers depending on the Bible version read: after his kind; according to its kind; according to their kinds

What does a fish reproduce? more fish

How do most fish reproduce? by laying eggs

- Direct attention to the picture of the salmon and its eggs.

What do the eggs of a fish look like? soft balls of jelly that stick together

🌸 Why do you think fish lay thousands of eggs at one time? Possible answers: Some fish eat the eggs; some people eat the eggs; some do not hatch.

- Display the *Life Cycles of Fish* visual. Discuss the stages in the life cycle of an egg-laying fish. (A newly hatched trout lives off an attached yolk sac. It has been removed for ease of discussion.)

- Direct attention to the picture of the guppies.

How do other fish reproduce? by giving birth to live offspring

What does a young fish look like when it is born or hatched? much like the adult fish, but smaller

Does it grow new parts later? no

- Direct attention to the *Life Cycles of Fish* visual. Discuss the stages in the life cycle of a livebearer fish.

- Direct attention to the pictures on page 127. Discuss the ways that the lemon shark and the angelfish reproduce. lemon shark: live offspring; angelfish: laying eggs

- To review the classification and features of fish, display the *Animal Classification 1* visual. Direct the students to complete the "Fish" and "Some Survival and Growth Features" cells on *Activities* page 95. Give help as needed.

✓ **Color helps a fish blend with its surroundings so that the fish can hide from other fish or its prey.**

👥 Differentiated Instruction

Enrichment: Spine Fish Feature
Materials: *Activities*, page 96, science notebook
The "Notebook Connection" on *Activities* page 96 extends the student's learning about the spines of fish.

Enrichment: Life Cycle of Fish
Materials: *Activities*, page 99, science notebook
The "Notebook Connection" on *Activities* page 99 extends the student's learning about the life cycle of a fish.

Preparation for Reading

- Preview and pronounce the vocabulary terms *amphibian*, *predator*, and *metamorphosis* and the words *newt* and *caecilian*.
- Direct attention to the big heading on page 130.
  What animal group will you be reading about? amphibians
- Direct the students to read pages 130–32 silently to find out what metamorphosis is.

Teach for Understanding

What is an *amphibian*? It is a cold-blooded vertebrate that lives in a water habitat when it is young and in a land habitat as an adult.

What does the name *amphibian* mean? double life

How does the name describe its habitats? It lives in a water habitat when it is young and a land habitat as an adult.

The characteristics of amphibians were studied in *Science 2* except for reproduction. Use the chart, pictures, and captions to review the characteristics of amphibians as needed. You may want to review by using the *Animal Classification 1* visual and the "Amphibian" cell on *Activities* page 95. Reproduction will be discussed in this lesson.

Amphibian Characteristics

Amphibians are cold-blooded vertebrates.

Amphibians live in water and on land.

Amphibians have thin, smooth, wet skin.

Young amphibians breathe with gills.

Adult amphibians breathe with lungs.

Amphibians reproduce by laying eggs.

Amphibians go through metamorphosis.

Amphibians
Characteristics

An **amphibian** is a cold-blooded vertebrate. The name *amphibian* means "double life." This name describes its habitats. A young amphibian lives in a water habitat. An adult amphibian lives in a land habitat.

eggs

lungs

tadpole of wood frog

vertebrate

gills

red newt

spotted salamander

Try It Yourself!

Amphibians that walk on land are able to "hear" without having ears. Turn an aluminum pie pan upside down on a table. Sprinkle a thin layer of salt onto the pan. Tap the pan with a metal spoon. Observe how the salt moves.

130

Activity

Try It Yourself!
Materials: aluminum pie pan, table salt, and metal spoon
Allow the students to try this activity on their own.

Helps

Metamorphosis
The students will learn about incomplete metamorphosis and complete metamorphosis in a later grade level.

Background

Amphibian Skin
Because amphibians have such thin skin, they are especially susceptible to pollution and other environmental problems. Biologists know that environmental problems are present when they find frogs or other amphibians that are missing limbs or are otherwise malformed.

Features to Survive and Grow

Amphibians have features to survive and grow. Most live on the ground in wetlands or forests. These habitats are damp. Here the sun does not dry out their skin. So being active at night helps keep their skin wet. At night they are also harder to see and can avoid becoming prey.

Adult amphibians have eyelids and lungs. Eyelids allow the amphibian to see outside of the water. Lungs allow adults to breathe on land. Some breathe through their skin as well.

Some amphibians have poison as a harmful feature. These amphibians are usually brightly colored. This warns predators to stay away. A **predator** is an animal that hunts and eats prey. The amphibians give off poisons from their skin. The poison tastes bad to predators. The poison is deadly to some predators. Poisonous amphibians are usually active during the day.

What feature of the crested newt is similar to a fish?

What feature is warning predators away from the poison dart frog?

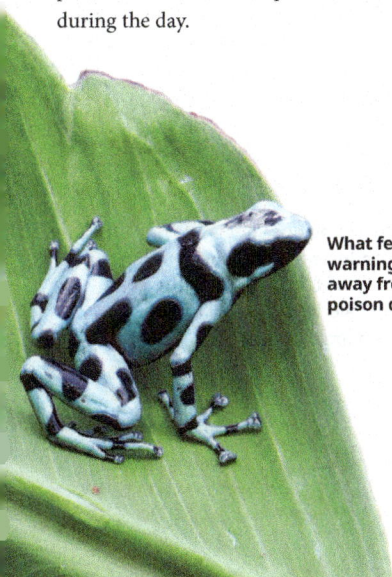

The caecilian lives mostly in the ground and mud of streams. What is different about the caecilian from the other amphibians?

131

Why do most amphibians live in wetlands and forests? The damp habitats help to keep their skin wet.

Why are some amphibians active at night? so the sun does not dry out their skin

- Direct attention to the newt picture and the caption.

What feature of the great crested newt is similar to a fish? Both have camouflage.

- Discuss the picture of the frog's eyes.

Why do adult amphibians have eyelids? They allow the amphibian to see outside of the water.

Why do you think the frog has its eyes and nose above the water? Possible answer: The frog is an adult. As an adult it has lungs and eyelids. Lungs do not help the frog breathe underwater.

What do young amphibians breathe with? gills

What do adult amphibians breathe with? Adults breathe with lungs; some breathe through their skin as well.

Why is it necessary to be able to breathe in two different ways? A young amphibian lives in the water where gills are needed. An adult amphibian lives on land where lungs are needed.

What is a *predator*? an animal that hunts and eats prey

What does a brightly colored amphibian usually have to help it survive? poison

What do the bright colors do? warn away predators

When are poisonous amphibians usually active? during the day

Why do you think that a poisonous amphibian is active during the day rather than at night? Possible answer: so that the bright colors can be seen

- Direct attention to the poison dart frog picture and its caption.

What feature is warning predators away from the poison dart frog? its bright colors

Is the poison dart frog active during the day or during the night? It is active during the day; poisonous amphibians are usually active during the day.

- Ask a volunteer to read the caption with the caecilian picture aloud.

What is different about the caecilian from the other amphibians on this page? It does not have legs.

- To review the classification and features of amphibians, display the *Animal Classification 1* visual. Direct the students to complete the "Amphibians" and "Some Survival and Growth Features" cells on *Activities* page 95. Give help as needed.

- Display the *Stages of Amphibian Metamorphosis* visual. Refer to it as you discuss the stages of a frog's life cycle.
  What does a fish look like when it is born? similar to an adult fish
  Does the fish change form as it grows? no
  What does a frog look like when it is born? a tadpole
  Does the tadpole change form as it grows? yes
  What is *metamorphosis*? the process of an animal changing form as it grows
  Does a frog give birth to live offspring, or does it lay eggs? It lays eggs.
  Where does the frog lay eggs? in the water
  Why do you think that a frog lays eggs in the water and not on land? Possible answers: When the tadpole is born, it lives in the water. When the tadpole is born, it is near the food that it eats.
  What does a tadpole have that an adult frog does not have? gills to breathe with and a tail to swim
  What happens to a tadpole as it grows? It changes form. It loses its tail and grows legs.
  What is this changing of form called? metamorphosis
  What does the adult frog have that the tadpole did not? legs and lungs to breathe with
- Direct attention to the tadpoles on page 130.
  Have the tadpoles just been born or are they almost ready to become adults? just been born because they still have their tails and no legs
- Discuss "Fantastic Facts."
  How is the poison dart frog different from most amphibians? It does not lay its eggs in water.
  Does the poison dart frog still go through metamorphosis? Yes, the male frog carries the tadpoles to water where they continue to go through metamorphosis.

✓ Possible answers: live in damp areas to keep skin wet; more active at night to keep skin wet; eyelids and lungs as adults; some are poisonous; bright colors to warn of poison

Apply

Activities

Animal Classification, page 95

The organizer was completed as part of the lesson. Direct the students to keep the page in their science notebooks to review for Test 6.

Study Guide, pages 97–99

These pages review the concepts taught in Lessons 41–42. After completion, direct the students to keep them for review for Test 6.

Some features in the fish and amphibians section of the *Student Edition* can also apply to the other animal groups. To avoid confusion, the features in the *Activities* book will follow the same animal groups as the *Student Edition*.

A Way to Reproduce

God designed frogs and other amphibians to grow in a special way. They change form as they grow. When an animal changes form as it grows, scientists call this **metamorphosis**.

Stages of Amphibian Metamorphosis

The first stage is when an adult frog finds a safe place in water and lays its eggs.

The adult frog is finally ready to move onto land.

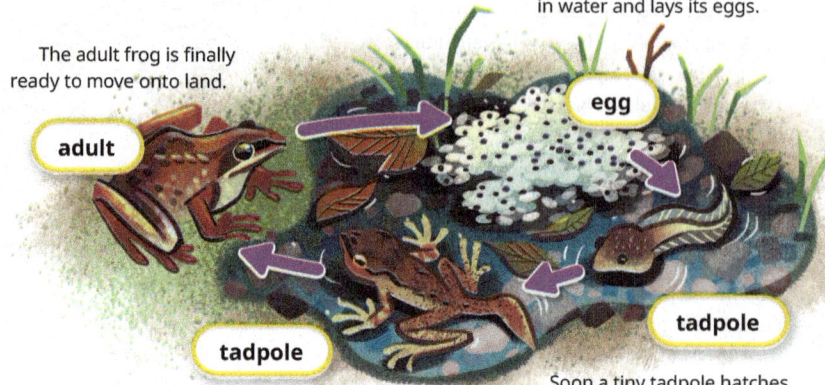

adult

egg

tadpole

tadpole

As the tadpole grows, it changes. It loses its gills and forms lungs. It also loses its tail and grows legs.

Soon a tiny tadpole hatches from each egg. A tadpole has gills to breathe and a tail to swim. It does not look like an adult frog.

Fantastic Facts

The poison dart frog does not lay its eggs in the water. Instead it lays eggs on dead leaves. After the tadpoles hatch, they wiggle onto the male frog's back. The male frog carries the tadpoles to water.

✓ What are two amphibian features that help them to survive and grow?

Assess

Quiz 6A

The quiz may be given at any time after completion of this lesson.

① Background

Frogs and Toads

Frog and *toad* are unscientific terms used as common names for convenience. There is no true classification distinction between frogs and toads. In fact, some species share characteristics of both groups.

Lesson 43

Reptiles
Characteristics

Reptiles and amphibians are sometimes together in books. But reptiles are quite different from amphibians. A **reptile** is a cold-blooded vertebrate with tough, dry, scaly skin. Compare the characteristics of a reptile with those of an amphibian.

Reptile Characteristics
Reptiles are cold-blooded vertebrates.
Reptiles live on land. Some can move back and forth between land and water.
Reptiles have tough, dry, scaly skin.
Reptiles breathe oxygen with lungs.
Most reptiles reproduce by laying eggs.

vertebrate

Does the alligator live in a water habitat or a land habitat?

tough, dry, scaly skin

A cottonmouth snake is venomous.

most lay eggs

lungs

coral snake

133

Lesson **43** begins here.

Objectives
* Identify the characteristics, structures, and features of reptiles.
* Describe how God has provided the structures and features that benefit reptiles for survival and growth. **BWS**
* Sequence the pattern of the reptile life cycle.
* Explain, using Scripture, why there will one day be no predators and no prey. **BWS**
* Infer in an Investigation how a cold-blooded animal depends on the temperature of its habitat for survival and growth.
* Explain why people should care for reptiles. **BWS**

Materials
* 2 thermometers
* tape
* timer
* leather belt or other piece of leather

Teacher Resources
* Instructional Aid 6.4: *Lizard*
* Visuals 6.4; 3.3; 6.5; 6.1–6.3: *Life Cycle of Reptiles*; *Science Inquiry Skills*; *Animal Classification 2*; *Animal Classification 1*; *Life Cycles of Fish*; *Stages of Amphibian Metamorphosis*

Copy and cut out 2 lizards from the *Lizard* Instructional Aid. To save time during the lesson, you will want to attach each lizard to a thermometer. The thermometer should be facing out so it can be read. The Investigation can be done as a class. If you choose to have groups do the activity instead of the class, you will need 2 copies of the lizard and 2 thermometers for each group.

Vocabulary
* reptile
* Fall

Engage

The Leaping Lizards Investigation is an excellent way to engage the students for this lesson. See *Teacher Edition* pages 167–69. You will need time in the day to read the thermometers after waiting 10 minutes and again after 2 hours.

Instruct
Preparation for Reading
* Preview and pronounce the vocabulary term *reptile*, and the words *Jaragua*, *Caribbean*, *leathery*, and *hatchling*.
* Direct the students to read pages 133–35 silently to find out how most reptiles reproduce.

Teach for Understanding
What is a *reptile*? a cold-blooded vertebrate with tough, dry, scaly skin

The characteristics of reptiles were studied in *Science 2* except for reproduction. Use the chart, pictures, and captions to review the characteristics of reptiles as needed. You may want to review by using the *Animal Characteristics 2* visual and the "Reptiles" cell on *Activities* page 101. The reproduction characteristic will be discussed in this lesson.

Describe a reptile's skin. tough, dry, with scales

What are five ways this type of skin is helpful to the reptile? It protects the animal from predators. It protects the animal as it moves across the ground. It prevents the body from drying out. It helps keep the heat in. The skin of many reptiles is camouflaged.

- Discuss the cottonmouth snake and that it sheds its skin as it grows as illustrated on the previous page.
- Explain that the coral snake and the cottonmouth are both venomous snakes.
- Discuss that some reptiles have legs and some do not. Explain that even the reptiles with legs move across the ground.

How would being camouflaged help a reptile? It would keep predators from seeing the reptile. It would help the reptile catch prey.

Are any of the reptiles on the picture on the previous page camouflaged? no

What determines how much a reptile moves? its temperature

How does temperature affect the reptile? It moves more during warm times and less during cold times.

How does the speed of some reptiles help them survive? Some have lightning speed and can move quickly away from predators.

- Direct attention to the picture of the hanging snake.

Describe the snake's tongue. It is slender and forked.

What does *slender* mean? thin

Why is it called a forked tongue? Elicit that the tongue is divided similarly to a fork.

How does the forked tongue help the snake to survive? It allows the snake to pick up scents.

- Explain that other reptiles also have forked tongues. Reptiles smell with the tip of their tongues and forked tongues allow them to sense which directions the smells are coming from.

How do you think the tongue helps a snake to survive? Possible answers: It helps a snake smell prey. It helps a snake recognize other snakes. It helps a snake choose a mate. It also helps the snake locate shelters and follow trails.

- Analyze both the picture of the snake swallowing the egg and its caption.

What feature allows the snake to swallow the egg? a movable upper jaw

- Direct the students to open their mouths wide and to point to the top jaw. This jaw is movable in snakes and lizards.
- Explain that jaws of lizards and snakes are attached with a stretchy material that allows them to open their mouths wide for their prey. Some snakes can open their jaws wide enough to swallow a crocodile.
- Explain that the upper jaw in humans is not movable.
- Discuss the picture of the lizard and its caption.

What special design did God create the lizard with? If grabbed by a predator, the lizard may lose all or part of its tail.

Features to Survive and Grow

Reptiles have features that help them survive and grow. A reptile's skin helps in five different ways. The tough, dry, scaly skin provides protection from other predators. It also protects a reptile as it moves across the ground. The skin prevents the reptile's body from drying out. It helps keep heat in. Many reptiles are camouflaged.

Because a reptile is a cold-blooded animal, the temperature determines how much a reptile moves. During the coldest parts of the year, a reptile does not move around much. When it is warmer, some reptiles are very fast. Some can scurry off at lightning speed.

Snakes and lizards have a movable upper jaw. This kind of jaw allows the snakes and lizards to swallow large prey. Snakes use their slender, forked tongues to pick up scents. If grabbed by a predator, some lizards can lose all or parts of their tails. This helps keep the lizards safe. The lizards will grow new tails.

What feature allows the snake to swallow the chicken egg?

What special design did God create the lizard with?

Fantastic Facts

The tiniest reptile is a Jaragua lizard. When it curls up, it fits on a dime. When it stretches out, it fits on a quarter. Scientists discovered these lizards on an island in the Caribbean Sea in 2001.

134

👥 Differentiated Instruction

Enrichment: Venomous Snakes

Materials: *Activities* page 102, science notebook

The "Notebook Connection" on *Activities* page 102 extends the student's learning about venomous snakes.

ⓘ Background

Poisonous and Venomous

According to biologists, the term *venomous* applies to organisms that bite (or sting) to inject their toxins, whereas the term *poisonous* applies to organisms that pass on toxins when you eat them. Very few snakes are truly poisonous. The vast majority of snake toxins are transferred by bite. One exception is the garter snake. It is small and harmless in terms of its bite, but it is toxic to eat. Its body absorbs and stores the toxins of its prey (newts and salamanders).

Jaragua Lizards

The Jaragua lizard is a member of the dwarf gecko family. The species was discovered in 2001 on Beata Island, a small Caribbean island. It shares the title of smallest reptile with the Virgin Gorda Least Gecko. This species was found in the British Virgin Islands in 1965. Both of these geckos measure only 16–18 mm when fully grown.

A Way to Reproduce

Fish and most amphibians lay eggs. What is different about a reptile laying eggs than a fish or an amphibian? Some reptiles dig a hole in sand or soil to bury their eggs. Others make a nest in grasses or hide their eggs in rotting logs. A sea turtle might swim thousands of miles to lay its eggs on the same beach every year.

A reptile can live a long time. Some live as long as most people do. Giant tortoises and alligators can live more than 80 years.

adult

young

The young reptile looks like its adult parents but is smaller. It does not change form. It does not grow new parts later. It just gets bigger and stronger.

Life Cycle of Reptiles

Most reptiles lay several eggs with a leathery shell at one time. Some lay 20 or 30 eggs in their nests. Others lay more than 100 eggs at a time.

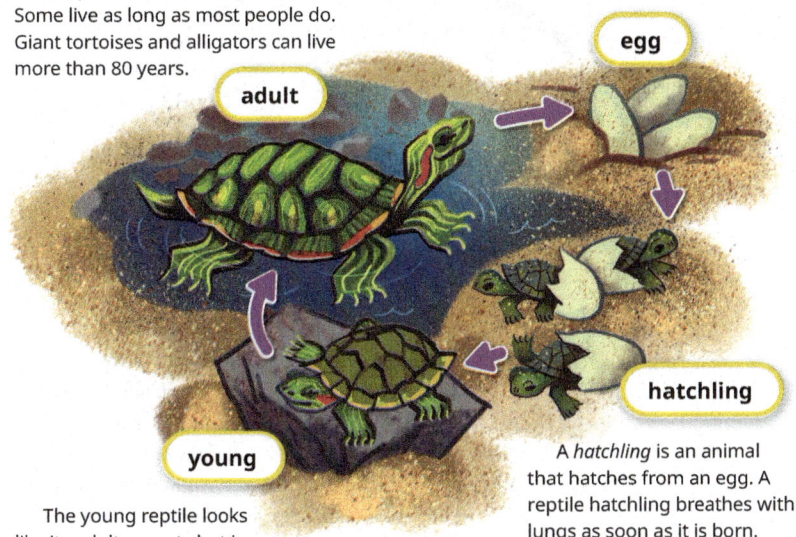

egg

hatchling

A *hatchling* is an animal that hatches from an egg. A reptile hatchling breathes with lungs as soon as it is born.

How do most reptiles reproduce?

135

- Where can you find reptiles? Possible answers: deserts, forests, swamps, the ocean
- Discuss "Fantastic Facts" and use the Background information about Jaragua lizards on the previous page about this tiny animal.

Discussion of page 135 begins here.

- Display the *Life Cycle of Reptiles* visual.
- Where do fish and most amphibians lay eggs? in water
  Where do reptiles lay their eggs? on land
- Discuss that there are exceptions to the characteristics being identified. Some reptiles, like sea turtles, live most of their lives in water, and some reptiles do give birth to live offspring.
  What kind of shell do reptile eggs have? leathery
- Twist, turn, and bend the leather belt.
- Why do you think that the eggs are leathery? Elicit that, like the leather belt, the eggs do not crack easily and it is not easy for predators to get through the outer shell.
  What is the reptile called when it hatches from an egg? a hatchling
  What does a hatchling breathe with? lungs
  Does the young reptile change form or grow new parts? No, it looks like the adult reptile except smaller.
- To review the classification and features of reptiles, display the *Animal Classification 2* visual. Direct the students to complete the "Reptiles and Some Survival and Growth Features" cells on *Activities* page 101. Give help as needed.
- Guide a discussion comparing and contrasting characteristics of amphibians and reptiles. You may choose to use the *Animal Classification 1* and *Animal Classification 2* visuals. You may want to list the characteristics in a Venn diagram. (See Helps.)

most lay eggs

⚠ Helps

Venn Diagram

Possible Venn diagram for the characteristics of amphibians and reptiles:

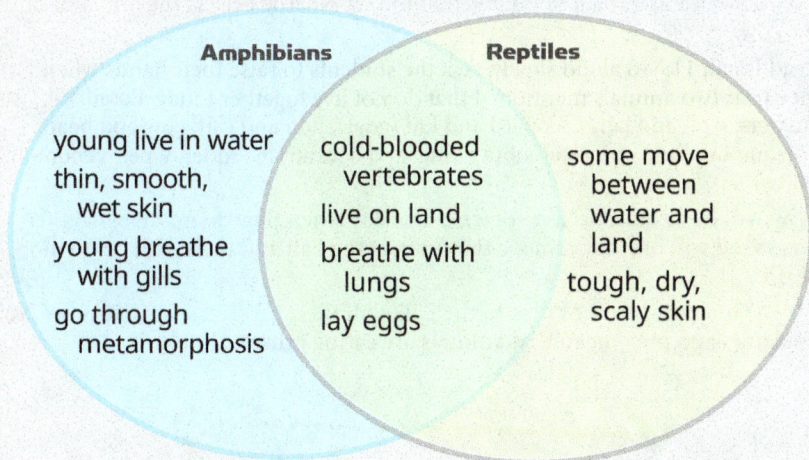

Amphibians

- young live in water
- thin, smooth, wet skin
- young breathe with gills
- go through metamorphosis

Reptiles

- cold-blooded vertebrates
- live on land
- breathe with lungs
- lay eggs

- some move between water and land
- tough, dry, scaly skin

Preparation for Reading

- Preview and pronounce the vocabulary term *Fall.*
- Direct attention to the heading at the top of page 136.
  What will you be reading about? the life cycles of cold-blooded animals
- Direct the students to read page 136 silently to find out the pattern of the life cycles of cold-blooded animals.

Teach for Understanding

What are two ways animals reproduce? Some lay eggs. Some give birth to live offspring.

What do you call this stage in the life cycle? birth

- Instruct a student to write the word *birth* for display. *Note:* The students will be writing and drawing the pattern for life cycles. The word *birth* will need to be written high enough to do this.

Do all animals look like their parents when they are born? No; some look like their parents, and some change after birth to look like their parents.

Name an animal that looks like its parents when it is born. Possible answers: snake, fish, turtle, lizard

Name an animal that does not look like its parents when it is born. Possible answers: frog, salamander, newt

After an animal is born, what does it do next? begins to grow

Do all animals grow after they are born? yes

What is this stage in the life cycle called? growth

- Direct a student to add an arrow and the step *growth* to the life cycle for display.

When an animal is finished growing what does it become? an adult

How do adult animals make more animals like them? They reproduce.

- Ask a volunteer to add an arrow and the stage *reproduce* in the life cycle for display.

What must all animals do at some point? die

- Ask a volunteer to add an arrow and the step *death* to the life cycle for display.
- Lead the students in saying the four stages in all life cycles together: birth, growth, reproduction, death.
- Display the *Life Cycles of Fish, Stages of Amphibian Metamorphosis,* and *Life Cycle of Reptiles* visuals one at a time. Guide the students as they find the four stages (birth, growth, reproduction, death) on each life cycle.

How did God create animals? perfect

What did God create animals to eat? plants

What happened in Genesis 3 that changed animals? Adam sinned. There are now predators and prey. There is now death. Some animals eat other animals instead of plants.

What is Adam's sin and God's curse because of that sin called? the Fall

- Read Revelation 21:4 aloud. Explain that this verse tells us what will happen when God removes the curse.

What does the verse tell us there will be no more of? death, sorrow (mourning, grief), crying, and pain

---

Life Cycles of Cold-Blooded Animals

God created all animals with a life cycle. But all animals do not have the same life cycle. Some animals reproduce from eggs the mother lays. Some grow from an egg inside the mother's body. All animals grow to be like their parents. Some animals look like the parents at birth. Some change to look like the parents as they grow.

All life cycles have the same stages. The life cycles repeat the same pattern: birth, growth, reproduction, and death.

Pattern of All Life Cycles

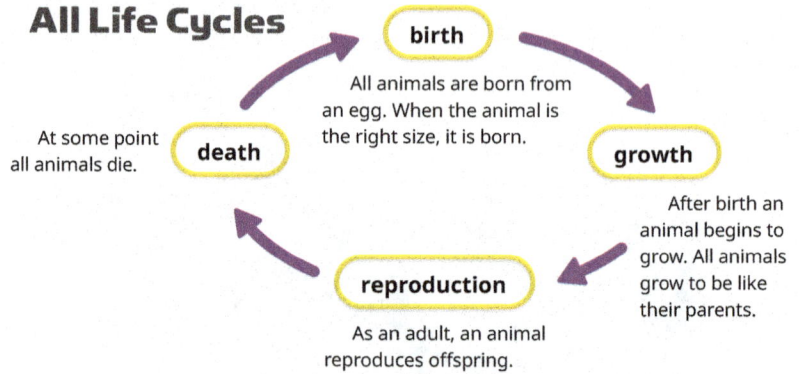

birth
All animals are born from an egg. When the animal is the right size, it is born.

growth
After birth an animal begins to grow. All animals grow to be like their parents.

reproduction
As an adult, an animal reproduces offspring.

death
At some point all animals die.

God created animals perfect. He created them to eat plants. Genesis 3 tells us that the animals changed after Adam sinned. God's curse, or punishment, for sin was death. What happened in Genesis 3 is called the **Fall.** Now there are predators that hunt and eat prey. Animals also die when they are old. All animal life cycles end in death. Death was not part of the way God created animals. One day God will remove this curse and death will be no more.

✓ What pattern do all animal life cycles have?

136

If there is no more death, can there be predators and prey? No, predators and prey mean that animals are eating other animals.

Do a wolf and a lamb live together today? No, the sheep can be the prey of the wolf.

Do a cow and a bear eat grass together today? No, the bear is the predator of the cow.

- Read Isaiah 11:6–8 aloud slowly. Ask the students to raise their hands when they hear two animals mentioned that do not live together today. Possible answers: wolf and lamb, leopard and kid (goat), lion and calf, cow and bear, lion and ox, child and asp (cobra), baby and cockatrice (adder, viper, venomous snake)

Why will these animals live together one day when they do not live together now? God will one day remove the curse of the Fall and there will be no more death.

When God removes the curse of the Fall, will there be predators and prey? No, predators and prey mean that animals are eating other animals.

✓ birth, growth, reproduction, death

INVESTIGATION

Leaping Lizards

Inquiry Skills
- Measure
- Infer

A cold-blooded animal's body temperature changes in different surroundings. For a lizard to be active, it must be warm. If you have a pet lizard, you need to give it both warm and cool places to live. You need to measure temperature to take care of your lizard.

In this Investigation, different temperatures will be measured. You will need to find a place where the "lizard" can be active. The "lizard" is active when its body temperature is about 21°C to 31°C (70°F to 88°F).

137

The Investigation can be used to engage the students at the beginning of the lesson.

- Choose volunteers to read the page aloud.
- What is a *cold-blooded animal*? an animal that has a body temperature that changes depending on its surroundings
- Direct the students to remove pages 103–4 from their *Activities* books. Display the *Science Inquiry Skills* visual. What inquiry skills will we be using? measure and infer
- Review what it means to *measure* and to *infer*.

⚠ Helps

Leaping Lizards
Try to find a location that will stay shady or sunny for two hours but move lizards as needed to keep them in sun or shade during the two hours.

If the temperature change is not sufficient to demonstrate a change in the lizard's activity, guide the students in concluding that the lizard's activity would not change under these conditions.

ⓘ Background

Ornate Tree Lizard
The paper lizards represent ornate tree lizards, which are found in the southwestern United States and in northern Mexico.

• Choose a student to read the **Problem** aloud.

Apply

• Direct attention to the **Hypothesis** to answer the problem.
  Decide whether the lizard will be most active in the sun or in the shade.

• Point out that the students will be doing the Investigation together and that you have already done Steps 1–2.

• Explain to the students that they will check off **Procedure** Steps 3–7 as each is completed.

• Ask two students to read the temperatures on the thermometers and say the temperatures aloud.

• Direct the students to record the temperatures on the **Observations** data chart.

What makes this investigation a controlled investigation? Elicit that although there are two paper lizards, they are representing only one lizard. They are the same color to make it a controlled investigation. The location of the lizard is what is changing.

• Discuss the outside locations where the students will be placing the lizards.

• Guide the students in placing the lizards outside, one in the sun and one in the shade.

• Set the timer for 10 minutes.

This Investigation follows the same format as other Investigations even though it is part of the Lesson 43 introduction. You may want to begin the reptile discussion during the waiting periods in the Investigation.

INVESTIGATIO

Leaping Lizards

Name _____

© 2021 BJU Press. Reproduction prohibited.

Problem

Where would a lizard be more active during the daytime?

Hypothesis

A lizard would be more active in the   **sun, shade** .

Procedure

☐ 1. Cut out the paper lizards.

☐ 2. Attach the thermometers to the lizards.

☐ 3. Complete the hypothesis and record the starting temperatures on the data chart.

☐ 4. Put the lizards outside in the morning. Place one in a sunny spot. This is your Sunny Lizard.

☐ 5. Place the other lizard in a shady spot. This is your Shady Lizard.

☐ 6. Wait 10 minutes. Read both thermometers. Record the temperatures.

☐ 7. Leave the lizards in the same places for at least two hours. After two hours read the thermometers again. Record the temperatures on the data chart.

Inquiry Skills
• Measure
• Infer

Materials
☐ 2 paper lizards
☐ scissors
☐ tape
☐ 2 thermometers

Observations

Time	Sunny Lizard Temperature	Shady Lizard Temperature
Starting temperature	___°C	___°C
After 10 minutes	___°C	___°C
After 2 hours	___°C	___°C

SCIENCE ❸ *Activities*

Lesson 43 • Page 137
Investigation

Conclusions

1. Why are both lizards the same color? *They are representing the same lizard. To have a controlled investigation, the lizards need to be the same color.*

2. Did the Sunny Lizard's temperature stay the same each time you checked? _____ If not, did it get cooler or warmer? _____

3. Did the Shady Lizard's temperature stay the same each time you checked? _____ If not, did it get cooler or warmer? _____

4. In which place would a lizard be more active? _____

5. Do you think a real lizard stays in the same place all day long? _____ Explain your answer. *Possible answers: A real lizard might move to follow the sun; it might move between sun and shade to keep from getting too hot or too cold; it might move to eat or avoid being eaten.*

6. Describe how what you learned can be helpful when caring for a pet reptile. *Possible answer: The home for the reptile would need to have warm and cool areas to be able to meet the animal's needs.*

7. Why is it important for people to care for reptiles? Use the Bible's teaching in your answer. *God created people to manage the earth. This means learning to take care of all animals, including reptiles.*

Lesson 43 • Page 137
Investigation

04

Differentiated Instruction

Remediation: Classify Reptiles

Materials: *Animal Cards*, on cardstock (Instructional Aids 6.1–6.3)

Direct the students to work in pairs to find the animal cards that represent the reptile group. When a student chooses a card, he should tell what characteristic of reptiles made him choose the card.

- Ask two students to read the thermometers and say the temperatures aloud.
- Direct the students to record the temperatures on the data chart.
- Set the timer for 2 hours.
- Ask two students to read the thermometers and say the temperatures aloud.
- Direct the students to record the temperatures on the data chart.
- Write the temperature range 21°C–31°C for display. Remind the students that this is the range at which the lizard would be most active.
- Direct the students to draw **Conclusions** using their observations. The conclusions may be done individually or with their science group.
- Guide a discussion of the students' conclusions.
- Discuss whether the results supported the hypothesis. Explain that having an incorrect hypothesis is not bad. An incorrect hypothesis just means that the experiment had different results from what you expected.

Activities

Animal Classification 2, page 101

This organizer was completed as part of the lesson. Direct the students to keep the page in their science notebooks for review for Test 6.

Investigation, pages 103–4

The student will infer how a cold-blooded animal depends on the temperature of its environment.

Study Guide, pages 105–6

These pages review the concepts taught in Lesson 43. After completion, direct the students to keep them for review for Test 6.

> Some features in the reptiles section of the *Student Edition* can also apply to the other animal groups. To avoid confusion, the features in the *Activities* book will follow the same animal groups as the *Student Edition*.

Assess

Rubric

Use the prepared rubric or design a rubric to include your chosen criteria.

Quiz 6B

The quiz may be given at any time after completion of this lesson.

Objectives
- Identify the characteristics of insects and spiders.
- Describe the structures and features that benefit insects and spiders in survival and growth. **BWS**
- Identify the ways that insects and spiders reproduce.
- Sequence the stages of ladybug metamorphosis.

Materials
- piece of scrap paper, per student
- timer or watch with a second hand
- piece of cotton clothing or other clothing made from plant or animal fibers
- piece of polyester clothing
- video of a ballooning spider

Teacher Resources
- Visuals 6.6–6.8: *Animal Classification 3*; *Life Cycle of an Insect*; *Life Cycle of a Spider*

Vocabulary
- insect
- spider

Engage
- Direct the students to write as many insect names as they can while you time them for one minute.
- Discuss the variety and number of names written.
- Explain that if the student could write 10 insect names every minute, it would take him more than two months of nonstop writing to list all the insects scientists know of. Scientists have discovered more than 900,000 insects and are still discovering new ones!

Instruct

Preparation for Reading
- Preview and pronounce the vocabulary term *insect* and the words *orchid mantis*.
- Direct the students to read pages 138–39 silently to find out some features that insects use to survive.

Teach for Understanding
What is an *insect*? a cold-blooded invertebrate with six legs and three body parts

> The characteristics of insects were studied in *Science 2* except for reproduction. Use the chart, pictures, and captions to review the characteristics of insects as needed. You may want to review by using the *Animal Characteristics 3* visual and the "Insects" cell on *Activities* page 107. The reproduction characteristic will be discussed in this lesson.

Lesson 44

Insect Characteristics

Insect Characteristics
Insects are cold-blooded invertebrates.
Insects have a hard, shell-like covering.
Insects have six jointed legs and three body parts.
Insects usually have four wings.
Insects have two antennae.
Insects reproduce by laying eggs.
Insects go through metamorphosis.

Insects
Characteristics

Do you remember any differences between an insect and a spider? Let's read to find out. An **insect** is a cold-blooded invertebrate with six legs and three body parts.

2 antennae
monarch butterfly
4 wings
praying mantis
3 body parts
6 jointed legs
blue and orange banded leaf beetle

138

Background

Insect Legs

The number of legs listed for insects refers to the adult insects. Many insect larvae have six legs as well, but there are some that do not. Caterpillars seem to have more than six legs, but the first six are the only ones considered true legs. Some larvae, such as fly larvae, have no legs at all.

Features to Survive and Grow

An insect is small, but it is not helpless. God created insects with features to survive and grow. You may remember that each insect has a hard, shell-like covering that gives it protection.

Many insects survive by camouflage. Some insects have colors or patterns that hide them. Other insects have shapes that look like their surroundings.

Some insects have a bitter taste or a bad smell. Predators avoid them. Insects with bright colors warn predators to leave them alone.

Speed is another survival feature. Have you tried to swat a fly or mosquito? You probably did not find it easy. Some insects move very quickly. Their speed helps them escape from predators and even from people.

Other insects survive by stinging or biting. One sting from a wasp causes a person to avoid a second meeting. Some animals do not eat wasps and bees. They avoid getting stung by killing the insect. Flea bites are unwelcome to animals and people. Some insects can bite and sting.

orchid mantis

milkweed locust

The fire ant bites to get a grip and then it stings to inject venom.

What are some insect features that help them to survive and grow?

139

ⓘ Background

Stingers

The stinger of a honeybee is barbed. These barbs cause the stinger and often other parts of the bee to remain attached to the victim after the bee pulls free. Any remaining venom in the stinger will continue to enter the victim, so the stinger should be quickly removed. A honeybee that has lost its stinger will die soon afterward.

Wasps and bees other than honeybees have straight stingers. As a result, their stingers do not remain in their victims. This means wasps and bees with straight stingers can sting repeatedly.

The bite of a fire ant is not what causes the pain. The pain comes from the stinger, which injects venom.

What kind of covering gives an insect protection? a hard, shell-like covering

- Analyze the orchid mantis picture.

  What does the orchid mantis have that helps it to survive and grow? It is shaped like a flower so that it blends in with the plant.

  What is the blending in with the surroundings called? camouflage

- Are insects the only group of cold-blooded animals to use camouflage? No, some fish, amphibians, and reptiles also use camouflage.

  Are any of the insects in the picture on the previous page using camouflage? the praying mantis

- Direct attention to the picture of the milkweed locust.

  What feature does the milkweed locust have to help it survive? bright colors

- How do the bright colors help insects? warns predators that the insect may have a bitter taste or a bad smell

- Analyze the picture of the mosquito.

  What survival features does the mosquito have? speed and biting

  How does speed help an insect? allows it to escape from predators and people

- Have you ever tried to swat a mosquito? Were you successful?

- What other insects use speed as a survival feature? Possible answers: bees, wasps, flies, dragonflies, cockroaches

  How does biting help the mosquito? causes an animal or person to avoid the biting insect

  What do some animals do to avoid getting stung by an insect? They kill it.

- Discuss the picture of the fire ant and its caption.

  What survival features does the fire ant have? biting and stinging

  How does the sting help a fire ant? It injects venom into its prey; it makes a predator want to leave it alone.

- Explain that a sting from a fire ant causes immediate pain. Then the skin starts to itch. This irritation can last from a few hours to a few days. Stings can be life-threatening for people allergic to the venom of fire ants.

- What other insects use biting or stinging as a survival feature? Possible answers: wasps, bees, fleas, hornets, chiggers, yellow jackets

- Allow several students to share their experiences with biting or stinging insects.

- After discussing the Quick Check question, ask the students to explain where the features that help insects survive and grow come from. God designed insects with features to help them survive and grow.

Possible answers: hard, shell-like covering, camouflage, taking the shape of their surroundings, bitter taste, bad smell, bright colors, speed, stinging, and biting

Preparation of Reading

• Preview and pronounce the words *larva* and *pupa*.
• Direct the students to read page 140 silently to find out how most insects reproduce.

Teach for Understanding

What process do most insects go through to become an adult insect? metamorphosis

What other animal group have you studied that can go through metamorphosis? amphibians

• Display the *Life Cycle of an Insect* visual. Explain that the insect used in this life cycle is a seven-spotted ladybug.

What is the first stage in the metamorphosis of an insect? egg

Do insects look like an adult insect when they hatch from eggs? no

Why does the ladybug choose a particular place to lay its eggs? to hide them from predators and to give the offspring food after they hatch

What other animal groups hatch from eggs? some fish, amphibians, and most reptiles

Which animal groups lay eggs on land? reptiles and insects

What is the second stage of metamorphosis of an insect? larva

Do insects look like adult insects when they are larva? no

Describe the stage when they are larva. The larva hatches after several days and looks more like a worm. It sheds its skin several times and eats a lot.

Why do you think the larva sheds its skin? Possible answer: because it eats large amounts of food and keeps growing bigger

Do people shed their skin when they grow bigger? No, God did not create people to shed their skin.

What is the third stage of the metamorphosis of an insect? pupa

What does the larva do in the pupa stage? attaches itself to a leaf and makes a covering

What kind of change happens at the pupa stage? a big change; grows body sections and wings

What is the last stage? adult

What happens to the cycle after the insect becomes an adult? It starts all over again. The female lays eggs.

What is a *female*? a girl, or in people, a woman or lady

What eventually happens to all adult insects? They die.

• Using the *Life Cycle of an Insect* visual, discuss the pattern of birth, growth, reproduction, and death found in the life cycle of an insect.

• To review the classification and features of insects, display the *Animal Classification 3* visual. Direct the students to complete the "Insects" and "Some Survival and Growth Features" cells on *Activities* page 107. Give help as needed.

✓ egg, larva, pupa, adult

Ways to Reproduce

Most insects do not look like an adult insect when they hatch from eggs. An insect becomes an adult through metamorphosis. Some insects such as bees, butterflies, and beetles go through all four stages of metamorphosis.

Life Cycle of an Insect

An adult ladybug lays a cluster of eggs on the underside of a leaf to hide them from predators. She chooses a place where there will be plenty of food when the eggs hatch.

egg

After several days, a *larva* hatches. The larva looks more like a worm. The larva sheds its skin several times and eats a large amount of food.

larva

Finally, the ladybug comes out of its covering as an adult. The cycle begins all over again when the female lays eggs. A ladybug's life may be only a month long.

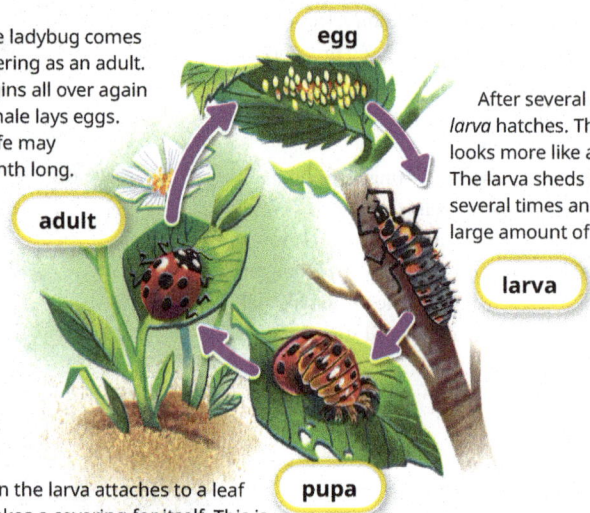

adult

Then the larva attaches to a leaf and makes a covering for itself. This is called a *pupa*. Here the insect makes a big change. It does not eat. It is busy growing its body sections and wings.

pupa

✓ What are the stages of metamorphosis for an insect?

140

⚠ Helps

Metamorphosis

The students will learn about incomplete metamorphosis and complete metamorphosis in a later grade level.

Spider Characteristics

Spiders are cold-blooded invertebrates.

Spiders have a hard, shell-like covering.

Spiders have eight legs and two body parts.

Spiders do not have wings.

Spiders do not have antennae.

Spiders reproduce by laying eggs.

Spiders

Characteristics

Can you spot some of the differences between an insect and a spider as you read? A **spider** is a cold-blooded invertebrate with eight legs and two body parts.

8 legs

daddy longlegs spider

The redback spider makes an egg sac around its eggs.

2 body parts

Australian peacock spider

no antennae

no wings

ground spider

141

Preparation for Reading

- Preview and pronounce the vocabulary term *spider* and the words *antennae*, *Australian*, *recluse*, *venom*, and *tarantula*.

  What group of animals will you be reading about? spiders

- Direct the students to read pages 141–44 silently to find out how spiders use silk.

Teach for Understanding

What is a *spider*? a cold-blooded invertebrate with eight legs and two body parts

- *Note*: All the spiders in the drawing can be found in Australia.

The characteristics of spiders were studied in *Science 2* except for reproduction. Use the chart, pictures, and captions to review the characteristics of spiders as needed. You may want to review by using the *Animal Classification 3* visual and the "Spiders" cell on *Activities* page 107. The reproduction characteristic will be discussed in this lesson.

👥 Differentiated Instruction

Enrichment: Spider Silk
Materials: *Activities* page 108, science notebook
The "Notebook Connection" on *Activities* page 108 extends the student's learning about spider silk.

❓ Misconception

Daddy Longlegs
You may have heard the saying that daddy longlegs are one of the most poisonous spiders, but their fangs are too short to bite humans. There is a daddy longlegs in the same classification as the daddy longlegs spider, but it is in a different order. It is sometimes called harvestmen. These creatures do not have venom and are not poisonous. The daddy longlegs spider does have short fangs, but there is no record of the spider's bite causing any adverse reaction for humans.

What do spiders eat? other animals, mostly insects

- Analyze the picture of the spider that is ballooning. Explain that this is a crab spider.

🔧 What is the crab spider doing? Elicit that it is using its silk to move from place to place; it is ballooning.

- Show the video of the ballooning spider. Share the Background information about how a spider balloons.
- Discuss the picture of the spider and its web. Explain that this is an orb spider.

Do all spiders spin webs? No, but all spiders make silk.

What do spiders that spin webs use their silk for? to catch food

🔧 What is a *fiber*? Elicit that it is a thin, thread-like structure.

- Display the cotton clothing. Explain that the fibers in the clothing are cotton. Point out that some fibers are natural and are made from animals or plants.

🔧 Where does cotton come from? a plant

🔧 What is a *manmade fiber*? Elicit that a manmade fiber is made by man, or people.

- Show the polyester clothing. Explain that some clothes are made from manmade fibers. The clothing is polyester, which is a manmade fiber. Many tents are made from manmade nylon fibers.

How does a spider's silk compare to most manmade fibers? The spider's silk is stronger.

What do some female spiders use their silk for? to make egg sacs

- Analyze the picture of the purse web spider and its caption. Explain that the spider is sitting on its tube-like web. The lighter tan or brown on either side is a tree.

How does the purse web spider uses the tube-like web? to catch prey

- Share the Background information about how a purse web spider catches its prey.
- Analyze the jumping spider and its caption. Explain that the spider is moving from acorn to acorn.

What does a jumping spider use the silk for? to help while it is in the air and when it lands

- Share the Background information about how a jumping spider uses the dragline of silk.
- Direct attention to the trapdoor spider and its caption. Explain that the trapdoor spider is active at night.

🔧 If the spider is waiting under its door and it is dark, how do you think the spider knows when prey is coming? If the students did the "Try It Yourself!" (*Student Edition* page 130) about how amphibians "hear," refer to the activity to help answer the question. Elicit that the spider feels the vibrations of the prey coming close to its burrow.

- Share the Background information on page 175 about how the trapdoor spider builds its trap.

How is the redback spider on the previous page using its silk? to make an egg sac around its eggs

✓ Possible answers: to make webs, to catch prey, to attach themselves to objects, to make egg sacs, to travel, to help in landing

---

Features to Survive and Grow

God created spiders to survive in different ways. Unlike insects, spiders do not eat plants. They eat other animals, mostly insects.

Not all spiders spin webs. But all spiders make silk. Those that spin webs use their silk to catch food. A spider's web looks weak, but it is actually quite strong. In fact, it is much stronger than most manmade fibers. Other spiders use their silk to climb. Some spiders use their silk to attach themselves to objects. This helps to keep spiders from falling. Female spiders use their silk to make egg sacs. Some spiders wrap their prey in silk.

ballooning

web

A jumping spider uses a dragline of silk to help while it is in the air and when it lands.

A purse web spider spins a tube-like sock to catch its prey.

A trapdoor spider builds a hidden door of silk and soil. It waits under the partly open door for its prey.

✓ What are some ways that spiders use silk?

142

ⓘ Background

Ballooning

Scientists at the University of Bristol studied spiders that balloon. The spiders were put in boxes. Then electric fields were generated similar to what the spiders experience outside. Tiny hairs stood up on the spiders' legs, just as a person's hair does after rubbing against a balloon. The electric fields also caused the spiders to start tiptoeing and raising their abdomens in the air. This happened right before the spiders released the silk to balloon. The scientists concluded that spiders sense Earth's electric field. They use the electric field and the air currents to travel through the air. The spiders can float up to 4.8 km (3 mi) above Earth's surface and as far as 1609 km (1,000 mi) out on the ocean.

Purse Web Spider

A purse web spider hides inside its tube-like web, waiting for its prey. When prey gets caught in the web, the spider bites through the web, pulling the prey inside the tube. The purse web spider eats insects and other spiders.

Jumping Spider

The jumping spiders are the largest family of spiders. They can jump over 50 times their body length. The jumping spider releases a line of silk, or a dragline, as it jumps. The dragline gives the spider balance as it lands. The dragline is also a safety net in case the spider needs to stop in the middle of its jump.

Almost all spiders use venom, or poison, to kill their prey. Once a spider catches its prey, the spider injects venom. The venom causes the prey to not be able to move. Most spiders' venom is not harmful to people.

Health and Safety

black widow spider

brown recluse spider

A black widow spider usually lives in dark and quiet places. A brown recluse spider lives in the same kind of places. You should stay away from both spiders. Their venom is harmful to people. Their bites can cause illness and sometimes death.

Technology Connection

Long ago spiderwebs were used to stop bleeding wounds. Some people even used the silk as fishing lines for small fish. Scientists know that a piece of spider silk is stronger than steel. It is lighter than cotton. It is thinner than your hair. It is tougher than a bulletproof vest.

There are problems with making spider silk as a product. Spiders cannot be raised together. They will eat each other. Spiders also spin small amounts of silk.

Scientists have experimented to produce more silk and keep the cost low. After many years, a way was found to make manmade silk that is like spider silk. Scientists continue to try to find an easier way to produce manmade silk.

The tarantula chases its prey at night. They eat insects but may eat other animals, such as toads, frogs, snakes, lizards, or small birds.

143

What do almost all spiders use to kill their prey? venom

- Explain that *venom* is a poison substance that a spider injects into its prey.

  What does the venom do to the prey? causes the prey not to be able to move

- Discuss "Health and Safety."

  What two spiders have venom that is harmful to you? black widow spider and brown recluse spider

  Where do these spiders like to live? in dark and quiet places

  What should you do if you see a black widow spider or a brown recluse spider? Elicit that the students should leave the spider alone, not touch the spider, and tell an adult about the spider.

- Direct attention to page 141. Explain that the redback spider is also known as the Australian black widow. It is a highly venomous spider.

- Discuss "Technology Connection."

  What were spiderwebs used for long ago? to stop bleeding wounds and as fishing lines for small fish

  How strong is a spider's web? stronger than steel and tougher than a bulletproof vest

  What are two problems with making spider silk as a product? Spiders eat each other and need to be raised separately. Spiders spin small amounts of silk.

  What way have scientists found to produce silk? to make manmade silk that is like spider silk

- Direct attention to the picture of the tarantula and its caption.

- Share the Background information about tarantulas.

Background

Trapdoor Spiders

A trapdoor spider lives in warm places and is active at night. It spends most of its time underground. The spider builds a trapdoor over a burrow. The burrow is about 30 cm (12 in.) deep and about 5 cm (2 in.) across. The soil mixed with the silk camouflages the trapdoor, making it difficult to see when it is closed. The trapdoor is hinged on one side with silk. When prey comes too close, the spider jumps out of its burrow and captures the prey. If a trapdoor spider makes its burrow beside creeks or rivers, it likes to catch little fish. Trapdoor spiders eat all types of insects, frogs, baby birds, baby snakes, mice, and small fish.

Tarantulas

Tarantulas look scary but are harmless to people. Their bite can be painful but their venom is weaker than a bee's. There are hundreds of species of tarantulas. They shed their hard coverings in a process called molting. Most tarantulas burrow into the ground. They move slowly and carefully during the night. They do not spin webs, although some spin a tripwire to alert them when prey is near their burrows.

- Display the *Life Cycle of a Spider* visual.

  How many stages are in the life cycle of a spider? three

  What is the first stage? egg

  Where does the female lay her eggs? in a silk egg sac or sacs

  Why do you think she lays so many eggs? Possible answer: so that some spiders will survive

  What is the second stage? spiderlings

  What are *spiderlings*? offspring; young spiders

  When does a spiderling leave the egg sac? after it loses its first outer skin

  What does a spiderling look like? an adult spider except smaller

  What is the last stage? adult

  What are some characteristics of the adult spider? It can take care of itself. Males are often smaller than females.

- Using the *Life Cycle of a Spider* visual, discuss the pattern of birth, growth, reproduction, and death found in the life cycle of a spider.

  Who cares for all organisms, even the smallest ones? God

  Why is it important to learn about cold-blooded animals? It is important in serving God and others.

  How is learning about cold-blooded animals important in serving God and others? Possible answers: helps you take better care of your cold-blooded pets; helps you obey God's command to take care of the earth; helps you know how to protect animals and other people from some animals

- To review the classification and features of spiders, display the *Animal Classification 3* visual. Direct the students to complete the "Spiders" and "Some Survival and Growth Features" cells on *Activities* page 107. Give help as needed.

- Guide a discussion comparing and contrasting characteristics of insects and spiders. You may choose to use the *Animal Classification 3* visual. You may want to list the characteristics in a Venn diagram. (See Helps.)

egg, spiderlings, adult

Apply

Activities

Animal Classification 3, page 107

This organizer was completed as part of the lesson. Direct the students to keep it for review for Test 6.

Study Guide, pages 109–10

These pages review the concepts taught in Lesson 44. After completion, direct the students to keep them for review for Test 6.

Some features in the insects and spiders section of the *Student Edition* can also apply to the other animal groups. To avoid confusion, the features in the *Activities* book will follow the same animal groups as the *Student Edition*.

A Way to Reproduce

A spider's life cycle has three stages: egg, spiderlings, and adult.

Female spiders lay several thousand eggs in a silk egg sac. Some female spiders spin several smaller egg sacs. Some carry the egg sac with them.

Life Cycle of a Spider

egg

adult

spiderlings

An adult spider can take care of itself. Males are often smaller than the females. Spiders mate and begin the life cycle again.

The spiderlings, or offspring, hatch in the egg sac. After a spiderling loses its first outer skin, it leaves the sac. A spiderling looks like its parents. It will lose its skin several times before becoming an adult spider.

God designed each part of His creation. He cares for even His smallest organisms. Learning about cold-blooded organisms is important in serving God and others.

What are the stages of a spider's life cycle?

👥 Differentiated Instruction

Remediation: Classify Insects and Spiders

Materials: *Animal Cards,* on cardstock (Instructional Aids 6.1–6.3)

Direct the students to work in pairs to find the animal cards that represent the insect and spider groups. When a student chooses a card, he should tell what characteristic of insects or spiders made him choose the card.

⚠ Helps

Venn Diagram

Possible Venn diagram of the characteristics of insects and spiders:

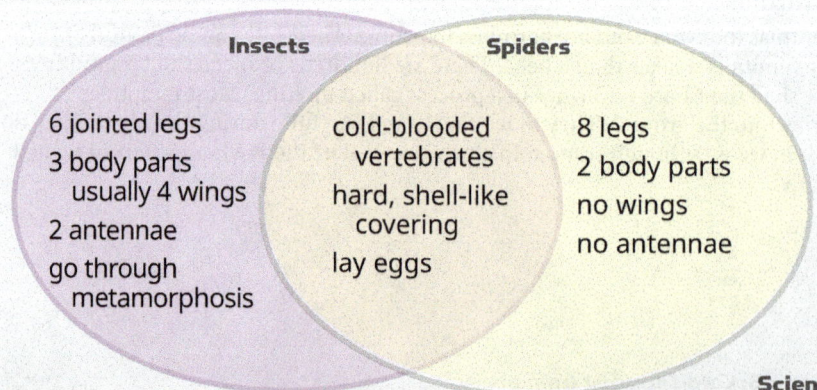

	Insects		Spiders
	6 jointed legs	cold-blooded vertebrates	8 legs
	3 body parts usually 4 wings	hard, shell-like covering	2 body parts no wings
	2 antennae		no antennae
	go through metamorphosis	lay eggs	

Changes in Dark Places

Inquiry Skills
• Predict
• Observe
• Communicate

A mealworm likes to live in dark places. It gets its water through some of the food that it eats. A mealworm goes through its life cycle better with warmer temperatures.

There are four stages of an insect's life cycle: egg, larva, pupa, and adult. The mealworm is part of a mealworm beetle's life cycle. What stage do you think the mealworm is?

In this Investigation, you will set up a living space for mealworms. You will predict and observe the stages of a mealworm beetle's life cycle.

145

Objectives
• Create a hypothesis about the stages in the life cycle of a mealworm.
• Construct a living space for mealworms.
• Observe and record the changes in a mealworm's life cycle.
• Draw conclusions from the data collected.

Materials
• See *Activities* page 111. (See Helps.)
• camera or cellphone camera (optional)

> This Investigation may be worked as individuals or in groups. You may choose to have one living space and allow each student to do his own observations of the mealworms. You will want to prepare the lid of each container with small air holes.

Teacher Resources
• Visuals 3.3, 6.9: *Science Inquiry Skills*; *Life Cycle of Mealworms*

Engage

Why are observations important for learning? Possible answers: to use all five senses when gathering information; to understand things better

Have you ever seen a beetle or a mealworm? Where?

• Explain that mealworms and beetles get into and harm grains and their products, such as flours and cereals. Some birds like to eat mealworms.

Instruct

Preparation for Reading
• Direct the students to remove pages 111–14 from their *Activities* books. Direct them to read page 145 and *Activities* 111–14 silently before beginning.

Teach for Understanding
• Display *Science Inquiry Skills* visual.
• Direct attention to the **Inquiry Skills**.

What do scientists use inquiry skills for? to gather and use facts

Which inquiry skills will you be using in this Investigation? predict, observe, and communicate

• Review what it means to *predict*, *observe*, and *communicate*.

⚠ Helps

Changes in Dark Places
Most pet or bait stores sell mealworms for an inexpensive price.

Some people with asthma tendencies may have respiratory difficulties when around mealworms. Students who have asthma may want to wear a mask when working with the mealworms. It is recommended that the students wash their hands before and after handling the mealworms, even though they are wearing gloves. Follow any other procedures that your school may have for handling live animals.

In place of the baby carrot you can use an apple slice, slice of raw potato, or a small wedge of cabbage. Remove and replace any of these if they get moldy or rotten. Depending on the length of the activity, more oatmeal may need to be added.

Choose one day a week for student observations. It should be the same day each week. Instead of the students drawing the changes each week, you may want to take pictures. Then direct the students to make a journal with the pictures or display the life cycle on posterboard.

Apply

- Choose a student to read the **Problem** aloud.
  How will you know what stages are in a mealworm's life cycle? by observing the mealworms
- Direct the students to complete the **Hypothesis** to answer the problem.
- Instruct the students to follow and check off **Procedure** Steps 1–11 as each is completed.

INVESTIGATIO

Changes in Dark Places

Name _____

Health & Safety

- Wear gloves when handling the mealworms.
- Wash your hands before and after handling the mealworms.

Inquiry Skills
- Predict
- Observe
- Communicate

Materials
- ☐ mealworms
- ☐ clear container with a lid
- ☐ disposable (latex-free) gloves
- ☐ uncooked oatmeal
- ☐ several baby carrots
- ☐ hand lens
- ☐ strips of construction pa[per] (optional)
- ☐ tape (optional)

Problem

What are the stages of a mealworm's life cycle?

Hypothesis

I will observe these stages of a mealworm's life cycle:
Accept any reasonable answer. The correct stages
are egg, larva, pupa, and adult.

Procedure

- ☐ 1. Prepare a living space for the mealworms.
- ☐ 2. Place a small amount of oatmeal in a corner of the clear container.
- ☐ 3. Add a carrot to the living space.
- ☐ 4. Wash your hands and put on the gloves.
- ☐ 5. Place the mealworms in the container.
- ☐ 6. Put the prepared lid with air holes on the container.
- ☐ 7. Wash your hands after handling the mealworms.
- ☐ 8. Observe the mealworms with the hand lens.
- ☐ 9. Record a description of the mealworms in the data chart.

SCIENCE 3 *Activities*

Lesson 45 • Page 145
Investigation

⚠ **Helps**

Living Space
Mealworms like temperatures around 25°C–27°C (77°F–80°F) and around 70 percent relative humidity. With temperatures below or above this the mealworms may take longer to go through the life cycle.

ⓘ **Background**

Mealworm Development
The darkling beetle, or mealworm beetle, lays hundreds of tiny, white, oval eggs. The eggs are about 2 mm (0.79 in.) in length and 0.9 mm (0.04 in.) wide. It usually takes from 4 to 19 days for the eggs to hatch.

In the larval stage a mealworm sheds the exoskeleton 9–20 times. The temperature and amount of moisture have some effect on the number of times shedding takes place. Also, the larva turns into a pupa more quickly in warmer temperatures.

If the pupal stage happens over winter, the pupal stage may last longer. The pupa does not eat while it is transforming into an adult.

A white adult darkling beetle emerges. It soon turns brown and then almost black. The adult lives for a few months, and then the entire life cycle starts again.

10. Draw a picture of the mealworms in the data chart.

11. Place the container in a dark place. If there is not a dark place, add strips of construction paper to the sides and top of the container. Cut air holes in the top piece of construction paper.

12. Choose the same day each week to observe the mealworms using the hand lens. Record your descriptions and draw a picture of the mealworms in the data table.

bservations

Week	Descriptions	Observations (Drawings)
Beginning		
1		
2		
3		
4		
5		
6		
7		

• Direct attention to Step 12. Ask a volunteer to read it aloud.

> You may want to choose the same day of the week for all students to do their observations. If you are using just one living space, you may want to divide the students, asking them to observe on different days. Each student needs enough time to observe and draw.

• Direct attention to the "Observations" data chart. Remind the students of the importance of accurately recording their **Observations**.
• Explain that a short description should be written, as well as a drawing done for each observation.

> The "Observation" data chart provides for 10 weeks of observations. You may want to extend the observations. You may also choose to observe for fewer than 10 weeks. Observing some changes in the mealworm is the important concept. The *Life Cycle of Mealworms* visual will allow you to show the complete life cycle whenever you decide to stop the observations.

- Direct the students to draw **Conclusions** using their observations.

Changes in Dark Places   Name _____

Week	Descriptions	Observations (Drawings)
8		
9		
10		

Conclusions

1. Why did you place the oatmeal and the baby carrot in the mealworms' living space? __
   The oatmeal was food. The baby carrot gave the mealworms water.

2. Why did you place the container in a dark place or add strips of construction paper to
   the container? *Mealworms like to live in dark places.* _____

3. Draw the life cycle of a mealworm. Use the drawings from the data table.

drawing box

4. Was your hypothesis correct?　　○ yes　　　　○ no

5. How did your findings agree or disagree with your hypothesis? _____

_____

_____

6. Compare your drawings and observations with other students'. How are they alike or

different?_____

_____

_____

• Guide a discussion of the students' conclusions.

Activities
Investigation, pages 111–14
Students will predict the life cycle of mealworms.

Assess

Rubric
Use the prepared rubric or design a rubric to include your chosen criteria.

Objective
• Recall terms and concepts from Chapter 6.

Apply

Review
• Material for Test 6 will come from the Study Guides on *Activities* pages 97–99, 105–6, and 109–10, as well as Quizzes 6A and 6B. You may review any or all of the material during the lesson.
• You may choose to review Chapter 6 by playing the Animal Puzzles review game or a game from the Game Bank (Teacher Resources).

🏆 Review Game

Animal Puzzles

Materials: full-page pictures of a fish, an amphibian, a reptile, an insect, and a spider

Note: Prepare as many different animal group pictures as there are teams.

Draw the same number of grid lines (strips or squares) on each picture and cut the pictures apart along the lines to form puzzles. Mix up the pieces and place them upside down on a table.

Group the students into teams and assign one puzzle to each team. For each correct answer to a review question, the team gets one piece of its animal picture. The first team to assemble its picture and identify and explain why the animal is a fish, an amphibian, a reptile, an insect, or a spider wins.

NOTES

NOTES

_____

_____

_____

_____

_____

_____

_____

_____

_____

_____

_____

_____

_____

_____

_____

_____

_____

_____

_____

_____

_____

_____

_____

_____

_____

_____

Objective

• Apply terms and concepts from Chapter 6.

Assess

• Administer Test 6.

Chapter Objectives

- Analyze the characteristics and features of warm-blooded animals.
- Formulate statements about warm-blooded animals and God's design.
- Analyze the life cycles of animals.
- Apply the inquiry skills of observe, predict, measure, and infer.
- Analyze inherited traits, instincts, and learned behaviors.

Lesson Objectives

- Recall the differences between invertebrates and vertebrates and between warm-blooded and cold-blooded animals.
- Identify the characteristics of birds.
- Identify and describe the structures and features that benefit birds in survival and growth. **BWS**
- Formulate a biblical explanation of why birds reproduce after their own kind. **BWS**
- Sequence the stages of a bird's life cycle. **BWS**
- Record observations of birds at a bird feeder.

Materials

- oral thermometer
- picture of a pelican with its throat pouch extended
- hanging bird feeder
- bird seed

> Prior to this lesson, if you have not already done so, hang a bird feeder where the students can observe it.

Teacher Resources

- Visuals 7.1–7.2: *Animal Classification 4; Life Cycle of a Bird*
- audio of a Yao hunter calling a honeyguide (TeacherToolsOnline.com)
- video of a male sandgrouse collecting water (TeacherToolsOnline.com)

Vocabulary

- bird

Chapter Introduction

- Direct the students' attention to the chapter title and the picture of the dolphin. Share any Background information you desire. Provide time for the students to leaf through the chapter, looking at the headings and pictures. Discuss what they think the chapter will be about.

> Use completed Looking Ahead, *Activities* page 115, to generate a discussion on what the students already know about birds and mammals.

Big Question
What are the characteristics of warm-blooded animals?

ⓘ Background

Chapter Opener Photo

The photo on pages 146–47 is a dolphin jumping out of the ocean. Dolphins are easily recognized by their curved mouths, which give them the appearance of smiling. They feed primarily on fish and squid, which they track using echolocation. A bottlenose dolphin can make as many as a thousand clicking noises per sound while using its echolocation. As mammals, dolphins are warm-blooded and feed milk to their offspring. An offspring usually stays with its mother for up to six years.

> Visit TeacherToolsOnline.com for resources to enhance the lessons.

Chapter 7

Warm-Blooded Animals

Suppose you take your temperature while you stand outside in the hot sun. Then you go inside a cool building and take your temperature again. Do you think they would be different or be about the same?

You may feel hot outside and cool inside. But your body temperature would not change much. Because you are warm-blooded, your body temperature stays about the same no matter where you are.

God has given your body different ways to keep the same body temperature. When you are hot, you sweat to cool off. When you are cold, you shiver to warm up. God made warm-blooded animals to do these things, too.

147

Big Question
- Ask a volunteer to read aloud the Big Question.
- Explain that the students will find the answer to the question as they read the chapter.

Engage

The Feathery Friends Exploration, *Student Edition* page 152, is an excellent way to engage the students for this lesson. You may want to use the Exploration on *Teacher Edition* pages 190–92 rather than the following Engage activity.

- Display the oral thermometer.
  What is this used for? to take a person's temperature
  When is this often used? when a person is not feeling well
- Explain that thermometers like these work to tell the temperature of people and some warm-blooded animals. Point out that the students will be learning more about different kinds of warm-blooded animals.

Instruct

Preparation for Reading
- Read aloud the first paragraph and poll the students as to whether the temperatures would be different or about the same.
- Direct the students to read page 147 silently to find out the answer.

Teach for Understanding
- Review the meaning of *warm-blooded*.
  What does it mean to say that you are warm-blooded? Your body temperature stays about the same no matter where you are.
  What happens to your body when you get too hot? You start to sweat.
  What happens when you get too cold? You start to shiver.

Preparation for Reading

- Preview and pronounce the vocabulary term *bird* and the words *egret* and *preening*.
- Direct attention to the headings, captions, and chart. Remind the students to read the captions and the chart as they read the text.
- Direct the students to read pages 148–50 silently to find out ways that birds use their beaks.

Teach for Understanding

- Guide a review of the main classifications of animals.
- Display the *Animal Classification 4* visual. Guide the students as they complete *Activities* page 117 during the lesson.

- What two groups that describe an animal's temperature can animals be classified into? warm-blooded and cold-blooded
- What is the main difference between the two groups? A cold-blooded animal's body temperature changes depending on its surroundings; a warm-blooded animal's body temperature stays about the same.
- What two groups can warm-blooded animals be classified into? vertebrates and invertebrates
- What is the main difference between vertebrates and invertebrates? Vertebrates have backbones and invertebrates do not have backbones.

- Discuss warm-blooded animals.
  What are two groups of animals that are warm-blooded? birds and mammals
  What do dogs do to keep cool? pant
- What does it mean to *pant*? Elicit that it means to breathe in short, fast gasps similar to what you do after hard exercise or running.
- Discuss the characteristics of birds.
  What is a *bird*? a warm-blooded vertebrate that has feathers and wings

> The characteristics of birds were studied in *Science 2*. Use the chart, pictures, and captions to review the characteristics of birds as needed. You may want to review by using the *Animal Classification 4* visual and the "Birds" cell on *Activities* page 117.

---

most fly

sandhill crane

breathe with lungs

blue jay

red-tailed hawk

vertebrate

lay eggs

2 wings

black-headed grosbeak

148

Warm-Blooded Animals

Many vertebrate animals are warm-blooded. Birds and mammals are warm-blooded animals. They have ways to keep their body temperature the same. Dogs pant to keep cool. Some mammals and birds shiver as we do.

Birds
Characteristics

A **bird** is a warm-blooded vertebrate that has feathers and wings.

Bird Characteristics

Birds are warm-blooded vertebrates.

Birds breathe air through lungs.

Birds have feathers and two wings.

Most birds fly.

Birds reproduce by laying eggs.

ⓘ Background

Feathers

Feathers are made of keratin, the same material that claws, hooves, human nails, animal fur, and human hair are made of. There are different types of feathers. These feathers look different and have different purposes. Some are used for flying, some act like sensors, and some simply help the bird keep warm and dry. A bird can have anywhere from 1,000 to 25,000 feathers.

Molting

Birds lose their feathers and grow new ones usually once or twice a year. This process is called *molting*. The feathers are lost in a specific order and at a specific rate depending on the species of bird.

Types of Flying

Birds can fly in different ways. Most birds flap their wings to fly. Many large birds rise high in the air and then soar on wind currents without flapping their wings. Hummingbirds are one of the few birds that can hover.

Features to Survive and Grow

Birds need to keep their body temperature the same. Some birds trap warm air next to their bodies by fluffing their feathers. Some birds also tuck their heads into their feathers. During the night some birds huddle together to keep warm. Some birds build up fat as cold weather approaches. This helps them to keep warm.

Birds have a need of food. Some birds that eat insects will follow animals that graze. Perhaps you have seen a cattle egret. The egret follows cattle. It eats insects that the cattle stir up as they move and eat. Some birds fly in groups to increase their opportunities of finding food.

Most birds use their feet for walking or perching. But God designed some of their feet for other purposes. A woodpecker's toes are arranged so that it can hang on to a tree trunk. Some birds have a fifth with a sharp claw. The claw is used to fight. Some birds, such as owls, use their sharp claws to kill their prey.

What does the duckling use its feet for?

Fantastic Facts

Several areas in Africa are home to small birds called honeyguides. The honeyguides and people communicate with each other. A person will call the honeyguide with a special sound. The honeyguide then guides the person to a bees' nest high up in a tree. The young man will collect the honey. The honeyguide will feed on the beeswax and bee larvae.

149

- Continue to display the *Animal Classification 4* visual and guide the students as they complete *Activities* page 117 during the discussion of the survival and growth features.

What does *survive* mean? to stay alive

- Guide students to analyze the first picture.

Why is the owl fluffing its feathers? to trap warm air next to its body to help it keep warm

What are some other ways that birds keep warm to survive? huddling together, tucking their heads into their feathers, building up fat as cold weather approaches

- Engage student discussion of the second picture.

Describe what the cattle egret is doing. The cattle egret is eating insects that the cow stirs up as the cow moves and eats.

How does flying in groups help some birds find food? Birds can take advantage of the same food supply; they can all use the same food source that one bird finds; their opportunities of finding food increase.

- Direct attention to the black-headed grosbeak on the previous page.

- Explain that toxins in the monarch butterfly make it poisonous to most birds. The black-headed grosbeak is one of the few birds that can eat it.

- Guide a discussion of the duckling picture and the caption on the page.

What does the duckling use its feet for? It uses its feet for swimming.

What kind of feet does the duckling have? webbed feet

What other purposes do birds use their feet for? walking, perching, hanging on a tree trunk

What do some birds use their claws for? fighting, killing their prey

- Discuss "Fantastic Facts."

What does the honeyguide feed on? beeswax and bee larvae

- Play the audio of a Yao hunter calling a honeyguide.

- Ask the students to try to imitate calling a honeyguide.

Why do people call to a honeyguide? so the bird will lead the person to a bees' hive

---

Differentiated Instruction

Enrichment: Birds That Do Not Fly

Materials: *Activities* page 118, science notebook

The "Notebook Connection" on *Activities* page 118 extends the student's learning about birds that do not fly.

- Guide a discussion of the different jobs of birds' beaks.
  What are some of the jobs of birds' beaks? eating food, building nests, killing prey, smoothing or cleaning feathers
  What is smoothing or cleaning a bird's feathers called? preening
- Explain that birds make a light oil near their tail feathers. As birds preen they spread the oil on their feathers. The oil helps keep birds dry in the rain or when they are in the water.
- Analyze the pictures of birds' beaks.
  What bird do you think the first beak belongs to? Elicit that the beak belongs to a pelican.
  How do you think a pelican uses its beak? It scoops up fish with its beak.
- Display the picture of the pelican. Explain that the pelican expands the throat pouch of its beak when it is fishing. At other times the pouch is folded up.
  What bird to you think the next beak belongs to? It belongs to a pied avocet.
- Explain that the pied avocet feeds in shallow water and mud flats. It moves its beak back and forth in the water searching for insects, crayfish, shrimp, and krill.
  What bird do you think the third beak belongs to? It belongs to a merganser.
  Look closely at the edges of the beak. What do they look like? like a saw or sharp teeth
- Explain that the merganser swims along the surface of fresh water. When the duck spots its prey or fish, it dives under water in pursuit. The saw-tooth beak helps the merganser hold on to its prey.
  What bird do you think the last beak belongs to? Elicit that it belongs to a hummingbird.
  Why do you think the beak is so long? Elicit that the hummingbird eats from flowers and needs to be able to reach the center of them.
- Discuss the designs of the birds' beaks on page 148.
- Analyze the picture of the eggs in the bird's nest.
  What feature helps hide the eggs? camouflage
- Explain that not only may the eggs be camouflaged but the nests themselves may also be camouflaged.
  Where are some places you can find birds' nests? trees, tall grasses, and cracks of rocks
- Discuss the Quick Check question. Then ask:
  Why do birds have these features to survive and grow? God designed birds with these features to help them with survival and growth.

✓ stay warm: fluffing feathers, tucking heads into feathers, huddling together, and building up fat;
find food: flying in groups to increase opportunities;
feet: walking, perching, and hanging on trees;
sharp claws: fighting and killing prey;
beaks: eating food, building nests, and smoothing or cleaning feathers;
camouflage: protecting eggs and nests

Designs of Beaks

God designed birds' beaks for different jobs. Birds use their beaks for eating food. Some beaks are designed to help them build nests. Some birds use their beaks to kill prey. Most birds also use their beaks to smooth or clean their feathers. This is called *preening*.

Bird eggs are often camouflaged. Speckled brown eggs may hide in nests of leaves and twigs. This helps protect the eggs from animals that eat them. Some nests are camouflaged in trees or in tall grasses. Others are hidden in the cracks of rocks.

Try It Yourself!

Penguins spend about half of their lives in the sea. How do they stay dry? Take a cutout of a penguin. Color the whole penguin completely with black, white, and orange crayons. Make sure you color each part twice so there is a thick coat of crayon. Take a spray bottle with water and spray the penguin two or three times. Observe what happens to the water.

✓ How do birds survive and grow?

150

👥 Differentiated Instruction

Remediation: Camouflage
Materials: jelly beans
Hide the jelly bean "eggs" around the room so that most are camouflaged.
Allow one student at a time to have 10 seconds to find all the eggs he can. Then allow another student to try. Continue until all the eggs are found.
Discuss which eggs were easier to find and why.

Enrichment: Bird Beaks
Materials: *Activities* page 122, science notebook
The "Notebook Connection" on *Activities* page 122 extends the student's learning about birds and their beaks.

🧪 Activity

Try It Yourself!
Materials: *Penguin* (Instructional Aid 7.1); black, white, and orange crayons; spray bottle with water
Allow the students to try this activity on their own.

Fantastic Facts

Sandgrouse live in dry habitats. A male sandgrouse will find water. He then stands in water. He rocks his body side to side, shaking his belly feathers in the water. The feathers collect water. This can take as long as 15 minutes. He can carry 25 mL (2 Tbsp) of water in his feathers back to the nest. The chicks use their bills to get the water from his feathers.

A Way to Reproduce

God created birds to reproduce after their own kind. Birds follow a life cycle to reproduce.

Life Cycle of a Bird

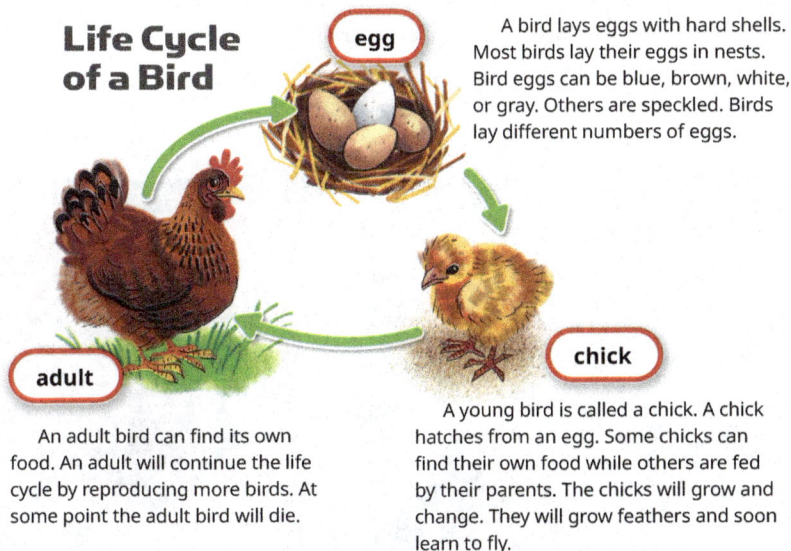

egg

A bird lays eggs with hard shells. Most birds lay their eggs in nests. Bird eggs can be blue, brown, white, or gray. Others are speckled. Birds lay different numbers of eggs.

adult

chick

An adult bird can find its own food. An adult will continue the life cycle by reproducing more birds. At some point the adult bird will die.

A young bird is called a chick. A chick hatches from an egg. Some chicks can find their own food while others are fed by their parents. The chicks will grow and change. They will grow feathers and soon learn to fly.

What pattern does a bird's life cycle have?

151

Preparation for Reading

- Direct the students to read page 151 silently to find out the stages in a bird's life cycle.

Teach for Understanding

- Direct attention to "Fantastic Facts."

  Who brings the water back to the sandgrouse chicks? the male sandgrouse

  Where do the chicks get the water? from his feathers

- Play a video of the male sandgrouse collecting water for the chicks.
- Read Genesis 1:24–25 aloud.
- Direct the students to collaborate with their science partners to write a statement about how God created animals to reproduce.
- Allow for several partners to share their statements. Write them for display.
- Discuss the statements. Guide the students as they make adjustments and write one statement.
- Display the *Life Cycle of a Bird* visual. *Note*: The chickens in the illustration are New Hampshire Red Chickens. They are one of the popular chickens being raised in backyards.

  What is the first stage in the life cycle of a bird? egg

  Do all birds lay the same number of eggs? no

  What colors are their eggs? blue, brown, white, or gray

  What is the second stage in the life cycle? chick

  What is a *chick*? a young bird that hatches from an egg

  How do chicks get food? Some find their own. Others are fed by their parents.

- Direct attention to the sandhill crane on page 148.
- Explain that the sandhill crane chicks can leave the nest within eight hours of birth and are capable of swimming.

  What is the last stage of the life cycle? adult

  How does an adult bird continue the life cycle? by reproducing more birds

- Using the *Life Cycle of a Bird* visual, guide the students in finding the life cycle pattern of birth, growth, reproduction, and death in the life cycle of a bird.

egg, chick, and adult or birth, growth, reproduction, and death

Differentiated Instruction

Remediation: Classify Birds

Materials: *Animal Cards,* on cardstock (Instructional Aids 6.1–6.3)

Direct the students to work in pairs to find the animal cards that represent the bird group. When a student chooses a card, he should tell what characteristic of birds made him choose the card.

Enrichment: Bird Eggs and Nests

Materials: *Activities* page 124, science notebook

The "Notebook Connection" on *Activities* page 124 extends the student's learning about bird eggs and nests.

Background

Birds and Camouflage

Not only do many birds build their nests in places where their nests and eggs are camouflaged, but often the bird itself is also camouflaged. For example, when a brown creeper feels threatened, it flattens its wings against a tree trunk and becomes motionless. The brown feathers blend in and hide the bird. Other birds, such as the ptarmigan, change colors as the seasons change. In the summer the ptarmigan has brown feathers, but in the winter it grows white feathers that help it blend in with its snowy environment.

Preparation for Reading

- Direct the students to remove pages 119–21 from their *Activities* books. Direct them to read page 152 and *Activities* pages 119–21 silently.

Teach for Understanding

- Direct attention to the **Inquiry Skills**.

  Which inquiry skills will you be using in this Exploration? observe and measure

  What does it means to *observe*? to use your senses to find out something

  What does it mean to *measure*? to find the size, number, or amount of something

- Choose a student to read the **Purpose** aloud.

Feathery Friends

Name _____

Purpose

I want to observe birds.

Procedure

☐ 1. Your teacher has placed a bird feeder where you will be able to observe birds.

☐ 2. Add birdseed to the bird feeder as needed.

☐ 3. Spend time each day for five days observing birds.

☐ 4. Record your observations in the data table. Put a tally mark for each bird you see under "How Many Seen" in the data table.

☐ 5. At the end of five days, mark the bar graph to show how many birds you saw each day.

Inquiry Skills
- Observe
- Measure

Materials
☐ bird feeder
☐ birdseed

Notebook Connection

Draw and color a picture in your notebook of one of the birds you observed at the bird feeder. Next to the picture, write the date you saw the bird.

SCIENCE 3 *Activities*

Lesson 48 • Page 152
Exploration

EXPLORATION

Feathery Friends

Inquiry Skills
- Observe
- Measure

You can learn about people by talking with them. Most birds cannot talk to you. But you can learn about them by observing. If you watch a bird feeder, you will see birds with different colors, sizes, and sounds. Your observations can help you identify the birds you are seeing.

In this Exploration, your teacher will hang a feeder near a window. You will observe the birds that visit it. When you complete your observations, you will graph how many birds you saw each day.

152

⚠ Helps

Bird Websites
Additional information, links, and apps you may find helpful can be found at TeacherToolsOnline.com.

Color of			Bird Name	How Many Seen				
Body	Wings	Beak		Day 1	Day 2	Day 3	Day 4	Day 5

Lesson 48 • Page 152
Exploration

20

Apply

- Discuss the type of food that will be used in the feeder. Guide the students to conclude that the type of food may determine the kinds of birds that come to the feeder.
- Direct the students to record their **Observations** in the data chart.

⚗ Activities

Bird Guide
Collect pictures of birds seen in your area. Label each picture with the bird's name and add a few facts about each bird. Possible facts to include would be information about its size, its song, what it eats, and where it lives. Organize the pictures in a notebook to make a local field guide.

Discuss Bird Sounds
Discuss the sounds that birds in your area make. Allow students to demonstrate several common bird sounds, such as a crow's caw, a duck's quack, a dove's coo, a hawk's screech, or an owl's hoot. You may choose to find audio recordings of bird songs on the internet.

- Direct the students to graph the number of birds they observed.
- Direct the students to draw **Conclusions** using their observations.
- Guide a discussion of the students' conclusions.
- Allow the students to share and compare their graphs.

Activities

Looking Ahead, page 115
This page assesses the students' knowledge prior to the chapter. *Note:* You may want the students to do the "Notebook Connection" at the bottom of the page.

Animal Classification 4, page 117
This page helps the students to classify animals. It was completed as part of the lesson.

Exploration, pages 119–21
The students will record and graph their observations of birds at a bird feeder.

Bird Puzzle, page 123
This page reinforces the concepts taught in Lesson 48.

Assess

Rubric
Use the prepared rubric or design a rubric to include your chosen criteria.

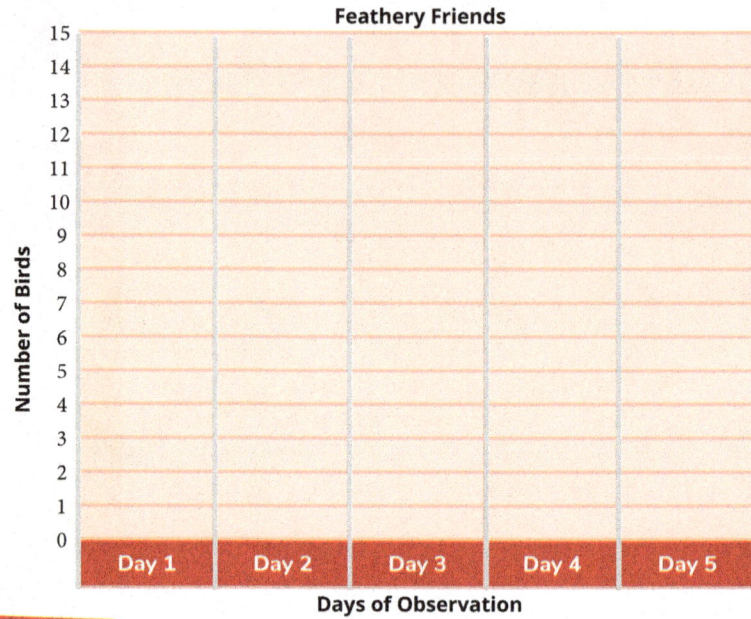

Feathery Friends   Name _____

Feathery Friends

Number of Birds

15 14 13 12 11 10 9 8 7 6 5 4 3 2 1 0

Day 1   Day 2   Day 3   Day 4   Day 5

Days of Observation

Conclusions

1. List some characteristics all birds that you observed have in common.

_____   _____
_____   _____
_____   _____

2. Which bird was your favorite? _____

3. What can you do for birds to obey God's command in Genesis to take care of the earth

Possible answers: Put out water for birds to drink; put out food for birds to eat; keep do

and cats away from birds' nests; take proper care of chickens being raised.

SCIENCE 3 *Activities*

Lesson 48 • Page 152
Exploration

Differentiated Instruction

Enrichment: Drawing a Bird
Materials: *Activities* page 119, science notebook

The "Notebook Connection" on *Activities* page 119 extends the student's learning by asking him to draw a bird that he observed at the bird feeder. You may want to extend the activity further by asking him to try to identify the bird.

Enrichment: Classify Birds
Materials: *Animal Cards*, on cardstock (Instructional Aids 6.1–6.3)

Direct the student to find the bird animal cards. After he has found the cards, ask him to classify the birds by other parts. For example, the birds could be grouped by their beak shape or the type of claws they have.

Mammals

Think of a name of an animal. You most likely named a mammal. Many pets are mammals. So are most farm animals. A **mammal** is a warm-blooded vertebrate that has fur or hair and feeds milk to its offspring.

Characteristics

A mammal may be as small as a bat that can weigh just 28.4 grams (1 ounce). A mammal may be as large as a blue whale that can weigh as much as 90.8 kilograms (200,000 pounds). Do you remember the characteristics of mammals?

Mammal Characteristics

- Mammals are warm-blooded vertebrates.
- Mammals breathe air through lungs.
- Mammals have fur or hair.
- Most mammals give birth to live offspring.
- Mammals feed milk to their offspring.

vertebrate

giraffes

elephant

cheetah with her cub

fur or hair

lungs

meerkat

warthog

oryx

153

Lesson **49** begins here.

Objectives

- Identify the characteristics of mammals.
- Identify and describe the structures and features that benefit mammals for survival and growth. **BWS**
- Sequence the stages of a mammal's life cycle. **BWS**
- Compare and contrast the views about egg-laying mammals. **BWS**
- Compare humans and mammals. **BWS**

Materials

- picture of Air Force One (plane)
- picture of a blue whale

Teacher Resources

- Visuals 7.3–7.5: *Fruit Bat*; *Animal Classification 5*; *Life Cycle of a Mammal*

Vocabulary

- mammal

Engage

- Ask volunteers to name ten vertebrates. Write the responses for display. Ask questions to eliminate any that are not mammals.
  Which animals on the list are cold-blooded? any fish, amphibians, reptiles, spiders, and insects

- Draw lines through those animals.
  What warm-blooded animals have we already discussed? birds
- Draw lines through any birds.
  The animals left on the list belong to another group of warm-blooded animals that we will talk about today.

Instruct

Preparation for Reading

- Preview and pronounce the vocabulary term *mammal* and the words *oryx*, *unique*, and *echidna*.
- Direct attention to the headings, captions, and chart. Remind the students to read the captions and chart as they read the text.
- Direct the students to read pages 153–55 silently to find out how an elephant uses its ears in addition to hearing.

Teach for Understanding

What classification are most farm animals? mammals
What is a *mammal*? a warm-blooded vertebrate that has fur or hair and feeds milk to its babies

- Discuss that some mammals have more hair than others.
- Most mammals give birth to live offspring. How is this different from most other animals? Most other animals lay eggs.
  What do mammal mothers feed their offspring? milk
- Display the pictures of the blue whale and Air Force One.
  How much can a blue whale weigh? 90.8 kilograms (200,000 pounds)
- Poll the class to see whether the students think the blue whale or Air Force One weighs more.
- Point out that the blue whale's weight is similar to the weight of Air Force One.
- Display the *Fruit Bat* visual.
  How small is a bat? 28.4 grams (1 ounce)
- Point out that most pencils weigh about 28.4 grams (1 ounce).
- Explain that some fruit bats weigh only an ounce but others can weigh as much as a couple of pounds. The smallest fruit bats are no more than 2 inches in length while the larger ones can be more than 16 inches in length.
- What characteristic does the bat have in common with most birds? It can fly.
- Explain that the bat is the only mammal that can fly.

The characteristics of mammals were studied in *Science 2* except for reproduction and feeding their offspring. Use the chart, pictures, and captions to review the characteristics of mammals as needed. You may want to review by using the *Animal Classification 5* visual and the "Mammals" cell on *Activities* page 125.

- Direct the students' attention to their shoes.
- Why do you wear shoes? to protect my feet
- Explain that God gave people the ability to think and to make clothing and shoes to cover and protect their bodies.
- Point out that animals have not been given those abilities. God has provided for them in other ways.
- Direct attention to the "Creation Corner." Discuss the unique eye features that God designed for some mammals.
- Analyze the picture of the camel's eye and the caption. Discuss the features of the camel's eye.
- Display the *Animal Classification 5* visual. Guide the students in completing the "Some Survival and Growth Features" section on the characteristics chart on *Activities* page 125 during the discussion.
- Name some places where mammals live. Possible answers: on farms, on mountains, in hot deserts, in wooded areas, in water, in cold climates

  How is the fur on a mammal that lives in a cold climate different from the fur on a mammal that lives in a warm climate? Those in cold climates have thicker fur to help keep their body temperature the same.
- How thick do you think the fur is on the animals on page 153? not very thick, because they are in a warm climate

  What else did God provide to help keep some mammals warm in cold climates? blubber
- Explain that blubber is a thick layer of fat found between the skin and muscles of marine mammals.
- Point out that marine mammals spend most of their time in marine surroundings. The word *marine* is used to describe something that lives in salt water or the ocean.
- Name some marine mammals. dolphins, whales, seals, walruses

  When the climate is cold and there is not much food, what do some mammals do? go into a deep sleep

  What features help some animals survive in hot and dry climates? Some are more active at night when it is cooler. Sweat glands help some to survive with little water. The camel has fat in its hump to use as energy. An elephant uses its ears to cool its body.
- Explain that a kangaroo rat is active in the desert for less than one hour each night.
- Remember that people have sweat glands too. Where are the sweat glands found in people? in the dermis, in the skin
- How do sweat glands help people? They help them to cool down.
- Analyze the picture of the seal and its caption.

  Why does the seal have blubber? to keep its body temperature the same
- Discuss the picture of the white-tailed antelope squirrel and its caption.
- How is the white-tailed antelope squirrel keeping cool? It is pressing its belly against the cool ground.
- Share the Background information about squirrels.

Creation Corner

God designed unique eye features for some mammals. An owl's eyes do not move. Instead, the owl can turn its head almost all the way around. A camel has three eyelids. These help protect its eyes from blowing sand. Horses and zebras both have eyes on the sides of their heads. This allows them to see all around their heads. A mammal that hunts food has both eyes facing forward like your eyes do. Prey can be viewed from two different angles.

Why does the seal have blubber?

How is the white-tailed antelope squirrel keeping cool?

Features to Survive and Grow

Mammals live in cold arctic climates. Some live in hot deserts. They live in all other areas as well. Mammals need to survive in different climates. Mammals that live in cold climates have thick fur. Their fur helps keep a warm body temperature. Some have a thick layer of blubber, or fat, that keeps their body temperature warm. When there is little food and the climate is cold, some mammals' bodies are able to go into a deep sleep. Mammals that live in hot climates are more active at night when it is cooler.

Deserts and savannas can be hot and very dry. The sweat glands of some mammals allow them to survive with very little water. The camel stores fat in its hump. The fat will be used as energy when the camel needs it. An elephant uses its ears to cool its body.

154

👥 Differentiated Instruction

Remediation: Classify Mammals

Materials: *Animal Cards*, on cardstock (Instructional Aids 6.1–6.3)

Direct the students to work in pairs to find the animal cards that represent the mammal group. When a student chooses a card, he should tell what characteristic of mammals made him choose the card.

Enrichment: Mammals with Blubber

Materials: *Activities* page 126, science notebook

The "Notebook Connection" on *Activities* page 126 extends the student's learning about mammals with blubber.

ⓘ Background

Squirrels

White-tailed antelope squirrels are out for brief periods on the hot daytime desert surface. Their chipmunk-size bodies heat up, even though their bushy tails hang like parasols over their backs. Squirrels may be seen pressing their bellies, with legs spread, against the cool soil or even the tiles of suburban patios. It has been thought that the squirrels use the cooler earth in their burrows in a similar way on hot days.

Mammals defend themselves from predators. Some escape by running very fast. A rabbit can hold very still when a predator is near. Hopefully the predator will not see it. Some mammals use camouflage to hide from predators. The porcupine and the echidna both have spines or quills to protect themselves. Horses, deer, and giraffes use their hooves for protection. A giraffe's large, heavy hooves can break the back of a lion.

echidna

Try It Yourself!

Wet a paper towel with water. Rub the wet towel over your arm. Hold an index card or a piece of cardboard about 10 cm (4 in.) above your wet arm. Fan the card back and forth about 10 times. Observe any cooling effect on the skin. An elephant uses its trunk to spray itself with water. Then it fans itself with its large ears.

How do mammals survive in different climates?

155

- Discuss ways that mammals can defend themselves.
  Some mammals escape predators by running very fast. How does a rabbit defend against a predator? by holding very still, hoping the predator does not see it
  What would a rabbit do if a predator saw it? run very fast in a zigzag pattern
- Point out that smaller brush rabbits can also climb trees to escape predators.
- Direct every other student to hold their books open at about 15 feet from the students behind them. Discuss what the students without books see in the picture at the top of the page.
  How is the kudu, or antelope, defending itself against predators? camouflage; hiding in the bushes
- Allow the students to switch places and follow the same procedure.
- Direct attention to the picture of the echidna.
  What does the echidna use to defend against predators? quills
- Point out the enlargement of the quills in the circle.
  Do you think you would like to meet the echidna on a walk?
  How do horses, deer, and giraffes defend themselves? with their hooves

in cold climates: thick fur or blubber for warmth, go into a deep sleep in some cold climates; in hot climates: more active at night, sweat glands to survive with little water, stores fat to use as energy, ears or tail to cool body

---

Activity

Try It Yourself!
Materials: paper towel, water, index card or piece of cardboard
Allow the students to try this activity on their own.

Bird or Mammal?
Materials: *Bird or Mammal?*, per student; *Bird or Mammal? Key* (Instructional Aids 7.2–7.3)
This page provides practice classifying birds and mammals in the Bible. It will also provide practice for finding references in the Bible.

Background

Mammal Species
Mammals do not have the largest number of species. Reptiles, birds, and fish each have more species than mammals do. However, mammals are typically the largest animals and are the most familiar to people.

Breathing
Most dolphins can hold their breath for 10–15 minutes. Some whales can stay under water for one to two hours.

Preparation for Reading

- Preview and pronounce the word *duck-billed platypus*.
- Direct the students to read pages 156–57 to find out how a duck-billed platypus gives birth.

Teach for Understanding

What book in the Bible tells us about Creation? Genesis

What does the Bible tell us in Genesis about how mammals will reproduce? God said that animals will reproduce after their own kind.

What does "their own kind" mean? It means the animal will reproduce more of the same kind of animal.

- Display the *Life Cycle of a Mammal* visual. Discuss the stages of the life cycle of a mammal.

What is the first stage in the life cycle? birth

What mammal is pictured in this life cycle? a horse

- Explain that the illustration is an American Saddlebred horse.

What is the second stage of the life cycle? newborn

What happens during the newborn stage? All newborns drink their mother's milk until they can eat other foods. Some mammals are born with their eyes open. Some are not. Some need their mothers for everything.

Is a foal helpless when it·is born? No, it can stand and run soon after birth.

What stage of the life cycle is it when a mammal can get and eat their food on their own? youngster

What are some of the changes that a mammal might have during the youngster stage? Possible answers: Some grow their own fur. Some leave their mothers. They grow bigger. Some find new homes or move on to different areas.

What is the last stage of the life cycle? adult

What can a mammal do as an adult? Possible answers: find its own food, take care of itself, find or make its own home, reproduce more of the same kind of mammal

What will happen to an adult mammal? It will die.

- Using the *Life Cycle of a Mammal* visual, guide the students in finding the life cycle pattern of birth, growth, reproduction, and death in the life cycle of a mammal.

A Way to Reproduce

God created mammals to reproduce after their own kind. Most mammals give birth to live offspring.

Life Cycle of a Mammal

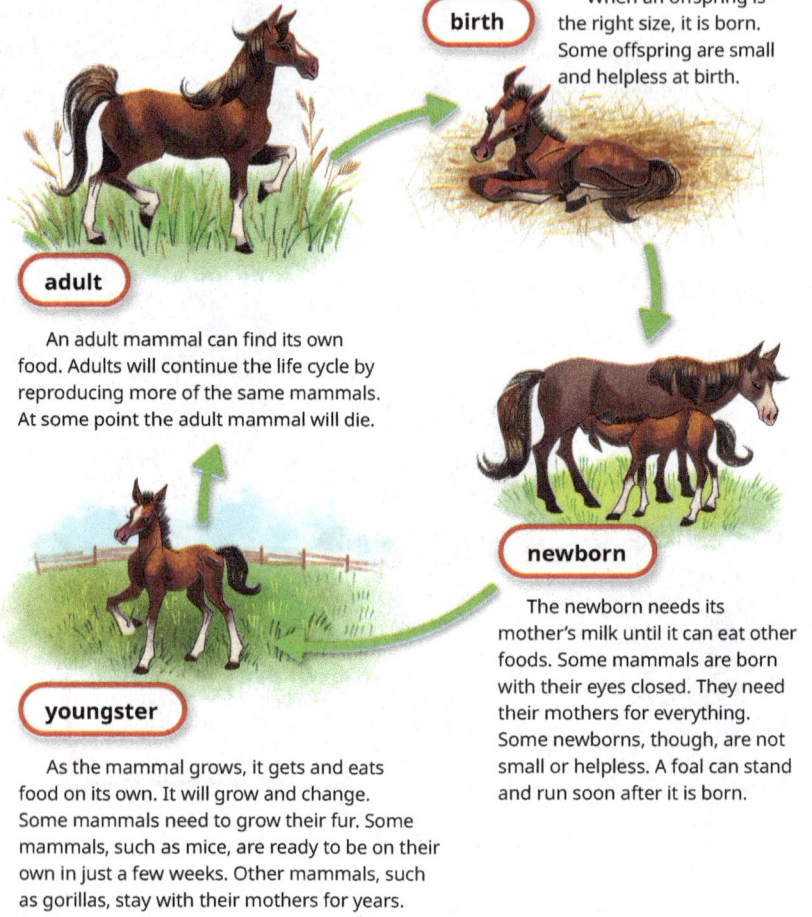

adult

An adult mammal can find its own food. Adults will continue the life cycle by reproducing more of the same mammals. At some point the adult mammal will die.

birth

When an offspring is the right size, it is born. Some offspring are small and helpless at birth.

newborn

The newborn needs its mother's milk until it can eat other foods. Some mammals are born with their eyes closed. They need their mothers for everything. Some newborns, though, are not small or helpless. A foal can stand and run soon after it is born.

youngster

As the mammal grows, it gets and eats food on its own. It will grow and change. Some mammals need to grow their fur. Some mammals, such as mice, are ready to be on their own in just a few weeks. Other mammals, such as gorillas, stay with their mothers for years.

156

👥 Differentiated Instruction

Enrichment: Life Cycle of a Bird or Mammal

Materials: *Activities* page 130, science notebook

The "Notebook Connection" on *Activities* page 130 extends the student's learning about the life cycle of a bird or a mammal.

Enrichment: Which Animal Group?

Materials: *Unusual Animals* (Instructional Aids 7.4–7.5), one per student; *Unusual Animals Key* (Instructional Aids 7.6–7.7)

These pages extend the student's learning about animal groups. Direct the students to read the paragraphs about four different animals. Instruct the students to decide what animal group each animal belongs to and give a reason why they chose that group. The animal groups include animals from both Chapter 6 and Chapter 7.

⚗ Activity

Mammal Posters

Direct the students to choose a mammal and make a poster that shows pictures and facts (where it lives, what it eats, how it protects itself, and any interesting facts) about that mammal. Allow each student to present his poster and then hang it for display.

Science and the Bible

The duck-billed platypus and the echidna do not give birth to live offspring. Both lay soft-shelled eggs. There are two reasons they are known as mammals. First, their bodies are covered with fur. Second, the mother feeds the offspring milk. Some people believe that all mammals laid eggs long ago. But the Bible tells us in Genesis 1 that God created all living creatures. He created them to reproduce after their own kind. He created most mammals to give birth to live offspring. He created some to lay eggs.

duck-billed platypus

Are Humans Mammals?

Humans have the same characteristics as mammals. Humans are warm-blooded. They are vertebrates. They breathe with lungs and have hair. Mothers feed milk to their babies.

Modern science considers man to be an animal. Many scientists believe that animals evolved to become humans. The Bible teaches that man is God's special creation. Man has common characteristics with mammals. But man was created in the image of God. This means that people are like God in certain ways and can communicate with Him. Animals are not like God and can not communicate with Him.

✓ How does the Bible say all mammals reproduce?

157

What does modern science consider man to be? an animal

How do scientists who believe in evolution think humans began? Animals evolved to become humans.

How do scientists who believe the Bible think humans began? Man was God's special creation.

- Read Genesis 1:27 aloud.

How did God create man and woman? in His image

What does being created in God's image mean for you? I am like God in certain ways, and I can communicate with God.

How are humans like God? Elicit that humans can talk or communicate, can reason, can know good and evil, and can think.

Are animals like God? No, animals were not created in God's image. Animals are not like God and cannot communicate with God.

✓ after their own kind

Apply

Activities

Animal Classification 5, page 125

This page helps the students classify animals. It was completed as part of the lesson.

Study Guide, pages 127–29

These pages review the concepts taught in Lessons 48–49. After completion, direct the students to keep them for review for Test 7.

Assess

Quiz 7A

The quiz may be given at any time after completion of this lesson.

Discussion for *Student Edition* page 157 begins here.

- Discuss "Science and the Bible," the picture of the duck-billed platypus, and the picture of the echidna on page 155.

What two characteristics make the duck-billed platypus and the echidna mammals? Their bodies are covered with fur, and the mother feeds the offspring milk.

What characteristic makes the duck-billed platypus and the echidna different from mammals? They both lay soft-shelled eggs.

What do some people believe about mammals long ago? that they all laid eggs

What does the Bible tell us in Genesis 1 about all living creatures? God created all animals to reproduce after their own kind. Some give birth to live offspring and some lay eggs.

- Discuss humans and mammals.

What characteristics do humans and mammals have in common? warm-blooded, vertebrates, breathe with lungs, have fur or hair, feed milk to their babies

How do human characteristics and mammal characteristics compare? They are the same.

How would you compare the classification of humans and mammals? They both have the same classification.

Objectives

* Explain why understanding the different kinds of animals is important for caring for them. **BWS**
* Identify the life cycle pattern that all cold-blooded and warm-blooded animals have.
* Compare and contrast life cycles of cold-blooded and warm-blooded animals.
* Identify similar and different patterns of inherited traits shared between offspring and their parents or among siblings.
* Identify a variation of an inherited trait.

Materials

* piece of scrap paper, one per student

Teacher Resources

* Visuals 7.2, 7.5: *Life Cycle of a Bird*; *Life Cycle of a Mammal*

Engage

* Direct attention to the picture of the cats on page 160.
  How are these cats alike? Possible answer: They all have eyes, ears, a nose, legs, and a tail.
  How are these cats different? Possible answer: different color of fur
* Explain that today the students will learn about why the same kinds of cats can look the same and different.

Instruct

Preparation for Reading

* Preview and pronounce the word *dominion*.
* Direct the students to read pages 158–59 silently to find out the life cycle pattern of warm-blooded animals.

Teach for Understanding

* Direct the students to collaborate with a science partner and write a definition for the word *dominion* on scrap paper.
* Allow several groups to share their definitions. Tell the students to turn to the Glossary. Ask a volunteer to read aloud the definition for *dominion*. "The job God gave to people to care for, manage, or rule over the world for the benefit of others and the glory of God."
  Where is God's command to have dominion over the earth found in the Bible? in Genesis 1:28

> Some Bible versions do not use the word *dominion*.

How does God expect you to use His command to have dominion over warm-blooded animals? God expects us to care for the birds and the mammals. He expects us to learn about them so we can use them and care for them properly.
How do we use animals? Possible answers: for food, for clothing, for enjoyment (pets and riding), for work (pulling wagons)
What pattern is repeated in cold-blooded animal life cycles? birth, growth, reproduction, and death

birth

Taking Care of Animals

In Genesis 1:28 God gave us the command to have *dominion* over the earth. He wants us to care for the earth. He told us to care for the fish of the sea. He told us to care for the birds of the air and every living thing on land.

Life Cycles of Animals

Think about the lives of fish, amphibians, reptiles, insects, and spiders. What pattern is repeated in cold-blooded animal life cycles? Now think about the lives of birds and mammals. What pattern do you see in their life cycles? Compare the two patterns. Are the patterns the same or different? All animals have the same life cycle pattern: birth, growth, reproduction, and death.

Pattern of All Life Cycles

death

158

* Display the *Life Cycle of a Bird* and *Life Cycle of a Mammal* visuals. Guide the students in discovering the pattern found in both life cycles. Write the pattern for display as the students name the stages in the pattern.
  What pattern is repeated in the bird and mammal life cycles? birth, growth, reproduction, and death
  How does the pattern of cold-blooded animals and the pattern of warm-blooded animals compare? They are both the same.
* Discuss the first stage: birth.
  Where does the birth of an animal begin? as an egg outside the mother's body or inside the mother's body
  How do animals look when they are born? Some look like their parents. Some do not look like their parents.
  What do some amphibians, insects, and spiders go through as they grow? metamorphosis
  As an animal goes through metamorphosis, does it look like its parents? No, not until it becomes an adult.
* Direct attention to the pictures with the birth stage. Discuss the birth of the monkey and alligator.
  How is a monkey born? It begins life inside its mother's body and then is born alive.
  What does the monkey look like at birth? like its parents only smaller
  How is an alligator born? It comes from an egg outside its mother's body.

growth

Birth

The birth of some animals begins as an egg outside the mother's body. Other animals begin their lives inside their mother's bodies. They are born alive. Some animals look like their parents when they are born. Some do not.

Growth

Offspring from cold-blooded and warm-blooded animals will grow. They grow and change until they become adults.

Reproduction

Some animals may die before they can reproduce. Most adult animals will obey God's command to reproduce. Each animal fills the earth with its own kind.

Death

All life cycles end in death. A life cycle ends in death because Adam and Eve sinned. God punished all His creation because of Adam's sin. The Fall brought death into the world.

reproduction

✔ What pattern do all life cycles have?

159

- Discuss the second stage: growth,
  How does the growth stage of cold-blooded and warm-blooded animals compare? In both animal groups the offspring will grow and change.
  How long will the animals grow? until they become adults
- Analyze the pictures with the growth stage.
  Does the salamander larva look like an adult salamander? No, it is going through metamorphosis and will look like a salamander as an adult. It looks like a tadpole now.
  Does the baby swan look like an adult swan? Yes, it is born looking the same as an adult only smaller.
- Discuss the third stage: reproduction.
- Read Genesis 1:21–22, 24–25 aloud to support the statements in the *Student Edition*.
  What job do most adult animals do that God gave them? to reproduce and fill the earth with animals after their own kind
- Analyze the pictures shown with the reproduction stage.
  How do birds or robins reproduce? by laying eggs
  How do wasps reproduce? by laying eggs
- Discuss the last stage: death.
  How do all life cycles end? with death
  Why do all life cycles end in death? God punished all His creation because of Adam's sin; the Fall brought death into the world.
- Discuss the pictures shown with the death stage.
- How do you think the whales died? Possible answers: died of old age, died because they were sick, washed on shore and were stranded
  How did the rabbit die? The rabbit became the prey of the fox.

✔ birth, growth, reproduction, and death

Preparation for Reading

- Direct the students to read pages 160–61 silently to find out what inherited traits in animals look like.

Teach for Understanding

What does the Bible tell us about offspring? The offspring will be the same kind as their parents.

- Direct the students to collaborate with a science partner and write a definition for the words *inherited trait* on scrap paper.

- Allow several groups to share their definitions. Tell the students to turn to the Glossary. Ask a volunteer to read aloud the definition for *inherited trait.* "A characteristic that is passed down from parents to offspring."

- Direct attention to the pictures of the cats and kittens.

Where did the kittens get their inherited traits from? their parents

- After each answer for the following question, ask the student if the parents have the trait also. *Note:* When a student mentions fur, be sure that he does not mention color. In these pictures color will be a variation that will be discussed next.

What inherited traits can you observe in the kittens? Possible answers: two eyes, two ears, four legs, one tail, one nose, whiskers, fur

- Direct attention to the pictures of the cardinals.

- After each answer for the following question, ask the student if the parents have the trait also.

What inherited traits can you observe in the baby cardinals? Possible answers: two eyes, two legs, beak, feathers, two wings, tail

- Explain that a male baby cardinal does not get its color until it is about a year old.

- Direct attention to the pictures of the dogs on page 161.

- After each answer for the following question, ask the student if the parents have the trait also. *Note:* When a student mentions fur, be sure that he does not mention color. In these pictures color will be a variation that will be discussed next.

What inherited traits can you observe in the puppies? Possible answers: two eyes, two ears, four legs, one tail, one nose, whiskers, fur

Do all cold-blooded and warm-blooded animals have inherited traits? Yes, they all have inherited traits that include color, size, and shape.

Animals and Their Offspring

Both cold-blooded and warm-blooded animals reproduce to make more living things. The Bible tells us in Genesis that each animal will reproduce the same kind of animal. The offspring will be the same kind as their parents.

Inherited Traits

Just as He did for plants, God gave animals inherited traits. Look at the adult cats and the kittens. What is the same about the parents and their offspring? The adult cats and the kittens have fur covering their bodies. Both parents and offspring have four legs. They all have two ears, two eyes, and a tail. Each of these is an inherited trait. Remember that an *inherited trait* is a characteristic that is passed down from parents to offspring. The adult cats passed down inherited traits to their kittens. The kittens' inherited traits are two eyes, two ears, four legs, a tail, and fur. The kittens inherited these traits from their parents.

Both cold-blooded and warm-blooded animals have inherited traits. The offspring receive inherited traits from the adult parents. Color, size, and shape are all inherited traits.

160

⑦ Misconception

Animal Size

Some students may think that males are always bigger than females. This is not always true. For some species the female is larger. For example, many female shorebirds are larger than the males. In many birds of prey, the females are larger than the males. Some female mammals that are larger than the males are some bats, common marmosets, jack rabbits, golden hamsters, and some whales.

God created animals with variations of inherited traits. Remember a *variation* is an inherited difference between two animals of the same kind.

Look at the dogs and the puppies. What is the same about the puppies and their parents? Each of the puppies has two eyes, two ears, four legs, a tail, and fur. These are inherited traits.

An offspring inherits its traits from both its father and its mother. What is different about the puppies and their parents? Sometimes the offspring may look more like one of its parents. For example, the father dog has curly fur and the mother dog has straight fur. Look at the puppies. Which ones inherited curly fur from their father? Which ones inherited straight fur from their mother?

Compare the puppies with each other. Do both of the offspring look the same? An offspring can also look the same as or different from its siblings. The puppies do not look the same because each one inherited different traits from a different parent. Variations can be seen in the offsprings' fur and color.

What are some inherited traits of animals?

161

- Direct the students to collaborate with a science partner and write a definition for the word *variation* on scrap paper.
- Allow several groups to share their definitions. Tell the students to turn to the Glossary. Ask a volunteer to read the definition for *variation*. "An inherited difference between two living things of the same kind."
- Emphasize that a variation is *an inherited difference*.
- Direct attention to the pictures of the dogs. Point out that the father is a poodle and the mother is a Yorkshire terrier.
  What inherited trait do you see that is different in the adult dogs and their puppies? their fur
  What kind of fur does the Yorkshire terrier have? long, straight, tan and brown fur
  What kind of fur does the poodle have? curly, short, brown fur
  Which puppy inherited its curly fur from its father? Accept any reasonable answer. Although both puppies appear to have at least some curly fur, the black one appears more curly.
  Which puppy inherited its straight fur from its mother? Accept any reasonable answer. Although the black and brown one appears to have some curly fur, it also has straight white fur and straight fur on its ears.
  Why do the puppies not all look the same? because each one inherited different traits from a different parent
- Discuss which puppies received their fur color from their father, their mother, or both parents.
  What variations can be seen in the puppies? The fur is different in its texture—straight or curly. The fur color is different.
- Direct attention to the pictures of the horses.
  What variation do you see between the adult parents and the colt? difference in color of the fur
  Did the colt inherit its fur or hair from its father or its mother? probably from both since it is brown
- Review the cat pictures on page 160, discussing the variations in the color of the kittens' fur.

color, shape, size, number of legs, ears, eyes, fur, tail

Apply

Activities
Inherited Traits, page 131
This page reinforces the concepts taught in Lesson 50.

Objectives
- Identify animal instincts.
- Explain where an animal's instincts come from. **BWS**
- Identify learned behaviors of some animals.
- Differentiate between inherited traits, learned behaviors, and instincts.

Materials
- nest-building materials such as clay, straw, grass, twigs, yarn, leaves, per student or group
- box or tray to make distribution of the nest-building materials easier, per student or group
- piece of paper with "Limited Food Supply" on it
- piece of paper with "Plenty of Food" on it

Teacher Resources
- Visual 7.6: *Compare and Contrast*

Vocabulary
- instinct
- migrate
- hibernate
- learned behavior

Engage
What kind of home do many birds have? nests

What materials are used to build nests? Possible answers: clay, straw, grass, twigs, yarn, leaves

- Direct attention to the nest-building materials.
- Give each student or group five minutes to use the nest-building materials to build a bird nest that will stay together when moved. *Note:* The bird nest needs to be built to move but not on cardboard. It will be used later in the lesson to show students how God gave birds the ability to build their nests better than people can.
- Ask the students to put aside their nests to be used later in the lesson.
- Explain that we will find out about this behavior and the name scientists give to it.

Instruct

Preparation for Reading
- Preview and pronounce the vocabulary terms *instinct, migrate,* and *hibernate* and the words *behavior* and *miracle.*
- Direct the students to read pages 162–63 silently to find out what *migrate* means.

Teach for Understanding
What is an *instinct*? a God-given behavior that an animal is born with

What is a *behavior*? the way an animal reacts to its environment

What do you do if you stub your toe? Possible answers: jump, cry, hold toe

Did someone teach you to behave or act that way? No, it is an instinct to react that way.

What are some examples of instincts? Possible answers: birds building nests, dolphins and whales knowing how to swim, a monkey with the ability to hang by its tail, elephants protecting their young

---

Lesson 51

hibernating dormouse

migrating elk

Instincts

An **instinct** is a God-given behavior that an animal is born with. A *behavior* is the way an animal reacts to its environment. Animals do not have to learn instincts. Birds are born with the ability to build a nest. Dolphins and whales are born with the ability to swim. A monkey is born with the ability to hang by its tail. These are instincts. Instincts are not taught. The animals are born with the ability to do these things.

Some birds and mammals have an instinct to migrate. A group of animals **migrates**, or moves from one area to another area. Some animals migrate to warmer climates. Some need to find food and water. Flycatchers, ruby-throated hummingbirds, ducks, geese, and swans are birds that migrate. Antelope, elk, whales, bats, and mule deer are mammals that migrate.

Not all animals migrate when the seasons change. Some mammals have the instinct to hibernate. An animal **hibernates**, or goes into a deep sleep, during the winter months. Before hibernating the animal usually eats large amounts of food in the summer and fall. The extra food is stored as body fat. This fat provides energy for the animal while it is hibernating.

162

Are instincts taught? No, instincts are given to animals by God.

- Direct attention to the bird nests the students built. Give the students the opportunity to try to move their nests.

Did your nest stay together?

Did God give you instincts to build a nest? No, God gave birds instincts to help them build their nests.

What do birds use to build their nests? their beaks and feet

- Explain that even with our hands it is difficult for us to build a sturdy nest, but God gave birds the instincts and abilities to build sturdy nests.
- Direct attention to the top two pictures.

What instinct is the spider showing? building a web

What instinct is the flying fish showing? The fish is jumping out of the water.

Why do you think the fish jumps out of the water? Elicit that it may be escaping predators.

Did the flying fish or the spider need to be taught to jump out of the water or build a web? No, God gave them instincts to do these.

- Direct attention to the elk at the bottom of the page.

What instinct do the elk have? to migrate

What is *migrate*? to move from one area to another area

What are some reasons why animals migrate? Possible answers: for warmer climates, to find food and water

Animals need instincts to survive and grow. God gave each animal certain instincts. Many animals have shared instincts, such as hunting or seeking shelter. But others have special instincts. When hunting, a cat may pounce on its prey. A spider may inject venom. God provided the exact instincts that each animal needs to survive and grow.

Science and the Bible

In Daniel 6 you read about a time when lions did not follow their instincts. A law had been made commanding people to pray only to the king. Daniel was a servant of God. He refused to stop praying to God. Because of this, Daniel was thrown into a lions' den. God kept the lions from following their instincts. They did not eat Daniel. The following day, the king commanded that Daniel should be taken up out of the lions' den. The men who had told the king about Daniel's praying were then thrown into the den. The hungry lions attacked the men before they even fell to the bottom. When God causes something to happen outside of nature's laws, we call that a miracle.

✔ **Where did animal instincts come from?**

163

- Place the two signs, "Limited Food Supply" and "Plenty of Food," in two separate areas of the room and some distance apart.
- Ask six to eight students to stand in the habitat with a limited food supply. Explain that they are "elk."
  What will the elk do? migrate
  Why will the elk migrate? because their food is in short supply
- Instruct the "elk" to migrate. Be sure the elk move as a group and not individually.
- Direct attention to the dormouse. Explain that the dormouse lives in Africa, Europe, and Asia.
  What instinct does the dormouse have? to hibernate
  What does *hibernate* mean? to go into a deep sleep during the winter months
  When do animals usually hibernate? in the winter
- Explain that a dormouse hibernates roughly from October to May, depending on the climate. It has one of the longest hibernation periods. *Note:* Hibernation is often associated with cold climates. Some animals do hibernate in tropical climates. In third grade only the animals that hibernate in cold climates will be discussed.
  What does an animal do before it hibernates? usually eats large amounts of food in the summer and fall
  Why do the animals eat large amounts of food? It is stored as body fat and provides energy for the animal while it is hibernating.

The discussion for page 163 begins here.

Why do animals need instincts? to help them survive and grow
Where do instincts come from? God
What are some instincts God has given animals? Possible answers: hunting, finding shelter, sleeping, making a shelter, avoiding or fighting enemies, migrating, hibernating
- Discuss the picture of the bees.
What are some instincts God has given bees? Possible answers: to protect themselves by stinging, to pollinate flowers, to make honey, to find their way back to the hive
- Direct attention to the picture of the turtle.
What instinct of the turtle does the picture show? making a shelter, making a nest
- Direct attention to "Science and the Bible."
Why was Daniel thrown into the lions' den? He refused to stop praying to God.
How did God protect Daniel? He kept the lions from following their instincts. They did not eat Daniel.
Did the lions follow their instincts when the other men were thrown into the den? Yes, they attacked and ate the men.

✔ **An animal's instincts came from God.**

👥 Differentiated Instruction

Remediation: Instincts and Learned Behaviors
Materials: *Activities* page 133, science notebook
The "Notebook Connection" on *Activities* page 134 allows the student to show his understanding of instincts and learned behaviors by drawing pictures.

ⓘ Background

Instincts and Learned Behaviors
Sometimes instincts and learned behaviors work together to help an animal survive. For example, lions are born with the instinct to kill and eat meat for food. But to be successful hunters, they need to learn from other lions. So lionesses teach the cubs how to hunt.

Ant Instincts
In Proverbs 30:25, the Bible speaks of the ant. The ant gathers food in the summer so that it can eat during the winter months when food is not available. The ant does not make a decision to do this. God gave the ant instincts that help it survive.

Preparation for Reading

- Preview and pronounce the vocabulary term *learned behavior* and the words *fiery, experience,* and *chimpanzee.*
- Direct the students to read pages 164–65 silently to find out which traits and behaviors God created animals with.

Teach for Understanding

What do you do if your shoes are untied? tie them

Did someone teach you how to tie shoes? yes

What is a *learned behavior*? something that an animal learns to do

Name some learned behaviors that you have seen an animal do. Possible answers: A dog shakes hands. A tiger jumps through a fiery hoop. A dolphin or orca does tricks. A dog catches a ball. A service dog helps its owner. A camel kneels to let the rider off. A horse pulls a wagon.

Is a learned behavior passed on from parent to offspring? no

- Explain that your parents may know how to ride a bike. This behavior is not passed on to you. You must learn to ride the bike yourself.
- Direct attention to the picture of the tiger.

Explain whether the tiger is showing an instinct or a learned behavior. It shows a learned behavior because tigers do not normally jump through hoops of fire. A tiger would be afraid of the fire and must learn to do the trick.

- Discuss the picture of the squirrel.

Do squirrels normally eat from a person's hand? No, it is a learned behavior.

What is a squirrel's instinct to do around people? run away from them

- Direct attention to the picture of the service dog.

How is the service dog helping his master? by opening the door

Is the service dog showing an instinct or a learned behavior? a learned behavior

- Analyze the picture of the cat.

How many of you have cats as pets?

What does your cat do when you open a can of food? Most cats will come running.

Why did the cat come running when it heard the can being opened? Elicit that it has learned that the sound of the can being opened means it is time to eat.

How do you think the cat learned this behavior? by hearing the can being opened over and over again and connecting the sound with being fed

- Tell the students to think about the Exploration that they are doing with the bird feeder.

Why do you think it can take several days for the birds to start coming to a bird feeder? Elicit that the birds must learn that the feeder is there and that it holds food for them.

Learned Behaviors

People can teach animals many things. Perhaps you have seen a dog shake hands or do other tricks. You may have seen a service dog do tasks for its owner. Have you seen a dolphin toss a ball? Maybe you have seen a tiger jump through a fiery hoop. These are all examples of learned behaviors. Someone taught the animals to do those things. A **learned behavior** is something that an animal learns to do.

Animals can learn by experience. Birds learn to come to a feeder for seeds and nuts. Dolphins learn to search for objects in the water. One animal can also teach another animal a behavior. A chimpanzee may learn to crack nuts by seeing another chimpanzee smash a nut with a rock.

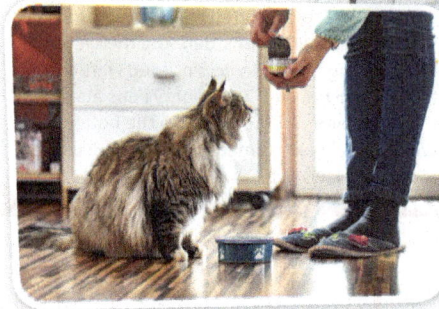

What learned behavior does the cat have?

164

⟨?⟩ Misconception

Learned Behaviors

Some students may think that learned behaviors are traits. But many of these are environmental rather than inherited traits. Some traits are environmental because they are traits that are chosen or learned, or that are influenced by the habitat.

Inherited Traits and Behaviors

Think about these characteristics: *inherited trait*, *instinct*, and *learned behavior*. What do these three characteristics have in common? Both cold-blooded animals and warm-blooded animals have these characteristics. Each characteristic is a type of animal behavior.

Animals have all of these characteristics. God created animals with inherited traits and instincts. Animals do not learn these characteristics. They are born with them. But animals also learn new behaviors. A learned behavior must be learned by an animal. The animal is not born with it.

Which Animal Shows a Learned Behavior?

✓ How are inherited traits, instincts, and learned behaviors different?

165

Discussion for *Student Edition* page 165 begins here.

- Direct attention to the *Compare and Contrast* visual. Instruct the students to remove page 133 from their *Activities* books.
- Guide the students in completing the chart as you compare and contrast inherited traits, instincts, and learned behaviors.
- Ask volunteers to help complete each definition in the first row of the chart.
- Direct attention to the second row in the chart.
  Think about these three characteristics. How do the animals that have inherited traits, instincts, and learned behaviors compare? Cold-blooded and warm-blooded animals have inherited traits, instincts, and learned behaviors.
- Direct attention to the third row.
  Are animals born with inherited traits? yes
  Are animals born with instincts? yes
  Are animals born with learned behaviors? no
  Compare and contrast which animals are born with inherited traits, instincts, and learned behaviors. Animals are born with both inherited traits and instincts. They are not born with learned behaviors.

- Direct attention to the fourth row in the chart.
  Where do inherited traits come from? Animals are created by God with inherited traits.
  Where do instincts come from? Animals are created by God with instincts.
  Where do learned behaviors come from? Animals learn behaviors from experience, people, or other animals.
- Name a learned behavior that comes from experience. Possible answers: birds eating at a bird feeder; squirrel eating from a person's hand
- Name a learned behavior that comes from people. Possible answers: tiger jumping through fiery hoop; service dog helping its owner; dolphins, dogs, and whales doing tricks
- Name a learned behavior that comes from other animals. Possible answers: A lion cub is taught how to capture prey; a chimpanzee learns to smash a nut with a rock by seeing another chimpanzee do it.
  Compare and contrast where inherited traits, instincts, and learned behaviors come from. Inherited traits and instincts are created in animals by God. Learned behaviors must be learned.
- Direct attention to the last row in the chart. Guide the students as they name examples of each trait or behavior.
  inherited trait: Possible answers: eyes, color, shape, size, legs, tail
  instinct: Possible answers: swimming, migrating, hibernating, hunting, injecting venom
  learned behavior: Possible answers: a dog shaking hands, birds coming to a bird feeder, dolphins searching for objects, service dogs doing tasks for owners
- Direct attention to the pictures under the heading "Which Animal Shows a Learned Behavior?"
  Which animal shows a learned behavior? dog catching a flying disk
- Discuss the trait or behavior that the other pictures show.
  butterfly: instinct
  octopus: instinct
  cows: inherited trait (body parts) or instinct (cuddling)
  turtles: inherited trait (body parts) or instinct (riding on parent's back)

✓ **Animals are born with instincts and inherited traits because God created animals with these. Animals are not born with learned behaviors because animals were not created with these behaviors. They must be learned.**

Preparation for Reading

- Direct the students to read page 166 silently to find out what God gave each animal for survival.

Teach for Understanding

- Direct attention to the picture.

What is happening in the picture? People are taking a ride through an area with animals in their natural environment (safari).

Have you ever taken a ride to see animals?

Name the animal you see and tell whether it is cold-blooded or warm-blooded. zebras, warm-blooded

How do you know that a zebra is warm-blooded? It has fur.

Would its fur be thick? No, it lives in a warmer climate.

- Discuss the power of God's Creation.

What animals did God create? all cold-blooded and warm-blooded animals; all animals

How did God create the animals? They were created by the power of His words; God simply spoke and they were made.

Were any two animals created exactly the same? No, God created each animal differently.

What did God provide for each animal to survive? Possible answers: their needs met, the right body parts, instincts, ability to protect themselves

- Read Genesis 1:28 aloud.

Who did God give authority or dominion over the earth to? people

Does this dominion include the animals? Yes, the verse says to have dominion "over the fish of the sea, and over the fowl of the air, and over every living thing that moveth upon the earth."

How should people have authority over animals? by taking care of them, by treating them properly

- Point out that God gave people the authority over animals and that it is up to each person to obey God.

What helps you to take care of animals properly? learning about them and understanding them

✓ It is important to learn about animals to use the authority over them that God gave people and to give them the proper care.

Apply

Activities

Compare and Contrast, page 133

This page was completed as part of the Lesson.

Study Guide, pages 135–37

These pages review the concepts taught in Lessons 50–51. After completion, direct the students to keep them for review for Test 7.

Studying Animals

Genesis 1 records God's power at Creation. God created all the cold-blooded and warm-blooded animals. He created animals by the power of His words. He simply spoke, and all the animals were made.

God made no two animals exactly the same. Even within the different kinds of animals there are many differences. He created each animal with the ability to survive in its environment. God planned and provided to meet the needs of each animal. He gave each animal the right body parts and instincts for survival.

God gave people authority over the earth. This includes authority over cold-blooded animals and warm-blooded animals (Genesis 1:28). We need to learn about and understand animals so we can give them proper care.

✓ Why is it important to learn about animals?

166

Assess

Quiz 7B

The quiz may be given at any time after completion of this lesson.

EXPLORATION

Blubber, Feathers, and Fur

Inquiry Skills
- Observe
- Measure
- Infer

Some animals are warm-blooded. Their body temperature stays about the same all the time. Birds and mammals are warm-blooded animals. What did God give birds that keeps their bodies warm? What did God give mammals that keeps their bodies warm?

In this Exploration, you will find out how blubber, feathers, and fur keep animals warm.

167

⚠ Helps

Stopwatch
If each science group will be using a stopwatch, instruct the students in its proper use and allow them time to practice starting, stopping, and reading the stopwatch. *Note:* You may display a stopwatch for the students to refer to during the Exploration, or you may direct them to keep track of time with the wall clock.

Kinds of Animals at Creation
God created all kinds of animals at Creation. All the variations of kinds of dogs, cats, or other animals were not created at the beginning.

Objectives
- Identify the purpose of blubber, feathers, and fur.
- Observe how blubber, feathers, and fur help keep animals warm.
- Record observations.
- Draw conclusions from data collected.

Materials
- 2 or 3 pictures of several different kinds of birds
- 2 or 3 pictures of several different kinds of mammals
- 2 or 3 pictures of several mammals that have blubber (sea lions, seals, whales, manatees, dolphins)
- See *Activities* page 139.

> To reduce the number of materials needed for this Exploration, you may choose to have the students work in groups.

Engage
- Display the pictures of the birds, mammals, and mammals with blubber out of group order.
- Instruct the students to classify the animals into groups.
  What groups did you classify the animals into? birds and mammals
  What kind of covering do the birds have? feathers
  What kind of covering do the mammals have? fur or hair
- Separate the pictures of the mammals with blubber from the other two groups.
  Can you think of something these mammals have in common that helps them keep warm? Elicit that all the animals have blubber.
- Explain that in this Exploration the students will use what they know about birds and mammals to predict whether blubber, feathers, or fur help keep animals warm.

Instruct
Preparation for Reading
- Direct the students to remove pages 139–40 from their *Activities* books. Direct them to read page 162 and *Activities* pages 139–40 silently before beginning.

Teach for Understanding
- Direct attention to the **Inquiry Skills**.
  What inquiry skills will you be using in this Exploration? Observe, Measure, and Infer
- Review the inquiry skills of *observe*, *measure*, and *infer*. Refer the students to the Inquiry Skills page in their science notebooks, if needed.

- Direct attention to the **Purpose**.

  What is the purpose of this Exploration? to find out how blubber and fur help to keep some mammals warm and to find out how feathers help to keep a bird warm.

- Direct attention to the **Materials**.

Apply

- Direct attention to the **Procedure**. Remind the students to follow each step. Encourage the students to check off each step as it is completed.

  What do the plastic bags represent? the epidermis

  What is the *epidermis*? the top layer of skin

- Point out that the students will record the temperatures of the ice water on the data chart.

EXPLORATIO

Blubber, Feathers, and Fur

Name _____

Purpose

I want to find out how animals keep warm.

Procedure

Inquiry Skills
- Observe
- Measure
- Infer

NOTE: This Exploration will require the help of a partner. The plastic bags used represent the epidermis of the skin.

☐ 1. Fill two containers with ice water.

☐ 2. Place the thermometers in the ice water containers.

☐ 3. Record the temperature of the ice water in both containers on the data chart.

☐ 4. Fill one plastic bag with shortening or soft butter.

☐ 5. Place your extended right finger in the bag of "blubber." Completely cover your ind finger with the fat. Do not let the tip of your finger poke through the shortening or butter.

☐ 6. With your left index finger extended, place your left hand in a clean, empty plastic bag. Let your science partner help you.

☐ 7. When your partner is ready to start the stopwatch, dip your index fingers, which ar in the bags, in the two containers of ice water.

☐ 8. Remove each finger from the ice water when that finger is too cold to keep in the water any longer.

☐ 9. Record on the data chart how long you were able to keep your fingers in the ice water.

☐ 10. Repeat steps 2–9 using feathers. If the water temperature begins to warm, add more ic

☐ 11. Repeat steps 2–9 using fur. If the water temperature begins to warm, add more ice.

Materials
- ☐ 2 containers of ice water
- ☐ 6 plastic sandwich bags, per group of 3 students
- ☐ 2 thermometers, per gro
- ☐ stopwatch
- ☐ feathers
- ☐ fur
- ☐ shortening or butter
- ☐ plastic spoon
- ☐ paper towels

SCIENCE 3 *Activities*

Lesson 52 • Page 167
Exploration

	Blubber		Feathers		Fur	
Finger	Left	Right	Left	Right	Left	Right
Temperature of Ice Water	___°C	___°C	___°C	___°C	___°C	___°C
Finger	Left	Right	Left	Right	Left	Right
Ending Time	___ min ___ sec	___ min ___ sec	___ min ___ sec	___ min ___ sec	___ min ___ sec	___ min ___ sec

Conclusions

1. What did you keep the same each time? _*containers, temperature of the water, the "skin"*_ _____

2. What changed each time? _*the material covering the right index finger; blubber,*_ _*feathers, fur surrounding the right index finger*_

3. Did the different coverings stay warmer than the empty bag?　○ yes　○ no

4. What can you infer about the coverings of birds and mammals? _*Possible answers:*_ _*The coverings help to keep an animal warm; the coverings help to keep an animal*_ _*warmer longer than skin alone would; the blubber, fur, and feathers help to protect*_ _*animals from the cold.*_

5. What do real animals have to help them stay warm that the materials in this Exploration did not have? _*Possible answers: Warm blood flowing through their bodies; skin*_ _*(epidermis and dermis) and fat covering their bodies.*_

- Direct the students to continue with the Procedure.
- Guide the students in choosing an appropriate outside area to place the cups.
- Remind the students to record the temperatures on the data chart after each ten minutes.
- Direct the students to draw **Conclusions** using their observations.
- Guide a discussion of the students' conclusions.

Activities
Exploration, pages 139–40
The students will predict whether blubber, feathers, or fur help keep animals warm.

Assess

Rubric
Use the prepared rubric or design a rubric to include your chosen criteria.

Objective

• Recall terms and concepts from Chapter 7.

Apply

Review

• Material for Test 7 will come from the Study Guides on *Activities* pages 127–29 and 135–37, as well as Quizzes 7A and 7B. You may review any or all of the material during the lesson.

• You may choose to review Chapter 7 by playing the Animal Mix-Up review game or a game from the Game Bank (Teacher Resources).

🏆 **Review Game**

Animal Mix-Up

Materials: *Animal Cards*, on cardstock (Instructional Aids 6.1–6.3)

Spread the cards face-down. Divide the students into two teams. Ask each team questions. If the team answers correctly, a member of the team chooses a card. If he can identify the animal group the animal belongs to, the team keeps the card. If not, he places the card face-down again. The goal is to get at least one card for each animal group (fish, amphibian, reptile, insect, spider, bird, and mammal).

If a student correctly identifies a card but his team already has that kind of animal, you may choose to let him replace the card and choose another. The student must correctly identify the new card to keep it, and he may not choose a third card.

NOTES

NOTES

_____

_____

_____

_____

_____

_____

_____

_____

_____

_____

_____

_____

_____

_____

_____

_____

_____

_____

_____

_____

_____

_____

_____

_____

_____

_____

_____

_____

Objective

- Apply terms and concepts from Chapter 7.

Assess

- Administer Test 7.

Chapter Objectives

- Explain how organisms have their needs met in an ecosystem.
- Analyze a food chain and a food web.
- Analyze ways ecosystems maintain balance.
- Apply the science inquiry skills of infer and predict to the relationship of predators and prey.
- Apply the steps of the engineering design process to solve a real-life problem.

Lesson Objectives

- Identify four characteristics of organisms.
- Identify what makes up an ecosystem.
- Explain the relationship between ecosystem, habitat, community, and population.
- Explain how resources meet the needs of organisms.
- Identify why there is competition between animals.

Materials

- pictures of plants, animals, and people

Teacher Resources

- Visual 8.1: *Fitting Together Organizer*

Vocabulary

- ecosystem
- population
- community
- habitat
- resource
- competition

Engage

Chapter Introduction

- Direct attention to the chapter title and the picture.
- What kind of animal do you think this is? Elicit that it is a jaguar.
- Share Background information at this time.
- Provide time for the students to leaf through the chapter, looking at the headings and pictures. Discuss what they think the chapter will be about.

> Use completed Looking Ahead, *Activities* page 141, to generate a discussion on what the students already know about ecosystems.

Big Question

- Ask a volunteer to read aloud the Big Question.
- Explain that the students will find the answer to the question as they read the chapter.

Big Question

How do organisms work together?

168

ⓘ Background

Chapter Opener Photo

The photo on pages 168-69 is a jaguar in a rainforest. The jaguar is the largest big cat found in the Western Hemisphere. Jaguars can be found in Central America and South America. They can be found in many habitats, including grasslands, deciduous forests, and rainforests. The rainforest habitat of some jaguars provides all they need to survive. The jaguars' spots help to camouflage them as they hunt their prey.

> Visit TeacherToolsOnline.com for resources to enhance the lessons.

Chapter 8

Ecosystems

How can you tell whether something is alive? Maybe you can see it breathing. Perhaps you notice it move. There are different ways to know whether something is alive.

Remember that all living things are made of cells. Cells make up tissues. Tissues make up organs. Several organs working together make up systems. Systems make up an organism.

All organisms have some of the same characteristics. Living things grow, develop, reproduce, and interact with things around them. For example, a tiny puppy grows quickly. Soon the puppy becomes an adult dog. Later this dog might have puppies of its own. Plants also follow this pattern. Plants make seeds, and new plants grow from those seeds.

169

Activity

Class Terrarium

During this chapter, you may want to set up a terrarium. A terrarium will help the students see connections between the environment (soil, water, air, and sunlight) and the living things (plants).

Lesson Introduction

- Display the pictures of the plants, animals, and people.
- What do these pictures have in common? They are of living things.
- What is a living thing? Something that must have food, water, air, and space to survive and grow.

  Are all of the living things in the pictures the same? No, there are plants, animals, and people.
- Ask volunteers to share characteristics of some of the living things pictured. List and display these characteristics.
- Explain that today they will be learning about living things, what they need, and how they live together.

Instruct

Preparation for Reading

- Preview and pronounce the word *interact*.
- Direct the students to read pages 169–70 silently to find out how plants and puppies are alike.

Teach for Understanding

What are some characteristics that all living things have in common? Possible answers: They grow and develop, reproduce, and interact with things around them.

- Refer to the list made during the introduction and compare new information with what the students listed.

  What does the word *reproduce* mean? to make more of the same kind of living thing
- You put a seed in the ground, and after a few months there is a plant with beautiful flowers. How does this show that the plant is a living thing? Possible answer: The plant grew and developed.
- How are plants and puppies alike? Possible answers: They are both living things; they follow the same pattern; they both were created by God to grow, develop, reproduce, and interact with things around them.

Lesson 55

What do you think it means to *interact* with something? to respond to it or to have an affect on it

- Discuss the examples of how dogs might interact with other animals or the rain.
- Ask volunteers to share how their pets interact with them.
- Ask the same volunteers to share how their pets interact with strangers.

Do your pets interact with strangers in the same way that they interact with you?

How do plants interact with sunlight? They grow toward the light.

✔ Living things grow, develop, reproduce, and interact with things around them.

Living things interact with things around them. A dog might bark and chase other animals. The other animals may hide or run away. They are reacting to the dog's behavior. The same dog might find shelter when it rains. The dog behaves a certain way because of what is happening to it.

Even plants interact with their surroundings. A plant grows toward sunlight. It is acting a certain way because it needs sunlight to meet its needs. A plant also takes in carbon dioxide and gives off oxygen.

✔ What are four characteristics of all living things?

This dog is interacting with an animal in the tree.

170

ⓘ **Background**

Characteristics of Living Things
Living things have five common characteristics. They grow and develop, reproduce, respond to their environments, use energy, and are made of cells. For something to be classified as a living thing, it must demonstrate all five characteristics. Nonliving things may demonstrate some of the characteristics, but they are not alive because they do not demonstrate all five characteristics.

Ecosystems

God created the world with both living and nonliving things. All the living and nonliving things and the environment in which they interact is called an **ecosystem**. An ecosystem can be big or small. It could be a pond or a desert. Even your backyard is an ecosystem! A tropical rainforest ecosystem includes animals and trees. It also includes nonliving things, such as water and soil.

171

Preparation for Reading

- Preview and pronounce the vocabulary terms *ecosystem*, *population*, *community*, and *habitat* and the word *cypress*.
- Direct the students to read pages 171–73 silently to find out what a community is.

Teach for Understanding

What is an *ecosystem*? all the living and nonliving things and the environment in which they interact

What size are ecosystems? Possible answers: any size; big or small

Name some examples of ecosystems. Possible answers: a pond, a desert, a backyard, a tropical rainforest

- Direct attention to the picture.

What kind of ecosystem is this? a pond

What kinds of animals are in a pond ecosystem? Possible answers: turtles, ducks, fish

- Display the *Fitting Together Organizer* visual. Guide the students as they complete the "Ecosystem" section on *Activities* page 142.

(i) Background

Ecosystem Boundaries

Ecosystems do not have set boundaries and distinct classifications. An ecosystem simply consists of any group of living and nonliving things that interact with each other. Smaller ecosystems, such as ponds, can exist within larger ecosystems, such as forests.

- Direct attention to the picture of the swamp ecosystem along the bottom of pages 172 and 173.
  What kinds of living things might be in a swamp ecosystem? Possible answers: grasses, cypress trees, catfish, wood ducks, alligators, deer, foxes

- Remind the students that all living things are made of cells. Many living things are made of more than one cell. Many cells working together make up tissues. Tissues working together make up organs. Several organs working together make up a system. Systems make up an organism. Many organisms make up the living things in an ecosystem.
  What kinds of nonliving things might be in a swamp ecosystem? Possible answers: water, mud, rocks, air, sunlight

- Introduce groups that live together within an ecosystem.
  What is a *population*? all the living things of one kind that live in an ecosystem
  Would a deer and a fox be part of the same population? No, they are two different kinds of animals.
  How big are populations? They can be different sizes.
  Why do you think there might be more catfish than alligators in a swamp? Possible answers: The alligators are larger and eat the catfish, which are smaller. Large animals require many smaller animals to meet their food needs. An ecosystem can meet the needs of only a few large animals but can meet the needs of many smaller ones.

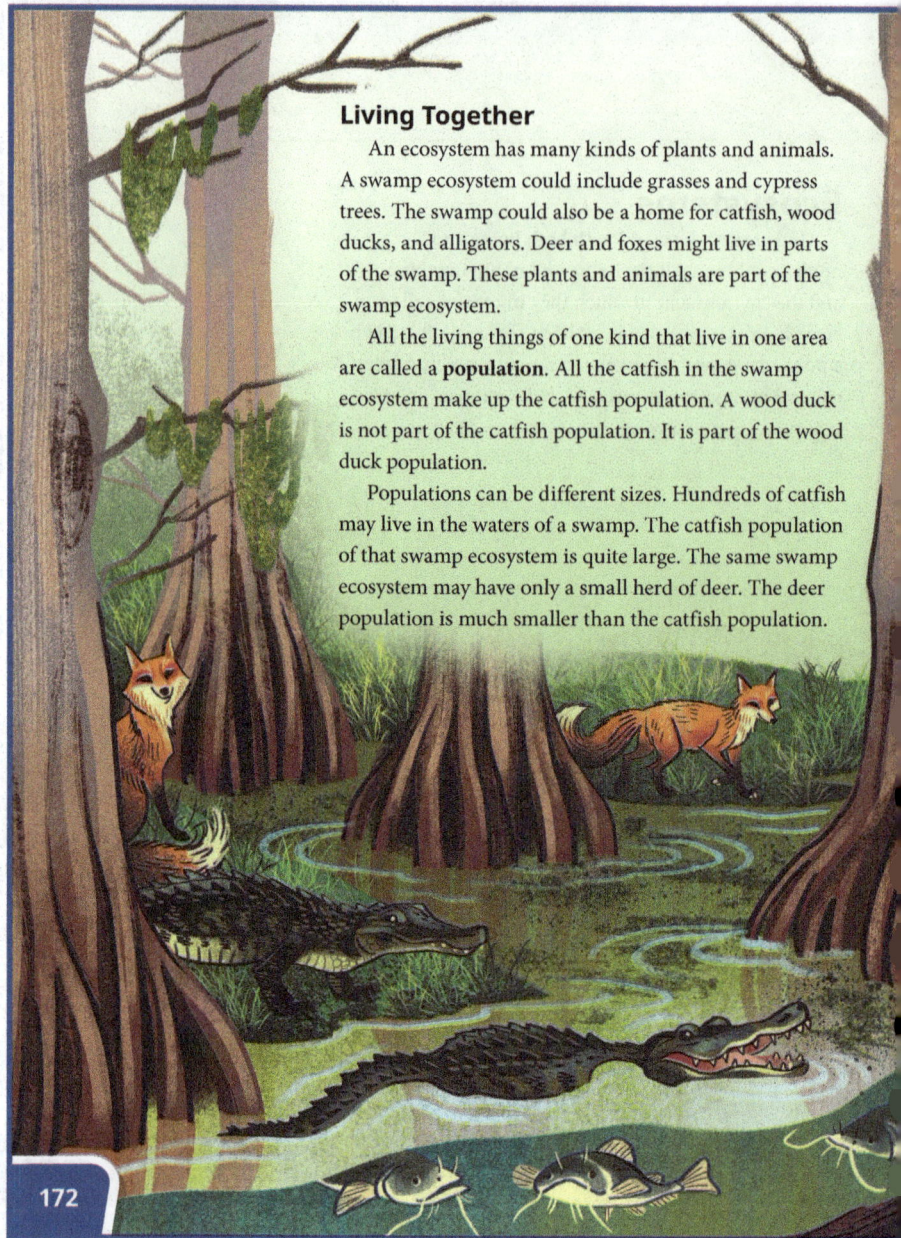

Living Together

An ecosystem has many kinds of plants and animals. A swamp ecosystem could include grasses and cypress trees. The swamp could also be a home for catfish, wood ducks, and alligators. Deer and foxes might live in parts of the swamp. These plants and animals are part of the swamp ecosystem.

All the living things of one kind that live in one area are called a **population**. All the catfish in the swamp ecosystem make up the catfish population. A wood duck is not part of the catfish population. It is part of the wood duck population.

Populations can be different sizes. Hundreds of catfish may live in the waters of a swamp. The catfish population of that swamp ecosystem is quite large. The same swamp ecosystem may have only a small herd of deer. The deer population is much smaller than the catfish population.

172

⚠ **Helps**

Understanding Ecosystem Relationships
Use the following analogy to help the students understand the relationships between a population, community, and habitat.
individual—student
population—family
community—neighborhood
habitat—town

All the different populations make up a community. A **community** includes all the living things in one area. For a swamp ecosystem, the community includes the cypress trees and other plants. It also includes the catfish population, the wood duck population, and the deer population.

Each population is part of a habitat in the ecosystem. A **habitat** is a place where a living thing lives. A habitat has the food, water, and shelter that an organism needs to live. An alligator spends most of its time in the water. The alligator's habitat is the water of the swamp. Since the water is also a habitat for cypress trees, the alligators and trees share a habitat. Many populations can share a habitat.

Deer live in the woods of the swamp. Their habitat is the woods. Foxes also live in the woods. The foxes share their habitat with the deer population.

> ✓ What is the difference between an ecosystem and a habitat?

173

- Discuss the relationship between a population and a community.

  In an ecosystem, what is a *community*? all the living things, or different populations, in that ecosystem

  Which is bigger: a population or a community? A community is made up of several different populations.

  What are some living things that make up a community in a swamp? Possible answers: cypress trees, other plants, catfish, wood ducks, foxes, deer

- Relate the vocabulary to the relationships of the students in a community. The people in a family are a population. The families who live in your neighborhood are part of your community. The many neighborhoods make up your town or habitat.

  What are some living things in your community? Possible answers: different kinds of plants and animals; people

  What word describes the place where a population lives? habitat

- Relate the term *habitat* with the town where the students live. Explain that the area where they live is their habitat. Point out it is where they eat and sleep, shop, go to church and school, and do many of their other activities.

  What is an alligator's habitat? Possible answers: water, most of the time; a swamp

  What are some populations that can share a habitat with alligators? Possible answers: cypress trees, catfish, wood ducks

  What is a deer's habitat? the woods of the swamp

  Name some living things that share a habitat with deer. Possible answers: grasses, foxes, trees

- Display the *Fitting Together Organizer* visual. Guide the students as they complete the "Population," "Habitat," and "Community" sections on *Activities* page 142.

> ✓ An ecosystem includes all the living and nonliving things and the environment in which they interact. A habitat is the place in the ecosystem where a living thing lives and finds food, water, and shelter.

Activity

Draw an Ecosystem

Materials: blank drawing paper, per student

Instruct the students to fold one third of the paper from the bottom, crease it, then unfold it. Direct them to create and draw an ecosystem in the larger top section of the page. Set a minimum for the number of living and nonliving things to include.

In the smaller, lower section of the page the students should write the name of their ecosystem and list the parts of the environment shown. They should also identify the populations and parts of the community.

Preparation for Reading

- Preview and pronounce the vocabulary terms *resource* and *competition*.
- Direct the students to read page 174 silently to find out what happens to animals when a resource in an ecosystem runs out.

Teach for Understanding

Where is the only place living things can survive in an ecosystem? in a place where their needs are met

What are the needs of all living things? air, food, water, shelter

What is a *resource*? Anything in an ecosystem that can meet a need of an organism.

What is an example of a living resource for a bear? Possible answer: food, such as fish and berries

What is an example of a nonliving resource that a bear needs? Possible answers: air, water, shelter

What might a bear use for shelter? a cave

What is *competition*? the struggle between two or more living things trying to use the same resource

What are some resources that animals may compete for? food, water, shelter

What food might a bear and a fox compete for? berries

What might the bear do if the berries in the ecosystem run out? Possible answers: The bear might eat other things; it might migrate to a new ecosystem where there are berries; it might not survive.

> Competition is the struggle between two or more living things trying to use the same resource.

Apply

Activities

Looking Ahead, page 141

This page assesses the students' knowledge prior to the chapter.

Fitting Together Organizer, page 142

This page was completed during the lesson.

Study Guide, pages 143–44

These pages review the concepts taught in Lesson 55. After completion, direct the students to keep them for review for Test 8.

Resources

Living things can survive only in an ecosystem where their needs are met. All organisms need air, food, water, and shelter.

Anything that can meet a need of an organism is a **resource**. Resources can be living or nonliving things. A bear's resources can include foods, such as fish and berries. Its resources also include water and air. A bear may use a cave for shelter. The struggle between two or more living things trying to use the same resource is called **competition**. Animals may compete for resources such as food, water, or shelter. For example, black bears may eat small animals, nuts, roots, and berries. Foxes may eat insects, eggs, and berries. If black bears and foxes live in the same area and eat the same kinds of berries, then we say they are competing for the berries.

Most ecosystems have a limited amount of resources. Sometimes a resource runs out. When this happens, the animals that depend on the resource must adjust or they will die. To avoid death they may change the way they meet their needs in the ecosystem, or they may migrate to a new ecosystem.

> What is competition?

174

STEM

Habitat Help!

One morning you watch the news before school. The meteorologist reports that your location is in a drought. The weather has been very dry. There has not been rain for many weeks.

Because of the drought, the small wild animals in your neighborhood cannot find enough food and water. You enjoy watching the animals in your backyard. You do not want them to leave your backyard ecosystem.

In this STEM activity, you will design a way to help provide food or water to a group of small animals during a drought. You will create a model of your design.

175

Objectives

- Plan and design a solution for providing food or water for an animal group during a drought, using the engineering design process.
- Create a model of the food system or water system.
- Test and compare models to improve the original design.
- Communicate to others how the model solves the problem.

Materials

- one 2-liter soda bottle, per student or group

> Students will use STEM to repurpose the 2-liter soda bottle as a means to help provide food or water to small wild animals. You may add items of your choosing to the materials list.

Teacher Resources

- Visual 2.7: *STEM: The Engineering Design Process*

Engage

- What small wild animals have you seen in your backyard? Possible answers: birds, chipmunks, squirrels, rabbits
- What is a drought? a period of time with very little, or no, precipitation
- What would happen to the animals in your backyard if there were a drought? Possible answers: They would leave the ecosystem to find food; they might not survive.
- How could you keep the animals in your backyard ecosystem? Provide them with resources, such as food and water.
- Why do you think it would be a good idea to help small wild animals get food and water? Possible answers: God expects people to care for the earth; providing resources for animals in a drought is part of caring for the earth.

Instruct

Preparation for Reading

- Direct the students to remove *Activities* pages 145–46. Direct the students to read page 175 and *Activities* pages 145–46 silently before beginning.

Teach for Understanding

- Display the *STEM: The Engineering Design Process* visual or direct the students to the copy of *Activities* page 35 in their science notebook.
- Review the engineering design process as needed.

- Ask a volunteer to identify the *Problem*.
- Direct attention to the blank lines in the statements in the Problem.
- Explain that the students are to identify a small wild animal they would like to help provide food or water for.
- Explain that in this STEM activity the students will design a way to use a 2-liter soda bottle to hold food or water for small animals.

Apply

- For this activity, each student will design his own solution to the problem, or the students may work in science groups to collaborate as they design a solution.
- Instruct the students to *Imagine* a solution to the problem and *Plan* a design using the empty 2-liter soda bottle.

STEM

Habitat Help!

Name _____

Ask

What is the problem?
There has not been rain for many weeks. There is a
drought and the _____
(an animal)
in my neighborhood cannot find enough food and water.

What can I design to help provide food or water to the
_____ during a drought?
(an animal)

Imagine

Think about possible answers to the problem.

_____

_____

_____

Plan

Draw and label your design. List and gather materials.

Accept any reasonable design.	
Design	**Materials**

SCIENCE 3 *Activities*

Create

ollow the plan and make a model.

est and Improve

What changes need to be made to the design?
Accept any reasonable answer.

_____

_____

_____

| |
| Improved Design |

hare

ll others how your design solves the problem.

- Direct the students to *Create* a model of their design.
- Instruct the students to *Test* their model.
- After testing their designs, direct the students to compare models and *Improve* their design.
- Invite the students to *Share* their designs with a partner or another science group. Instruct the students to explain how their design solves the problem.

Activities

STEM, pages 145–46

The students will design a way to provide food or water to a small wild animal during a drought.

Assess

Rubric

Use the prepared rubric or design a rubric to include your chosen criteria.

Objectives

- Explain how producers get their energy.
- Classify organisms as producers, consumers, or decomposers.
- Describe the three types of consumers.
- Explain, using biblical teaching, why there were no carnivores before the Fall. BWS
- Analyze the advantages of animals living together in groups.

Vocabulary

- producer
- consumer
- decomposer
- herbivore
- carnivore
- omnivore

Engage

- Identify and discuss different jobs that people in your community have.

  What does a doctor or nurse do? Possible answers: gives medicine; takes care of sick or injured people

  What does a chef at a restaurant do? Possible answers: makes food that people like to eat; tries out new recipes

  What does a pastor do? Possible answers: preaches God's Word; counsels and encourages people; visits sick people

  What does a police officer do? Possible answers: protects people; helps people

- Point out that different people have different jobs, and each job is important. People work together to help and serve others in their community. The living things in an ecosystem also have different jobs but work together.

Instruct

Preparation for Reading

- Preview and pronounce the vocabulary terms *producer*, *consumer*, and *decomposer*.
- Direct the students to read pages 176–77 silently to find out what the job of a decomposer is.

Teach for Understanding

What do all living things need? energy

Where do plants get energy from? sunlight

Where do animals get energy from? plants and other animals they eat

What are the three main groups of living things? producers, consumers, and decomposers

What is a *producer*? a living thing that makes its own food

Where does a producer get its energy from? directly from the sun

What do plants use that energy for? to make food during photosynthesis

- Review the process of photosynthesis as needed. Photosynthesis is discussed in Chapter 5.

  Name an example of a producer. any kind of plant

- Direct attention to the pictures and caption. Ask volunteers to identify the producers.

Lesson 57

Eating for Energy

Living things need energy to live. Plants get energy from sunlight. Animals get energy from plants and other animals. Plants and animals can be put into three main groups: producers, consumers, and decomposers.

Producer

A **producer** makes its own food. It gets its energy directly from the sun. Plants are producers. They use energy from the sun to make their own food during photosynthesis. Plants use some of the food for their own needs. They store the rest of the food.

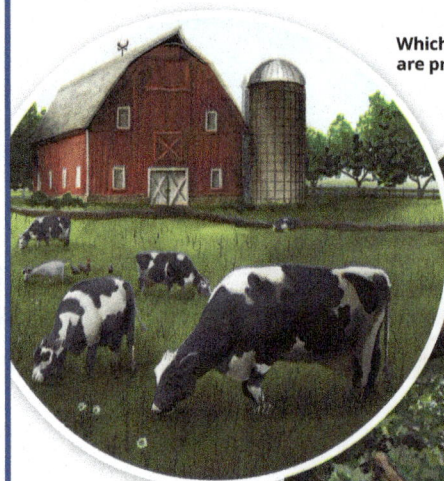

Which living things are producers?

176

ⓘ Background

Other Producers

Plants are not the only producers. Algae and some bacteria are also producers.

Kinds of Consumers

Decomposers and scavengers could also be considered consumers since they do not make their own food. Scavengers are a group of consumers that help keep ecosystems clean. Scavengers usually do not kill their prey but instead eat things that have already died. Cockroaches, vultures, and hyenas are common scavengers. Most scavengers are carnivores.

ⓘ Misconception

People Making Food

Although people can combine food products to make a meal, they cannot make food using energy from the sun.

Consumer

Many living things cannot get their energy directly from the sun. An organism that gets its energy by eating other living things as food is called a **consumer**. Animals and people are consumers. They depend on producers for food. Consumers must eat plants or animals to get energy.

Decomposer

The last main group is called decomposers. A **decomposer** helps break down dead organisms and decaying matter. Breaking down these dead things adds nutrients to the soil. Plants then use the nutrients to grow and produce more food.

There are many kinds of decomposers. Some are very small. Bacteria are so small that they cannot be seen without a microscope. Other decomposers include mold, mushrooms, and earthworms.

Which living thing is the consumer?

How do consumers get their energy?

What are *consumers*? living things that get their energy by eating other living things

Name an example of a consumer. Possible answers: any kind of animal; people

Why are people considered consumers? Possible answers: We must eat plants or animals to get energy; we cannot use the sun to make food for ourselves.

What are *decomposers*? living things that help break down dead organisms and decaying matter

Why do you think decomposers are important to an ecosystem? Possible answers: They add nutrients to the soil; they help plants to grow, which makes more food for consumers; they help keep ecosystems clean; they keep dead things from piling up.

Name an example of a decomposer. Possible answers: bacteria, mold, mushrooms, earthworms

- Direct attention to the pictures and caption. Ask a volunteer to identify the consumer.

by eating other living things as food; by eating producers and other consumers

177

Preparation for Reading

- Preview and pronounce the vocabulary terms *herbivore*, *carnivore*, and *omnivore*.
- Direct attention to the pictures on the next three pages. Guide the students to conclude that all the animals shown are consumers based on the heading for this section. Review that headings help the reader know what a section is about.
- Direct the students to read pages 178–80 silently to find out what the three types of consumers are.

Teach for Understanding

What are *herbivores*? consumers (animals) that eat only plants

What do herbivores eat? parts of plants such as bark, branches, leaves, roots, stems, grass, seeds, fruit, and nectar

- Direct attention to the picture at the top of the page.

What animal is this? a koala

What is the koala eating? leaves; eucalyptus leaves

What are the horse and sheep eating? grass

Name some herbivores. Possible answers: rabbits, elephants, giraffes, koalas, zebras, horses, sheep, sparrows, butterflies

How did God design each herbivore to get the energy it needs? from the parts of plants that it eats

What part of the plant does a butterfly use for food? nectar from flowers

- Explain that nectar is the sweet liquid produced by many flowers.

Types of Consumers
Herbivore

Animals eat a variety of foods. Some consumers, such as rabbits and elephants, eat only plants. Consumers that eat only plants are called **herbivores**.

Herbivores eat different parts of plants. Some herbivores, such as elephants, eat bark, small branches, leaves, grasses, roots, and the fruit of plants. Giraffes and koalas eat leaves. Zebras and sheep eat grasses.

Sparrows eat the seeds and fruit of plants. Butterflies drink nectar from flowers. God designed each herbivore to get the energy it needs from the parts of plants that it eats.

178

ⓘ Background

Koalas

Koalas are very selective eaters. Almost all their food comes from eucalyptus trees. They even get most of their water from eating the leaves of the plant.

Herbivores

Some additional herbivores include beavers, bees, bison, deer, grasshoppers, porcupines, and snails.

Carnivore

Some consumers eat only other animals. These consumers are called **carnivores**, or meat eaters. Carnivores get their energy by eating other consumers. Wolves, weasels, tigers, and some large birds are carnivores.

Animals were not always carnivores. When God first created the world, all animals were herbivores. The Bible tells us in Genesis 1:29–30 that God created people and animals to eat only plants. But Adam and Eve sinned. Because of their sin people and animals can die. After the Fall God allowed people and animals to begin to eat animals as well as plants for food.

Fantastic Facts

Not all carnivores are animals. Plants such as butterworts and Venus flytraps "eat" insects and spiders. These plants still get energy from the sun, but they grow in places where the soil has few nutrients. The insects and spiders provide some of the nutrients the plants need.

179

What kind of consumer eats only meat? a carnivore
How do carnivores get the energy they need? by eating other animals

- Direct attention to the picture of the bird.
- What kind of bird is this? an eagle; a bald eagle
  What is the eagle carrying? a fish
- What kind of consumer is the eagle? It is a carnivore.
- Direct the students to Genesis 1:29–30 in the Bible. Ask a volunteer to read the verses aloud.
  When God created animals and people, what did He create them to only eat? plants
- Why did animals and people eat only plants? because there was no sin; because there was no death
  When did God allow animals and people to begin to eat both plants and animals? after the Fall
- Discuss "Fantastic Facts" on page 179.
  What are some plants that eat insects and spiders? Possible answers: butterworts, Venus flytraps
- Point out that the picture is a Venus flytrap.
  Do these plants get energy from the sun? yes
  Why do the plants need to eat insects as well? The soil where the plants grow does not have many nutrients. The insects and spiders provide some of the nutrients the plants need.

ⓘ Background

Carnivores

Some additional carnivores include crocodiles, frogs, leopards, lions, octopuses, owls, penguins, polar bears, salmon, sharks, snakes, and spiders.

Carnivorous Plants

Although insects do provide minerals needed by the plant, the insects provide little energy for the plant. Carnivorous plants still depend on the sun for the energy to make food.

- Discuss omnivores.

  What do omnivores eat? both plants and animals

  Name some omnivores. Possible answers: bears, skunks, robins

  Why do some omnivores change their eating habits? Possible answer: What they usually eat is not available.

- Direct attention to the pictures at the bottom of the page.

  What kind of omnivore is shown? a bear

  What do the photos show the bears eating? fish, berries

  How do these pictures show us that bears are omnivores? The bears are eating the meat of another animal and the fruit of plants.

- Classify the type of consumer people are.

  Which type of consumer are people? People are omnivores, because people eat both plants and animals.

- Remind the students that, since the Fall, God allowed people to eat both plants and animals for energy.

> ✓ Herbivores eat only plants. Carnivores eat only other animals. Omnivores eat both plants and animals.

Omnivore

Other consumers eat both plants and animals. These consumers are called **omnivores**. Bears, skunks, and robins are omnivores. Bears eat other animals, such as fish, insects, and small mammals. But they also eat grass, berries, roots, and nuts.

Many omnivores change their eating habits when the seasons change. For example, a skunk eats whatever is available in its habitat. Skunks often eat rats and small mammals in the winter months. During spring and summer, they eat plants and insects. In the fall skunks add fruit and berries to their diet.

> ✓ What do herbivores, carnivores, and omnivores eat?

180

ⓘ Background

Omnivores

Some additional omnivores include chickens, crows, pigs, raccoons, rats, seagulls, squirrels, and foxes.

Living Together in Groups

Some animals live together in a group to meet their needs. Living together gives animals advantages. Living in a group can help some animals find food more easily. Being in a group can help protect animals and their offspring from predators. Animals who live in a group can also be protected from the cold.

Some animals who live in groups can hunt bigger animals because the group can overpower the prey more easily. When large groups of animals see predators, they sometimes form tight circles around their offspring and move as a group. Some animals in the group will also take turns watching for predators and sleeping. Others may fight predators to keep the rest of the group safe. When it is cold, large groups of animals can stay warm by staying very close together.

What advantage do these fish have in a group?

What advantage do these elephant seals have in a group?

✓ **What are some advantages of animals living together?**

What advantage do these elephants have in a group?

181

Preparation for Reading

- Preview and pronounce the word *advantage*.
- Direct the students to read page 181 silently to find out why it is beneficial for some animals to live in groups.

Teach for Understanding

Why do some animals live in a group? to meet their needs

What does living together give these animals? advantages

⚙ What do you think an *advantage* is? a benefit

Why can some animals who live in groups hunt bigger animals? The group can overpower the prey more easily.

How can living in a group be an advantage to offspring? The larger adult animals can form a tight circle around the smaller offspring, protecting them from predators.

When it is cold, how is being in a group an advantage? The animals can stay warm by staying very close to one another.

- Direct attention to the picture of the school of fish. Explain that a large group of the same kind of fish is called a school.
- Choose a student to read the caption aloud.

⚙ What advantage do fish in a school have? Possible answers: protection from predators; easier to find food

- Direct attention to the pictures of the elephants and elephant seals.

⚙ What advantages do these animals have by living together in a group? Possible answers: finding food more easily; protection for themselves and their offspring from predators; staying warm when it is cold

✓ **finding food more easily; protection from predators; protection from the cold**

Apply

Activities
Study Guide, pages 147–48
These pages review the concepts taught in Lesson 57. After completion, direct the students to keep them for review for Test 8.

Assess

Quiz 8A
The quiz may be given at any time after completion of this lesson.

Objectives

- Recall what a food chain is.
- Sequence on a food chain the transfer of energy from one organism to another. **BWS**
- Identify what a food web is.
- Describe the relationship between a food chain and a food web. **BWS**
- Interpret a food web.
- Explain what happens when one part of a food web changes.

Materials

- mouse, picture of mouse or live pet mouse
- paper chain with 4 links, per student

> Make one paper chain for display. Each student should make his own.

Teacher Resources

- Visuals 8.2–8.3: *Food Chain*; *Food Web*

Vocabulary

- food chain
- food web

Engage

- Display the mouse.
- What food do you think a mouse eats? Possible answers: oats, seeds, fruit, grains, insects
- Are mice herbivores, carnivores, or omnivores? omnivores, because they eat plants and other animals
- What animals might eat a mouse? Possible answers: cats, owls, hawks, snakes, foxes
- Owls eat mice. What other things do you think an owl might eat? Possible answers: rabbits, squirrels, snakes, frogs, insects
- Point out that most animals eat more than one kind of food. Explain that today the students will learn more about the foods animals eat.

Instruct

Preparation for Reading

- Preview and pronounce the vocabulary term *food chain*.
- Direct the students to read pages 182–83 silently to find out what the movement of energy from one living thing to another is called.

Teach for Understanding

What is the process called in which a plant uses the sun to make its own food? photosynthesis

What provides energy for grass and other plants? the sun

What happens to a plant's energy when an animal eats the plant? The energy passes from the plant to the animal.

What is the movement of energy from one living thing to another living thing called? a food chain

Food Chains

A blade of grass produces its own food through photosynthesis. A grasshopper gets food as it nibbles on the blade of grass. Later a garter snake catches and eats the grasshopper.

The movement of energy from one living thing to another living thing is called a **food chain**. The sun provided energy for the grass. That energy then passed from the grass to the grasshopper to the garter snake.

A food chain begins with the sun. The first living thing in a food chain is always a producer. A producer uses sunlight to provide energy for itself. The next link in the food chain is usually a herbivore. The animals that make up the other links of a food chain are predators. Predators can be carnivores or omnivores.

= Movement of energy

182

- Display the paper chain.
- What are the loops that make the chain called? links

What is the first link of every food chain? the sun

Why is the sun the first link on every food chain? The sun provides the energy for plants.

- Write "sun" on the first link of the paper chain and direct the students to write it on theirs.

What are the first living things on a food chain? plants

- What type of living thing is a plant—a producer, consumer, or decomposer? producer, because it uses sunlight to provide energy for itself
- Direct the students to write "producer: plants" on the second link.

Why is the next link in the food chain usually a herbivore? Herbivores are consumers that eat only producers (plants) for energy.

- Direct the students to write "consumer: herbivore" on the third link.
- What other kind of consumer eats producers? omnivores
- What are predators? animals that hunt and eat other animals
- What do we call the animals that predators hunt and eat? prey

What type of consumer are the predators that make up the other links in the food chain? carnivores or omnivores

- Direct the students to write "consumer: carnivore or omnivore" on the fourth link.

A food chain is only a few links. Each living thing in each link uses some of the energy and stores the rest. The grass received energy from the sun and used most of that energy to grow. Only some of the energy was stored.

The grasshopper ate the grass and gained the stored energy. The grasshopper used most of that energy for its own needs. Some of the energy was stored in the grasshopper's body. The garter snake then came along and ate the grasshopper. When it did, the garter snake received the grasshopper's stored energy.

Each organism uses most of the energy it gets for its own needs. That is why a grasshopper has to eat more than just one blade of grass. It needs the energy from many plants. In the same way, a garter snake needs to eat many grasshoppers.

> **How is energy passed down in a food chain?**

183

- Display the *Food Chain* visual and explain that the arrow shows the movement of energy from one link of the chain to the next.
- Ask volunteers to summarize the flow of energy through the links shown in the food chain.
  What moves from one living thing to another as animals eat? energy
  What happens to the energy that a plant or animal does not use? It is stored to use later.
- Review that God created all living things to grow and develop, reproduce, and interact with things around them. All of these characteristics require energy. Living things are constantly using the energy they eat and store. Explain that this is why living things need to eat every day to stay healthy.
  Where did the grasshopper get its stored energy from? the grass that it ate
  How did the garter snake receive its energy? When the garter snake ate the grasshopper, it received the grasshopper's stored energy.
  Why do grasshoppers eat more than one blade of grass and garter snakes eat more than one grasshopper? to get enough energy to meet their own needs
- Discuss that since people are consumers, they are also part of food chains.
- Direct attention to the picture on page 183.
  Which animal is the predator and which is the prey? The bird is the predator and the rabbit is the prey.
- Point out that the bird is a hawk and that it eats only meat.
  Is the bird a herbivore, carnivore, or omnivore? It is a carnivore.
- Ask for volunteers to construct a food chain for display that includes the rabbit and the hawk. Answers should include: sun, plants/grass, rabbit, hawk

> **Energy in a food chain is passed down from one living thing to another living thing.**

Differentiated Instruction

Remediation: Build Food Chains

Materials: *Food Chain Links*, one set per student or group, on cardstock and cut apart (Instructional Aids 8.1–8.2)

Give each student or group of students a set of cards. Direct them to arrange some of the cards into a food chain using the descriptions on the cards. Then instruct them to list the order of the chain on paper. Repeat to form more chains.

Preparation for Reading

- Preview and pronounce the vocabulary term *food web* and the word *source*.
- Direct the students to read pages 184–85 silently to find out how a food chain and a food web are connected.

Teach for Understanding

In a food chain how many sources of food are shown for each animal? one

Did God design most animals to usually eat only one kind of food? No, most animals eat more than one kind of food.

Are most animals part of one food chain or many food chains? many food chains

What do we call several food chains linked together? a food web

- Display the *Food Web* visual and explain that the arrow shows the movement of energy from one link of the chain to the next.

What part of a food chain is not shown in this web? the sun

- Explain that some food-web diagrams include the sun, and others do not.
- Direct attention to the frog.

What does the frog's food chain include? (sun), plants, grasshopper

- Direct attention to the fox.

What does the food chain of the fox include? (sun), plants, grasshopper, frog, garter snake

Do you think a food chain or a food web is a more accurate scientific model of the relationships in an ecosystem? Elicit that a food web is more accurate in an ecosystem because a food chain shows only one possible connection, but plants and animals are usually part of many food chains.

What does a food web show us about how God created plants and animals? God made it possible for plants and animals to have their needs met from several sources of food.

What happens when one part of a food web changes? The whole food web is affected.

What would happen to the grasshopper population if there were fewer frogs? The grasshopper population would increase; there would be more grasshoppers because the frogs eat the grasshoppers.

What would happen to the plants in the ecosystem if there were more grasshoppers? There would be fewer plants because grasshoppers eat only plants.

Food Webs

A food chain shows only one source of food for each animal. Most animals, though, eat more than one kind of food. God made it possible for animals to have their needs met from several sources of food. This is the way that God created animals. To show animals having more than one source of food, we use a food web. A **food web** is several food chains linked together.

For example, a grasshopper eats plants, and a garter snake might eat the grasshopper. That is one food chain. In the same community, another food chain might also start with plants and a grasshopper. But in that food chain, a frog might eat the grasshopper. The frog might be eaten by a fox. The grasshopper is prey for both frogs and snakes. It is part of more than one food chain.

The fox is a predator of frogs. But it is also a predator of snakes and grasshoppers. It will even eat plants. The fox is part of a different food chain for each food it eats. A food web shows many prey and predator links.

When one part of a food web changes, it affects the entire web. Frogs eat grasshoppers. If there were fewer frogs, fewer grasshoppers would be eaten. The grasshopper population would increase. More grasshoppers means more plants would be eaten. If there were fewer frogs, it would also mean that some of the frog's predators would not have enough to eat.

184

Food Web

→ = Movement of energy

✓ What happens when one part of a food web changes?

- Direct attention to the diagram and display the *Food Web* visual. Discuss what the diagram is showing.

  What three animals eat grasshoppers? wood frogs, garter snakes, and red foxes

  What is shown in the diagram that helps you to know what animals eat grasshoppers? The arrows from the grasshopper point to those animals.

  What kinds of plants are included in this web? grass and a berry bush

  What might happen to the grasshoppers if many of the plants were taken away? Possible answers: The grasshoppers would not have enough to eat, and some might die. The grasshoppers might have to move to find food.

  Is the garter snake the predator or the prey of the red fox? the prey

  What animal in the food web has no predators? the fox

  What would happen to the garter snake population if there were more foxes? There would be fewer garter snakes because foxes eat snakes.

  What would happen to the red fox population if there were fewer frogs? The foxes would have to eat more snakes, grasshoppers, and berries to receive the energy they need; if the red foxes could not get the energy they need from the snakes, grasshoppers, and berries, they may move to find more food.

✓ It affects the entire food web.

Apply

Activities
Connecting Links, pages 149–50
These pages reinforce the concepts taught in Lesson 58.

👥 Differentiated Instruction

Remediation: Model Changing a Food Web
Materials: 5 pieces of paper, marker, hole punch, yarn

Choose a basic food web to demonstrate, such as the one pictured on *Student Edition* page 185 or *Activities* page 150. Write the name of one part of the food web on each sheet of paper. Punch two holes at the top of each piece of paper, and string a piece of yarn through the holes to make a hanging sign. Give the signs to students to wear.

Direct the students to use the yarn to make connections to match the food web. Emphasize the movement of energy by making connections one at a time. Repeat until the web is complete as pictured.

Choose one part of the food web to change by directing one member of the web to drop his pieces of yarn and move away from the web.

Discuss how the connections are broken and how they affect the other parts of the web.

Objectives

- Name some causes of change in an ecosystem.
- Explain a way people should exercise dominion when making changes to ecosystems. **BWS**
- Give examples of characteristics that help organisms survive in a changed ecosystem.
- Analyze why some organisms survive well and some do not in a changed ecosystem.
- Identify how an ecosystem maintains balance.

Engage

- Discuss how the students handle changes.

  What are some changes that have happened in your life during the past few months? Possible answers: becoming a third grader; moving to a new house or a new school; having a new sibling born; getting a new pet; having new chores or responsibilities at home

  How did you feel about those changes? Elicit that the students might have been excited about some changes and fearful of or unconcerned about other changes.

- Point out that changes happen to us all the time. Changes also happen in ecosystems. Explain that today the students will find out about some changes that can happen in ecosystems.

Instruct

Preparation for Reading

- Direct the students to read pages 186–88 silently to find out what helps living things survive in different environments.

Teach for Understanding

What was the biggest change that happened to any ecosystem? death

What caused that change? the Fall, or Adam's sin

What do herbivores, carnivores, and omnivores help us to understand about the Fall? Possible answers: There was no death before Adam sinned; before the Fall, people and animals were herbivores; after the Fall, some living things ate other living things for food.

Do ecosystems still change today? yes

How might disease change an ecosystem? Possible answers: A plant population may be destroyed by disease; the animal population that eats the plants may not have enough food; the animals may move to another ecosystem; more animals in the other ecosystem may cause a lack of food for all the other animals.

What are three things that can cause changes in an ecosystem? the environment, animals, and people

What are some ways the environment in an ecosystem could change? Possible answers: Too much rain could flood an ecosystem. Not enough rain could cause a drought, or dry soil. The weather could become much hotter or much cooler than normal.

Animals also can change an ecosystem. How can beavers change an ecosystem? by building a dam that turns a stream into a pond and the pond into a meadow

---

Changes in an Ecosystem

Ecosystems are always changing. The biggest change to any ecosystem took place when Adam's sin brought death into the world.

Ecosystems today continue to change. A disease may destroy a certain plant population. Animals that eat that plant may not get enough food. Then they might move to another ecosystem. The increase in animals would then affect that ecosystem.

Sometimes it is the environment that causes change. Heavy rains can flood an ecosystem. The extra water might destroy plants and cause animals to move to another place. Not enough rain can cause dry soil and can keep plants from growing.

Sometimes animals can also cause an ecosystem to change. Beavers build dams across streams. The dam stops the flow of water and turns the stream into a pond. After a while, the pond dries out and becomes a meadow.

How did this ecosystem change?

186

(i) Background

Changes Made by People

People change ecosystems in both negative and positive ways. It is easy to notice the negative ways, such as over-fishing, littering, and destroying habitats. People help ecosystems by setting aside areas for wildlife, cleaning up pollution, and planting trees and other vegetation in urban areas.

Sometimes people can change an ecosystem. If fishermen catch a lot of one kind of fish, its population gets smaller. Other animals that depend on that fish for food will not have enough to eat. They may have to move to another area. In some places, fishing laws set limits on the number of fish that can be caught. The limits help keep the ecosystem from being changed too much.

God wants us to wisely use the resources He has given us. The Bible tells us in Genesis 1:28 that God told Adam and Eve to take care of the earth. When people make wise changes to ecosystems, they glorify God.

When an ecosystem changes, the plants and animals are affected. Some can survive well. Some cannot survive well. Some animals move to new ecosystems, and some may even die.

When a small pond dries up because of a drought, tadpoles do not survive.

When a wildfire spreads through a woodland forest, deer have to move to another ecosystem to survive.

187

What are some ways people could change ecosystems? Possible answers: catching too much of one kind of fish; building things; cutting things down

- Point out that pollution is another way people change an ecosystem. Explain that people can help control pollution by observing the three Rs of conservation: reduce, reuse, and recycle. Use the Background information to remind the students what each of these means.

God has given people the responsibility to have dominion, or rule, over His creation. People should use the earth's resources while being good stewards of God's creation.

How does God want us to use the resources He has given us? wisely

Why should you use the resources wisely? God told Adam and Eve in Genesis 1:28 to take care of the earth. Using resources wisely is one way to take care of the earth.

What is one way people can use resources wisely in an ecosystem? Possible answers: When trees are cut down, new trees can be planted; when the number of fish in a river or pond runs low, we can bring in more fish or, for a time, the number of fish caught can be limited (or returned to the water as "catch and release") so the fish have time to reproduce.

How are plants and animals affected when an ecosystem changes? Some plants and animals can survive well. Some cannot survive well. Some animals move to new ecosystems. Some die.

- Direct attention to the pictures and captions. Ask volunteers to read the captions aloud.
- Discuss how the plants and animals have been affected by the changes to their ecosystems.

Why do tadpoles not survive when the pond dries up during a drought? Possible answers: Tadpoles need water to survive; tadpoles breathe with gills in water.

Why do deer have to move to another ecosystem after a wildfire has happened in the woodland forest? Possible answer: They move to find food; the fire has destroyed their food source (shoots of plants, fruit, nuts, acorns, etc.).

(i) Background

Reduce, Reuse, Recycle

The three Rs of conservation are reduce, reuse, and recycle. Reducing involves decreasing the resources used, such as using less water or energy. Reusing is simply using something over again, such as giving clothing you have outgrown to someone who can use it. Recycling is processing material to make a new product out of an old one. Items such as aluminum cans, plastics, and paper can be recycled.

⚗ Activity

List Seasonal Changes

Divide the students into groups. Direct them to identify and list changes that animals and plants in a forest or other ecosystem might experience as winter approaches.

Possible changes to animals include hibernating, growing thicker coats of fur, migrating, and fur changing to a different color.

Possible changes to plants include trees and bushes dropping leaves and seeds, parts of some plants dying, and some whole plants dying.

What helps living things to survive when their ecosystem changes? **the ability of living things to adapt; their created characteristics that help them to survive**

How did living things get the ability to adapt? **God created them with the ability.**

What are some ways plants and animals adapt? **Possible answers: animals growing thicker coats of fur in cold weather and storing fat in their bodies; plants having longer roots**

- Explain that some plants and shrubs in the desert have adapted by growing longer roots, but that the roots of some of these plants spread out from the plant, near the surface, not deep into the ground. Because the roots are near the surface, they can quickly absorb the little water that does fall when it rains. Point out that the amount of rainfall in a desert is about 10 inches or less per year.

- Direct attention to the picture of the camel and discuss the purpose of its humps.

- Direct attention to the pictures of the weasels.

How does a weasel like this one adapt during the winter? **It grows white fur.**

How does having white fur help the weasel during the winter? **Possible answers: The white fur helps the weasel blend in with the snow. It helps the weasel hide from predators and not be as noticeable to its prey.**

What word could you use to describe the weasel's ability to not be seen easily in its surroundings? **camouflage**

- Point out that we can observe the weasel's fur change color. But this does not mean that the weasel will change its fur for scales or for feathers at some future point. While God designed animals with some ability to adapt, there is no evidence that animals can change into other kinds of animals.

changes in population sizes; changes in environment; changes caused by animals; changes caused by people

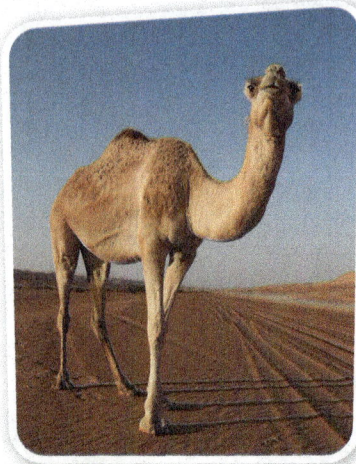

God created living things with the ability to adapt, or change. Some characteristics help plants and animals survive when ecosystems change. Certain animals, like deer, can grow thicker coats of fur. They do this when the temperature gets colder in the woodland forest ecosystem. Trees and shrubs that live in deserts usually have long roots. The long roots help the plants get water from the ground. Camels store fat in their bodies. Their bodies can use the fat for food. This characteristic helps them survive when food is hard to find in the desert ecosystem.

Long-tailed weasels are usually brown, but some weasels that live in areas where it snows grow white fur during the winter.

What kinds of changes can happen in an ecosystem?

Preparing Ahead

Lesson 65 Materials
Look ahead to the Materials list in Lesson 65 for required preparation.

Balance in an Ecosystem

Changes happen to an ecosystem. Even a small change can cause many other changes in an ecosystem. But God is a wise Creator. He designed ecosystems to change and to balance each other.

Lightning can start a fire that burns a forest. Many habitats are destroyed in the fire. Plants and animals might be killed. Changes like this are sad and even scary, but they also show God's amazing design. Out of the ashes of a forest fire, a new ecosystem comes to life. Soon new plants start to grow. Animals that left may return to the ecosystem. Other animals also come to live in the new ecosystem. Over time another forest grows. It is somewhat like the forest it replaced, but it is also different.

A fire changes an ecosystem.

Preparation for Reading

- Direct the students to read pages 189–90 silently to find out what predators help to control.

Teach for Understanding

What happens when something in an ecosystem changes? Other things also change.

- Discuss that even though ecosystems change, God designed them to be balanced. God is in control of the changes.
- Direct attention to the picture. Ask a volunteer to read aloud the caption.

What are some changes to an ecosystem that are caused by a forest fire and might seem sad or scary? Possible answers: Habitats are destroyed. Some plants and animals die.

How can God use a forest fire for good? He can use it to renew the ecosystem.

In what ways can an ecosystem come to life again? Possible answers: New plants start to grow. Animals come back to the ecosystem. A new forest grows to replace the old.

Is the new forest exactly like the forest it replaced? No, in some ways it is like the forest it replaced, but it is also different.

What are some things that might be different in the forest? Possible answers: Different kinds of trees and other plants may grow; different animals may move into the forest looking for water, food, and shelter.

189

ⓘ Background

Stresses and Succession

Disasters such as floods, fires, and droughts change the environment of an ecosystem and are often referred to as stresses. A *stress* is any physical hazard to life caused by having too much or too little of things needed for life. Stresses are often a normal part of the changes happening in an ecosystem. After a stress, the ecosystem is different. Populations may have gotten smaller, and even the types of living things may have changed. Eventually, the ecosystem renews itself. The gradual changes in the populations of an ecosystem are called *succession*.

Invasive Species

Invasive species are organisms that are deliberately or accidentally introduced into an ecosystem where there are not natural checks to keep them in balance. Some examples include kudzu vines, fire ants, gypsy moths, and European starlings.

God's Control

Remind the students that God is always in control of the changes that happen. This also applies to natural disasters.

What does God use to keep an ecosystem in balance? predators

What do predators help control? population sizes

- Direct attention to the pictures.

What population is the leopard helping to balance? antelope; bohor reedbuck

What population is the praying mantis helping to balance? lizard; wall lizard

- Discuss things that might happen if a population grows too big. Explain that a larger population will eat more and use up its food resources.

What determines how many living things an ecosystem can support? whether the needs of the living things are being met

What are some needs that animals have? Possible answers: food, water, shelter

What might animals do if they cannot find all the resources they need? move to a new ecosystem or die

Why is death part of our world? Adam disobeyed God, and death is a result of that sin.

- Read and discuss Revelation 21:1–6. Point out that God has revealed to John a vision of the new heaven and the new earth (verse 1).

What is different about death in the new heaven and new earth? There is no more death (verse 4).

To whom does God give the water of life? to the one who thirsts

- Explain that God is promising to give eternal life to all those who come and ask for it (cf. John 6:37). These are the ones who will experience no more death (Revelation 21:4). Share *Explaining the Gospel* (see Teacher Resources, A2).

move to a new place or die

Apply

Activities

Study Guide, pages 151–54

These pages review the concepts taught in Lessons 58–59. After completion, direct the students to keep them for review for Test 8.

Assess

Quiz 8B

The quiz may be given at any time after completion of this lesson.

God also uses predators to keep an ecosystem in balance. Predators help control population sizes. Ecosystems can meet the needs of only a certain number of living things. Resources will run out if too many animals compete for food, water, or shelter. Animals that do not have all their needs met will either move to a new place or die.

The Bible tells us in Revelation 21:4 that someday there will be no more death. This will be when God reveals His new heaven and new earth. God promises us that He will make all things new and perfect. God can make you part of that new and perfect world.

What do animals do when their needs are not met in an ecosystem?

190

EXPLORATION

Ecosystem Tag

Inquiry Skills
• Measure
• Infer

Animals in a community interact with each other. Some animals are the prey. Other animals are the predators. But all are important to the balance of an ecosystem.

In this Exploration, you will represent different animals that interact in an ecosystem.

191

⚠ Helps

Life Card Record
Copy the chart onto posterboard.

Life Card Record			
	Chipmunks	**Snakes**	**Owls**
Start per player			
Start total			
Round 1 Total			
Round 2 Total			
Change that will affect animal interactions:			
Round 3 Total			
Round 4 Total			

Objectives

• Identify the roles of predator and prey in a food web.
• Infer changes in an ecosystem.
• Draw conclusions from data collected to make inferences about a population after its predator dies off in an ecosystem.

Materials

• See *Activities* page 155.
• safety pins, one per student
• posterboard
• marker
• large plastic hoops
• stopwatch
• whistle

Teacher Resources

• Instructional Aids 8.3–8.5: *Tag Identity Badges*; *Tag Life Cards*

Make an identity badge for each student. For each owl, there should be 2 snakes and 6 chipmunks. Make enough life cards for each owl to have 1, each snake to have 2, and each chipmunk to have 5. Copy the "Life Card Record" chart onto the posterboard. See the Helps information.

Engage

• Relate the players in a game of tag to a predator-and-prey relationship.
• Direct the students to think about playing a game of tag. Explain that the person who is "it" chases the others to try to tag them. The game usually includes a safe zone or base.

  What is the safe zone or base for? to give players a place to rest or get away from the person who is "it"

• Explain that in tag, the person who is "it" is like a predator chasing his prey. The prey must stay alert for danger and may run to shelter for rest and safety.
• Point out that in this activity the students will model some interactions between predators and prey. Explain that they will also be able to see the effects of changes in populations.

Instruct

Preparation for Reading

• Direct the students to remove pages 155–56 from their *Activities* books. Direct them to read page 191 and *Activities* 155–56 silently before beginning.
• Explain that it is helpful to look over all the pages before starting an Exploration. Point out that it will help them to see what they are going to do and when they will be writing information down.

Teach for Understanding

• Direct attention to the **Inquiry Skills**.
• Remind the students that scientists use science inquiry skills to gather and use facts.

  Which inquiry skills will you be using in this Exploration? Measure and Infer

What does it mean to *infer*? to use what you know to tell why things happen; to interpret observations and draw conclusions based on previous knowledge and experience

- Choose a student to read the **Purpose** aloud.
- Discuss ways that the students have seen animals, such as dogs and cats, interact. Remind the students that animals have different reactions or behaviors around other animals.

Apply

- Instruct the students to follow and check off **Procedure** Steps 1–9 as each is completed.

> This Exploration may be set up inside in a gym or outside on an open field.

- Distribute the identity life cards. Direct the students to pin the identity badges to their shirts and carry the life cards in their hands.
- Explain that each life card represents one animal in that population. For example, chipmunks and snakes will have multiple life cards, representing multiple animals. Group the students by kind of animal (chipmunk, snake, owl).
- Discuss the game rules. Identify the boundaries and where students who have lost all their life cards will go. Point out the "den" hoops where chipmunks can rest for 10 seconds. Explain that each round will last two minutes, and the whistle will signal the beginning and end of a round.
- Start Round 1 by releasing the chipmunks. Once they scatter, release the snakes. Wait a few seconds and then release the owls. At the end of the round, use the "Life Card Record" chart to record the number of life cards each group has.
- Play Round 2 with the same number of animals and life cards as Round 1 and record the results.

Ecosystem Tag

Name _____

Purpose

I will model animal interactions in an ecosystem.

Inquiry Skills
- Measure
- Infer

Procedure

Materials
- ☐ Tag Identity Badge
- ☐ Tag Life Cards

- ☐ 1. Collect the identity badge and the correct number of life cards for your animal from your teacher.

- ☐ 2. Identify others who share the same kind of identity badge. Form an animal group.

- ☐ 3. Read the rules of the game.
  - Chipmunks and snakes must avoid being "eaten," or tagged. Snakes can eat chipmunks. Owls can eat snakes and chipmunks.
  - Chipmunks are safe from their predators only when they are in a "den." Only one chipmunk can be in a den at a time, and he can be there for only 10 seconds.
  - When a chipmunk or snake is tagged, he must give up one of his life cards to the predator who tagged him.
  - If a chipmunk or snake loses all his life cards, he must sit out the rest of the round.

- ☐ 4. Play the game until your teacher tells you to stop.

- ☐ 5. Count your group's life cards. Tell that number to your teacher and turn in your identity badge and life cards.

- ☐ 6. Play the game again with the same number of animals and life cards.

- ☐ 7. Count your group's life cards. Tell that number to your teacher and turn in your identity badge and life cards.

- ☐ 8. Think about what you learned about predators and prey from the first two rounds. Infer what change will affect the animal interactions in the ecosystem.

- ☐ 9. Play the game two more times in the same way, making the change.

SCIENCE 3 *Activities*

Lesson 60 • Page 191
Exploration

onclusions

1. Did each kind of animal have the same number of life cards left each time you played?

○ yes   ● no

2. What are some reasons each kind of animal did or did not have the same number of life cards left? _Possible answers: Owls might catch more snakes in one round, leaving fewer_ _snakes to catch chipmunks._

How do you think the ecosystem could be affected if the entire population of snakes died? _Possible answer: The chipmunk population would likely increase._

Lesson 60 • Page 191
56 Exploration

- Guide the students in making inferences about the roles of the animals in the food web and the number of life cards each animal had at the end.
  Which animal(s) ended with fewer life cards?
  Was it the same in both rounds?
  What caused the change?
- Direct the students to make inferences using what they have learned from the first two rounds to make a change for the next round.
- Discuss what could be changed to increase the number of life cards for the animals discussed above, such as the starting amount of a kind of animal or the number of life cards an animal has.
- Choose one variable to change. Write the change on the posterboard.
- Play Rounds 3 and 4 the same way, making the change. Record the results.
- Direct the students to draw **Conclusions** using their observations.

Activities
Exploration, pages 155–56
Students will model animal interactions in an ecosystem.

Assess

Rubric
Use the prepared rubric or design a rubric to include your chosen criteria.

Objective
- Recall terms and concepts from Chapter 8.

Apply

Review
- Material for Test 8 will come from the Study Guides on *Activities* pages 143–44, 147–48, and 151–53, as well as Quizzes 8A and 8B. You may review any or all of the material during the lesson.
- You may choose to review Chapter 8 by playing the Word Chains review game or a game from the Game Bank (Teacher Resources).

🏆 Review Game

Word Chains

Materials: *Word Chains Cards*, on cardstock and cut apart (Instructional Aids 8.6–8.8); 15 envelopes

Place the cards for each term into separate envelopes. Group the students into teams. Give each team an envelope.

For each correct answer to a review question, allow the team member to draw one card (letter). Once the team has drawn all the cards for their term, the team unscrambles the word to earn a point. Give additional points if a team member can define or explain the meaning of the term. Give the team a new envelope for a new term.

The following terms are included on the *Word Chains Cards* pages.

- Ecosystem
- Environment
- Population
- Habitat
- Community
- Producer
- Consumer
- Decomposer
- Herbivore
- Omnivore
- Carnivore
- Food Chain
- Predator
- Prey
- Food Web

NOTES

NOTES

_____

_____

_____

_____

_____

_____

_____

_____

_____

_____

_____

_____

_____

_____

_____

_____

_____

_____

_____

_____

_____

_____

_____

_____

_____

_____

_____

Objective

• Apply terms and concepts from Chapter 8.

Assess

• Administer Test 8.

Unit Introduction

- Direct attention to pages 192 and 193.

  What is the title of Unit 3? Let's Connect with Physical Science

- What does the word *connect* mean? Possible answers: to join together, to fasten together

- What do you think the word *connect* means in the unit title? Elicit that *connect* means to see how physical science is meaningful to your life.

- Explain that the picture is a sports car at night.

> The unit opener art is a digital rendering of a sports car. It is not a specific make or model.

- Direct the students to look through Chapters 9–11, pages 192–261. Guide the students through several pages, stopping to look at headings and pictures.

- What do you think we are going to be talking about in Unit 3? matter, sound, forces, motion, electricity, and magnetism

- Direct the students to look back at the picture of the sports car.

  What do you see in the picture? Possible answers: race car, lights, tires, darkness, colors

- How does each part of the picture connect with physical science? Elicit that the car is matter; the engine is loud; the car moves when forces act on it; electricity and magnets provide power for the lights and the car.

Let's connect with

Physical Science

Chapter Objectives

- Analyze the characteristics, properties, and states of matter.
- Analyze chemical and physical changes of matter.
- Analyze the relationship between sound and matter.
- Apply inquiry skills of predict, measure, observe, classify, and infer.

Lesson Objectives

- Identify physical properties of matter.
- Explain the difference between mass and volume.
- Identify what an atom is.
- Compare and contrast the properties of solids, liquids, and gases.

Materials

- 5 clear containers of various shapes and sizes, each large enough to hold 250 mL
- beaker that holds at least 250 mL
- water
- decorative pillow
- toy truck
- whole apple
- knife
- plate or cutting board
- 9" latex balloon, one per student

Vocabulary

- physical property
- mass
- volume
- matter
- atom
- solid
- liquid
- gas

Chapter Introduction

- Direct attention to the chapter title and the picture of blown-glass balloons.
- Explain to students that blown glass comes from a process where glass is melted at a very hot temperature and then air is blown into the hot glass to form it into a shape, like these balloons.

  What do the blown glass balloons look like? Possible answers: round, colorful

  What happened to the glass before it could be formed into the shape of a balloon? Possible answers: It had to change; it had to be melted; it had to get hot.

- Provide time for the students to leaf through the chapter, looking at the headings and pictures. Discuss what they think the chapter will be about.

> Use completed Looking Ahead, *Activities* page 157, to generate a discussion on what the students already know about matter, atoms, and sound.

Big Question

- Ask a volunteer to read aloud the Big Question.
- Explain that the students will find the answer to the question as they read the chapter.

Big Question

What changes to matter do I observe around me?

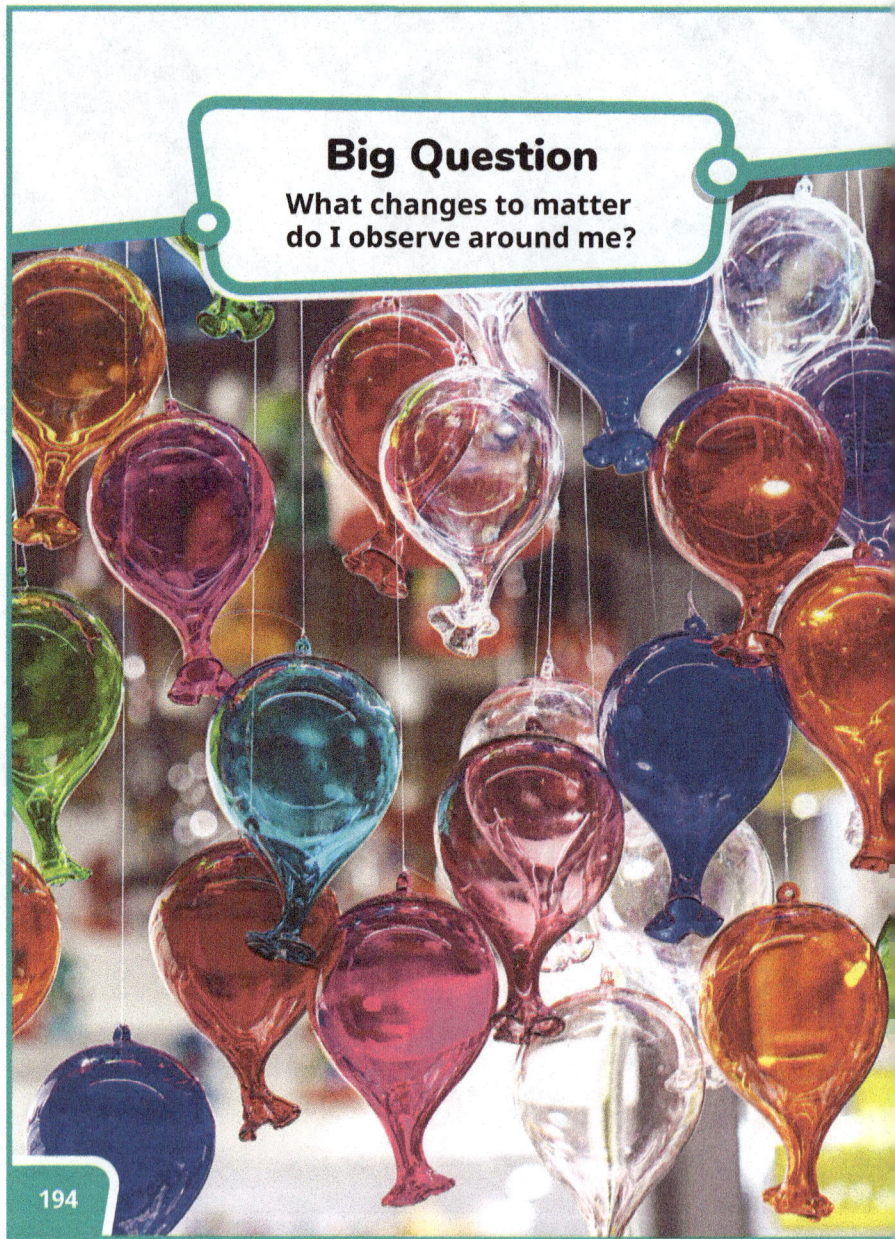

194

Visit TeacherToolsOnline.com for resources to enhance the lessons.

Chapter 9

Matter and Sound

You live in a world full of objects. Think about your bedroom. You have a bed and clothes there. You might have toys, books, or even a pet fish or hamster. Your room has many things in it.

Each object has characteristics that you can observe and describe. A book is hard and shaped like a rectangle. A sweater might be soft and fuzzy. A blanket might be bright red. A fishbowl is smooth and the water inside is cool. Each object has a color, size, and shape that make it different from other things.

195

Preparing Ahead

Lesson 68 Materials

Look ahead to the Materials list in Lesson 68 for required materials. You may want to ask students to bring in one or two small plastic containers such as yogurt, butter, or sour cream containers.

Engage

- Fill each of the 5 clear containers with 250 mL of water. If possible, fill them without the students seeing.
- Display the clear containers.
  Which container has the most water?
- Record students' answers on a tally chart.
- Ask several volunteers to explain why they think the container they selected has the most water.
- After students have shared their answers, pour the water from the container that received the most votes into the beaker. Choose a student to record the amount for display. Empty the beaker.
- Repeat for the water in the remaining containers, measuring the amount one at a time. Choose a student to record the amount of each one.
- Invite students to share ideas about why the containers looked like they had different amounts of water.

Instruct

Preparation for Reading

- Direct the students to read page 195 silently to find out what makes one object different from another object.

Teach for Understanding

What do objects have that makes them different from other objects? different characteristics such as color, size, and shape

- Display the pillow and the toy truck.
- Guide a discussion comparing the characteristics of the pillow and truck, using a Venn diagram.

How are the pillow and the toy truck the same? Possible answers: The pillow and truck might be the same color. They could be the same shape. They might be the same size.

What makes the pillow different from the toy truck? Possible answers: The pillow is soft, and the truck is hard. The pillow and truck might be different shapes. The truck can roll; the pillow doesn't move. The pillow and truck might be different colors.

- Ask several volunteers to describe the same object, such as a blanket, from their rooms at home. Write the descriptive words for display.

Is everyone's blanket the same? no

How are the blankets the same? Possible answers: They are soft; they are the same shape.

How are the blankets different? Possible answers: They are different colors; they are different shapes.

All the objects described are blankets, but each has different characteristics.

Preparation for Reading

- Preview and pronounce the vocabulary terms *physical property*, *mass*, *volume*, *matter*, and *atom*.
- Direct the students to read pages 196–97 silently to find out what all matter is made up of.

Teach for Understanding

What is another name for a characteristic? property

The closet door in the picture is matter. What are some properties of that door? Possible answers: hard, rectangular, white, tall

What is a *physical property*? anything about an object that can be observed with your senses

What are our senses? sight, hearing, touch, taste, and smell

What are some physical properties? Possible answers: color, size, shape, hardness, texture, temperature, mass, and volume

- Direct attention to the lists of characteristics made earlier in the lesson. Emphasize that each characteristic is a physical property of the object.
- Direct attention to the picture of the bedroom. Ask volunteers to describe objects pictured using the term *physical property* in the following sentence structure.
- A physical property of the ___ (pillow) is that it is ___ (soft).

What is *mass*? the amount of matter in an object

Does a large object always have more mass than a small object? no

What example is used in the text to show this idea? A baseball is smaller than a beach ball, but the baseball has more mass.

What physical properties can you observe in this picture?

196

Physical Properties

Everything around you has certain characteristics, or properties. You can observe those properties and use them to describe objects. For example, baseballs are round and hard. Pillows can be square and soft. Shape and hardness are two physical properties. A **physical property** is anything about an object that can be observed with your senses. Color, size, shape, hardness, texture, temperature, mass, and volume are physical properties.

Mass

Mass is one physical property. **Mass** is the amount of matter in an object. Some objects have more mass than others do. Often a larger object will have more mass than a smaller one, but mass does not depend on size. A baseball is smaller than a beach ball, but the baseball has more mass.

Volume

Volume is another physical property. **Volume** is the amount of space that an object takes up. A large object, such as a dump truck, takes up a lot of space. The volume of a dump truck is quite a large amount! Smaller objects, such as a toy truck, take up only a little space.

Matter

The dump trucks and the things in your room are matter. **Matter** is anything that has mass and takes up space. All matter has physical properties, including mass and volume. All matter is made up of atoms. An **atom** is a tiny particle that makes up matter. Atoms are so tiny that you cannot see them. Even you are made up of atoms.

God gave you five senses that help you observe the properties of matter. Your sense of sight can tell you about color and shape. Your sense of touch can tell you about hardness, temperature, and texture. Your senses of taste, smell, and hearing can also help you learn about matter around you.

What is matter?

197

Teach for Understanding

What is *volume*? the amount of space that an object takes up

- Direct attention to the picture of the dump trucks.

What is the picture of the trucks showing? that the volume of matter can be large or small

What is the difference between mass and volume? Mass is the amount of matter in an object. Volume is the amount of space that an object takes up.

What is *matter*? anything that has mass and takes up space

What is all matter made up of? atoms

What is an *atom*? a tiny particle that makes up matter

What are the tiny parts that you learned about in Chapter 4 that make up living things? cells

- Explain that atoms are tiny parts that are even smaller than cells. Cells are made up of atoms.

- Direct attention to the picture at the bottom of the page.

What might the girl be observing with her senses? Possible answers: She can *see* the flowers. She can *smell* the flowers. She can *touch* the flowers. She can *feel* the warmth of the sun.

anything that has mass and takes up space

ⓘ **Background**

Cells and Atoms

Cells are the smallest living parts of living things. While atoms make up cells, atoms themselves are not living things. It takes about 100 trillion atoms to make up a single cell.

Preparation for Reading

- Preview and pronounce the vocabulary term *solid*.
- Direct the students to read page 198 silently to find out what the three states of matter are.

Teach for Understanding

What are the three states of matter? solid, liquid, and gas
How is matter grouped into these three states? by its properties

- Discuss properties of solids.

What is a *solid*? anything that has a definite shape and volume

- Direct attention to the pictures and caption.
- Display the whole apple on the plate or cutting board, then use the knife to cut the apple into slices.

Does the apple still have the same volume after it is sliced? yes

- Demonstrate that the volume has not changed by putting all the slices back together. Use a rubber band, if necessary, to hold the apple slices together.

What are some examples of solids? Possible answers: book, desk, water bottle

Are all solids hard? no

What are some solids that are not hard? Possible answers: a blanket, a stuffed animal, cotton candy

- Direct attention to the picture of the teddy bear.

What does the teddy bear feel like? Possible answers: soft, squishy, fuzzy

What does the chair feel like? Possible answers: hard, smooth

Are the bear and the chair both solids? yes

✔ solid, liquid, gas

States of Matter

Matter can be a solid, a liquid, or a gas. These are called the states of matter. Each state of matter has certain properties. These properties help us group matter into the states.

Solid

Blocks, books, and baseballs are solids. A **solid** has a definite shape and volume. You can put an apple inside a box, but the apple is still round. It does not change its shape to fit the shape of the box. It keeps its own shape.

The volume of a solid also stays the same. Even when you cut the apple into pieces, all the pieces together have the same volume as the whole apple had. The apple's volume does not change.

Some solids, such as glass and tables, are hard. But solids do not have to feel hard. A blanket and a teddy bear are also solids.

Does the apple have the same volume after it has been sliced?

Solid
Keeps its shape
Keeps its volume

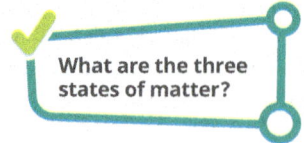

✔ What are the three states of matter?

ⓘ **Background**

Plasma

Plasma is a fourth state of matter. It exists naturally in very hot conditions, such as in lightning bolts and on the surface of the sun. It can also exist in manmade conditions that are cooler, such as in fluorescent lights and plasma screen technology.

Liquid

Water, milk, and oil are kinds of liquids. A **liquid** has a definite volume, but not a definite shape. A liquid takes the shape of whatever container it is in.

Milk in a tall, thin glass takes the shape of the glass. But if the glass of milk is poured into a short, wide bowl, the milk changes shape. It takes the shape of the bowl. The volume of the milk does not change, but the shape of the milk does. The same volume of a liquid can change to different shapes.

Does the liquid in both containers have the same shape?

Solid	Liquid
Keeps its shape	Takes the shape of its container
Keeps its volume	Keeps its volume

Gas

The air in your bicycle tires is a gas. The helium in a balloon is also a gas. A **gas** has no definite volume or shape.

Like a liquid, a gas takes the shape of its container. The air in a tire takes the shape of the tire. If you pump air into a soccer ball, the air takes that shape. If you pump the same amount of air into a football, the air takes that shape.

199

Preparation for Reading

- Preview and pronounce the vocabulary terms *liquid* and *gas* and the word *helium*.
- Direct the students to read pages 199–200 silently to find out which state of matter does not have a definite volume or shape.

Teach for Understanding

- Discuss properties of liquids.

  What is a *liquid*? anything that has a definite volume but not a definite shape

- Allow volunteers to name examples of liquids.

  How is the shape of a liquid different than the shape of a solid? A liquid takes the shape of whatever container it is in, but a solid keeps its shape.

- Remind students of the demonstration at the beginning of the lesson.

  When the water was in the containers, what was its shape? the shape of the container

  When the water was poured into the beaker, what was its shape? the shape of the beaker

  Did the volume of the water stay the same? yes

  Did the shape of the water stay the same? no

- Direct attention to the pictures of the milk. Choose a volunteer to read and answer the caption.

  How is the volume of a liquid like the volume of a solid? The volume of a liquid and the volume of a solid stay the same.

  How is the shape of a liquid different from the shape of a solid? The shape of a liquid changes to fit the container it is in. The shape of a solid does not change.

- Discuss the properties of gases.

  What is a *gas*? matter that has no definite volume or shape

  What shape does gas have? the shape of its container

- Direct attention to the picture at the bottom of the page.

  What shape does the gas in this picture have? the shape of the bicycle tire

  What are some gases we have already talked about this year? oxygen, gases from cars and factories that burn fuel, carbon dioxide as part of photosynthesis

- Distribute one 9" latex balloon to each student. Instruct students to blow up the balloon and tie it off.

  If you blow up a balloon, what gas are you putting inside it? Possible answers: air, carbon dioxide

  The gas takes the shape and size of the balloon. What shape does the air in this room have? the shape of the room

- Compare the properties of gases and liquids.

  How are gases like liquids? They take the shape of their containers.

  How are gases different from liquids and solids? Gases expand and stretch out to fill the container. Liquids and solids do not.

  What stops a gas from expanding? a container or a closed space

- Direct attention to the picture of the balloons.

  What gas is in the balloons? air

  Can you see the gas? no

  A balloon is an example of a container that we put air in. What are some other containers that we put air in? Possible answers: air mattresses, basketballs, beach balls, inflatable swim floats and pools, soccer balls, tires

  What determines the shape or volume of a gas? its container or the closed space it is in

- Direct attention to the picture of the girl with the chocolate milk.

  What is in the glass? milk and bubbles

  Can you see the air in the glass? no

  What is the girl adding to the glass when she blows into the straw? air

  How can you see that she is blowing air into the milk? Possible answer: You can see the air making bubbles in the milk. The bubbles break and air goes into the glass.

  Does the air stay in the glass? No, air escapes into the room.

- Direct attention to the "Fantastic Facts" box. Explain that helium is a gas that is lighter than air, so when the balloons are filled with helium they are able to float.

  How many balloons did we fill today?

- Explain that it would take as much helium to fill 48,000 of these small balloons as it would to fill a large parade balloon.

  takes the shape of its container, takes the volume of its container

Apply

Activities

Looking Ahead, page 157
This page assesses the students' knowledge prior to the chapter.

Study Guide, pages 159–60
These pages review the concepts taught in Lesson 63. After completion, direct the students to keep them for review for Test 9.

What shape is the gas?

Air comes to the top of the glass and escapes.

Gases are different from solids and liquids. A gas expands, or stretches out. It takes the volume of its container. That means that it fills up all the space inside a closed container. A gas keeps expanding until something stops it. That something may be any kind of container or closed space. A tied balloon and a bicycle tire are examples of closed spaces.

You can use jars to compare how a liquid and a gas fill containers. When you pour water into a jar, all the water goes to the bottom. If you do not tip the jar, the water does not come out. When you blow air, which is a gas, into a jar, the air does not go all the way to the bottom. Instead, the air fills all the space in the jar. Without a lid on the jar, the air will expand and escape.

Solid	Liquid	Gas
Keeps its shape	Takes the shape of its container	Takes the shape of its container
Keeps its volume	Keeps its volume	Takes the volume of its container

Fantastic Facts

Helium is a gas that people use to fill balloons. Small balloons need just a small amount of helium. But some balloons are large. The character balloons used in some parades are very, very large. The amount of helium needed to fill a parade balloon is about 340 cubic meters (12,000 cubic feet). That much helium would fill about 48,000 small balloons!

What are the properties of a gas?

ⓘ Background

Air
Air is a mixture of nitrogen, oxygen, carbon dioxide, hydrogen, helium, and other gases. Animals and humans keep some of the inhaled oxygen and exhale excess carbon dioxide, but all the gases are present in both inhaled and exhaled air.

Atmosphere
The gases in the atmosphere are not in a closed space. They continue to expand until the molecules are so far apart that they are really no longer "air." Gravity keeps the gases near the earth's surface denser.

Shrinking Gases
The pressure exerted by a gas on its container is caused by the collisions of moving gas molecules. When thermal energy is removed (the gas is cooled), the molecules slow down and strike the walls of their container less frequently. This causes an elastic container to shrink. For example, a balloon inflated with room temperature air will shrink after being placed in a freezer.

EXPLORATION

Which Kind of Matter?

Inquiry Skills
• Observe
• Measure
• Classify

Matter is found in different forms. The food you eat and the juice you drink are kinds of matter. The air you breathe is also a kind of matter. If you list their properties, you will find each kind is different.

In this Exploration, you will compare properties of different kinds of matter. Then you will classify the matter based on its properties.

201

⚠ Helps

Making Graduated Containers

Extra graduated containers can be made from clear plastic cups. Measure a certain amount of water, such as 100 mL, in a metric measuring cup. Pour the water into a cup. Mark the water line and write the measurement on the cup. Continue adding water and marking the measurement lines in the increments you choose. Using this method, you can make as many measuring cups as you need for your class.

Lesson Organization

This lesson may be taught in several formats. One option allows for volunteers to demonstrate to the entire class one type of observation. Guide a discussion of each observation as it is made. Another option allows for you to set up a separate station for each type of observation. The students would then rotate between the stations as you guide their work. Afterward lead a discussion of each observation.

Objectives
• Compare properties of different kinds of matter.
• Measure mass and volume of different kinds of matter.
• Classify matter based on properties.

Materials
• See *Activities* page 163.

> This activity will work best with a balloon that is partially inflated.

Engage

• Read the following poem to review the states of matter.

What is a solid?
That's easy to say.
(It keeps its same volume,
and keeps its same shape.)
You can turn it or slice it
Any which way!

What is a liquid?
That's trickier yet.
(It keeps its same volume,
but changes its shape.)
Don't spill it or splash it—
'Cause you might get wet!

What is a gas?
That's the hardest of all.
(It doesn't keep volume,
and doesn't keep shape.)
It fills its container
Whether bubble or ball!

Instruct

Preparation for Reading

• Direct the students to remove pages 163–66 from their *Activities* books. Direct them to read page 201 and *Activities* 163–66 silently before beginning.

  What states of matter do you see in the picture? Possible answers: The coffee and juice are liquids. The fruit, eggs, bacon, and table are solids. The steam is a gas.

• Explain that it is helpful to look over all the pages before starting an Exploration. Point out that it will help the students to see what they are going to do and when they will be writing information down.

Teach for Understanding

• Direct the students to remove pages 161–62 from their *Activities* books.

• Direct the students to place the pages in their science notebook. These pages will be used as reference throughout the course.

• Read and discuss *Measure Up: Mass and Volume* on *Activities* pages 161–62. *Measure Up* questions may be completed as a class or assigned as time allows.

- Direct attention to the **Inquiry Skills**.
- Remind the students that scientists use inquiry skills to gather and use facts.
  Which inquiry skills will you be using in this Exploration? Observe, Measure, and Classify
- Review what it means to *observe*, *measure*, and *classify*.
- Choose a student to read the **Purpose** aloud.

Apply

- Direct attention to the **Procedure** list on *Activities* page 163-64. Instruct the students to follow and check off Steps 1–5 as each is completed.
- Give help as needed as the students complete the steps in the Procedure.
- Direct students to mark their observations on the chart on *Activities* page 165.

Which Kind of Matter?

Name _____

Purpose

I want to compare the properties of different kinds of matter. I want to classify the different types of matter based on their properties.

Procedure

Observe a Solid

☐ 1. Write the name of your objects in the data chart.

☐ 2. Using the balance, measure and record the mass of the first object in the data chart.

☐ 3. Observe whether your object takes up space. Mark the chart with an *X* if it does.

☐ 4. Place the object in the container to see whether its volume changes to fill up the entire container. Mark whether the volume stays the same or changes to fill the entire container.

☐ 5. Look to see whether the object keeps its shape or takes the shape of the container. Mark your observation.

☐ 6. Repeat steps 1–4 with the second object.

Observe a Liquid

☐ 7. Measure the mass of the empty beaker. Add 200 mL of water to the beaker and measure the mass again. Record the difference between the two measurements.

Work Space	
mass of container with water	_____ g
mass of empty container	_____ g
difference	_____ g

☐ 8. Observe whether the water takes up space. Mark the chart if it does.

Inquiry Skills

- Observe
- Measure
- Classify

Materials

☐ 2 solid objects of different shapes and sizes

☐ ___*object 1*___

☐ ___*object 2*___

☐ balance

☐ clear container

☐ beaker or metric measuring c

☐ water

☐ balloon, not fully inflated

SCIENCE 3 *Activities*

9. Look at the shape of the water in the beaker. Pour the water into the container and look at its shape. Mark your observation for its shape.

bserve a Gas

10. Squeeze the balloon and then press it into the container. Mark whether the shape of the air in the balloon stays the same or takes the shape of the container.

11. Observe whether the air in the balloon takes up space. Mark the chart if it does.

12. Press on the balloon and observe whether the air continues to fill the entire balloon. Mark whether the volume stays the same or changes.

13. Measure the mass of the filled balloon. Open the clip and measure the mass of the clip and the empty balloon. Record the difference between the measurements.

Work Space	
mass of filled balloon	_____ g
mass of clip and empty balloon	_____ g
difference	_____ g

- Direct attention to the Procedure list on *Activities* page 163-64. Instruct the students to follow and check off Steps 6-13 as each is completed.
- Give help as needed as the students complete the steps in the Procedure.
- Direct students to mark their observations in the chart on *Activities* page 165.

Lesson 64 • Page 201
Exploration

What is a physical property that does not change unless matter is added or removed? mass

What physical property does an object have if it takes up space? volume

What do you know about the object if it has mass and takes up space? It is a kind of matter.

Why did you need to measure the mass of the empty container? The container is a kind of matter, so it also has mass. Its mass is not part of the water's mass.

What did you use to find the water's volume? a beaker

Does the water's volume change when it is poured from the container into the beaker? no

You tested air by using an inflated balloon. Why do you think this is a good way to test air? The air needs to be held in some kind of container.

Why did you need to squeeze the balloon when you did not squeeze the objects or the water? to test whether the air inside the balloon can take the shape of its container

When you opened the balloon, what happened to the air? Most of it left the balloon. Why do you think that happens? Possible answer: The air is always trying to fill its container and take its shape, so the air leaves the balloon to do this. Air takes the shape of the room when it leaves the balloon.

Why did you need to measure the mass of the clip and the empty balloon? They are kinds of matter, so they also have mass. Their mass is not part of the air's mass.

EXPLORATIO

Which Kind of Matter?

Name _____

Write the mass of each item of matter.
Mark an X in the boxes that match your observations.

	Property	Object 1 *object name*	Object 2 *object name*	Water	Air
Mass	measure on a balance				
	takes up space				
Volume	stays the same (keeps its volume)				
	changes to fill the container (takes the volume of its container)				
Shape	keeps its shape				
	takes the shape of the container				

SCIENCE 3 *Activities*

Lesson 64 · Page 201
Exploration

What states of matter do you see represented in the picture?

onclusions

. Look at the first two rows of the chart. How do you know that each item is matter?
Each item has mass and takes up space.

. Do all types of matter have the same mass? *no* the same volume? *no*

. Look at the information for volume and shape for both solid objects. What can you
infer about their volume and shape? *Both the volume and shape stay the same.*

. Look at the information for volume and shape for the water. What can you infer about
the water? *Its volume stays the same but its shape changes to the shape of the container*
it is in.

. Look at the information for volume and shape for air. What can you infer about air?
Its volume and shape change to fill the container it is in.

66 Lesson 64 • Page 201
Exploration

Do all the kinds of matter have the same properties?
- Discuss differences between the kinds of matter.
- Direct the students to draw **Conclusions** using their observations. The conclusions may be done individually or with their science group.
- Guide a discussion of the students' conclusions.
- Guide the students in making inferences using the information about the properties they observed.
- Guide a discussion of the picture on *Activities* page 166.

Activities
Measure Up: Mass and Volume, pages 161–62
These pages were discussed in the lesson.

Exploration, pages 163–66
Students will observe, measure, and classify different types of matter.

Assess
Rubric
Use the prepared rubric or design a rubric to include your chosen criteria.

Objectives

- Describe how heating and cooling can cause matter to change states.
- Infer why there is moisture inside a window on a cold day.
- Describe the three states of water.

Materials

- ice cube, per student
- paper towel, per student
- unpopped popcorn kernels
- small cup or bowl, per student
- hot air popcorn popper
- melted butter, optional
- salt, optional
- additional popcorn toppings, optional
- empty drinking glass or jar
- ice
- water

Teacher Resources

Visual 9.1: *Changing States*

Vocabulary

- Review the vocabulary terms *evaporation* and *condensation* from Chapter 2.

Engage

- Distribute one paper towel and one ice cube to each student. Instruct the students to fold the paper towel in half and place it on their desks. Students should place the ice cube on the paper towel.

  What are some observations you can make about the ice cube? Possible answers: It is cold. It is wet. It is hard.

  Is the ice cube a solid or a liquid? solid

- Direct students to leave the ice cube and paper towel on their desks.

Instruct

Preparation for Reading

- Direct the students to read pages 202–4 silently to find out what is formed when water is heated.

Teach for Understanding

- Direct students to hold the ice cube in their hands.

  What has happened to the ice cube since you put it on the paper towel? Possible answers: It is melting. It has gotten smaller.

- Discuss reasons why the students think the ice is melting.

  At what temperature does ice melt? above 0°C (32°F)

  Is your hand at 0°C? no Why is the ice cube melting? My hand is warmer than 0°C.

  What forms as the ice melts? Possible answers: water, a liquid

  What needs to be done in order to turn the water back into ice? Possible answers: Cool it; put it in the freezer; drop its temperature to 0°C (32°F).

- Collect the ice and paper towels.

  Does matter stay in the same state all the time? no

- Explain that many solids are melted and formed into other shapes to make them useful.

Changes in States

God wants people to use the things He has provided in the world. One way we do that is by changing matter from one state to another. Think about the foods that you enjoy. Some of them are frozen. When you freeze fruit juice to make an ice pop, the liquid juice becomes a solid frozen treat. Freezing also helps keep some foods from spoiling.

Sometimes states change without help from people. A chocolate candy bar left in a hot car melts. The solid chocolate changes to liquid chocolate.

Solids and Liquids

Heating some solids can cause them to melt and become a liquid. Solids melt at different temperatures. Chocolate melts at about 36°C (96.8°F), which is just a little cooler than your body temperature. That is why chocolate melts on your hands. Ice melts at 0°C (32°F).

Some solids melt at very high temperatures. Solid gold will not become liquid unless it is heated to a temperature of over 1000°C. That is over 1800°F!

Gold bars are made by pouring liquid gold into molds. When the gold cools, the bars are solid.

202

- Solids melt at different temperatures. Do you think liquids also become solids at different temperatures or at the same temperature? at different temperatures
- Direct attention to the pictures of the fruit pop, chocolate candy bar, and ice cream cone.

  What causes some solids to become liquid? heat

  What does it mean when we say something melts? It changes to a liquid.

  Do all solids melt at the same temperature? no

- Which melts at a higher temperature—ice or chocolate? chocolate
- Direct attention to the picture at the bottom of the page.

  What liquid is being poured into the mold? gold

- Is this the state that gold is usually found in? no
- What is gold used for? Possible answers: jewelry, statues, gold bars

  How hot does gold need to be heated to for it to melt? over 1000°C (over 1800°F)

Some solids do not become liquids when they are heated. When a log is burned in a campfire or fireplace, it becomes ash. Its shape and texture have changed, but it is still a solid. Another example is popcorn. When popcorn kernels are heated, they pop! The popcorn is still a solid. But its shape and texture have changed.

Cooling a liquid can cause it to become a solid again. A candy maker melts a block of chocolate. Then he pours the chocolate into pans that are different shapes. As the liquid chocolate cools, it becomes a solid again. When the chocolate is taken out, it is in the shape of the pan.

Liquids and Gases

Solids can change to liquids, and liquids can change to solids. Liquids can also change to gases. Liquids become gases when they are heated.

You have seen this happen when water boils in a pot or teakettle. The boiling water bubbles. Steam rises from the water. The steam is a gas called *water vapor*. As the water vapor spreads out in the room, it seems to disappear. But it is still there.

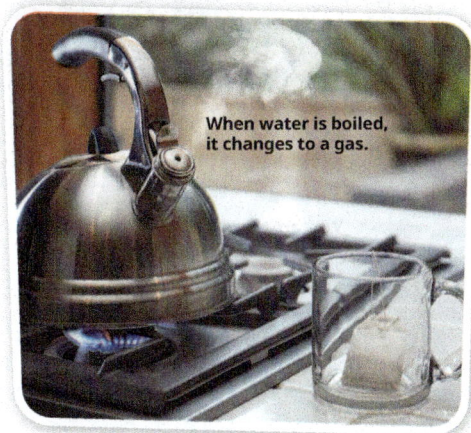

When water is boiled, it changes to a gas.

203

Do all solids melt when they are heated? no

What are some examples of solids that do not melt when heated? burning logs, popcorn

- Pop the popcorn and distribute some to each student.
- Share Background about popcorn as time allows.
- Direct attention to the pictures of the logs and popcorn.
- Explain that some solids burn at a lower temperature than they melt. Logs burn before they are hot enough to melt. Popcorn pops before it gets hot enough to melt.

If heating some solids causes them to melt, what happens when they are cooled? They become a solid again.

- Direct attention to the picture of the chocolate mold.

Why do you think this chocolate was melted? Possible answer: to make it into another shape

- Direct attention to the kettle on the stove.

What happens when a liquid is heated? The liquid turns into a gas.

What gas forms from water? water vapor

⚠ Helps

Popcorn

The popcorn in this lesson is used to demonstrate that not all solids become liquids when they are heated. It is best if the students are able to watch the popcorn as it is popping. If resources are not available for this, you may choose to provide already popped popcorn and show the video *Popcorn* (TeacherToolsOnline.com).

ⓘ Background

Why Does Popcorn Pop?

Popcorn kernels have a small droplet of water on the inside, surrounded by a hard shell. That shell is called a hull. When the popcorn kernel is heated, the water inside the kernel turns into steam. The steam builds up pressure inside the hull until the kernel explodes in a pop.

- Direct attention to the picture of the water puddle. Choose a student to read and answer the caption.
  Boiling water creates water vapor. How else can water vapor form? Possible answer: when water is heated by the sun
  What is the process called when a liquid changes to a gas? evaporation

- Display the word *evaporation* and circle the word *vapor* in it. Discuss this root word and how it can help the students remember that water vapor is formed when water evaporates.
  When you paint with watercolor paints, the paint is wet when you put it on the paper. Later the paint on the paper is dry. Why? The liquid in the paint evaporates into the air.
  Sometimes you might hang up clothes or towels to dry. When the clothes are dry, what has happened to the water that was in them? Possible answers: It dripped out; it evaporated; it became water vapor.

- Discuss examples of other liquids evaporating, such as hand sanitizer. Explain that hand sanitizer dries quickly because it contains a liquid, such as alcohol, that evaporates faster than water does.

- Allow the students to examine and feel the outside of the empty jar. Fill the jar with ice and water and set it in a warm location.
  What is it called when a gas changes to a liquid? condensation
  For a gas to change to a liquid, does the gas need to be heated or cooled? cooled
  What are clouds formed from? water vapor in the sky that cools and condenses
  Why do little drops of water sometimes form on the outside of a cold glass? The cold drink cooled the glass and the air around it, causing the water vapor in the air to change from a gas to a liquid.

- Display the jar and allow the students to feel the outside. Discuss the changes they observe.
  What is different about the outside of the jar? It is wet.
  What caused this? Water vapor from the air cooled and condensed on the glass.

- Direct attention to the picture of the window at the bottom of the page.
  Why are there drops of water on the inside of the window? Possible answer: The cold air outside is causing condensation on the inside of the warm window.

It condenses.

What is happening to the water in the puddle?

Water vapor forms when water is heated. Steam is water vapor from boiling water. Water vapor also forms as the sun warms bodies of water. The sun's heat causes the water in lakes, ponds, and even mud puddles to change to water vapor. When a liquid changes to a gas, it is called *evaporation*. The water that evaporates stays in the air for a while.

When a gas changes to a liquid, it is called *condensation*. This can happen to water vapor in the air. When the water vapor cools, it condenses and forms clouds and rain. You can also see water vapor condense when you pour a cold drink into a glass on a hot day. Little drops of water form on the outside of the glass. The cold drink cools the glass and the air around it. The water vapor in the air condenses, or changes from a gas to a liquid.

Why are there drops of water on the inside of the window?

What happens to a liquid when it cools?

204

Differentiated Instruction

Enrichment: Compare Evaporation Rates
Materials: 2 clear plastic cups, permanent marker, beaker, rubbing alcohol, water
Label one cup *A* and the other *W*. Measure and pour some rubbing alcohol into cup *A*. Measure and pour the same amount of water into cup *W*. Mark the top of the level of liquid on both cups. Set the cups in a sunny place for several days. Periodically observe and mark the top of the level of liquid. Discuss changes that the students observe and compare which liquid is evaporating faster.

States of Water

Any type of matter can change to another state, but most do not change easily. We usually think of each kind of matter in the state that we usually see it. Wood is a solid. Milk is a liquid. Oxygen is a gas in the air. They almost never change.

However, water is different. Water can easily change from one state to another. We usually see water in its liquid state. Ice, frost, and snow are solid forms of water. You cannot see water vapor in the air, but the water is still there as a gas.

What is the state of matter of the water?

The Celsius temperature scale is based on the changing states of water. At 0°C (32°F), water freezes and changes to a solid. At 100°C (212°F), water boils and changes to a gas.

Most types of matter contract, or get smaller, as they freeze. Water is one of the few kinds of matter that does the opposite. Water expands, or gets bigger, when it freezes and changes from a liquid to a solid. Frozen water, or ice, takes up more space than liquid water does.

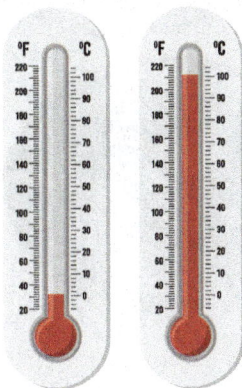

✓ **What does frozen water do that is different from most other frozen matter?**

205

ⓘ Background

Freezing Water

Like other liquids, water shrinks as the temperature nears 4°C. However, after the temperature reaches 4°C, the freezing water begins to expand a little. As it reaches the freezing point, 0°C, it expands greatly. This happens because of the shape of its molecules and how they bond together. Because of this expansion, ice is less dense than liquid water and can float.

Preparation for Reading

- Direct the students to read page 205 silently to find out what is different about water than most other types of matter.

Teach for Understanding

What substance can change easily from one state to another? water

What state do we most often see water in? liquid

What are some solid forms of water? Possible answers: ice, frost, snow

- Display the *Changing States* visual and guide the students in completing *Activities* page 167 during the discussion.

What is the form of water that is a gas? water vapor

- Direct attention to the pictures of the bird bath and water faucet.

What is the state of matter of the water coming from the faucet? liquid

What is the state of matter of the water in the bird bath? solid

What is the Celsius temperature scale based on? the changing states of water

- Direct attention to the pictures of the thermometers.

What happens to water at 0°C? Water freezes and changes to a solid.

What happens to water at 100°C? Water boils and changes to a gas.

What do most types of matter do as they freeze? contract, or shrink

What does water do when it freezes? expands

- Explain that because water expands, it is able to float. The fact that frozen water floats instead of sinks demonstrates God's design. As a body of water freezes, the ice forms on top. This layer of ice insulates the water underneath and helps keep the living things in the water below from freezing.

If the "Try It Yourself" in Lesson 18 (page 59) was not completed during that lesson, you may choose to have students complete it at this time.

✓ It expands.

Apply

Activities
Changing States, page 167
This page was completed during the lesson.

Objectives

- Identify what physical and chemical changes are.
- Compare and contrast physical and chemical changes.
- Give one example of a physical change and one example of a chemical change.
- Explain how people can use changes in matter to serve God and other people. **BWS**

Materials

- self-sealing sandwich bag, per student
- 3 or more ingredients to make a snack mix, per student
- serving spoon for each ingredient

Vocabulary

- physical change
- mixture
- chemical change

Engage

- Display the snack mix ingredients and allow volunteers to identify each ingredient. Discuss properties that make each a different type of matter. Distribute the bags and allow the students to put some of each ingredient into their bags. Direct the students to seal their bags and shake to mix the ingredients.

  What name can we call this? Possible answers: snack mix or mixture

  Can you separate the ingredients? yes

- Explain that the ingredients were combined and changed into something called a snack mix, but each ingredient can still be identified. They did not make a new kind of matter.

Instruct

Preparation for Reading

- Preview and pronounce the vocabulary terms *physical change* and *mixture*.
- Direct the students to read pages 206-7 silently to find out what a mixture is.

Teach for Understanding

What are some ways matter can change? Possible answers: It can change states. Its color and shape can change. It can combine with other matter to form new kinds of matter.

What is a *physical change*? a change in matter that does not form something new

- Direct attention to the picture of the cake.

  If you cut a cake into slices, are you making a new kind of matter? no

  What are you changing? the size, the shape

  What kind of change is this? a physical change

Physical and Chemical Changes

God made matter to change in different ways. It can change from one state to another. Some of its physical properties, such as color and shape, can change. Some kinds of matter can combine to form new kinds of matter. All changes to matter are either physical changes or chemical changes. Learning about changes to matter helps us serve God and other people.

Physical Changes

A **physical change** happens when matter changes but does not form something new. If you have a cake to share with friends, you cut it into pieces. The cake's size and shape have changed. But the cake did not turn into something new. If you put all the pieces back together, they would have the same mass as the whole cake did.

Physical changes also happen as matter changes state. Ice cream straight from the freezer has a definite shape. It is a solid. But as it melts, it changes to a liquid. The melted ice cream takes the shape of the bowl it is in. The ice cream has had a physical change, but it is still ice cream.

Which bowl of ice cream shows a physical change?

206

When ice cream melts, it changes states and becomes a liquid. Does it become a new kind of matter? no

What changes? the state

What kind of change is this? a physical change

- Direct attention to the pictures of the ice cream. Choose a volunteer to read the caption aloud.

  Which bowl of ice cream shows a physical change? Possible answers: the bowl on the right; the bowl with the melted ice cream

  How do you know that the melted ice cream has had a physical change? It changed from a solid to a liquid.

Physical changes can also happen when matter is mixed together. A **mixture** is two or more kinds of matter that are combined. The parts of most mixtures are easy to see and separate. A fruit salad can be made with sliced bananas and strawberries and topped with yogurt. The fruit salad is a mixture. Because it is a mixture, the fruit salad can be taken apart. Cutting the fruit did not change the kind of matter. The banana is still a banana. The strawberry is still a strawberry. The fruit has had a physical change. The pieces are smaller.

The parts of some mixtures are not easy to see or separate. A smoothie might have bananas, strawberries, and yogurt in it. The smoothie is a mixture. To make a smoothie, you would cut the fruit into smaller pieces. The fruit and yogurt are placed in the blender. The blender chops and mixes the fruit and yogurt to form a thick drink. Now the fruits are not easy to see. But you can still taste the fruits and the yogurt. Combining the ingredients together does not make a new kind of matter. It makes a mixture.

What is a physical change?

207

What is a *mixture*? two or more kinds of matter that are combined

What kind of change happens when objects are combined into a mixture? a physical change

What is an example of a mixture? Possible answers: a snack mix, a fruit salad, a fruit smoothie

- Direct attention to the picture of the fruit at the top of the page.

  What kinds of fruit are shown in the picture? strawberries and bananas

- Direct attention to the picture of the fruit salad.

  What is shown in the picture? a fruit salad; a mixture

  How do we know that this is a mixture? The fruit did not become a new kind of matter. It is still the same fruit that it was.

- Refer to the snack mix made in the lesson introduction.

  When you made your mixture, what kind of change happened to the ingredients? a physical change

  How do you know? The ingredients still have their same properties; they did not make a new kind of matter.

  What kind of physical change did you see happen in the last lesson (Lesson 65)? popcorn popping

  How do you know that popping popcorn is a physical change? It did not make a new kind of matter. It is still corn.

  What are some other examples of physical changes? Possible answers: cutting paper, crushing a soda can, melting chocolate or ice cubes

- Direct attention to the picture of the fruit smoothie.

  Did the fruit make a new kind of matter when it was mixed in the blender? No, it is still fruit.

a change in matter that does not form something new

Preparation for Reading

- Preview and pronounce the vocabulary term *chemical change* and the word *borax*.
- Direct the students to read pages 208–9 silently to find out what new kind of matter forms when iron tools get wet.

Teach for Understanding

What is a *chemical change*? a change in matter that makes a new kind of matter

How is a chemical change similar to a physical change? They are both a change in matter.

How is a chemical change different from a physical change? A chemical change forms a new kind of matter. A physical change does not.

- Direct attention to the pictures of the two eggs.

When you make a snack mix, you are combining matter. When you boil an egg, you are changing matter. Why is making a snack mix a physical change, but boiling an egg is a chemical change? The snack mix ingredients keep their own properties. The raw egg changes and becomes a new kind of matter.

How does an egg change when it is cooked? It changes from a liquid to a solid; it is a chemical change.

> Remind the students that in second grade they learned that cooking an egg is an irreversible change. Chemical changes are irreversible changes.

- Direct attention to the picture of the slime.

What happens to glue, water, and borax when they are mixed together? They make slime.

What kind of change is that? a chemical change

- Share the Background information at this time.

What are some other examples of chemical changes? baking cookies, burning wood, milk turning sour

Chemical Changes

A **chemical change** happens when matter changes into a new kind of matter. Sometimes two or more kinds of matter combine to form a new kind of matter. The different kinds of matter lose their own properties and get new ones. A chemical change can also happen when some kinds of matter change state.

For example, think about the properties of a raw egg. It is a sticky, clear liquid with a thick, yellow center. However, these properties change when you cook the egg. The clear part becomes a white, rubbery solid. The yellow part becomes a crumbly solid. The egg's matter had a chemical change because it was heated. The egg changed and can never be liquid again. A chemical change cannot be undone.

Slime is another example of a chemical change. To make slime, you combine glue, water, and borax. When you first stir the water and glue together, the mixture stays a liquid. But when you add borax and more water, the matter becomes very thick and stretchy. The liquid mixture and the solid borax lost their own properties. They took new properties. They had a chemical change.

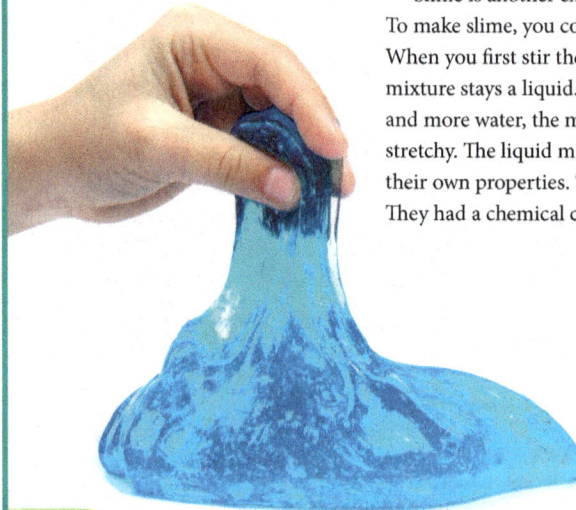

208

ⓘ Background

Slime

Slime was first introduced by a major American toy company in 1976. It was green, and it was sold in small green trash cans. Other toy companies also began making slime and slime-based games. In the 2000s, slime became a popular do-it-yourself activity.

Slime is considered a non-Newtonian fluid, which means it is neither a solid nor a liquid.

⚠ Helps

Borax

Borax is an alkaline mineral salt. It is a white powder that dissolves in water. It is natural and is nontoxic. Borax is not harmful to the body under normal, external use.

Try It Yourself!

Make your own chemical change! Mix 50 mL (1.75 oz.) of white glue and 50 mL (1.75 oz.) of water in a plastic cup. In another plastic cup, mix 25 mL (0.88 oz.) of water with 1 g (1/4 tsp) of borax. Pour a small amount of the borax mixture into the glue mixture and stir with a wooden craft stick. Keep pouring the borax mixture into the glue mixture a little at a time. Stir until the matter is stiff and mostly stays on the wooden craft stick. Remove the matter from the craft stick with your fingers. Keep squeezing the matter with your hands until it is no longer sticky. Now you have slime!

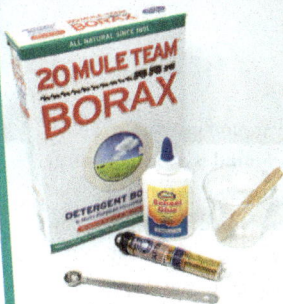

Another chemical change happens when iron rusts. Nails, garden tools, and other objects made of iron can rust when they get wet. Water contains oxygen. The oxygen in the water mixes with the iron. When oxygen and iron combine, they form a new kind of matter called rust.

What kind of change happened to the rusty nail?

Serving with Matter

Matter is all around you. Your food, shelter, and clothing are all different kinds of matter. The air you breathe is a kind of matter. Matter is very changeable, and God made it that way for your good.

Sometimes we make matter change from one state to another. We may cause physical or chemical changes. We can change matter to meet our needs. Cooking food is one common way we change matter every day. We can change matter to help others.

✓ **What is a chemical change?**

209

- You may choose to do the "Try It Yourself" activity at this time.

  When iron and oxygen combine, what new kind of matter is formed? rust

- Direct attention to the pictures of the nails.

  What happened to the nail that is rusted? It got wet. The nail had a chemical change.

  How can understanding the chemical change of rusting nails help you to serve others? Possible answers: to keep bikes out of the rain so they don't rust; to put away metal tools so they don't rust

  How can changes to matter be helpful? People can use and change matter to meet their needs.

- Direct attention to the picture of the family.

  How is this family changing matter? Possible answers: Mixing cookie dough, slicing vegetables, and boiling water are physical changes. Baking cookies is a chemical change.

  How can you use changes in matter to serve other people? Possible answers: cooking, baking, recycling aluminum cans or plastic to make other items

  How can you use changes in matter to serve God? Changing matter to meet the needs of others is serving God.

Apply

Activities

Study Guide, pages 169–70

These pages review the concepts taught in Lessons 65–66. After completion, direct the students to keep them for review for Test 9.

Assess

Quiz 9A

The quiz may be given at any time after completion of this lesson. This quiz covers material from the Study Guides in Lessons 63 and 66.

✓ **when matter changes into a new kind of matter**

Activity

Try It Yourself!

Materials: white glue, water, borax, wooden craft stick, 2 plastic cups, beaker or metric measuring cup, metric or standard measuring spoon

Guide the students as they try these steps: Mix 50 mL of white glue and 50 mL of water in a plastic cup. In another plastic cup, mix 25 mL of water with 1 g (1/4 tsp) of borax. Pour a small amount of the borax mixture into the glue mixture and stir with a wooden craft stick. Keep pouring the borax mixture into the glue mixture a little at a time. Stir until the substance is stiff and mostly stays on the craft stick. Remove the substance from the craft stick with your fingers. Keep squeezing the substance with your hands until it is no longer sticky.

Objectives
- Recall what sound is.
- Identify causes of sound vibrations.
- Explain how sound travels.
- Describe three ways sound waves interact with matter.

Materials
- bell
- 2 wooden pencils
- small rock or pebble
- large shallow container, partially filled with water

Teacher Resources
- Instructional Aids 9.1–9.2: *Instrument Cards*
- Visual 9.2: *Speed of Sound*

Vocabulary
- sound
- vibration
- reflect
- echo
- absorb

Engage

- Direct the students to close their eyes and remain still while you time 15 seconds. Do not tell them to think about listening. Direct students to open their eyes at the end of 15 seconds.

  What did you hear while your eyes were closed?

- Direct the students to close their eyes while you time another 15 seconds and direct them to think about listening this time. Direct students to open their eyes at the end of 15 seconds.

  What did you hear this time?

- Discuss whether they heard more when they thought about it.

Instruct

Preparation for Reading

- Preview and pronounce the vocabulary terms *sound* and *vibration* and the words *vocal cord*.
- Direct the students to read pages 210–11 silently to find out what part of your body vibrates to make sound.

Teach for Understanding

What is *sound*? a form of energy that you can hear
What causes a sound? a moving object that is vibrating
What is a *vibration*? a rapid back-and-forth movement
Can you see a sound? no
What might you be able to see and feel? a vibration

- Hold up the bell, tap it with a pencil, and allow it to vibrate.

  How is the bell making sound? by vibrating

- Tap the bell again. Direct a volunteer to touch the outside of the bell.

  What happened to the sound when the bell was touched? It stopped.

  Why did the sound stop? The hand caused the vibrations to stop, which caused the sound to stop.

Sound
Vibrations

We know that matter is all around us. We also know that sounds are all around us. **Sound** is a form of energy that you can hear. You can hear sound as it travels through matter. Every sound is caused by a moving object that is vibrating. A **vibration** is a rapid back-and-forth movement. You cannot see sound, but you may be able to see and feel the vibrations.

You can cause an object to vibrate by blowing, hitting, shaking, plucking, or rubbing. Musicians use these actions to produce sounds. A trumpeter blows into a trumpet. The blowing causes the air inside the trumpet to vibrate. The top of a drum vibrates when it is hit. The strings of a guitar vibrate when they are plucked. A violinist rubs a bow across the strings of a violin to make them vibrate.

rubbing

blowing

hitting

210

- Now ring the bell. Explain that when a bell is rung, the clapper on the inside of the bell is creating the vibrations.
- Direct attention to the students playing instruments on pages 210–11.

  Look at the boy with the trumpet. What is he doing to make vibrations? blowing air through the trumpet

  What vibrates inside the trumpet? air

  Look at the boy with the guitar. What is he doing to make vibrations? plucking the strings of a guitar

  Look at the girl with the violin. What is she doing to the strings to make them vibrate? rubbing them with a bow

  Look at the girl with the drum. What is she doing to cause vibrations? hitting the drum

  Look at the girl with the maracas. What is she doing to cause vibrations? shaking the maracas

⚠ Helps

Tapping the Bell
Tapping the bell will help the students see that it is the vibration of the bell that causes sound, not the back-and-forth hand movement usually associated with ringing a bell.

The sounds of your voice are also caused by vibrations. When you speak or sing, the air you breathe out makes your vocal cords vibrate. Place your fingers on the front of your neck. Now hum or say a word to feel the vibrations of your vocal cords.

The Bible tells us in Psalm 150:6 that everything that breathes should praise the Lord. Some people can play an instrument. Some people can sing very well. But even if you cannot play an instrument or sing well, you can still praise God. The words you say can give God glory. We can all use our voices to praise God.

left vocal cord

right vocal cord

larynx

vocal cords

✓ **What are five ways an object can vibrate?**

shaking

plucking

211

Look at the picture of a boy holding a book. What is he doing? singing

When you placed your fingers on your neck and hummed or said a word, what did you feel? vibrations

What did the vibrations make? the humming sound or a word spoken

What part of your body vibrates to make sounds? the vocal cords which are in the throat

- Direct attention to the diagrams of the vocal cords.
- Explain that air passes through the throat when you talk, sing, or hum, and makes the vocal cords vibrate. These vibrations make the sound that is heard.

We have talked about five ways vibrations can be made. What are they? hitting, plucking, rubbing, blowing, and shaking

When you talk, sing, or hum, which of these ways are you using? Elicit that air blowing through the throat makes the vibrations.

- Direct attention to the picture of the students singing.

How are these students making sound? Air blowing or moving through the throat makes the vocal cords vibrate.

- Choose a volunteer to read Psalm 150:6 aloud.

What are some ways you can praise God with sound? Possible answers: play an instrument, sing, praise God with your words

✓ **blowing, plucking, rubbing, hitting, shaking**

ⓘ Background

Wind Instruments

For brass instruments (trumpet, tuba, French horn, etc.) the vibrations start as the player vibrates his lips. For woodwind instruments with reeds (saxophone, clarinet, oboe, etc.) the air from the player causes the reed(s) to vibrate, producing sound.

⚗ Activity

How Instruments Make Sound

Materials: Prepare Instructional Aids 9.1–9.2: *Instrument Cards* on paper or cardstock. Make as many copies as needed for each student to have at least one card. Cut cards apart and distribute.

Call out a way that sound is made, such as by blowing or plucking. Instruct each student who has a card with an instrument that makes sound that way to raise his card. You may choose for the students to stand and pretend to play their instruments.

Preparation for Reading

- Preview and pronounce the vocabulary terms *reflect*, *echo*, and *absorb*, and the word *source*.
- Direct the students to read pages 212–14 silently to find out what kind of matter sound waves travel through fastest.

Teach for Understanding

- Display the shallow dish of water. Allow the surface of the water to become still. Drop the pebble into the center of the water, and allow the students to observe that ripples radiate in all directions on the surface of the water.

  What happened to the water when the pebble was dropped into it? It made ripples, or waves, that spread outward in circles from where the pebble went into the water.

- Explain that the pebble is called the "source" of the waves in the water.
- Discuss how sound also travels in waves.

  Can you see sound waves like water waves? no

- Direct attention to the picture of the water puddle.

  How are sound waves like the waves in water? They travel outward from their source.

- What is different about the way sound waves and water waves travel? Answer should include that sound travels in all directions, not just on one surface.

- Direct attention to the picture of the fire truck.

  Have you ever heard a siren from a fire truck before you could see it? yes

- Why can you hear the siren before you see the fire truck? Possible answer: The sound travels outward from the source.

- What do the curved lines around the fire truck show? that the sound waves move in all directions

  If you are shouting across a ball field, which friend will hear a louder sound: the one next to you or the one across the field? the one next to you Why? Sound waves spread out and get quieter as they move farther away.

  What do most of the sound waves that we hear travel through? the air

- What state of matter is air? gas

  Can sound waves travel through other states of matter? yes

- What are the other two states of matter? liquid and solid

Sound Waves

What happens to the water when you drop a rock into a calm pond? Small waves, or ripples, spread out from the place where the rock went into the water. The waves move outward.

Sound also travels in waves through matter. But these waves cannot be seen. A vibration causes the sound waves. Even though you cannot see the sound waves, they act like the waves on the pond. The sound waves move outward from their source in all directions.

Because sound travels in all directions, the source of a sound does not need to be facing you for you to hear the sound. You can hear sounds behind, beside, and in front of you. You can also hear sounds above and below you.

Sound Passes through Matter

Sound waves travel through the air. Air is one kind of matter. Sound waves can also travel through other kinds of matter. You can sit in a room with the doors and windows closed. But you still may hear sounds from outside or from the next room. The sounds travel through the air. Then they also travel through the solid wall, door, or window.

212

Sound waves travel at different speeds through solids, liquids, and gases. Sound waves travel slower through gases, such as air. They travel faster through liquids. They travel the fastest through solids.

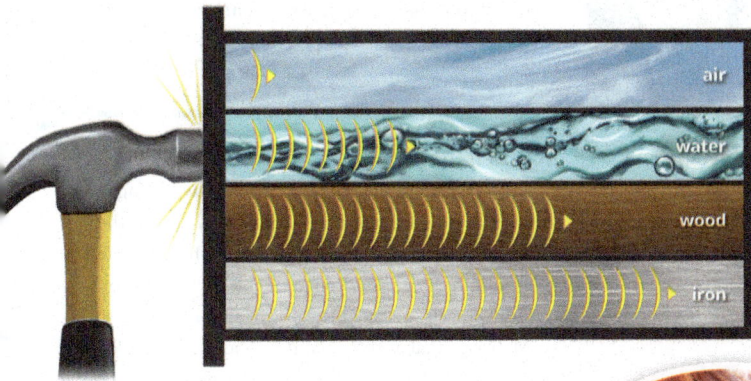

Sound Is Reflected by Matter

Sound waves travel outward from a source. They keep moving outward until something blocks their path.

When a sound wave hits an object, the object reflects the sound wave. **Reflect** means to bounce off an object. Large, hard, smooth surfaces reflect sound waves better than other surfaces do. A large building or room might have these surfaces. Canyon walls also have large, smooth surfaces.

Perhaps you have heard an echo. An **echo** is a sound wave that reflects clearly enough to be heard again. To reflect strongly, the wave must bounce off a large, hard surface. The smoother the surface, the more sound is reflected. That is why you hear echoes best when you are surrounded by hills, cliffs, or large buildings.

Sound reflects well in a canyon.

213

- Display the *Speed of Sound* visual.

  Do sound waves travel at the same speed through all states of matter? no

  Do sound waves travel faster through air or a liquid? a liquid

  Do sound waves travel faster through a liquid or a solid? a solid

  Through which state of matter does sound travel the fastest? solid

- What solids are listed on the diagram? wood and iron
- If wood and iron are both solids, why doesn't sound travel through both at the same speed? Iron is harder, or more dense, than wood.
- How can you tell from the diagram that sound travels more slowly through air than through wood? The arrow in the air does not go as far as the arrow in the wood.

  Do sound waves travel toward or away from their source? away from

  What causes a sound wave to stop moving? Something blocks its path.

  What do we call it when a sound wave bounces off an object? The wave reflects.

  What kinds of surfaces reflect sound waves the best? large, flat, smooth surfaces

  Name some examples of places where these surfaces would be found. Possible answers: large buildings, large rooms, canyon walls

- Direct attention to the picture of the canyon.
- What characteristics of this canyon would cause sound waves to reflect? The canyon's walls are large, smooth surfaces.

  What is a sound wave that reflects clearly enough to be heard again? an echo

- When you are in a crowded gymnasium, auditorium, or other large, open room, why do you think it is harder to hear someone talking? Possible answer: There are many sound waves reflecting off objects and other sound waves. This makes it harder to hear one sound.

- Direct attention to the "Fantastic Facts" box.

  How do bats see to travel in the dark? They send sound waves out and listen for them to echo back.

  What happens to some sound waves that are not reflected? They are absorbed.

  What does *absorb* mean? to take in

  What kinds of surfaces absorb sound? rough, soft, uneven surfaces

- A room that has many large, flat, smooth surfaces reflects a lot of sound. Would this room tend to be quiet or noisy? noisy

- Direct attention to the pictures of the two rooms.

  Which room will absorb more sound? Possible answers: the one on the left; the one with the bed and curtains and pillows

  Which room would be noisier? the one with the wood floor

- What are the three ways that sound waves act with matter? Sound waves travel through matter, they are reflected by matter (echo), and they are absorbed by matter.

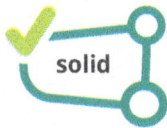

solid

Apply

Activities

Study Guide, page 171

This page reviews the concepts taught in Lesson 67. After completion, direct the students to keep it for review for Test 9.

Enrichment, page 172

This enrichment activity may be used to reinforce the concepts taught in Lesson 67.

Fantastic Facts

Some animals use echoes as a way to "see" where they are going. Bats can fly through the dark sky by sending very high-pitched sound waves out around them. These sound waves are so high pitched that people cannot hear them. But bats have very sensitive ears. They listen for the sound waves to echo, or bounce, back to them. The bats can then "see" what is around them because of the way the sound waves are reflected.

Sound Is Absorbed by Matter

Sometimes sound waves are not reflected. When sound waves hit an object, the object may absorb the sound waves. **Absorb** means to take in. The sound wave is taken into the object. Rough or soft surfaces absorb sound. Uneven surfaces also absorb sound.

Materials that absorb sound can be used to make a room quieter. Rooms that have hard, smooth surfaces can be very noisy. The walls, ceiling, and floor are all places that may reflect sound. Echoes are easily created in rooms like these. Rough, uneven materials can help keep sounds from being reflected. Carpet, fabric, and ceiling tiles are materials used to absorb some of the sounds in a room.

Which room will absorb more sound?

What kind of matter do sound waves travel through the fastest?

214

Characteristics of Sound

Pitch

Pitch is how high or low a sound is. Every sound has a pitch. Pitch depends on how fast the source of the sound is vibrating. A cello has a low pitch. It has long strings, and they vibrate slowly. A small violin has a high pitch. The strings are shorter and vibrate faster.

A string player changes the pitch by moving the position of his fingers on the strings. When he makes the strings shorter, the instrument plays a higher pitch.

Volume

Volume is how loud or soft a sound is. When you shout to a friend across a ball field, you make a loud sound. When you whisper a secret to your friend, you make a soft sound. Volume depends on how much force is used to make an object vibrate. When a stronger force is used, the vibration makes a louder sound. A weaker force makes a softer sound.

Sound waves spread out as they move farther away from the source of their vibrations. The vibrations become weaker. The sound becomes quieter.

Which picture shows more volume?

215

Lesson **68** begins here.

Objectives

- Describe the characteristics of pitch and volume.
- Explain how the physical properties of a musical instrument affect pitch and volume.
- Create musical instruments to demonstrate pitch and volume.
- Identify ways that sound is used to communicate. BWS

Materials

- sound clips of alarm clock, siren, flute, and lawnmower
- plastic containers of various shapes and sizes, one per student
- aluminum foil
- balloons, not inflated
- rubber bands
- squares of cardboard
- whistle

Teacher Resources

- Visual 9.3: *Parts of the Ear*

Vocabulary

- pitch
- volume
- audiologist

Engage

- Play the different sound clips (TeacherToolsOnline.com).
- Discuss ways those sounds are useful or helpful.

What are some other sounds that are helpful? Possible answers: doorbell, phone ringing, car horn, music

How can you tell an alarm clock from a flute? They sound different.

- Explain that sounds have different characteristics that help us tell them apart.

Instruct

Preparation for Reading

- Preview and pronounce the vocabulary terms *pitch* and *volume* and the word *communicate*.
- Direct the students to read pages 215–16 silently to find out what kind of force makes a louder sound.

Teach for Understanding

What is *pitch*? how high or how low a sound is

What does pitch depend on? how fast the source of the sound is vibrating

Does a sound with a low pitch vibrate quickly or slowly? slowly

A sound that vibrates quickly has what kind of pitch? a high pitch

- Direct attention to the pictures of the boys playing instruments.

Which instrument do you think has the higher pitch? the violin

Why does the violin have a higher pitch than the cello? The strings are shorter and they vibrate faster.

What is *volume*? how loud or how soft a sound is

> You may wish to remind students that they learned in *Science 1* that the thickness of the strings also affects the pitch of an instrument. The concept being taught in this lesson deals with the length of the string.

- Direct attention to the picture of the lawnmower and the girls whispering.

What kind of volume does a lawnmower have? a loud volume

What kind of volume does a whisper have? a soft volume

Which picture shows more volume? the lawnmower

What does volume depend on? how much force is used to make the sound

How can you make a soft sound louder? use more force

- Complete the "Musical Instrument" activity at this time so that students can create an instrument that can be played by hitting or striking it.

  What materials did you use to make an instrument?

  Does your instrument have a low pitch or a high pitch?

  What made your instrument have a low (or high) pitch?

  How can you make the pitch of your instrument higher or lower?

  How can you make the volume of your instrument louder? softer?

- Direct attention to the pictures of the two drums.

  Which drum has more volume? the large one

  Why does the larger drum have more volume? It has a larger surface, so it has more volume.

- Direct attention to the picture of the marching band.

  Which drums have a higher pitch? the small ones

- Direct attention to the "Fantastic Facts" box.

  Why does the beating of a mosquito's wings make a high-pitched sound? The wings beat quickly and cause a fast vibration.

✓ Pitch is how high or low a sound is.
Volume is how loud or soft a sound is.

Which drum has more volume?

There are many different types of drums. Some drums are very large. When a large drum is struck, it vibrates slowly. It has a low pitch. Because it is large, it has a loud volume. You can usually hear a large drum from a long distance. You might have even felt the vibration of a large drum at a parade or a school band concert!

Other drums are small. When a small drum is struck, it vibrates quickly. A small drum has a higher pitch because it vibrates quickly. A small drum has a softer volume than a large drum. The physical characteristics of the drums change their pitch and volume.

Fantastic Facts

A mosquito beats its wings 300 to 600 times a second. The fast vibration of its wings causes the high-pitched buzzing sound you hear.

✓ **What is the difference between the pitch and the volume of a sound?**

216

Activity

Musical Instruments

Materials: plastic containers, squares of aluminum foil, balloons, rubber bands, squares of cardboard

Direct students to make a musical instrument using the given materials, as well as items the students may have in their desks. The instrument should be played by hitting or striking it. Students should be able to demonstrate how to change either the pitch or the volume of their instruments.

Uses of Sounds

Some sounds are pleasing to us. Most people enjoy listening to good music. We like to hear the voices of our friends and family.

Other sounds are harmful. Loud, harsh sounds can damage our ears. That is why people who work with loud machines should wear ear protection.

Sounds can communicate as well. A ringing alarm clock tells you it is time to get up. A stove timer buzzes when it is time to take something out of the oven. A siren on a police car, a fire truck, or an ambulance warns drivers to move out of the way.

We also use sound to communicate with each other. We hear the sounds others make, and we make sounds for others to hear. The words we say are very important. The Bible tells us that our speech should please God. The Bible tells us in Psalm 19:14 that the words that come from our mouth should be acceptable to God.

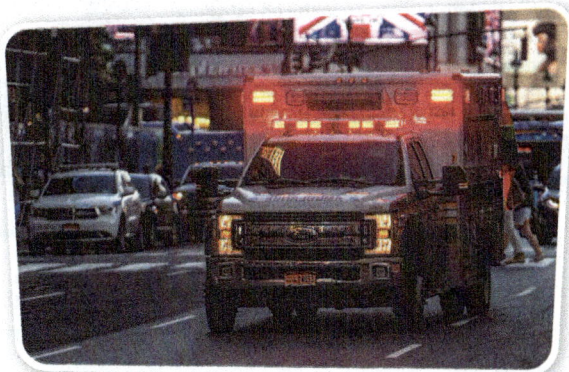

217

Preparation for Reading

• Direct the students to read pages 217–18 silently to find out some ways trumpets were used in the Bible.

Teach for Understanding

• Refer to the sounds heard during the lesson introduction and discuss the variety of sounds that you can hear.

What are some sounds that are pleasant to hear? Possible answers: singing, music, laughter, birds chirping

What are some sounds that are unpleasant to hear? screaming, loud banging, arguing, loud machines

What can happen if sounds are too loud or harsh? They can damage your ears.

How can you protect your ears from loud sounds? Cover them; wear ear protection.

• Direct attention to the picture at the top of the page.

What sounds are the people in this picture enjoying? talking, laughing, noisemakers

What does the word *communicate* mean? to give information to others

What are some things that sounds can communicate? Possible answers: They can tell about the things around us; they can warn of danger.

• Demonstrate the sound made by the whistle.

If you are playing in a game and you hear the sound made by this whistle, what does it communicate to you? Possible answers: All play should stop. A rule has been broken.

• Choose volunteers to name other sounds and explain the messages they communicate.

We use sound when we speak with each other. What does the Bible tell us about the kinds of words we should use to communicate with others? Our words should be pleasing to God.

• Direct attention to the pictures at the bottom of the page.

What does the alarm clock communicate? Possible answer: It is time to get up.

What does the siren on the ambulance communicate? Possible answers: There is an emergency. Cars need to move out of the way.

(i) **Background**

Loud Sounds

The amount of force (intensity) used to make a sound is measured in decibels. The lowest level that a person can hear is 0.0 decibels.

Any sounds above 85 decibels can cause hearing damage over time. Precautions should be taken when exposed to these sounds frequently or for prolonged periods.

For sounds above 120 decibels (the threshold of pain), even short-term exposure can cause hearing damage. Sounds such as the explosion of a firecracker, the pop of a balloon, the firing of a gun, and the takeoff of a jet plane are above 120 decibels.

- Direct attention to the picture at the top of the page.
  What is this picture showing? Possible answers: a man blowing a trumpet (shofar); a man from long ago
- Ask a volunteer to read Joel 2:15 aloud.
  What is the sound of the trumpet communicating in Joel 2:15? Possible answers: to call the people together, to assemble the people
- Ask a volunteer to read Psalm 98:6 and Psalm 150:3 aloud.
  What is the sound of the trumpet communicating in Psalm 98:6 and Psalm 150:3? Possible answers: to praise God, to worship God
- Ask a volunteer to read Judges 7:18 aloud.
  What is the sound of the trumpet communicating in Judges 7:18? that men were going into battle
- God made us able to hear the sounds around us as the sounds pass through matter. Why do you think God gave us the gift of sound? Possible answers: to help us observe the world around us; to help us communicate with others
  How can we use the gift of sound to bring glory to God? to serve others by playing music; to serve others by speaking kindly to them; to worship God
- Direct attention to the "Health and Safety" box.
  What does the ear do? collects sound waves
  What carries the sound wave messages to the brain? the nerves
  What is one thing that could harm the parts inside your ear? pushing objects too far into the ear
  What is the best way to keep your ears healthy and safe? to wash the outer ear and keep it clean
- Direct attention to the picture of the father and son.
  How is this boy using his ears? He is listening to the words his father is saying.

✓ possible answers: speech, an alarm clock, a stove timer, a siren

In Bible times the sound of trumpets was often used to give information to many people. The priests blew trumpets to tell the people to come together. Trumpets were used in the temple for worship. Trumpets were used to send men into battle.

You glorify God when you use sound to serve other people. You can serve people by playing beautiful music. You can also serve people by speaking kind words to them. In 1 Corinthians 6:20, the Bible tells us that God wants you to use your body to give Him glory. Your voice is part of your body. You can use sound to worship God and serve others.

Health and Safety

The amazing design of the ear should cause you to give glory to God as your Creator. The ears collect sound waves. The nerves carry the sound wave messages to the brain. This allows you to hear.

Sticking things in your ears can be unsafe. It is easy to push things too far into the ear. Then the object could harm the parts inside your ear. This would be painful and could harm your hearing.

Only a doctor should put anything into your ear. It is best to wash the outer ear with a washcloth to keep the ear clean. This will keep your ears healthy and safe.

✓ **What are some ways sound is used to communicate?**

218

STEM Career

Audiologist

An **audiologist** is a doctor who treats patients with hearing problems. You may have had a hearing screening at school or at your doctor's office. This test tells the audiologist whether you are hearing correctly. If you are not hearing correctly, your doctor may send you to an audiologist.

The audiologist can do more tests to know what the problem is. Sometimes the parts inside the ear are not working correctly. An audiologist might use an otoscope. An otoscope is a tool that shines a light in the ear to allow a doctor to see well.

An audiologist might also use a test called *tympanometry*. This test blows small puffs of air into your ear. This can measure how well your eardrum moves. You may not be hearing well if your eardrum is not moving like it should.

Sometimes the audiologist will want to do a hearing test. You would probably wear earphones to listen to different sounds and words. This test lets the audiologist know what sounds you can hear.

219

Preparation for Reading

- Preview and pronounce the vocabulary term *audiologist* and the words *otoscope* and *tympanometry*.
- Direct the students to read pages 219–20 silently to find out what an audiologist does.

Teach for Understanding

- Choose volunteers to share a time when they have had a hearing test.

  What does an audiologist do? treats patients with hearing problems
- Direct attention to the first picture.

  What is an *otoscope*? a tool that shines a light into the ear

  Why would an audiologist need to shine a light into someone's ear? Possible answer: to see if the parts are working correctly
- Direct attention to the second picture.

  What is *tympanometry*? a test that measures how well the eardrum moves
- Direct attention to the last picture.

  Why do you think you would need to wear earphones during a hearing test? Possible answers: so you can't hear other noises; so you can hear only the sounds the audiologist makes

ⓘ Background

Hearing Frequency

Most people can hear a sound with a frequency (how fast the sound is vibrating) between 20 and 20,000 hertz. But there are sounds that humans cannot hear. Many animals can hear sounds at much higher frequencies. As humans get older, the ear becomes less sensitive to higher frequencies, so overall hearing diminishes.

Why would a graph be useful for an audiologist to see the test results? It can show exactly what a person is hearing.

- Direct attention to the picture in the middle of the page.
  What can a person wear to hear better? hearing aids
  What important job do audiologists have? to help people to hear better

- Display the *Parts of the Ear* visual. Discuss the parts of the ear shown on the diagram.

> The sponge-like area around the ear is the temporal bone.

Where do sound waves enter the ear? the ear canal
Where do the sound waves go next? to the eardrum

- Explain that the eardrum is like a very tiny drum inside the ear. When the sound waves hit the eardrum, it vibrates. As the eardrum vibrates, it moves the sound waves to the next part of the ear.
  Where do the sound waves go after they move past the eardrum? to the hammer, anvil, and stirrup

- Explain that the cochlea is full of liquid.
  Do sound waves travel through liquid? yes

- Explain that the sound waves travel in the liquid to the auditory nerve.
  What do nerves do? take messages to the brain and parts of the body
  What does the auditory nerve do? takes sound wave messages from the ear to the brain

Apply

Activities
Study Guide, page 173
This page reviews the concepts taught in Lesson 68. After completion, direct the students to keep it for review for Test 9.

Assess

Quiz 9B
The quiz may be given at any time after completion of this lesson. This quiz covers the Study Guides from Lessons 67 and 68.

STEM Career

When the tests are finished, the audiologist will know all about your hearing. The audiologist will know whether you can hear loud sounds and soft sounds and high pitches and low pitches. The audiologist can put all the test results into a graph. The graph shows exactly what you can hear.

Sometimes hearing aids will help you to hear better. The audiologist will measure your ear and choose a hearing aid to fit your ear. But you get to choose the color!

Audiologists must know a lot about the parts of the ear. They need to know exactly how the ear works so they can help people. An audiologist does an important job to help others hear better.

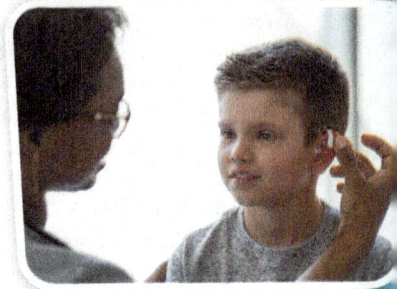

Parts of the Ear

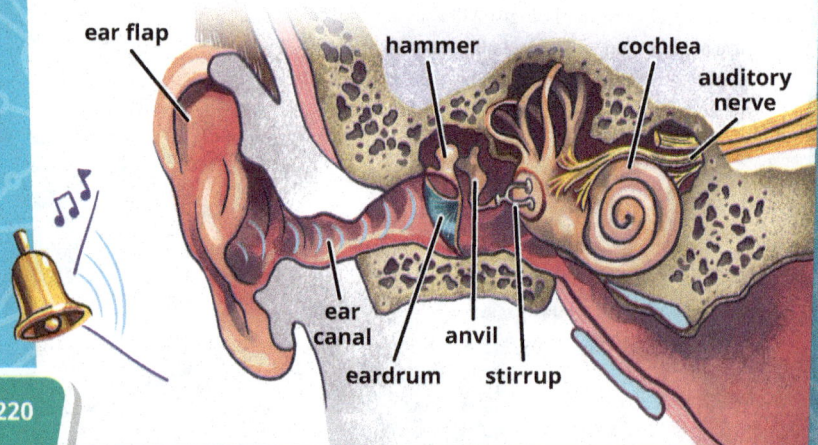

ear flap, hammer, cochlea, auditory nerve, ear canal, anvil, eardrum, stirrup

220

Musical Bottles

Inquiry Skills

- Predict
- Measure
- Infer

The pitch of some instruments is related to the length of the part that vibrates. Long strings make low sounds. Short strings make high sounds. The pitch of other instruments is related to the amount of air that vibrates. Some instruments, such as flutes, can control the amount of air that vibrates. This gives those instruments a high or low pitch.

In this Investigation, you will predict how the amount of vibrating air affects pitch.

Let's inquire into . . .
Vibration and Pitch

Repeat the Investigation using one 2 L (67.6 oz.) bottle and one 500 mL (16.9 oz.) bottle.

In this Inquiry, you will predict how changing the amount of air in the bottle will change the pitch.

221

Objectives

- Create a hypothesis to predict how air affects pitch.
- Observe and record data.
- Draw conclusions about how the amount of air in a bottle affects the pitch.

Materials

- See *Activities* page 175.
- *Musical Bottles* video

Engage

- Play *Musical Bottles* (TeacherToolsOnline.com).
  When the boy was tapping the bottles, what was vibrating? the glass
  When the boy blew across the top of the bottle, what was vibrating? the air inside the bottle
  How did he make the pitch change? He poured water out of the bottle.

Instruct

Preparation for Reading

- Direct the students to remove pages 175–77 from their *Activities* books. Direct them to read page 221 and *Activities* 175–77 silently before beginning.
- Explain that it is helpful to look over all the pages before starting an Investigation. Point out that it will help the students to see what they are going to do and when they will be writing information down.

Teach for Understanding

- Direct attention to the **Inquiry Skills**.
- Remind the students that scientists use inquiry skills to gather and use facts.

  Which inquiry skills will you be using in this investigation? Predict, Measure, and Infer
- Review what it means to *predict*, *measure*, and *infer*.

Apply

- Choose a student to read the **Problem** aloud.

  What are you going to use to test how air affects pitch? bottles and water
- Direct the students to complete the **Hypothesis** to answer the problem.
- Instruct the students to follow and check off **Procedure** Steps 1–5 as each is completed.
- Remind the students of the importance of accurately recording their **Observations**.

Musical Bottles

Name _____

Problem

How does the amount of air in a bottle affect its pitch?

Hypothesis

The **more, less** air there is in a bottle, the higher the pitch will be when you blow across the top of the bottle.

Procedure

☐ 1. Label one bottle **A** and the other **B**.

☐ 2. Measure 500 mL of water. Use a funnel to pour the water into bottle B. Leave bottle A empty.

☐ 3. Draw a line to mark the water level on each bottle below. Write the amount of wate in each bottle. Color the space the water takes up in each bottle blue. Color the spa the **air** takes up in each bottle red.

_0_ mL   _500_ n

☐ 4. Gently blow across the top of each bottle. Listen carefully to the pitch of the sound.

☐ 5. Circle the bottle that makes the higher pitch.

Inquiry Skills
- Predict
- Measure
- Infer

Materials
- ☐ 2 narrow-necked plastic bottle 2 liters (67.6 oz) each
- ☐ 3000 mL water
- ☐ beaker or metric measuring c
- ☐ funnel
- ☐ colored pencils, blue and red

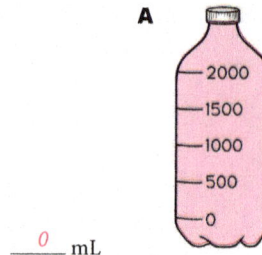

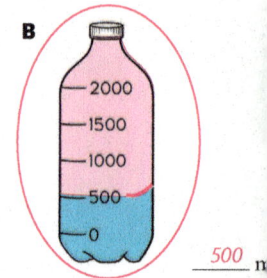

6. Measure 1500 mL of water. Use a funnel to pour the water into bottle A.

7. Draw a line to mark the water level on each bottle below. Write the amount of water in each bottle. Color the space the water takes up in each bottle blue. Color the space the **air** takes up in each bottle red.

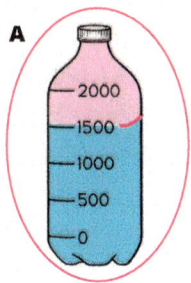 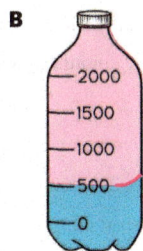

_1500_ mL   _500_ mL

8. Gently blow across the top of each bottle again. Listen carefully to the pitch of the sounds.

9. Circle the bottle that makes the higher pitch.

10. Measure 1000 mL of water. Use a funnel to pour the water into bottle B.

11. Draw a line to mark the water level on each bottle below. Write the amount of water in each bottle. Color the space the water takes up in each bottle blue. Color the space the **air** takes up in each bottle red.

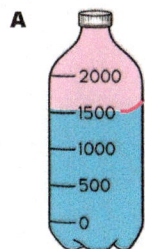

Neither bottle should be circled; they should both have the same pitch.

_1500_ mL   _1500_ mL

12. Gently blow across the top of each bottle again. Listen carefully to the pitch of the sounds.

13. Circle the bottle that makes the higher pitch.

• Instruct the students to follow and check off Procedure Steps 6–13 as each is completed.

- Direct the students to draw **Conclusions** using their observations. The conclusions may be done individually or with their science group.
- Guide a discussion of the students' conclusions.

Activities

Investigation, pages 175–77

Students will predict how the amount of air in a bottle affects its pitch.

Assess

Rubric

Use the prepared rubric or design a rubric to include your chosen criteria.

Musical Bottles

Name _____

Conclusions

1. Did your observations support your hypothesis?   ○ yes   ○ no

2. Did the pitch change when the amount of air in the bottle changed?   ● yes   ○ n

3. How did the amount of water affect the amount of air that was in the bottle? _Answers should include that more water in the bottle means less air can be inside the bottle._

4. How could you make the pitch lower in bottle A? _Pour some of the water out so that there will be more air._

5. How could you make the pitch higher in bottle A? _Add more water so there will be less air._

6. How does the amount of air that vibrates affect pitch? _The pitch is lower when more a is vibrating and higher when less air is vibrating._

7. How did the same amount of air in each bottle affect the pitch? _The pitch was the sam when both bottles had the same amount of air._

SCIENCE 3 *Activities*

Musical Bottles

Name _____

Problem

How can I make the pitch of a 500 mL bottle and a 2 L, or 2000 mL, bottle be the same?

Hypothesis

Write a hypothesis that will answer the problem.
Hypothesis should include that both bottles must have the same amount of air in them.

Procedure

☐ 1. Label the larger bottle **A** and the smaller bottle **B**.

☐ 2. Gently blow across the top of each empty bottle. Listen carefully to the pitch of the sound.

☐ 3. Measure _____ mL of water. Use a funnel to pour the water into bottle _____.

☐ 4. Draw a line to mark the water level on each bottle below. Write the amount of water in each bottle. Color the space the water takes up in each bottle blue in each bottle. Color the space the **air** takes up in each bottle red.

A

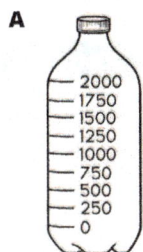

2000
1750
1500
1250
1000
750
500
250
0

_____ mL

B

500
400
300
200
100
0

_____ mL

SCIENCE 3 *Activities*

Lesson 70 • Page 221
Inquiry **179**

Inquiry Skills

• Predict
• Measure
• Infer

Materials

☐ 1 narrow-necked plastic 500 mL (16.7 oz.) bottle
☐ 1 narrow-necked plastic 2 liter bottle, or 2000 mL bottle (67.6 oz.)
☐ water
☐ beaker or metric measuring cup
☐ funnel
☐ colored pencils, blue and red

Lesson **70** begins here.

Objectives

• Create a hypothesis to predict how the size of a bottle will affect the pitch.
• Develop the steps of the procedure in the Inquiry.
• Observe and record data.
• Draw conclusions about how the size of a bottle will affect the pitch.

Materials

• See *Activities* page 179.

Engage

• Discuss observations from the previous day's lesson.
  🎧 How did you change the pitch in the bottle? added or removed water
  🎧 How did adding water affect the pitch? When more water is in the bottle, there is less air. Less air vibrating makes a higher pitch.
  🎧 How did removing water affect the pitch? When less water is in the bottle, there is more air. More air vibrating makes a lower pitch.

Instruct

Preparation for Reading

• Direct the students to remove pages 179–81 from their *Activities* books. Direct them to read *Activities* 179–81 silently before beginning.

• Explain that it is helpful to look over all the pages before starting an Inquiry. Point out that it will help the students to see what they are going to do and when they will be writing information down.

Teach for Understanding

This is an Inquiry lesson. Inquiry lessons allow the students the opportunity to form the hypothesis and design more of the investigation. The *SCIENCE 3 Activities Answer Key* provides possible answers to help the teacher guide the students to success. Encourage the students to collaborate as they work through the Inquiry.

• Direct attention to the **Inquiry Skills**.
• Remind the students that scientists use inquiry skills to gather and use facts.
  Which inquiry skills will you be using in this Investigation? Predict, Measure, and Infer
• Review what it means to *predict*, *measure*, and *infer*.

Apply

• Choose a student to read the **Problem** aloud.
  What are you going to use to test how air affects pitch? bottles and water

• Guide the students in a discussion as they write the **Hypothesis** together as a class. So that each student is testing the same hypothesis, display the students' suggested hypotheses and have the class agree on a single hypothesis to test.

• Direct attention to the **Materials** list and make the materials available to the students.

• Instruct the students to follow and check off **Procedure** Steps 1–4 as each is completed.

• Remind the students of the importance of accurately recording their **Observations**.

• Explain that information is missing in several steps and that the students are to determine how to best complete the steps of the Procedure. Point out that they can refer back to the Procedure in the Investigation in Lesson 69.

- Instruct the students to follow and check off Procedure Steps 5–10 as each is completed.
- Point out that if the pitch is not the same after Step 10, students should continue to try to make the pitch the same. Observations should be recorded in Step 12.

☐ 5. Gently blow across the top of each bottle. Listen carefully to the pitch of the sound.

☐ 6. Is the pitch of both bottles the same? If not, continue to step 7.

☐ 7. Measure _____ mL of water. Use a funnel to pour the water into bottle _____.

☐ 8. If needed, measure _____ mL of water. Use a funnel to pour the water into bottle _____.

☐ 9. Draw a line to mark the water level on each bottle below. Write the amount of water in each bottle. Color the space the water takes up in each bottle blue in each bottle. Color the space the **air** takes up in each bottle red.

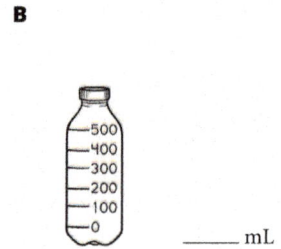

A

```
— 2000
— 1750
— 1500
— 1250
— 1000
— 750
— 500
— 250
— 0
```
_____ mL

B

```
—500
—400
—300
—200
—100
—0
```
_____ mL

☐ 10. Gently blow across the top of each bottle again. Listen carefully to the pitch of the sounds.

☐ 11. Is the pitch of both bottles the same? If not, repeat steps 7–10 as needed.

☐ 12. Continue recording your measurements in the chart below until the pitch of both bottles is the same.

Bottle A	Bottle B
_____ mL	_____ mL
_____ mL	_____ mL
_____ mL	_____ mL
_____ mL	_____ mL

Musical Bottles

Name _____

Conclusions

1. Did your observations support your hypothesis? ○ yes ○ no

2. Did the pitch change when the size of the bottle changed? ○ yes ○ no

3. How did the size of the bottle affect the amount of air that was in the bottle? *Answers should include that a smaller bottle means less air can be inside the bottle.*

4. How does the amount of air that vibrates affect pitch? *The pitch is lower when more air is vibrating and higher when less air is vibrating.*

5. Were you able to make the pitch of both bottles the same? ○ yes ○ no

6. Explain how you could or could not make the pitch of both bottles the same. *Answers should include that the pitch will be the same if there is the same amount of air in both bottles.*

• Direct the students to draw **Conclusions** using their observations. The conclusions may be done individually or with their science group.

• Guide a discussion of the students' conclusions.

Activities

Investigation, pages 179–81

Students will predict how the amount of air in a bottle affects its pitch.

Assess

Rubric

Use the prepared rubric or design a rubric to include your chosen criteria.

Objective

• Recall terms and concepts from Chapter 9.

Apply

Review

• Material for Test 9 will come from the Study Guides on *Activities* pages 159–60, 169–70, 171, and 173, as well as Quizzes 9A and 9B. You may review any or all of the material during the lesson.

• You may choose to review Chapter 9 by playing the States of Matter review game or a game from the Game Bank (Teacher Resources).

🏆 Review Game

States of Matter

Divide the students into two teams. Designate three locations in the classroom for solids, liquids, and gases. Direct all the students to start at the area for solids.

Alternate questions between teams. If a team member answers the question correctly, he moves to the area for liquids. If the next person on his team answers correctly, that person moves to the area for liquids, and the first team member advances to the area for gases. When a third team member answers correctly, the one in the area for gases returns to his seat, and each of the other members advances one position.

If a team member answers incorrectly, any team members in the gas or liquid areas move back one position. But any team members already seated remain seated. The goal is to get all the team members through the states of matter and seated.

Once all team members have advanced past the area for solids, direct questions to students in the gas or liquid areas.

NOTES

NOTES

_____

_____

_____

_____

_____

_____

_____

_____

_____

_____

_____

_____

_____

_____

_____

_____

_____

_____

_____

_____

_____

_____

_____

_____

_____

_____

Objective

• Apply terms and concepts from Chapter 9.

Assess

• Administer Test 9.

Chapter Objectives

- Analyze matter in terms of force and motion.
- Formulate a biblical explanation for God's design and order about how forces and motion work together.
- Apply the steps of an investigation to predict, observe, and infer balanced and unbalanced forces.
- Apply the steps of the engineering design process to solve a real-life problem.

Lesson Objectives

- Recall what force, friction, gravity, weight, and magnetism are.
- Formulate a biblical explanation for how we know that God created force. BWS
- Identify contact and noncontact forces.
- Compare and contrast contact and noncontact forces.

Materials

- dart board game with balls using hook-and-loop fasteners
- jump rope
- soccer ball
- scrap of paper, per student
- scale
- object to be weighed
- strong magnet
- iron filings, nails, or paper clips

Teacher Resources

- Visuals 10.1–10.2: *Cause and Effect*; *Kinds of Forces*

Vocabulary

- force
- exert
- contact force
- friction
- noncontact force
- gravity
- weight
- magnetism

Chapter Introduction

- Direct the students' attention to the chapter title and the picture of the fair games and a Ferris wheel spinning at night. Provide time for the students to leaf through the chapter, looking at the headings and pictures. Discuss what they think the chapter will be about.

  Where was the picture taken? at the fair, at a carnival, at a theme park

  What do you see in the picture? Possible answers: spinning Ferris wheel, stuffed teddy bears and lions, signs

  Name some things that are in motion. Ferris wheel, possibly swinging stuffed animals and signs

  Use completed Looking Ahead, *Activities* page 183, to generate a discussion on what the students already know about forces and motion.

Big Question

- Ask a volunteer to read aloud the Big Question.
- Explain that the students will find the answer to the question as they read the chapter.

Big Question

How are force and motion connected?

222

Big Question
How are force and motion connected?

⚠ Helps

Recalling Concepts

The students learned what force, friction, gravity, weight, and magnetism are in *Science 2*. They also learned about the cause-and-effect relationships listed in the chart on *Student Edition* page 225. These concepts are reviewed at the beginning of this lesson so the students will be able to build on them later in the chapter.

Scale

A bathroom scale will work if a scale is not available. If using a bathroom scale, find an object heavy enough to weigh on the bathroom scale, other than a student.

ⓘ Background

Chapter Photo

The photo on pages 222–23 is a picture of games that can be played at a fair and of a Ferris wheel spinning at night.

Visit TeacherToolsOnline.com for resources to enhance the lessons.

Chapter 10

Forces and Motion

As Mei-yook and her family arrive at the fair, the merry-go-round music welcomes them. Mei-yook is drawn to a booth by the smell of buttery popcorn. She munches the popcorn and sips lemonade as she watches her dad win several games. He pulls tickets from his pocket to pay for three darts. Taking careful aim, he hits the target with each throw. Mei-yook excitedly adds a pink bear to her armload of prizes.

Think about what Mei-yook was doing at the fair. She moved herself and other objects. Her feet moved as she walked. Her hands moved as she carried the popcorn and prizes. Her mouth and tongue moved as she ate.

223

Engage

- Display the dart board.
- Allow several students to throw the balls at the dart board. Ask the students to watch carefully as each volunteer throws a ball.

  How did the ball get from the student's hand to the dart board? The ball was thrown.

  Did the student push the ball or pull the ball? pushed the ball

  Do you remember what a push is called? a force

- Explain that today the students will learn about two different kinds of forces.

Instruct

Preparation for Reading

- Preview and pronounce the name *Mei-yook*.
- Direct the students to read page 223 silently to find out what kinds of motions are at the fair.

Teach for Understanding

What moved as Mei-yook munched popcorn and sipped lemonade? Possible answers: Her mouth and her tongue moved as she munched and sipped. Her hands moved as she carried the popcorn and lemonade.

What moved as Mei-yook's father played darts? His hands moved as he picked up the darts and threw them at the target.

Have you ever been to a fair?

What movement did you see at the fair?

⚠ Helps

Pronunciation

Mei-yook is pronounced *me yuk*. The name is common among Cantonese from Hong Kong.

Preparation for Reading

- Preview and pronounce the vocabulary terms *force* and *exert*.
- Direct the students to read pages 224–25 silently to find out when forces began.

Teach for Understanding

The practical aspects of forces, motion, and work are usually very evident to the students. However, since the wording may be difficult, regularly give additional practical examples to help the students understand the terminology throughout the chapter.

What is a *force*? a push or a pull

Use the word *exert* as much as possible when talking about applying an effort so that students will become familiar with the term.

- Direct several students to exert a force using objects in the classroom. Possible ways: pull out or push a chair, push a book, open a cabinet door, roll a ball, roll a pencil, erase the whiteboard
- Display the *Cause and Effect* visual.
  What does the effect tell you? what happened
  What does the cause tell you? why it happened
- Direct a volunteer to read the first example aloud.
  What question do you ask to find an effect? "What happened?"
- Instruct the students to read the example again silently, asking themselves the question, "What happened?"
  What happened? The water became a gas.
  What question do you ask to find the cause? "Why did it happen?"
- Instruct the students to read the example again silently, asking themselves the question, "Why did it happen?"
  Why did it happen? because the water was heated
- Follow the same procedure with the second example. Fill in the last chart on the visual as you discuss the effect and cause. Effect: The water vapor in the air condensed and changed to a liquid.
  Cause: because a cold drink cooled the glass and the air around it
- Discuss cause-and-effect relationships with force.
  What are the effects of force on an object? can make the object go faster, go slower, or change direction
  What does the word *effect* mean? something happens because of something else
  What question can you ask to find the effect? "What happened?"
  What does the word *cause* mean? why something happens
  What question can you ask to find the cause? "Why did it happen?"
- Give the jump rope to two students and ask them to turn it.

Force

Mei-yook's movements at the fair were pushes and pulls. A push or a pull is called a **force**. When you push open a door, you exert a force. When you **exert** a force, you apply effort. When Mei-yook moved her hands and chewed her food, she exerted a force. You exert a force to kick a ball or turn a jump rope. Matter cannot move unless a force causes it to move.

Fast, Slow, and Direction

Force also acts on objects that are already moving. Force causes objects to go faster, slow down, or change direction. Think about a jump rope. You exert a force to make a jump rope go faster, go slower, or change its direction.

224

- Instruct the students to demonstrate turning the jump rope slowly.
  What was the cause for the jump rope moving slowly? A light force was exerted.
- Instruct the students to demonstrate turning the jump rope faster.
  Why did the jump rope move faster? A strong force was exerted.
  How can the students change the direction of the jump rope? Possible answer: turn the jump rope in the opposite direction
- Direct the students to demonstrate changing the direction of the jump rope.
  Why did the jump rope change direction? The force exerted changed direction.
- Analyze the playground picture.
  Are the boy and girl on the swings exerting a push or a pull? Both; they are pushing when going forward and pulling when going backward.
  Are they exerting a strong or a light force? Accept either with a reasonable explanation.
- Continue asking questions about the force (push or pull) shown by other children in the picture and whether it is a strong or light force.
  Possible answers: Accept forces as strong or light with a reasonable explanation. boy and girl on the seesaw: push; boy and girl swinging the jump rope: push and pull; girl jumping rope: push; boy sliding: push; boy sitting at top of slide: no force exerted

When you kick a soccer ball, you exert a force to the ball. The ball goes in the same direction as the force. An object will move in the same direction as the force that moves it.

You may pass the ball to someone else on your team. Your teammate kicks the ball in another direction. When the direction of the force changes, the direction of the ball changes.

Effect	Cause
Object moves fast	Strong force
Object moves slowly	Light force
Object changes direction	Objects touch
Object moves in the same direction	Direction of the force
Object changes direction	Direction of the force changes

Creation Corner

On the first day of Creation, God created light. "God called the light Day, and the darkness he called Night. And the evening and the morning were the first day" (Genesis 1:5). Because Earth rotates, it has day and night. Because of Earth's motion, Earth has evening and morning. What causes motion? A force causes motion. Earth cannot be in motion without force. When God spoke on the first day of Creation, He created forces.

✔ **What does it mean to exert a force?**

225

- Gently kick the soccer ball to a student.
  Why did the soccer ball travel? A force was exerted.
  What direction did the ball travel? the same direction as the force
- Direct a student to place the soccer ball in front of him. Change your position slightly to the side of the student.
- Instruct the student to pass the ball to you.
  Why did the ball travel to me? A force was exerted.
  What direction did the student exert the force? in the same direction that he wanted the soccer ball to travel; in the same direction as you were standing
  What is the cause and effect between force and direction? An object moves in the same direction (effect) as the direction of the force exerted (cause). If the force changes direction (effect), the object changes direction (cause).
- Guide a discussion of the picture of the soccer players.
  Will a light force or a strong force be exerted? a strong force because you want the ball to travel fast
  Which direction is the soccer ball going to travel? the direction of the force, toward the teammate
- Discuss the playground picture on the previous page.
  Which students are showing a change in the direction of force? the seesaw students, the swinging students, the students turning the rope, the student jumping rope, the student sliding
- Review cause-and-effect relationships as needed using the chart.
- Discuss "Creation Corner." Ask a volunteer to read Genesis 1:5 aloud from the box.
  When did God create light and darkness? on the first day of Creation
- *Note:* This is an unknown light created by God. This is not the sun. The sun was created on Day Four.
  What causes the daytime and nighttime on Earth? the rotation of Earth
  The rotation of Earth is motion. What causes motion? a force
- Instruct the students to pair with their science partners. Direct the pairs to collaborate and write a biblical explanation on a scrap piece of paper for how we know that God created force.
- Remind them that *biblical* means to use the Bible's teaching.
- Allow some of the partners to share their explanations. As a class, discuss the explanations and write an explanation for display. Possible answer: God created light on the first day of Creation. Earth has day and night because the earth rotates. Rotating is motion. Earth cannot be in motion without force. When God spoke on the first day of Creation, He created forces.

✔ to apply effort

Differentiated Instruction

Remediation: Pantomime a Force

Direct the students to pretend to do the actions listed below without talking or making noises. Instruct them to identify each force in the action as a push, a pull, or both. Begin with the following actions and add more as time allows.

- shooting an arrow pull
- sweeping or raking both
- kicking a ball push
- rowing a boat both
- raising a flag pull
- swinging a bat push

Helps

Teaching Cause-and-Effect Relationships

It is easier for students at this age to first find the effect in a cause-and-effect relationship and then find the cause. In a later grade, cause-and-effect relationships will be taught as finding the cause first and then the effect.

Background

Forces

Some forces exist even when an object does not appear to be moving. A book resting on a table is exerting a force on the table, and the table is pushing back against the weight of the book.

Preparation for Reading

- Preview and pronounce the vocabulary terms *contact force, friction, noncontact force, gravity, weight,* and *magnetism*.
- Direct the students to read pages 226–27 silently to find out what weight is.

Teach for Understanding

What does the word *contact* mean? Elicit that it means to touch.

- Ask the students to demonstrate each of the following sentences. Emphasize the word *contact* in each.

  My elbow contacted the desk.
  My right foot is contacting my left foot.
  My hand contacts the book.
  My grip is strong as it contacts the pencil.

- Display the *Kinds of Forces* visual. Direct the students to remove *Activities* page 185 from their books. Guide them as they complete the organizer about forces during the discussion.
- Discuss the first kind of force on the organizer.

  What is a *contact force*? a force that acts only when one object touches another object

- Gently roll the soccer ball.

  What is the moving of the ball called? motion

  What is *motion*? what happens when an object moves

  Does the ball continue to roll on forever? No, it will begin to roll more slowly and gradually come to a stop.

  What force slows or stops motion? friction

- Gently roll the soccer ball again.

  Where is there friction? between the ball and the floor

  Are the floor and the ball in contact with one another? Yes, there is a contact force because one object (floor) touches another object (ball).

- Discuss the picture of the snowboard.

  Are two objects in contact? Yes, the snowboard is in contact with the snow.

  What force is the contact force? friction

  What does friction do? causes the snowboard to slow down and stop

- Discuss the second kind of force on the organizer.

  What is a *noncontact force*? a force that acts between objects that are not touching

- Gently toss the soccer ball in the air.

  Does the soccer ball keep going up forever? No, it falls back down.

  Did the ball touch anything to make it fall back down? no

  What force pulls objects toward the floor or the center of the earth? gravity

  Did gravity touch or contact the ball? No, it is a non-contact force.

- Guide a discussion of the geyser picture.

  Explain what you see. Possible answer: A geyser shoots water into the air. The water falls back to the ground.

  What noncontact force is being exerted? Gravity; the water is not in contact with anything but is being pulled toward the center of the earth.

What causes the water to fall back to the earth?

226

Kinds of Forces

Think about kicking a ball. You can see how the force from your foot causes the ball to move. But your kick is not the only force acting on the ball. There are other forces at work as well.

Contact Forces

Some forces act only when two objects touch each other. A force that acts only when one object touches another object is called a **contact force**. A rolling ball will stop rolling at some point. Friction causes the ball to roll more slowly and gradually stop. **Friction** is a contact force that slows or stops *motion*. As the ball rolls, there is friction between the ball and the ground. When the ball and the ground come in contact with each other, there is friction.

What causes the snowboard to slow down and stop?

Noncontact Forces

A force that acts between objects that are not touching is called a **noncontact force**. If you kick a ball into the air, it does not keep going up forever. It falls back to the ground. The noncontact force, called gravity, pulls the ball down. **Gravity** is the force that pulls all objects toward the center of the earth. Gravity pulls the ball down without contacting the ball.

ⓘ Background

Friction

Sometimes people want to reduce the amount of friction. A bowling lane is usually waxed or polished to reduce friction. The parts inside machines are often oiled to reduce friction. However, sometimes more friction is needed. The tread on tires helps them grip the road better. A baseball player wears cleats to keep from sliding on the ball field. Workers sometimes add sand to icy roads to help prevent accidents.

Gravity

All objects, even small ones, exert gravity. However, the greater the mass of an object, the greater the gravitational force it exerts.

The earth's gravity keeps us anchored to the earth's surface and keeps the earth's atmosphere in place. The sun's gravity keeps the earth in its orbit around the sun. The sun's gravity and the moon's gravity both pull on the earth and cause the tides.

When you weigh an object, you are measuring the pull of gravity. **Weight** is the measure of the force of gravity on an object. Scientists use a scale to measure weight. Weight and mass are not the same. Remember that *mass* is the amount of matter in an object.

What is used to measure the force of gravity?

Fantastic Facts

Your weight is the force of gravity acting on your mass. Your mass will always stay the same. But your weight can change on different planets. For example, you weigh almost three times more on Earth than you would on Mars.

Contact Forces	Noncontact Forces
Friction	Gravity
	Magnetism

The pushing or pulling force caused by a magnet is called **magnetism**. Is magnetism a contact force or a noncontact force? Does a magnet need to touch an object to work? Think of using a magnet to pick up a paper clip. You might hold the magnet over the paper clip. Suddenly the magnetic force pulls the paper clip up to the magnet. Magnetism is a noncontact force.

Try It Yourself!

Stack 2 books on a table. Tie a string around both books. Leave at least 5 cm (2 in.) of string hanging from the knot. Attach a rubber band to the loose ends of the string. Move the books by pulling on the rubber band. Measure how far the rubber band stretches. Place 6 pencils under the books. Move the books again. Measure how far the rubber band stretches. Which surface caused more friction?

✔ **What is the difference between contact forces and noncontact forces?**

227

When you weigh an object, what are you measuring? the force of gravity

What do scientists use to measure weight? a scale

- Display the scale and demonstrate how to weigh an object. *Note:* To give the students opportunities to weigh objects, you may display the scale and allow the students to weigh different objects during their free time.

- Analyze the picture of the scale.

  Where have you seen a scale similar to this one? in the grocery store, in the kitchen

  What is being weighed on the scale? fruit

  What is being measured? the force of gravity

- Discuss "Fantastic Facts."

  Which one always stays the same—your weight or your mass? your mass

  Where can your weight change? on different planets

  Where would you weigh more—on Earth or on Mars? on Earth

  Why would you weigh more on Earth than on Mars? Elicit that the force of gravity is stronger on Earth than on Mars.

- Discuss magnetism as a noncontact force.

  What is the pushing or pulling force caused by a magnet called? magnetism

- Display the iron filings on a solid wood surface.

- Ask a student to slowly lower the magnet toward the iron filings. Stop the student when the magnet begins to pull the filings toward it.

  Does the magnet need to touch the metal to work? no

  What is a force that acts without touching or contacting the object called? noncontact force

- Explain that the paper clips in the picture are metal that is covered with plastic. The magnet is strong enough to work through the colorful plastic coverings.

- Review which forces are contact or noncontact forces, using the chart.

✔ Contact forces act when an object touches or contacts another object. Noncontact forces act between two objects that are not touching or in contact.

Apply

Activities

Looking Ahead, page 183

This page assesses the students' knowledge prior to the chapter. *Note:* You may want the students to do the "Notebook Connection" at the bottom of the page.

Kinds of Forces, page 185

This page helps the students to classify forces. It was completed as part of the lesson.

Contact and Noncontact Forces, page 186

This page reinforces the concepts in Lesson 73.

Activity

Try It Yourself!
Materials: 2 books; string, the length of the circumference of the books plus at least an additional 5 cm; rubber band; 6 pencils; ruler
Allow students to try this activity on their own.

Objectives

- Describe direction, speed, distance, and position as they relate to motion. **BWS**
- Identify balanced and unbalanced forces.
- Analyze whether the forces acting on an object are balanced or unbalanced.
- Explain why it is important to know about balanced and unbalanced forces. **BWS**

Materials

- foam ball
- chair
- book
- table
- rope for tug of war
- toy truck

Teacher Resources

- Visual 10.3: *Balanced or Unbalanced?*

Vocabulary

- position
- motion
- direction
- distance
- speed
- strength
- balanced forces
- unbalanced forces

> This lesson incorporates demonstrations to aid the students' understanding. You may choose to teach the lesson outside to be able to use baseball equipment with the examples given from a baseball game.

Engage

- Choose several volunteers to pantomime different movements at the same time, such as eating soup, climbing a ladder, and bouncing a ball.
- What is each student doing? Answers should describe what each is pantomiming.
- Guide the students in concluding that all are moving.
  We can also say that they all are "going through the motions of _____."
- Point out that we frequently use terms like *move, moving,* and *motion,* but that we might not think of them the same way that scientists do.

Instruct

Preparation for Reading

- Preview and pronounce the vocabulary terms *position, motion, direction, distance,* and *speed.*
- Remind the students that captions also give us information.
- Direct the students to read pages 228–29 silently to find out three ways to describe the motion of an object.

Teach for Understanding

> In *SCIENCE 2* the students learned that *motion* was "what happens when an object moves." The definition has developed with the grade level to mean "a change in position."

What is *motion?* a change of position
What does *position* mean? the place where an object is

- Direct attention to the picture of the bat and the ball.

Lesson 74

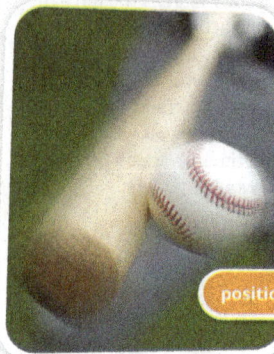
position

What is the position of the ball?

Motion

In a baseball game, the pitcher throws the ball toward home plate. The ball's position changes from the pitcher's hand to the air. **Position** is the place where an object is. The pitcher exerts a force to move the ball. The force puts the ball in motion. **Motion** is a change of position.

Direction, Distance, and Speed

You can describe an object's motion in three ways. You can describe the *direction* the object travels, the *distance* it travels, and the *speed* it travels.

Direction is the way an object faces. The object may face east, west, or to the right of your desk.

When a baseball is hit, its motion changes. The ball's direction changes from moving toward the bat to moving away from the bat. When hit, a baseball can travel in many directions. It might go left over third base or straight toward second base. It might even bounce up behind the batter.

motion

How is the baseball showing motion?

direction

Motion can be described by the direction an object travels.

228

- What is the ball's position? Possible answers: just hitting the bat, beginning to leave the bat
- Toss a ball to several volunteers. Guide the students in identifying when the ball changes position.
  What causes motion? a force
- What is causing the force that the volunteers are using to move the ball? their hands and arms
- Direct attention to the picture of the boys throwing the ball.
- How is the baseball showing motion? The position of the baseball is changing.
- Discuss how motion can be described.
  What are three ways we can describe the motion of an object? the direction it travels, the distance it travels, and the speed it travels
- Allow volunteers to describe the kinds of motion they have experienced while playing baseball or other ball games.
  What does *direction* mean? the way the object faces
- Which direction are you facing? Possible answers: facing toward the teacher, facing toward the front of the room, facing west
- Direct attention again to the picture of the bat and the ball.
- How will the direction of the ball's motion change? from moving toward the bat to moving away from the bat
- What direction is the motion of the bat? toward the baseball

Distance is how far an object travels. The baseball can travel different distances. It might travel far into the outfield. Or it may go only a few meters in front of home plate. If you measure from where the ball was hit to where it stopped, you will know the distance it traveled.

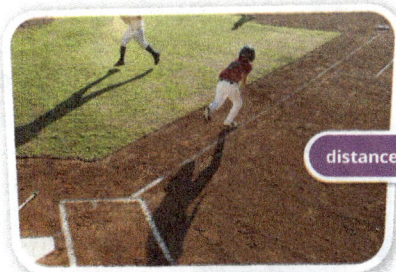

Speed is the measurement of how fast or slow an object moves. A baseball can move at different speeds. If the ball slowly rolls to the first baseman, he can easily pick it up. But if the ball quickly whizzes into the outfield, a player may not be fast enough to catch it.

distance

Motion can be described by the distance an object travels.

speed

Motion can be described by the speed an object travels.

Fantastic Facts

The time it takes you to travel a distance depends on your speed. The speed is followed by the letters km/h (mph). The letters *km/h* mean "kilometers per hour." The letters *mph* mean "miles per hour." Fast speeds cover more distance in a shorter time. Look at these speeds!

cheetah	112 km/h (70 mph)
hare	55 km/h (35 mph)
man	25 km/h (15 mph)
student	13.5 km/h (8.4 mph)

✓ **What are three ways to describe the motion of an object?**

229

- Discuss motion and distance.
  What is *distance*? how far an object travels
- Analyze the picture of the baseball player running.
  How could you find the distance of the baseball player's motion? by measuring the distance between the bases
- Refer to the second picture on the previous page.
  How could you find the distance of the baseball's motion? by measuring from where the ball was thrown to where the boy caught the ball
- Discuss the second baseball picture, designated speed.
  What is the last way that motion can be described? by the speed it travels
  What is *speed*? the measurement of how fast or slow an object moves
  What is happening in the picture? The ball is being thrown from the pitcher to the batter.
  What do you think the speed of the baseball's motion is? Possible answer: Most pitchers throw very fast.
- Explain that there are devices that can measure the speed of moving baseballs and that technology is used in places such as batting cages and in minor- and major-league baseball games.
- Allow the students to gently toss a ball to one another. Explain that for this demonstration, the speed of the ball should be slow each time.
- Each time the ball is tossed, ask the student tossing the ball to describe the direction of motion and the distance of motion.

You may choose to play a game of kickball or softball outside, continually reinforcing the ways motion can be described.

- Why do speed, motion, and position work together? Because God designed them to work together.
- Discuss "Fantastic Facts." Point out the meaning of the letters *km/h* and *mph*.
- Discuss the speeds of the animals and people pictured.
  How fast do you think you can run?

Teaching for *Student Edition* page 228 continues here.

- Direct attention again to the picture of the two boys tossing the ball.
- In what direction is the ball's motion? from one boy to the other boy
- Refer to the boy and his father throwing the ball.
  What direction is the baseball's motion? Possible answers: toward the baseball glove, toward the boy

✓ **the direction an object travels, the distance an object travels, and the speed an object travels**

Preparation for Reading

- Preview and pronounce the vocabulary terms *strength*, *balanced forces*, and *unbalanced forces*.
- Direct the students to read pages 230–31 silently to find out what a spring scale is.

Teach for Understanding

- Place a chair in the front of the room. Ask a student to sit in the chair.

  What force pulls the student down toward the chair seat? gravity

  What is *gravity*? a force that pulls objects toward the floor or the center of the earth

  What force pushes up on the student? the chair's force

  How are gravity and the chair's force pushing on the student? with the same strength, with equal strength

  What would happen if gravity and the chair's force were not equal in strength? The student would move upward or downward.

  Why would the student move downward? Gravity's force is stronger than the chair's force. The chair might break.

  Why would the student move upward? The chair's force is stronger than gravity's force.

  What is *strength*? the amount of force exerted on an object

  What direction are gravity and the chair's force pushing on the student? in opposite directions

  What can gravity and the chair's force be called? balanced forces

  What are *balanced forces*? two forces that are equal in strength and opposite in direction as they act on an object

  Do you observe any motion? No, balanced forces do not cause motion. The object or student keeps doing what he is doing.

- Follow the same procedure with a book on a table as was done with the student and chair, guiding the students to see that the forces (gravity and the table's force) have equal strength and are opposite in direction.

- Ask a student to push on one side of a cabinet door or the classroom door as you push on the other side.

  What kind of force do I exert on the door? a big force

  What kind of force does the student exert on the door? a smaller force

  What is the strength of the forces? unequal

  What are two forces that are unequal called? unbalanced forces

  Do you observe any motion? Yes, when the forces are unbalanced, there is motion.

- Instruct five students who are about the same size to grab the rope to play tug of war. Place three students on one end of the rope and two on the other end.

- Follow the same procedure with the tug of war as was done with the door, guiding the students to see that the forces (the force of the three students and the force of the two students) are unequal in strength and are unbalanced.

How Forces Affect Motion

Balanced and Unbalanced Forces

You are probably sitting in a chair. Gravity's force pulls you down toward the chair seat. The chair's force pushes up on you. The chair's force and gravity's force are both pushing with the same strength. **Strength** is the amount of force exerted on an object. The chair's force and gravity's force are pushing in opposite directions. Gravity's force and the chair's force are balanced forces. **Balanced forces** are two forces that are equal in strength and opposite in direction as they act on an object. If the chair's force pushing up did not equal gravity's force pulling down, you would move upward or downward. With balanced forces, the object keeps doing what it is doing. Gravity keeps pushing you down and the chair's force keeps pushing you up. Balanced forces do not cause a change in the motion.

A man and a third-grade boy are exerting force on a door. The man exerts a big force on the door. The boy exerts a small force against the door. Are the two forces equal or unequal in strength? The two forces are unequal in strength. **Unbalanced forces** are two forces that are unequal in strength as they act on an object. Unbalanced forces do not need to be opposite in direction. When the forces are unbalanced, there is motion. The man's force will move the door toward the boy.

Balanced forces are shown by two equal-sized arrows.

Unbalanced forces are shown by two unequal-sized arrows.

230

- Direct attention to the two pictures.

  How can balanced forces be shown? by two equal-size arrows

  What do the two equal-size arrows show? The forces are equal in strength.

  Which direction do the arrows point? in the direction the force is being exerted

  How can unbalanced forces be shown? by two unequal-size arrows

  What do the two unequal-size arrows show? The forces are unequal in strength.

  Which arrow is bigger? the one showing the stronger strength

  Which direction do the arrows point? in the direction that the force is being exerted

Activity

Balanced and Unbalanced Forces

Materials: drawing paper or science notebook

Direct the student to draw a picture of balanced forces and a picture of unbalanced forces. Tell him to add arrows, showing how the forces are balanced or unbalanced. Students may also take digital pictures of students demonstrating balance or unbalanced forces.

Are the forces acting on the toy truck balanced or unbalanced?

A toy truck is sitting on a table. The forces pushing on the truck are gravity and the table's force. The forces are equal in strength. What kind of forces are equal in strength? Balanced forces are equal in strength.

Balanced Forces	Unbalanced Forces
Equal in strength	Not equal in strength
No change in an object's motion	Change in an object's motion

How can you tell that the forces acting on the toy truck are unbalanced?

What happens if you give the toy truck a push? The toy truck moves forward. The truck has motion. What kind of forces cause motion? Unbalanced forces cause motion. The forces acting on the toy truck are unbalanced.

A *spring scale* measures the strength of a force.

✔ **What is the difference between balanced and unbalanced forces?**

231

• Is there motion with the two students seesawing? Yes, there is a change in the seesaw's motion.

• Which forces have a change in motion? unbalanced forces

• Discuss the picture of the game board.
  Describe what you see. a game board with a player's piece or marker
  Are forces balanced or unbalanced? balanced

• Explain why they are balanced. There is no change in the motion of the game piece. Gravity and the game board's force are equal in strength.

• How could the forces become unbalanced? The game piece could be picked up and moved.

• Follow the same procedure with the last two pictures on the visual.
  soccer ball: There is no motion. The forces are balanced. Gravity and the ground's force are equal. The forces could become unbalanced by the foot kicking the soccer ball.
  dominoes: There is motion. Gravity and the finger's force are unequal in strength. The forces are unbalanced.

• Lead a discussion about how balanced and unbalanced forces are important.

• Why would learning about balanced and unbalanced forces be helpful? Possible answers: I could know that I need unbalanced forces to move something. Balanced forces would be needed to keep something in place. Knowing how these forces work would help me better serve other people.

• I want to help my elderly neighbor by sweeping her sidewalks. Will I be using balanced forces or unbalanced forces? unbalanced forces

• Give an example of how you can serve other people better by knowing about these forces. Possible answer: Unbalanced forces would be needed to weed someone's garden. Knowing about how the forces work would help me to use them properly to work better.

✔ **Balanced forces are equal in strength and unbalanced forces are unequal in strength. Balanced forces have no change in an object's motion. Unbalanced forces have a change in an object's motion.**

Apply

Activities

Study Guide, pages 187–90

These pages review the concepts taught in Lessons 73–74. After completion, direct the students to keep them for review for Test 10.

Assess

Quiz 10A

The quiz may be given at any time after completion of this lesson.

• Place the toy truck on a table.
  What forces are pushing on the truck? gravity and the table's force
  What kind of strength do the forces have? equal strength
  What kind of forces are equal in strength? balanced forces
  Do balanced forces have motion? No, there is no change in the truck's motion.

• Give the toy truck a little push.
  What happened to the truck when I gave it a little push? The truck moved forward. The truck has motion.
  Do balanced or unbalanced forces cause motion? unbalanced forces
  Are the forces acting on the toy truck balanced or unbalanced? unbalanced, because there is motion

• Guide a discussion of the picture of the spring scale.
  What is used to measure the strength of a force? a spring scale

• Direct attention to the chart on the page and the first picture on the *Balanced or Unbalanced?* visual.
  Describe what you see in the first picture. Two students are seesawing.

Objectives

- Predict how balanced and unbalanced forces will affect the motion of a ball.
- Observe and record data in a chart.
- Infer how balanced and unbalanced forces affect the motion of a ball.
- Apply to the game of bowling the concepts learned about balanced or unbalanced forces.

Materials

- See *Activities* page 191.
- science notebook, per student

> This Investigation will need to be done in groups of at least two students.

Teacher Resources

- Visual 3.3: *Science Inquiry Skills* (optional)

Engage

- Place a Ping-Pong™ ball on a table.
  What is the position of the ball? The position is on the table.
- Blow on the Ping-Pong ball so that it begins to move.
  Is the Ping-Pong ball in motion? Yes, the position of the ball changed.
  Today we will learn about balanced and unbalanced forces using a Ping-Pong ball.

Instruct

Preparation for Reading

- Direct the students to remove pages 191–92 from their *Activities* books. Direct them to read page 232 and *Activities* pages 191–92 silently before beginning.
- Explain that it is helpful to look over all the pages before starting an Investigation. Point out that it will help the students to see what they are going to do and when they will be writing information down.

Teach for Understanding

- What is a controlled investigation? It is an investigation that changes only one thing at a time; except for one thing, all parts of the investigation stay the same.
- What makes this investigation a controlled investigation? The one thing that changes is the forces acting on the ball.
- Direct attention to the **Inquiry Skills**.
  Which inquiry skills will you be using in this investigation? Predict, Observe, and Infer
- Review what it means to *predict*, *observe*, and *infer*. You may want to display the *Science Inquiry Skills* visual for this review.

Lesson 75

INVESTIGATION

Inquiry Skills
- Predict
- Observe
- Infer

Balanced and Unbalanced Forces

You have learned that *position* is the place where an object is. You have also learned that *motion* is a change of position. For example, a soccer ball is placed on the ground. Its position is the ground. If you kick the ball, the position changes. The ball is in motion because the ball's position changed.

In this Investigation, you will find out how balanced and unbalanced forces affect the motion of a Ping-Pong™ ball.

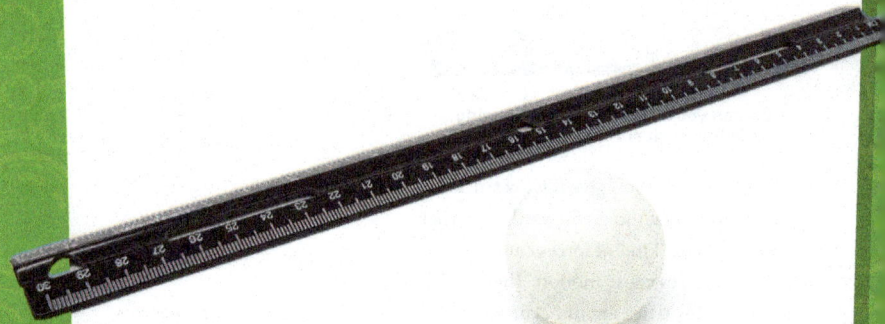

232

Balanced and Unbalanced Forces

Name _____

Problem

How do balanced and unbalanced forces affect the motion of a ball?

Hypotheses

If the forces are balanced, the ball

will have motion, will not have motion .

If the forces are unbalanced, the ball

will have motion, will not have motion .

Inquiry Skills

- Predict
- Observe
- Infer

Materials

- ☐ Ping-Pong™ ball
- ☐ ruler

Procedure

☐ 1. Place the ball on the floor.

☐ 2. Observe and record the motion of the ball in the data chart.

☐ 3. Complete the data chart for this position of the ball.

☐ 4. Push the ball gently with one finger while a team member pushes back gently with one finger in the opposite direction.

☐ 5. Observe and record the motion of the ball. Complete the data chart for this position.

☐ 6. Push the ball gently when a team member is not pushing back.

☐ 7. Observe and record the motion of the ball. Complete the data chart for this position.

☐ 8. Balance the ball on a ruler.

☐ 9. Observe and record the motion of the ball. Complete the data chart for this position.

☐ 10. Balance the ball on the ruler again. Tilt the ruler.

☐ 11. Observe and record the motion of the ball. Complete the data chart for this position.

- Discuss the **Problem**.

  What problem are you solving? How do balanced and unbalanced forces affect the motion of a ball?

  What ball will you test with? a Ping-Pong ball

- Direct the students to complete the **Hypotheses** to answer the problems. Instruct the students to circle their prediction for each force that will be tested. Remind the students that more than one hypothesis is called *hypotheses* with a long *e* at the end.

Apply

- Instruct the students to follow and check off **Procedure** Steps 1–11 as each is completed.

- Direct attention to the "Observations" data chart. Remind the students of the importance of accurately recording their **Observations**.
- Direct the students to draw **Conclusions** using their observations. The conclusions may be done individually or with their science group.
- Instruct the students to use their science notebooks and complete the "Notebook Connection," applying balanced or unbalanced forces to the game of bowling.
- Guide a discussion of the students' conclusions.
- Allow the students to present their examples and drawings of balanced or unbalanced forces in the game of bowling. Possible answers: balanced forces: pins sitting upright on the bowling alley, bowling ball resting in the ball return, person standing still at the beginning of the alley; unbalanced forces: person releasing a bowling ball, pins being knocked down by a bowling ball, pins scattering, bowling ball rolling down the alley

Activities
Investigation, pages 191–92
Students will predict the motion in balanced and unbalanced forces.

Assess

Rubric
Use the prepared rubric or design a rubric to include your chosen criteria.

Observations

	Motion of the Ball	Forces Acting on the Ball	Balanced or Unbalanced	How I Know It Is Balanced or Unbalanced
Sitting on the floor	no motion	gravity and the ball's force	balanced	because there is no motion
Pushing on either side with a finger	no motion	forces from both fingers	balanced	because there is no motion
Pushing with one finger	moves forward	finger's force and ball's force	unbalanced	because there is motion
Ball balanced on a ruler	no motion	gravity and the ball's force	balanced	because there is no motion
Ball on tilted ruler	downward motion	ball's force and ruler's force	unbalanced	because there is motion

Conclusions

1. When you pushed the ball on one side, what caused the ball to stop? _friction_
2. Did your observations support your hypothesis? ○ yes ○ no
3. I can conclude that balanced forces cause a ball to _not have any motion_.
4. I can conclude that unbalanced forces cause a ball to _have motion_.

Notebook Connection

Think about the game of bowling. What is an example of a balanced or unbalanced force that happens during the game? Draw a diagram of your example.

192 Lesson 75 · Page 232
Investigation

Lesson 76

Equations of Balanced and Unbalanced Forces

An equation, or number sentence, can show whether forces are balanced or unbalanced. What do these math symbols mean?

$$< \quad > \quad =$$

< less than
> greater than
= equal

These same symbols show whether two forces are balanced or unbalanced. Which symbol do you think you would use to show balanced forces? If you said the equal sign, you are correct. Notice the forces in the example with the two boys.

The boys are exerting the same strength of force. They are exerting force in opposite directions. Are the forces balanced or unbalanced? The forces are balanced. Look at the equation for these balanced forces below the picture.

Notice the forces in the example with the baby and the dog.

Are the forces balanced or unbalanced? The strength of both forces is unequal. The forces exerted by the baby and the dog are unbalanced. What symbol is used to show these unbalanced forces? Was your equation like the one below the picture?

10   10

10 = 10

5   15

5 < 15

233

Lesson **76** begins here.

Objectives
- Write equations for balanced or unbalanced forces on an object.
- Identify observed patterns of motion and unseen patterns of motion.
- Predict how an object will move next, based on observed patterns of motion.
- Identify patterns of motion that God made at Creation. **BWS**
- Explain why some motion cannot be predicted.

Materials
- 2 small balls with different masses (e.g., Ping-Pong™ ball and rubber ball; large and small marbles)
- large plastic or metal bowl
- stapler
- globe

Teacher Resources
- Visual 10.4: *Equations*

Engage
- Write ◇□○○◇□○○◇□○○ for display.
- What is this called? a pattern
- What is a *pattern*? a design or natural order that happens over and over again
- What pattern is repeated in the symbols? ◇□○○

- Write these numbers for display: 0, 5, 10, 15, 20.
- What number pattern do you see? counting by 5s, adding 5 to each number to get the next number
  Today we will learn more about patterns of motion.

Instruct

Preparation for Reading
- Direct the students to read pages 233–35 silently to find out what some unseen patterns of motion are.

Teach for Understanding
- What are *balanced forces*? two forces that are equal in strength and opposite in direction as they act on an object
- What are *unbalanced forces*? two forces that are unequal in strength as they act on an object
- Ask a volunteer to give an example of an equation or number sentence. Possible answers: 2 + 2 = 4, 20 − 5 = 15, 13 > 5
- Discuss the meaning of <, >, and =, using the box on the page.
- Why would you use the equal sign to show balanced forces? Balanced forces exert the same strength of force.
- Point out that the numbers in the picture of the boys playing tug of war show that the boys are exerting the same strength of force.
  What kind of forces have the same strength of force and are opposite in direction? balanced forces
- Write 10 ○ 10 for display.
  What symbol can be used to complete this equation? the equal sign
- Complete the equation. Direct attention to the picture of the baby and the dog.
  What is the strength of the force being exerted by the baby? 5
  What is the strength of the force being exerted by the dog? 15
  Are the forces equal in strength? No, they are unequal and unbalanced forces.
- Write 5 ○ 15 for display.
- What symbol can be used to complete this equation? the less than sign
- Direct the students to tear out *Activities* page 193. Display the *Equations* visual.
  What is happening in the first picture? A fish is being pulled from the water by a fishing pole.
  What is the fishing pole's strength of force? 20
  What is the fish's strength of force? 10
- What equation should we write to show the strength of these forces? 10 < 20 or 20 > 10
  What phrase should we mark to show the kinds of forces? unbalanced forces showing motion
- Instruct the students to complete the page, giving help as needed. Discuss the equations.

What is a *pattern*? a design or natural order that happens over and over

What are *patterns of motion*? The same motion happens over and over again.

- Discuss observations of motion.
- Analyze the picture across the top of the page.
- What pattern of motion do you observe in the ball practice drills? Possible answers: a zigzag pattern or an *S* pattern
- Ask the students to turn to page 224 and to observe the students seesawing.
- What pattern of motion do you observe with a seesaw? up and down
- Direct attention back to the picture of the girl playing hopscotch.
- What pattern of motion do you observe with the girl jumping in the hopscotch squares? one leg, two legs, one leg, two legs; one-leg hop, two-leg hop, one-leg hop, two-leg hop
- Discuss prediction of motion.
- What does the word *predict* mean? to make a careful guess at what may happen after observing; using what you know to say what may happen
- Display a large plastic or metal bowl. Place the lighter ball at the bottom of the bowl. Move the bowl in a circular motion while holding it. *Caution:* Moving the bowl too fast may cause the ball to fly out of the bowl.
  What is the pattern of the ball's motion? around and around the inside of the bowl
- Display the ball with a greater mass.
  What do you predict this ball's pattern of motion will be? Poll the students.
- Follow the same procedure with the heavier ball as you did with the lighter ball.
  How many correctly predicted the ball's pattern of motion? Poll the students.
- How were you able to predict the ball's pattern of motion? from observing the lighter ball
- Continue the discussion of predications of motion.
  Have you observed a ball at the top of a hill?
  What happened to the ball? It rolled down the hill.
- Direct attention to the picture of the human hamster ball.
  What pattern of motion do you predict the human hamster ball will have? roll down the hill
  How can you make this predication? because I have observed a ball rolling down the hill before
- Guide a discussion using an analog clock in the room.
  What pattern of motion do you predict the hands on a clock have? a circle
  Why can you make this prediction? because I have observed a clock's hands moving in a circle before

Patterns of Motion
Observations

Remember that a *pattern* is a design or natural order that happens over and over again. Patterns of motion happen when the same motion happens over and over again. You can observe patterns of motion. A merry-go-round goes around in a circle. The circle pattern happens over and over again. Some of the animals on a merry-go-round go up and down. The up-and-down pattern happens over and over.

Predictions

Have you played with a ball on a hill? If you place a ball at the top of a hill, you expect the ball to roll down the hill. You expect this motion to happen because you have observed a ball rolling down a hill. You can use what you observe to predict how an object will move next.

Look at a clock with hands. How do you predict the clock's hands will move? The motion of the hands will be in a circle. Why can you make this prediction? You have observed a clock's hands moving in a circle at an earlier time.

234

Differentiated Instruction

Enrichment: Balanced and Unbalanced Forces
Materials: *Activities* page 194, science notebook
The "Notebook Connection" on *Activities* page 194 extends the student's learning about balanced and unbalanced forces by observing more than one kind of force.

Enrichment: A Change of Forces
Materials: *Activities* page 194, science notebook
The "Notebook Connection" on *Activities* page 194 extends the student's learning about balanced and unbalanced forces by describing how forces change from unbalanced to balanced.

You are sitting on a swing on the playground. What pattern will the swing's motion be? The swing's motion is back and forth. You can predict the motion of the swing, because you have observed the swing move back and forth.

Unseen Patterns

These motions have been easy to observe. Some patterns of motion are harder to observe. Sound is made when something vibrates. The vibration has a back-and-forth pattern. You cannot observe this pattern. Sound also vibrates from the TV. The sound from the TV has a pattern, but you cannot observe the pattern.

You cannot see the wind blow. If you observe a flag waving, you can see the pattern of the wind. If it is a gentle breeze, the flag's motion will be slow. If it is a strong wind, the flag's motion will be fast. The flag helps you to observe the pattern of the wind.

What are some patterns of motion?

wind chimes

235

- Analyze the picture of the tire swing.
- What will the tire swing's pattern of motion be? Possible answers: back-and-forth, a circle

  How can you predict the tire swing's motion? because I have observed a swing before
- Display a stapler.
- What pattern of motion is the stapler going to have? up and down
- How do you know that the stapler has an up-and-down motion? I have observed the stapler having this motion.
- Point to the globe.
- What pattern of motion can you predict that the globe will have? a circle, round and round

  How were you able to predict the globe's motion? because I have observed a globe's motion before
- Point out that we have been able to observe all the motions we have discussed.

  Can you observe all patterns of motion? No, some are unseen patterns.
- Guide a discussion of the picture of the boy with the megaphone.
- Can you observe the pattern of sound coming from the megaphone? No, sound makes a back-and-forth pattern that cannot be seen.

  Can you observe the pattern of sound through a walkie talkie? No, sound makes a back-and-forth pattern that cannot be seen.
- Discuss the pictures of the wind blowing.

  What is happening in each of the bottom pictures? The wind is blowing.

  Can you see the pattern of the wind blowing? No, I can observe the wind chimes and flag being blown by the wind.

Possible answers: circle, zigzag, back-and-forth, up and down, rolling down a hill

Preparation for Reading

- Direct the students to read page 236 silently to find out what motions do not form a pattern.

Teach for Understanding

- Discuss created patterns of motion.

  What two patterns of motion did God design at Creation? revolution and rotation

  What pattern of motion is the revolution of the planets? The planets revolve around the sun.

  How do you know this is a pattern? It happens over and over again.

  What pattern of motion is the rotation of the planets? Each planet rotates on its axis.

  How is rotation a pattern? It happens over and over again.

- Discuss motion with no pattern.
- Guide an analysis of the predator and prey picture.

  Can you predict the pattern of motion for the jaguar and the antelope? No, you cannot predict how a predator or its prey will move or act.

  Why can you not predict the pattern of motion for a predator and prey? The forces of the predator and prey are acting in different directions. The motion does not have a pattern.

- Discuss each of the remaining pictures, asking whether there is a pattern of motion and why the pattern can or cannot be predicted. The motion of the girl with chalk and the motion of the player with the football do not have patterns.

✓ revolution and rotation

Apply

Activities

Equations, page 193

This page was completed as part of Lesson 76.

Created Patterns

God designed two patterns of motion in the solar system at Creation. Can you remember what they are? The two patterns of motion are *revolution* and *rotation*. God created the planets to revolve around the sun. The revolution of each planet around the sun happens over and over again. God created the moon to revolve around Earth. The moon's revolution happens again and again. God also designed the rotation of each planet. Each planet rotates on its axis. The rotation of each planet happens over and over again.

Motion with No Pattern

There are some motions that do not form a pattern. Because there is no pattern, the motion cannot be predicted. Think of a leopard chasing an antelope. Can you predict the predator's or the prey's motion? A toddler is splashing with his hands at a water play table. Can you predict the toddler's motion? Some motion does not have a pattern. When forces act in different ways or in different directions on an object, patterns are hard to predict.

predator and prey

✓ What patterns of motion did God make at Creation?

Lesson 77

Work

Suppose you were told to read five pages in your book. You read the pages and put down the book. You might think that reading the pages was hard work. A scientist, however, would say that reading is not work.

Scientists say **work** is done when a force moves something. Turning the pages of a book is work because you are moving something. Even moving your eyes across the pages is work. But understanding the words is not work. You are not moving anything.

sweeping

climbing stairs

folding laundry

237

⚠ Helps

Identifying Work

It is very difficult to find an activity that a child would do that does not involve some minor work, such as moving the eyes. For this reason, emphasize the big ideas rather than picking apart each activity to determine whether work is occurring or not. Asking a student to explain his answers will help you assess his understanding.

Objectives
- Identify what work is.
- Identify the living things or nonliving things that can do work.
- Determine what is doing the work in a scenario.
- Explain, using examples, how forces help in doing work better.

Vocabulary
- work
- machine

Engage

- Direct the students to stand by their desks and do jumping jacks until you tell them to stop.

  Have you ever heard or used the expression "that's hard work!" before?

  Do you think the jumping jacks were hard work?

  Name some things you do that you think are hard work. Possible answers: studying for a test, working in the yard, cleaning a room, doing homework, doing jumping jacks

- What do you call things that are not work? Possible answers: playing, resting, goofing off

  Work is a word that scientists use differently from the way we usually use it.

Instruct

Preparation for Reading
- Preview and pronounce the vocabulary term *work*.
- Direct the students to read pages 237–38 silently to find out when work is being done.

Teach for Understanding

Would a scientist say that thinking about a page in your science book is work? No, because nothing is moving.

What are some things you do when you are reading that are work to a scientist? Possible answers: turning or swiping the pages, eyes moving across the pages, picking up or putting down the book

When is work being done? when a force moves something

- Discuss the pictures, asking the students to identify when work is being done and explain what is moving. Possible answers: The boy in the hammock is not doing any work until he turns the page. The girl is sweeping, and the broom, her arms, and her legs are moving. The boys are climbing the stairs, and their legs are moving. The mother and son are folding laundry. Their hands and the clothes are moving.

- Direct attention to the picture of the girl in the library.
  If you are holding several books, are you doing work?
  No, because nothing is moving the books.
  If you drop the books, is work done? Yes, because the books move as they fall.
  What force causes the books to fall? gravity
  Is the girl in the picture doing work? Yes, she is walking. She is not doing any work with the books.
- Direct attention to the other pictures on the page. Allow volunteers to identify and describe the work being done.
  girl watering: watering plants, tilting the water can
  boy with the tablet: swiping screen
  parachutist: moving as he falls from the force of gravity
- Allow volunteers to describe other examples of work.

✔ **when a force moves something**

If you carry some books across the room, you are doing work. If you only hold the books, you are not doing work. Nothing is moving. Sometimes other forces do work. If you drop the books, you do not apply a force, but work is still done. The force of gravity is doing work and causes the books to move.

✔ **When is work done?**

parachutist

238

⚠ **Activity**

Work Charades
Allow several students to take turns silently acting out a person doing some kind of work. Allow the other students to guess what work is being done.

People can do work. You work when you move yourself or an object. Some animals work by carrying or pulling things.

People and animals are not the only ones who do work, though. Wind and moving water can also do work. Wind works when it moves wind mills, flags, or leaves. Flowing water moves logs, rocks, boats, and water wheels.

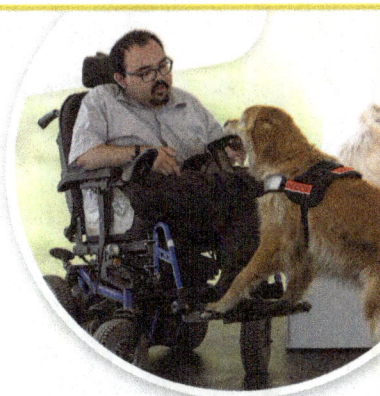

Hoover Dam

239

Preparation for Reading

- Preview and pronounce the vocabulary term *machine*.
- Direct the students to read pages 239–40 to find out five things that can do work.

Teach for Understanding

- Choose volunteers to name work that people can do. Possible answers: cleaning a sink, running the track, swinging, pushing a lawnmower, building with connecting blocks, completing a puzzle

  What are four other things that can do work? animals, wind, water, and machines

  What are some examples of work that animals can do? Possible answers: pulling a plow, carrying people or packs, pulling a dogsled

  What work can wind do? Possible answers: moving turbines, flags, leaves, sailboats, kites, or balloons

  What work can water do? Possible answers: moving logs, rocks, boats, water wheels, or soil

- Direct attention to the pictures on the page.

  Which picture shows a person doing work? The man in the wheelchair is handing the dog something. The man in the carriage moves the reins of the horses.

  Which picture shows water doing work? The turbines are being turned by the water at Hoover Dam.

  Which pictures show an animal doing work? The dog is bringing the wallet to the man, and the horses are pulling a carriage.

  Which picture shows wind doing work? The wind turbines and windmills in the fields are turned by the wind.

Differentiated Instruction

Enrichment: Five Things Doing Work

Materials: *Doing Work*, per student (Instructional Aid 10.1); *Doing Work Key* (Instructional Aid 10.2)

Direct each student to read the Bible verses and tell what is doing the work in each verse. This activity will help the student to determine what is doing the work in a scenario.

🔩 Why do we say that machines do work? They have moving parts and move things.

What provides the force to make machines work? People provide force for some machines, but most machines get their force from engines.

- Direct attention to the pictures. Discuss the name of the machine doing the work and the force doing the work. first picture: bikes, people; second picture: train, engine; third picture: construction crane, engine; last picture: tractor, engine

- Discuss other machines that make work easier.

- Draw a T chart for display, labeling the sides "Living" and "Nonliving."

What are the five things that can do work? As each one is mentioned, ask the next question.

Is this a living thing or a nonliving thing? (Write it under the correct category.) Living: people, animals; Nonliving: machines, wind, water

- Discuss how forces help in doing work better.

- Ask the students to share some examples of work they have done today. After a student gives an example, ask questions such as the following.

🔩 What forces did you use to brush your teeth? Possible answer: the force of my hand and arm

🔩 How did these forces help you work better? Possible answer: The forces made my work easier. The forces helped me clean my teeth.

🔩 What forces did you use to ride your bicycle to school? Possible answers: the force of a machine, friction, force of my legs

🔩 How did these forces help you work better? Possible answers: Friction helped me slow down and not run into things. The forces of a machine (bicycle) and of my legs helped me get to school faster than walking.

- Guide the students' understanding about the reason for studying forces and motion.

Why do you need to understand how forces and motion work? to work wisely in God's world

What is one way the knowledge of forces and motion will help you? Possible answers: to serve other people, to glorify God, to have dominion over the earth

✓ They help move people and things and make work easier.

Apply

Activities
Study Guide, pages 195–97

These pages review the concepts taught in Lessons 76–77. After completion, direct the students to keep them for review for Test 10.

Assess

Quiz 10B

The quiz may be given at any time after completion of this lesson.

Machines do a lot of work. Cars, trucks, tractors, and airplanes move people and things every day. These **machines** make work easier. But they still need forces to cause them to run. People provide the force for some machines, but most machines get their forces from engines. Engines help make the work easier for us.

The world is full of forces and motion. We need to understand each so we can work wisely in God's world. Learning about forces and motion will also give us the knowledge we need to serve other people.

✓ How are machines helpful?

240

👥 Differentiated Instruction

Enrichment: Forces
Materials: *Activities* page 198, science notebook

The "Notebook Connection" on *Activities* page 198 extends the student's learning about forces, combining different concepts together.

STEM

Lesson 78

A-Maze-ing Marble Game

Have you played a game using marbles? Did you move the marbles? Did the marbles move in some other way? Did the marbles move in a pattern? Have you tried to get a marble through a maze? A maze is an arrangement of pathways that a marble moves through.

In this STEM activity, you will design a maze that has the marble moving in two different patterns of motion. Your teacher will give you the patterns to include in your maze. To test the maze, allow someone else to try it. Can he see the patterns of motion?

241

Objectives

* Design a game path, using the engineering design process, that gives a marble two different patterns of motion.
* .Create a path with two patterns of motion for a marble.
* Test and compare models to improve the original design.
* Communicate how the redesign solves the problem.

Materials

* marble, per student
* variety of paper tubes
* ball of string
* ball of yarn
* paper clips of different sizes
* sheets of cardboard or posterboard
* sheets of paper
* tape
* glue
* dowels of different sizes
* boxes of different sizes

Teacher Resources

* Visual 2.7: *STEM: The Engineering Design Process*
* Instructional Aid 10.3: *Pattern of Motion Cards*

Copy the *Pattern of Motion Cards* enough times so that each student has two cards.

Engage

Have you ever played a game that used a maze?

* Discuss the picture of a maze.
  What do you see in the picture? Possible answers: corn, a maze
  What patterns do you see in the maze? Accept any reasonable answer with an explanation.
  Have you ever run through a cornfield maze?
* Explain that in today's lesson the students will design a maze with a marble moving in two different patterns of motion.

Instruct

Preparation for Reading

* Direct the students to remove *Activities* pages 199–200. Direct the students to read page 241 and *Activities* pages 199–200 silently before beginning.
* Remind the students that it is helpful to look over all the pages before starting a STEM activity. It will help them to see what they are going to do and when they will need to write information.

Teach for Understanding

* Display the *STEM: The Engineering Design Process* visual or direct the students to the copy of *Activities* page 35 in their science notebook.
* Review the design process, as needed.

⚠ Helps

Materials

Other materials may be added to the STEM activity. Lay out all the materials in one area rather than giving each student some of each. This will give the student the ability for greater creativity instead of the student trying to use each item.

Variation: You could substitute dominoes for this activity.

Apply

- Ask a volunteer to read the *Problem* aloud.
- Distribute two pattern of motion cards to each student.
- For this activity, each student will design his own solution to the problem, or the students may work in science groups to collaborate as they design a solution.
- Instruct the students to *Imagine* a solution to the problem and to *Plan* a design.
- Remind the students that the design needs to include two patterns of motion.
- Display the materials. Some materials will work better than others. Allow the students to choose materials that they believe will best solve the problem for the maze to have two patterns of motion.
- Direct the students to list and gather the materials needed for their design.

A-Maze-ing Marble Game   Name _____

Ask

What is the problem?

> A *maze* is an arrangement of pathways that a marble can move through. I want to design a maze that has the marble moving in two different patterns of motion.

How can I design a maze for a marble to move in two different patterns of motion?

Imagine

Think about possible answers to the problem.
Accept any reasonable answer.

_____

_____

Plan

Draw and label your design. List and gather materials.

The design should show a maze that allows a marble to move in the two different patterns of motion that were given to the team.

Design	Materials

SCIENCE 3 *Activities*

Create

Follow the plan and make a model.

Test and Improve

What changes need to be made to the design?
Accept any reasonable answer.

_____

_____

Improved Design

Share

Tell others how your design solves the problem.

Lesson 78 • Page 241
00 STEM

- Direct the students to *Create* a model of their design.
- Instruct the students to *Test* their model.
- After testing their designs, direct the students to compare models and *Improve* their design.
- Invite the students to *Share* their designs with a partner or another science group. Instruct the students to explain how their design solves the problem.

Activities
STEM, pages 199–200
The students will design a maze that has a marble moving in two different patterns of motion.

Assess

Rubric
Use the prepared rubric or design a rubric to include your chosen criteria.

Objective

• Recall terms and concepts from Chapter 10.

Apply

Review

• Material for Test 10 will come from the Study Guides on *Activities* pages 187–90 and 195–97, as well as Quizzes 10A and 10B. You may review any or all of the material during the lesson.

• You may choose to review Chapter 10 by playing the Tug of War review game or a game from the Game Bank (Teacher Resources).

🏆 **Review Game**

Tug of War

Materials: 20 ft string, 18 strips of paper, masking tape

Prepare the string by attaching the strips of paper at 12 in. increments. This is the tug-of-war "rope."

Divide the students into two teams. Direct the teams to stand on opposite sides of the room. Mark a dividing line on the floor between the teams with a strip of masking tape. Lay the string on the floor, from one team to the other, placing the center of the string on the dividing line.

For each correct answer to a review question, the team member pulls the rope one marker over the dividing line toward his team. At the end of the game, the team with the most markers on its side wins.

NOTES

NOTES

_____

_____

_____

_____

_____

_____

_____

_____

_____

_____

_____

_____

_____

_____

_____

_____

_____

_____

_____

_____

_____

_____

_____

_____

_____

_____

Objective

• Apply terms and concepts from Chapter 10.

Assess

• Administer Test 10.

Chapter Objectives

- Analyze the characteristics, properties, and benefits of electricity.
- Apply science inquiry skills of observe, predict, and infer.
- Analyze the characteristics, properties, and benefits of magnetism.
- Apply the steps of a scientific investigation to predict, observe, and infer the relationship between magnetic force and matter.
- Apply the steps of the engineering design process to solve a real-life problem.

Lesson Objectives

- Identify what electricity is.
- Relate atoms to electricity.
- Identify when an object is positively or negatively charged or when the object is neutral.
- Explain what static electricity is.
- Describe what happens when like charges and opposite charges are brought near each other.
- Analyze the effects that a charged object has on other objects.

Materials

- picture of socks (or other articles of clothing) attached to clothing by means of static electricity, for display
- resealable sandwich bag, per student for *Charged Objects* instructional aid cards. *Note:* The resealable bag will make for easier distribution of the cards.
- 9" balloons, 1 for teacher and 1 per student
- paper confetti, small amount per student
- salt, small amount per student
- pepper, small amount per student

Teacher Resources

- Instructional Aid 11.1: *Charged Objects*, per student

Vocabulary

- electricity
- electric charge
- static electricity
- repel
- attract

Engage

Chapter Introduction

- Direct attention to the chapter title and the picture. Ask the students to describe what they observe.
- Provide time for the students to leaf through the chapter, looking at the headings and pictures. Discuss what they think the chapter will be about.

> Use completed Looking Ahead, *Activities* page 201, to generate a discussion on what the students already know about electricity and magnetism.

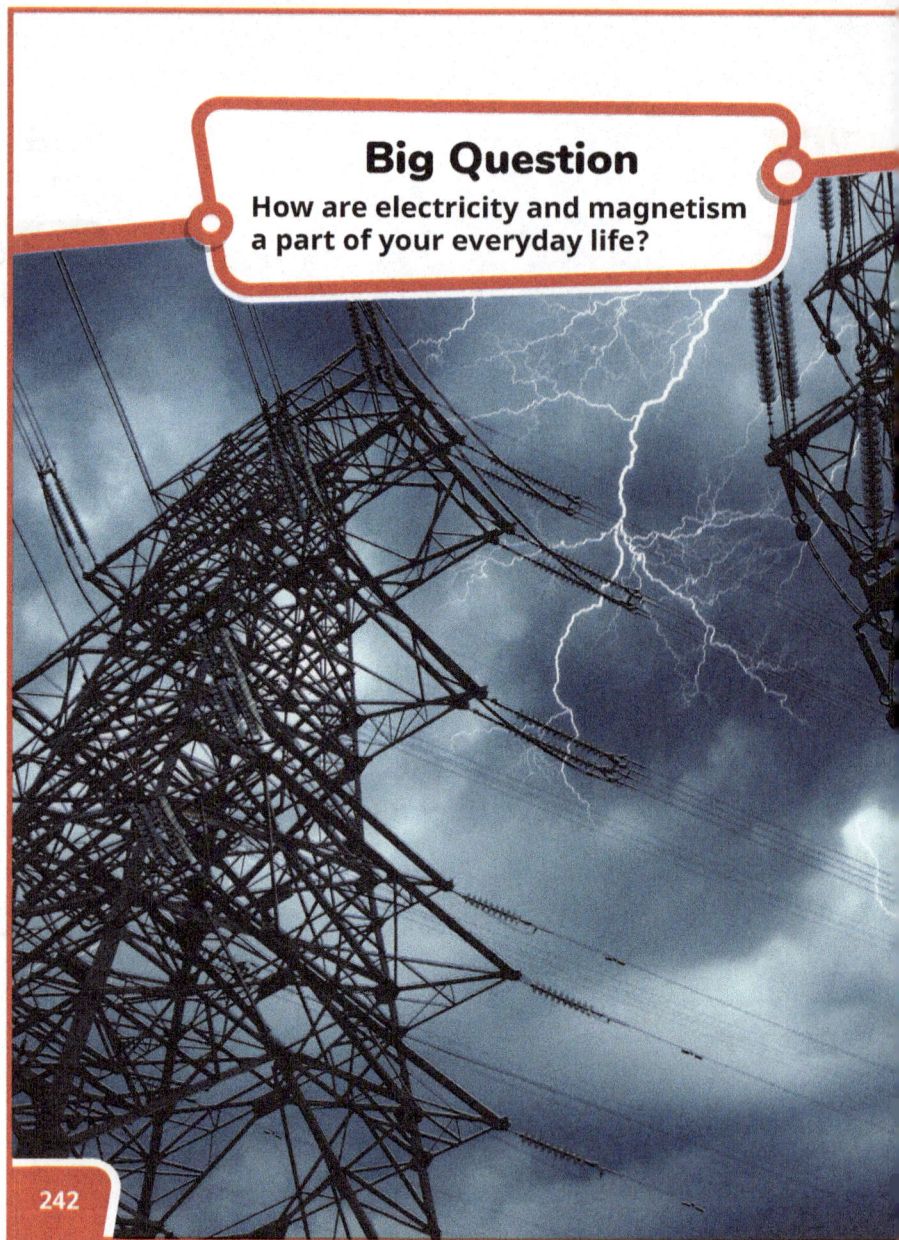

Big Question
How are electricity and magnetism a part of your everyday life?

242

Big Question

- Ask a volunteer to read aloud the Big Question.
- Explain that the students will find the answer to the question as they read the chapter.

> Visit TeacherToolsOnline.com for resources to enhance the lessons.

Chapter 11

Electricity and Magnetism

Think about some things you do every morning. Maybe you wake up to the sound of an alarm clock or a radio. You probably turn on a light. Perhaps you use a toaster as you fix your breakfast. You might listen to a weather report on the radio or television. All these things use electricity. Electricity is an important part of our daily lives.

Lesson Introduction

- Display the picture of the socks attached to clothing by means of static electricity. Do not use the term "static electricity" at this time.

  Have you ever had one piece of clothing stick to another piece of clothing?

  Do you think the power lines and lightning in the picture on pages 242–43 have anything in common with the socks that are attached to the clothing?

- Explain that today the students will learn how power lines, lightning, and the articles of clothing that stick together are all connected.

Instruct

Preparation for Reading

- Preview and pronounce the vocabulary terms *electricity* and *electric charge* and the word *neutral*.
- Direct the students to read pages 243–45 silently to find out what electricity is.

Teach for Understanding

- What are some ways you have used electricity since you woke up this morning? Possible answers: heard the sound of an alarm clock, turned on lights, used water that was heated by an electric water heater, used an electric stove or microwave oven to cook or heat up breakfast, dried hair quickly with a blow dryer, turned on the TV or radio to receive weather reports or news, charged cell phone battery
- What would your day be like if there were no electricity? Possible answers: no alarm clock, would use candles for light, water would be heated in small amounts on a wood stove or over a fire in a fireplace, hair would dry slowly, no TV or radio for weather reports or news, no cell phone
- Point out how important electricity is to our daily lives.

243

(i) Background

Chapter Opener Photo
The photo on pages 242–43 is high voltage towers and electric power lines in a lightning storm.

⚠ Helps

Paper Confetti
Paper confetti can be made by tearing a sheet of tissue paper or notebook paper into small pieces. A hole punch may also be used to make paper confetti.

What is *electricity*? a form of energy that involves the flow, or movement, of electric charges

What do many items use electricity for? to do work

What are *electric charges*? the particles of electricity that are found in everything

What are the tiny particles that make up matter called? atoms

What does each atom have? electric charges

What are the two types of electric charges? positive and negative

What does it mean that something is neutral? The object has the same number of positive charges and negative charges.

- Direct attention to the picture and caption. Explain that each plus sign (+) is a positive charge and each minus sign (–) is a negative charge. Ask a volunteer to read the caption.

  Answer the question in the caption. Neutral, because the pencil has the same number of positive and negative charges.

- Distribute the *Charged Objects* cards (Instructional Aid 11.1). *Note:* The cards will be used throughout the lesson.

- Direct the students to look at the cards.

  Hold up the cards that picture things that are neutral. the glass of water; the soccer ball

Electricity

Electricity is a form of energy that involves the flow, or movement, of electric charges. It provides a force that can make things move. Many items use electricity to do work. The particles of electricity are called **electric charges**.

Electric Charges

Remember that an atom is a tiny particle that makes up matter. Remember that even you are made up of atoms. Each atom has electric charges. An electric charge can be positive (+) or negative (–). A neutral object has the same number of positive and negative charges.

Count the number of positive charges and negative charges. Is the pencil positively charged, negatively charged, or neutral?

ⓘ Background

Atoms and Charges

The small particles that make up all matter are atoms. Each atom consists of electrons, protons, and neutrons. The protons have a positive charge, the electrons have a negative charge, and the neutrons have no charge.

Positive charges do not move. But negative charges can move from object to object. Rubbing an object causes it to gain or lose negative charges. When you rub your hair with a balloon, your hair loses negative charges. The balloon gains the negative charges from your hair.

Because your hair lost negative charges, it is no longer neutral. It now has more positive charges than negative charges. This means your hair is now positively charged.

The balloon is also no longer neutral. It gained negative charges. It now has more negative charges than positive charges. The balloon is now negatively charged.

Negative charges can move any time two objects rub together or touch each other. Negative charges can move when you comb your hair with a plastic comb. They can also move when you rub your feet across carpet.

Count the number of positive charges and negative charges. Why are the measuring cup and towel neutral?

Rubbing causes negative charges to move from the glass measuring cup to the dishtowel. Count the number of positive charges and negative charges. Which one is positively charged? Which one is negatively charged?

Which charges move, positive or negative?

245

Activity

Charged Objects

Using the *Charged Objects* cards, direct the students to match a negatively charged object with a positively charged object so that both become neutral. Encourage the students to explain whether the negative charges will move. For example, extra negative charges from the sock can be shared with the sweater so that both become neutral. Extra negative charges from the hairbrush can be shared with the girl's hair so that both become neutral.

Which kind of charge does not move? a positive charge

Which kind of charge can move from object to object? a negative charge

When can negative charges move? when two objects rub together; when two objects touch each other

When you rub your hair with a balloon, does your hair have a positive charge, a negative charge, or is it neutral? It is positive.

When you rub your hair with a balloon, which charges move from your hair to the balloon? The negative charges move to the balloon, because only negative charges can move.

The students will have the opportunity to use a balloon to create static electricity and observe the results that static electricity has on several objects, including their own hair, in the "Try It Yourself!" at the end of the lesson.

What does it mean for an object to be positively charged? It has more positive charges than negative charges.

- Direct attention to the *Charged Objects* cards (Instructional Aid 11.1).

Hold up the cards that picture things with positive charges. the girl's hair; the sweater

If an object has more negative charges than positive charges, what kind of charge does the object have? a negative charge

- Direct attention to the *Charged Objects* cards.

Hold up the cards that picture things with negative charges. the sock; the hairbrush

- Direct attention to the pictures of the bowl and towel on page 245. Choose a volunteer to read the caption of the first picture.

In the first picture, how many positive and negative charges do the bowl and towel each have? Each has five positive and five negative charges.

How would you describe the bowl and towel in the first picture? They are both neutral.

- Direct attention to the caption with the second picture.

What is different about the charges of the bowl and towel in this picture? The bowl has three negative charges, and the towel has seven negative charges.

Which one is positively charged, the bowl or the dishtowel? the bowl

Which one is negatively charged? the dishtowel

Where did the towel get the two additional negative charges? from the bowl

Why did the negative charges move from the bowl to the towel? The bowl and towel were rubbed together.

- Use the Activity to continue discussing charges and attractions of the objects on the cards.

negative

Preparation for Reading

- Preview and pronounce the vocabulary terms *static electricity*, *repel*, and *attract*.
- Direct the students to read pages 246–47 silently to find out when objects attract each other.

Teach for Understanding

When does static electricity occur? when electrical charges build up on the surface of an object

What are some activities that can cause static electricity? Possible answers: walking across a carpet, combing hair, rubbing things together

Would static electricity be more likely to happen on a warm, rainy day or on a cold, dry day? cold, dry day

What do the electric charges on an object cause? an electric force

What is a force? a push or a pull

- Direct attention to the pictures of the balloons and their captions.

What are the charges doing when they *repel* each other? pushing away from each other

What are the charges doing when they *attract* each other? pulling toward each other

Under what conditions do objects repel each other? when they have the same charge

- Direct attention to the picture of the girl on the slide.

Why are the hairs on the girl's head pushing away from each other? The positively charged hairs are repelling each other.

Why are the hairs on the girl's head being pulled toward the slide? The positively charged hairs are being attracted to the negatively charged slide.

> Teaching for page 247 begins here.

Objects that have positive charges will repel each other. If both objects have negative charges, will they attract or repel each other? They will repel each other because the charges are the same, and like charges repel.

Under what conditions do objects attract each other? when the charges are opposite, or different

- Display the picture used in the lesson introduction of the socks attached to the clothing.

Why do clothes sometimes stick together when they come out of a dryer? Some pieces of clothing lose negative charges and become positively charged. Other pieces of clothing gain negative charges and become negatively charged. When opposite charges come near each other, they attract.

Why do you sometimes feel tiny shocks or see tiny flashes of light when you pull apart clothes that are full of static electricity? The negative charges jump from the negatively charged clothes to the positively charged ones.

- Point out that lightning is a giant spark of static electricity.

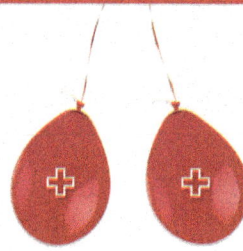

Like charges repel.

Opposite charges attract.

Static Electricity

When you walk across the carpet, your shoes collect negative charges. These charges build up on your body. Now, when you touch a doorknob, the negative charges jump from your hand to the doorknob. This jump can cause a small spark and mild shock. The spark and shock are caused by static electricity. **Static electricity** occurs when electric charges build up on the surface of an object. This happens more often when the air is cold and dry.

The electric charges on an object cause an electric force. Sometimes the charges **repel**, or push away from, each other. At other times, the charges **attract**, or pull toward, each other.

Objects with the same charge repel each other. When you comb your hair, the comb collects negative charges. The hairs lose negative charges to the comb and the hairs become positively charged. Each positively charged hair repels the other positively charged hairs. This repelling is what makes your hairs move apart. The negatively charged comb attracts the positively charged hairs. This attraction is what makes the hairs move toward the comb. If you rub two balloons with a piece of wool, each balloon gains negative charges. When you then try to bring the balloons close to each other, they repel each other.

246

Differentiated Instruction

Remediation: Explore Static Electricity

Materials: several plastic or nylon combs, wool (optional), several table-tennis balls

Give a student a plastic comb and a table-tennis ball. Direct him to comb his hair quickly about 15 times. If his hair is too short, he can rub the comb on a piece of wool. Direct the student to place the ball on a table, hold his comb near the ball, and then slowly move the comb away from it. The ball should be attracted to the comb and follow it.

Why does the ball follow the comb? The comb has become negatively charged and is trying to gain positive charges from the ball.

Allow several students to race their balls to see which comb has the most static electricity.

Enrichment: Build an Electroscope

Materials: bendable drinking straw, lump of clay, 1" x 6" strip of tissue paper or lightweight aluminum foil, tape, inflated balloon, piece of wool or fur

A calibrated electroscope can be used to measure the amount of static electricity in an object. The simple electroscope in this demonstration shows whether static electricity is present but cannot measure the amount.

Objects with opposite charges attract each other. The results of this can sometimes be seen when clothes come out of a dryer. The clothes touch each other as they tumble in the dryer. Some pieces of clothing lose negative charges and become positively charged. Other clothes gain negative charges and become negatively charged. The clothes with opposite charges stick to each other.

As you pull apart clothes that are full of static electricity, you may feel tiny shocks. You may even see little sparks. These effects happen as the negative charges jump from the negatively charged clothes to the positively charged ones. Once this occurs, both pieces of clothing have the same number of positive and negative charges. The clothes are now neutral.

Try It Yourself!

Blow up a balloon. Rub the balloon on your hair. Observe what happens when you hold the balloon near your hair.

Rub the balloon on your hair. Hold the balloon near a wall. Let go of the balloon. What happens?

Tear a piece of paper into tiny pieces. Rub the balloon on your hair. Hold the balloon over the small pieces of paper. What happens?

Sprinkle some salt and pepper on a piece of paper. Rub the balloon on your hair. Hold the balloon over the salt and pepper. What happens?

✔ **What happens when two objects with different charges are brought near each other?**

247

Once the clothes have been pulled apart, why are they no longer attracted to each other? Both pieces of clothing have the same number of positive and negative charges. The clothes are now neutral.

- Direct attention to the "Try It Yourself!" box. Distribute the materials needed to each student. Direct the students to follow the directions carefully. Discuss with the students their observations.

Why is your hair attracted to the balloon? The hair and balloon have opposite charges. When the balloon rubbed the hair, negative charges moved from the hair to the balloon. The positively charged hair is attracted to the negatively charged balloon.

Why did the balloon stick to the wall? The balloon and the wall have opposite charges. When the balloon rubbed the hair, negative charges moved from the hair to the balloon. The negatively charged balloon is attracted to the positively charged wall.

Why did the balloon lift the paper confetti? The balloon and the paper confetti have opposite charges. When the balloon rubbed the hair, negative charges moved from the hair to the balloon. The negatively charged balloon is attracted to the positively charged paper.

Why did the balloon lift the salt and pepper? The balloon and the salt and pepper have opposite charges. When the balloon rubbed the hair, negative charges moved from the hair to the balloon. The negatively charged balloon is attracted to the positively charged salt and pepper.

- Explain to the students that the balloon may not pick up as much salt as it did pepper because the salt is heavier than the pepper.

✔ **They attract each other.**

Apply

Activities
Looking Ahead, page 201
This page assesses the students' knowledge prior to the chapter.

Charge It!, page 202
This page reinforces the concepts taught in Lesson 81.

👥 Differentiated Instruction

Enrichment: Build an Electroscope (continued)
Push the bottom of the straw into the lump of clay. Press the clay onto a flat surface so that the straw stands upright. Bend the top of the straw parallel to the flat surface. Fold the tissue paper or foil in half and drape it over the bent part of the straw so that the ends of the paper are even. Tape the tissue paper to the straw. This is a simple electroscope.

Rub the balloon with the wool or fur. Once the balloon is charged, bring it close to the tissue paper or foil from underneath.

What happens to the tissue paper or foil? The tissue paper or foil will move toward or away from the balloon, indicating a charge is present.

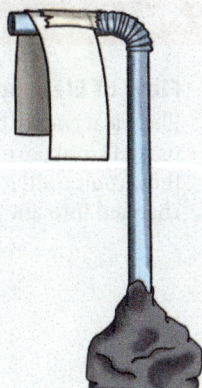

⚠ Activity

Try It Yourself!
Materials: balloon, paper confetti, salt, and pepper
Allow the students to try this activity on their own.

Objectives

- Identify what current electricity is.
- Differentiate between static electricity and current electricity.
- Compare and contrast conductors and insulators.
- Differentiate between a closed simple electric circuit and an open simple electric circuit.
- Explain, using biblical teaching, why it is beneficial to understand conductors and current electricity. **BWS**

Materials

- toaster
- bread
- insulated wire with some insulation removed
- pliers or wire cutters with insulated grips

Teacher Resources

- Instructional Aids 11.2–11.3: *Indoor Safety Counts!*; *Outdoor Safety Counts!*, per student

Vocabulary

- current electricity
- conductor
- insulator
- circuit
- closed circuit
- open circuit

Engage

- Display the toaster and bread. Try toasting the bread without plugging in the toaster.
- Ask volunteers to explain why the toaster will not toast the bread. The toaster needs to be plugged into an electric outlet to work.
- Explain that today they will learn about the kind of electricity that allows a toaster and other electric appliances to work.

Instruct

Preparation for Reading

- Preview and pronounce the vocabulary terms *current electricity*, *conductor*, and *insulator*.
- Direct the students to read pages 248–49 silently to find out what materials are called that do not allow electricity to flow through them easily.

Teach for Understanding

Why does static electricity last for only a short period of time? The negative charges move all at once.

- Explain that once the negative charges in static electricity are balanced (neutral), they stop moving.

What is *current electricity*? the continuous movement, or flow, of negative charges through a material

What are some ways you have used current electricity today? Possible answers: turning on lights, charging a cell phone battery, running a computer, toasting bread, heating food in a microwave

What is a *conductor*? a material that allows electricity to flow through it easily

What are some good conductors? Possible answers: metals, such as copper and silver; our bodies

Current Electricity

All electricity happens because negative charges move. In static electricity negative charges build up on one material and jump to another material all at once. This happens for brief periods of time. In **current electricity** the negative charges have continuous movement, or flow, through a material.

Energy produced by current electricity can be used in many ways. It can light a bulb in a lamp. It can charge the battery in a cell phone. It can also run a computer or toaster.

Current electricity is the continuous flow of negative charges through a material. Electric current can flow, or move, through many types of materials. A **conductor** is a material that allows electricity to flow through it easily. Metals, such as copper and silver, are good conductors.

However, metals are not the only conductors. Your body can also conduct electricity. That is why you need to be careful when you use electricity.

248

Why do we need to be careful when we use electricity? Our bodies are good conductors. Electric shocks can harm or even kill a person.

- Explain that water contains dissolved minerals and other substances that conduct electricity. The high amount of water and minerals in our bodies is what makes them such good conductors. Discuss the importance of not using electricity around water.
- Display the wire and pliers. Discuss how the insulation helps to make them safer. Kinds of insulation on tools include rubber, plastic, and fiberglass. If possible, identify the materials used on your tools.

⚲ Misconception

Flow of Electricity

Electrical energy is not carried by individual moving particles. Electrical energy is passed from particle (electron) to particle (electron). The energy is passed along the circuit until it reaches an electrical device, where the electrical energy is changed into another form of energy.

Sometimes, though, electricity cannot flow easily through a material. An **insulator** is a material that does not allow electricity to flow through it easily. Wood, glass, rubber, fiberglass, and plastic are all good insulators.

Whenever current electricity is moving through a wire, that wire is called a live wire. Touching a live wire might shock you. If the shock is strong enough, it might even cause death. So a live wire is usually covered with an insulator to prevent injury. The covering is called insulation. Most electric cords have plastic or rubber insulation covering the wires.

conductors

insulator

Power lines use insulators.

What is current electricity?

• Explain to the students that although many materials are good insulators, nothing completely insulates the flow of electricity. Extra care is always needed when working with electrical items.

the continuous flow or movement of negative charges through a material

249

Preparation for Reading

- Preview and pronounce the vocabulary terms *circuit*, *closed circuit*, and *open circuit*.
- Direct the students to read pages 250–51 silently to find out what happens to electricity if a circuit is open.

Teach for Understanding

What two things are needed for electricity to flow? a power source and a path

What is a *circuit*? an unbroken path for negative charges to flow through

- Direct attention to the pictures of the closed and open circuits on page 251.

What is the power source for each of these circuits? a battery

Why is the light bulb lit in the closed circuit? There is an unbroken, or a complete, path for the electricity to flow through.

Why is the light bulb not lit in the open circuit? Electricity cannot flow through an open, or incomplete, circuit.

Electric current cannot flow without a power source. A battery is an example of a power source. The power source keeps the negative charges flowing. Electricity also needs a path. The path must be unbroken for the negative charges to flow through. This path is called a **circuit**.

What if you want to light a bulb using a simple electric circuit? You will need to connect wires to the positive and negative ends of a battery. You will also need to connect the wires to the light bulb. The current electricity will flow through the wires and the bulb will light up in a **closed circuit**. If one wire along the circuit is not connected, the circuit is open. Electricity cannot flow through an **open circuit**. The bulb will not light up when the circuit is open.

Health and Safety

People are made in God's image (Genesis 1:26–27). We are important to God. Learning about conductors and circuits helps to keep people safe. Children must not play near electric outlets because an electric outlet is an open circuit. What would happen if a child put something that conducts electricity into the circuit? He would close the circuit and receive an electric shock. Studying circuits makes people aware of the dangers. It can help to keep us safe.

250

Activity

Model Static Electricity and Current Electricity

Materials: 11 craft sticks, a group of two students, a group of three students

Explain that for this activity, the craft sticks represent negative charges.

In the group of two students, give one student six craft sticks and give the other student two sticks. Direct the students to stand and simultaneously shake hands and exchange sticks until both have the same number. As soon as they have the same number of sticks, they should stop shaking hands and stand still.

In the group of three students, give one student three craft sticks. This student is the "battery." Direct the students to stand in a circle with one hand touching the next person. Direct them to pass the craft sticks around the circle at a steady pace. Direct them to continue passing the sticks until you remove the "battery" from the circle. Then the others should stop and stand still.

Which group represents static electricity? The group of two students represents static electricity because their charges stopped moving when they got the same amount, or became neutral.

Which group represents current electricity? The group of three students represents current electricity because their charges kept moving through a complete path (circuit) until the battery (power source) was removed.

A circuit might have a switch. The switch is used to open or close the circuit. The switch is a conductor that moves. When the switch is open, it breaks the path. This stops the flow of electricity. When the switch is closed, the circuit is complete and electricity can flow.

Simple Electric Circuit

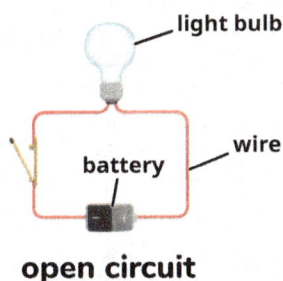

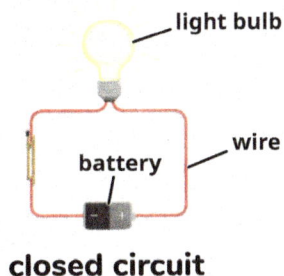

light bulb — light bulb

wire — wire

battery — battery

open circuit — **closed circuit**

Fantastic Facts

Have you ever noticed a squirrel running across electric power lines? Have you ever wondered why the squirrel does not get an electric shock? The answer is that the squirrel is touching only one wire. By touching only one wire, the squirrel is not providing a path for the electricity to travel through. The squirrel is not completing the circuit. If the squirrel were touching two wires or touching the wire and the ground, the electricity would flow through the squirrel. If electricity flowed through the squirrel's body, it would very likely die from the electric shock.

✓ **What happens to electricity if a circuit is closed?**

251

What might a circuit have? a switch

How is a switch useful? It opens or closes a circuit.

Describe how a switch works. When a switch is open, it breaks the path and stops the flow of electricity. When the switch is closed, the circuit is complete and electricity can flow.

- Point out that the switch is a movable conductor. Remind the students of what a conductor is and how the switch completes the circuit when it is closed.
- Direct attention to "Health and Safety" on page 250.

⚙ Why is it important to understand how current electricity works? Possible answer: People are made in God's image (Genesis 1:26–27), and learning about conductors and circuits helps to keep people safe. Keeping people safe pleases God. The human body is a conductor. Knowing this will help to avoid electric shock and will keep people safe around current electricity.

- Display and distribute the *Indoor Safety Counts!* and *Outdoor Safety Counts!* instructional aids. Review the safety rules for inside and outside the home with the students. Allow time for the students to contribute additional ways to be safe around electricity.
- Discuss "Fantastic Facts."

Why can a squirrel run across an electric power line and not be shocked? The squirrel is touching only one wire. The squirrel is not completing the circuit.

✓ **Electricity will flow through the wires of a closed circuit.**

Apply

Activities

Study Guide, pages 203–4

These pages review the concepts taught in Lessons 81–82. After completion, direct the students to keep them for review for Test 11.

Assess

Quiz 11A

The quiz may be given at any time after completion of this lesson.

Objectives

- Predict whether objects are conductors or insulators.
- Build a circuit tester.
- Identify which objects complete the circuit and which do not.
- Infer which materials are conductors and which are insulators.

Materials

- See *Activities* page 205.

Engage

How do you turn on the lights in your bedroom? by using a switch

What kind of circuit lights the bulb in the lamp—an open or a closed circuit? a closed circuit

What is a circuit? an unbroken path for negative charges to flow through

What does the switch in the lamp act as to make a closed circuit? a conductor

What is a conductor? a material that allows electricity to flow through it easily

- Explain that in this Exploration the students will use what they know about circuits, conductors, and insulators to predict and infer which items are conductors.

Instruct

Preparation for Reading

- Direct the students to remove pages 205–6 from their *Activities* books. Direct them to read page 252 and *Activities* pages 205–6 silently before beginning.

Lesson 83

EXPLORATION

Conductors Needed

Inquiry Skills

- Predict
- Observe
- Infer

It is a dark and stormy night! You and your family are at home. The power goes out. Your father lights a small candle to help you find your flashlight. When you turn on the flashlight's switch, you discover that it will not come on. You open the flashlight and notice that a wire to the switch is broken. The ends of the wire cannot be reattached. In order for the flashlight to work, you must complete the circuit. Can you find something to complete the circuit?

Knowing about conductors can help you complete the circuit and have light during the storm. In this Exploration, you will identify which items are conductors.

252

Conductors Needed

Name _____

Purpose

I want to complete a circuit. I want to test several objects to find out which objects will complete the circuit.

Procedure

☐ 1. You have 5 items to test. Choose 2 additional items to test. Record the additional items on the Materials list and the Observations chart.

☐ 2. Predict whether each item will complete the circuit. Record your predictions on the chart by checking the box.

☐ 3. Build a circuit tester using the following steps:

Inquiry Skills

- Predict
- Observe
- Infer

Materials

☐ 3 pieces of insulated wire, each 20 cm long
☐ miniature lamp and socket with 2 terminals (1.2 volt bulb)
☐ battery (1.5 volt C or D cell)
☐ masking tape
☐ aluminum foil
☐ button
☐ craft stick
☐ metal key
☐ penny
☐ _____
☐ _____

Exploration Setup

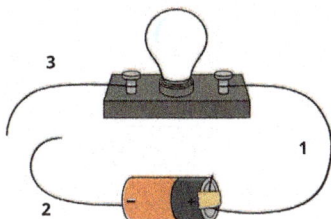

☐ Cut 3 pieces of insulated wire into 20 cm lengths.

☐ Remove 1 cm of insulation from each end of each wire.

☐ Attach wire 1 to the positive end of the battery and wire 2 to the negative end of the battery, using masking tape.

☐ Attach the other end of wire 1 to the light bulb.

☐ Attach wire 3 to the light bulb.

Teach for Understanding

- Direct attention to the **Inquiry Skills**.

  What inquiry skills will you be using in this Exploration? Predict, Observe, and Infer

- Review the inquiry skills of *predict*, *observe*, and *infer*. Refer the students to the Inquiry Skills page in their science notebook, if needed.

- Direct attention to the **Purpose**.

  What is the purpose of this Exploration? to use a circuit tester to test several objects to find out which objects will complete the circuit

- Direct attention to the **Materials**. *Note:* To reduce the number of materials needed for this Exploration, you may choose to have the students work in groups.

- Point out that the students are to choose two additional items to test in the circuit tester and they are to write each item on a blank line. You may wish to have each group test two different items.

Apply

- Direct attention to the list of **Procedures**. Remind the students to follow each step. Encourage the students to check off each step as it is completed. Direct the students to continue with the Procedure.

- You may need to demonstrate how to assemble the circuit tester. Explain that the noninsulated copper part of the wires should touch the battery, light, and objects to test.

- Caution the students to work carefully. Pulling too hard may pull the tape off the battery.

- When the student is testing the foil, you may suggest that he roll the foil into a ball or twist it into a rope.
- Point out that the students will record their **Observations** on the data chart.
- Direct the students to draw **Conclusions** using their observations.
- Guide a discussion of the students' conclusions.

Activities

Exploration, pages 205–6

Using a circuit tester, the students will predict, observe, and infer which objects will act as conductors by completing the circuit.

Assess

Rubric

Use the prepared rubric or design a rubric to include your chosen criteria.

☐ 4. Gently touch the 2 unattached ends of the wires together. The light bulb should glow. This is your circuit tester. To test an item, you will place it between the 2 unattached ends of the wires. If the light bulb glows, the item completes the circuit and is a conductor.

☐ 5. Place the aluminum foil between the 2 wires. Each end of the wire should be touching the foil. Determine whether it completes the circuit and is a conductor. Record your observation.

☐ 6. Repeat step 5 with the button, craft stick, metal key, penny, and two objects of your choice. Record your observations after testing each item.

Observations

Items	Predictions		Results	
	Will the object complete the circuit?		Did the object complete the circuit?	
	Yes	No	Yes	No
aluminum foil ball				
plastic button				
wooden craft stick				
metal key				
copper penny				
paper clip, staples,				
pencil, pen, marble, etc.				

Conclusions

1. Which item(s) completed the circuit? *Answers should include foil, key, penny, and possibly one of their chosen items.*

2. What are the items called that completed the circuit? *conductors*

3. What material are all the conductors that were tested made of? *metal*

4. What are the items called that did not complete the circuit? *insulators*

5. What materials are the insulators that were tested made of? *Answers should include wood, plastic, and any of the chosen items.*

Magnetism

What do you think keeps your refrigerator door closed? It is the same thing you may use to hold pictures or notes on the outside of the door. Most refrigerators have magnets in the seal around the inside edge of their doors. These magnets help keep the door closed so the cool air stays inside the refrigerator.

Magnets

A **magnet** is an object that attracts certain metals. It attracts metals such as iron, steel, and nickel. Objects made of wood and paper are not attracted to magnets.

Magnets can be different shapes and sizes. You may be most familiar with bar magnets. These magnets are shaped like a rectangle. However, there are also magnets that are horseshoe-shaped or circular in shape.

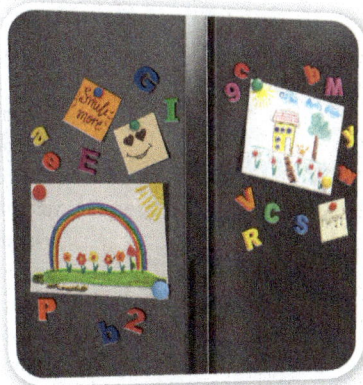

magnets

253

Objectives

- Recall what a magnet is.
- Explain why magnets can act at a distance.
- Explain how the invisible magnetic field can be "visible."
- Identify the areas of a magnet with the strongest magnetic force.
- Analyze the magnetic forces of attract and repel.
- Explain how magnets are beneficial. **BWS**

Materials

- 2 bar magnets, per group of students
- paper lunch bag with bar magnet inside
- paper clips
- iron filings
- sheet of white paper
- plastic bag or plastic wrap
- document camera (optional)
- variety of magnets (bar, horseshoe, disc)

Vocabulary

- magnet
- magnetism
- magnetic field
- pole

Engage

- Without the students seeing, place the bar magnet in the paper bag. Scatter several paper clips on a table. Direct the students' attention to the paper clips as you move the bag toward them.

  What happened to the paper clips? Some of them moved toward the bag.

  What do you think is in the bag? a magnet

- Explain that in today's lesson the students will learn about magnets and their force. *Note:* The paper bag with the bar magnet and the paper clips will be used again in Lesson 85.

Instruct

Preparation for Reading

- Preview and pronounce the vocabulary terms *magnet*, *magnetism*, and *magnetic field* and the word *circular*.
- Direct the students to read pages 253–54 silently to find out what a magnetic field is.

Teach for Understanding

What is a *magnet*? an object that is able to attract certain metals

What kinds of metals are attracted to magnets? iron, steel, nickel

What kinds of things are not attracted to magnets? Possible answers: wood, paper, plastic, rubber, cotton

- Review with the students the various types of magnets in the picture. Explain that some magnets are rectangular (bar magnets), while others are horseshoe-shaped or even round (disc magnets). Point out that the disc magnets are red on one side and blue on the other side to identify the north and south poles.
- Provide time for the students to test items around the room to observe whether or not they are magnetic. Guide the students in making generalizations about the materials the items are made of.

? Misconception

Magnetic Metals

Magnets do not attract all types of metal. Iron, nickel, and cobalt are magnetic metals. *Nickel* refers to both the element nickel and the five-cent coin. The coin is made from a mixture of metals, however, and does not contain enough of the element nickel to be magnetic.

ⓘ Background

Peter Peregrinus of Maricourt

Peter Peregrinus of Maricourt was a French scholar who wrote about magnetism in 1269. He wrote a letter describing everything that was currently known about magnetite, a magnetic rock. He was the first to identify the poles of a magnet, realizing that unlike poles attract and like poles repel. He established that fragments of magnetite could act like magnets. He also designed two compasses. His letter was later published under the title *Epistola de magnete* and used as a resource by scientists.

What force does every magnet have? a pushing or pulling force on certain objects; magnetism

What is the pushing or pulling force caused by a magnet called? magnetism

Why does a piece of iron not need to touch the magnet to be pulled toward the magnet? because of the magnet's magnetic field

What is a *magnetic field*? the area around a magnet where the force of the magnet can act

Can the magnetic field be seen? No, it is invisible.

What can you do to make the invisible magnetic field visible? Sprinkle iron filings around the magnet and observe the pattern the iron filings make.

• Place a bar magnet under a sheet of white paper. Sprinkle some iron filings on the paper. You may use a document camera to display the demonstration. *Note:* Iron filings are difficult to remove from magnets. To make cleanup easier, place the magnet in a plastic bag or plastic wrap before you use it with the iron filings.

What does the pattern made by the iron filings on the paper show? the magnetic field

the area around a magnet where the force of the magnet can act

Every magnet has a pushing or pulling force on certain objects. The pushing or pulling force caused by a magnet is called **magnetism**. If you place a magnet near a piece of iron, the magnet and the metal pull toward each other. Neither object needs to touch for the metal to be pulled toward the magnet. The area around a magnet where the force of the magnet can act is called the **magnetic field**.

A magnetic field cannot be seen. It is invisible. But you can use small pieces of iron to see the magnetic field formed by the magnetic force. You could put a bar magnet on a sheet of white paper and sprinkle iron filings on the paper. The iron is attracted to the magnet. The pattern made by the iron filings shows the magnetic field for the magnet.

magnetic field

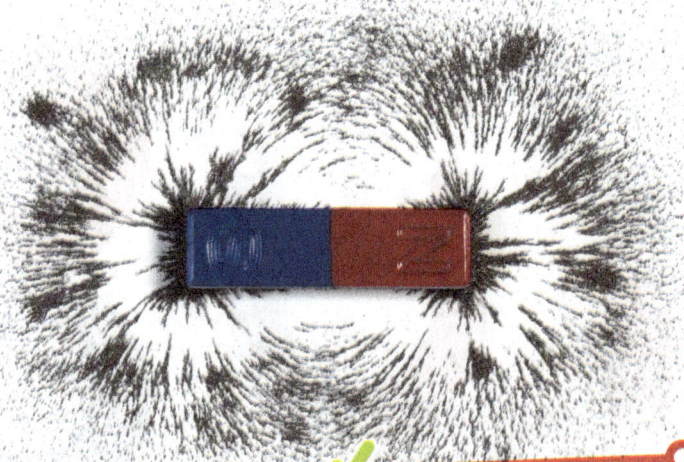

What is the magnetic field?

Most of the iron sticks to the **poles**, or ends of the magnet. This is because a magnet's magnetism is strongest at its poles. Every magnet has a north pole (N) and a south pole (S). If you put two magnets near each other, the opposite poles attract. The north pole of one magnet will pull toward the south pole of the other magnet. But if you try to put two south poles together or two north poles together, the magnets repel each other.

Science and History

Magnets can be used to find direction. The earth is like a giant magnet. It has a magnetic north pole and a magnetic south pole. There are also magnetic rocks on the earth. Ships once sailed across the oceans with a compass that used a piece of magnetic rock. This rock is called lodestone. The lodestone pointed to the magnetic north pole. The sailors used the lodestone compasses to find direction. Today, a compass has a small magnetic needle and a dial. The magnetic needle always points to the magnetic north pole.

Try It Yourself!

Using two bar magnets, place the two south poles (S) near one another. What happens?

Turn one bar magnet around and place the north (N) and south (S) poles near one another. What happens?

Place the two north (N) poles near one another. What happens?

255

Preparation for Reading

- Preview and pronounce the vocabulary term *pole*, as well as the word *lodestone*.
- Direct the students to read pages 255–56 silently to find out what some magnets are used for.

Teach for Understanding

What do you notice about the pattern of iron filings? Some areas are darker than others. There are more filings in some places.

Where are most of the filings? around the poles

What are the *poles* of the magnet? the ends of the magnet; the north pole and the south pole

Why are most of the filings around the poles? That is where the magnet's magnetism is the strongest.

What happens when the north poles of two magnets are put together? The magnets repel each other.

- Direct attention to the "Try It Yourself!" box. Using two bar magnets, instruct the students to analyze the magnetic forces of *attract* and *repel*.

What force is used when magnets attract? pulling force

What force is used when magnets repel? pushing force

- Discuss "Science and History."

Why can magnets be used to find direction? The earth is like a giant magnet. A magnetic needle on a compass is pulled toward the magnetic north pole of the earth.

What material did sailors once use to help them sail the oceans? lodestone

Why could the sailors tell their direction using lodestone? It is a piece of magnetic rock. It always points to the north.

What do sailors use today to tell direction? a compass

Differentiated Instruction

Enrichment: Magnetism at a Distance
Materials: 2 toy cars, 2 disc magnets, electrical tape
Direct the student to attach the disc magnets to the cars so one car can push the other without touching it.

Background

Earth's Magnetic North Pole
The geographic North Pole and magnetic north pole are not in the same place on the earth, but they are relatively close to each other. The magnetic north pole is the point on the earth where the earth's magnetic field is directed vertically downward. This point continuously moves and shifts position. The magnetic north pole is located several hundred miles south of the geographic North Pole. A compass needle points toward the earth's magnetic north pole, not the geographic North Pole.

Activity

Try It Yourself!
Materials: two bar magnets
Allow the students to try this activity on their own.

How does learning about magnets help you and others? Possible answers: It helps me to know how magnets can be used; it helps me to know how to use magnets to help others.

What are some ways magnets are used? Possible answers: pick things up; sort things; hold screws on tip of screw driver; identify counterfeit money; hear sound from speakers; help motors, doorbells, phones, and some toys work

- Discuss "Fantastic Facts."

How can magnets help cows stay healthy? When a cow is given a magnet to swallow, any wire or metal in the cow's stomach is attracted to the magnet and does not cut through the cow's stomach.

What instrument can a farmer use to check for a magnet in a cow's stomach? a compass

✓ at the poles

Apply

Activities

Study Guide, pages 207–8

These pages review the concepts taught in Lesson 84. After completion, direct the students to keep them for review for Test 11.

Uses of Magnets

Learning about magnets can help you know how to use magnets. Magnets can be used to pick things up. You can use a magnet to pick up dropped pins or nails that have spilled. Some screwdrivers have magnets to help hold screws in place. Some paper money has magnetic properties. This helps to identify bills that are not real. Magnets are used to help us hear sound from the speakers in cars, TVs, and computers. Magnets also help the motor in a washing machine to work. There are magnets in some doorbells, phones, and toys. Magnets are an important part of our everyday life. We can show love for God and love for others by using magnets to help others.

Fantastic Facts

Did you know magnets can help cows stay healthy? As a cow grazes, it sometimes eats bits of metal or wire along with the grass. So some farmers have the cow swallow a strong magnet. The magnet stays in the cow's stomach. Any bits of wire or metal that the cow eats are attracted to the magnet. This helps keep the metal from cutting through the cow's stomach and hurting the cow. A farmer can use a compass to check whether a cow already has a magnet in its stomach.

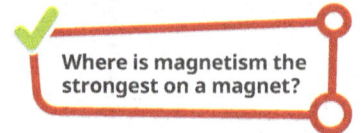

✓ Where is magnetism the strongest on a magnet?

256

ⓘ Background

Caring for Animals

Some people place the same value on animals as they do on humans. While humans have a much higher value in God's eyes (Genesis 1:26–27), this does not mean animals are unimportant. Proverbs 12:10 tells us that we should take care of the animals we own. Using magnets to help cows is one way people can take care of animals.

Magnetism and Matter

Name _____

Problem

Does the amount of matter between a magnet and an object affect the magnetism of the magnet?

Hypothesis

The **thicker/thinner** the amount of matter a magnet is pulling through, the **stronger/weaker** the magnetic force.

Inquiry Skills

- Predict
- Observe
- Infer

Materials

- ☐ horseshoe magnet or bar magnet
- ☐ paper clip
- ☐ 100 sheets of paper

Procedure

☐ 1. Place a paper clip on top of 1 sheet of paper. Hold the paper flat and still.

☐ 2. Hold the magnet against the underside of the paper. Move the magnet. Record your observations.

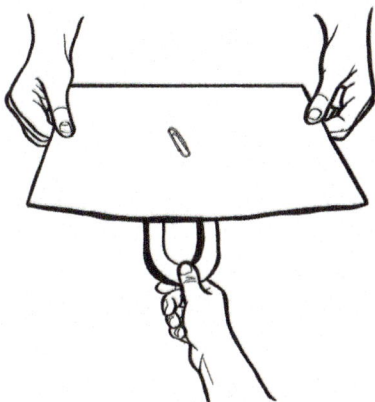

☐ 3. Repeat step 2, adding 10 sheets of paper each time. Record your observations. Stop recording when the magnet no longer moves the paper clip.

SCIENCE ③ Activities

Lesson 85 • Page 257
Investigation **209**

INVESTIGATION

Lessons 85–86

Magnetism and Matter

Inquiry Skills
- Predict
- Observe
- Infer

Magnetism can pass through matter. It can pull on a metal object even through matter. Magnetic force can move an object. The magnet and the object do not have to touch for the object to move.

In this Investigation, you will predict how the thickness of matter affects magnetism.

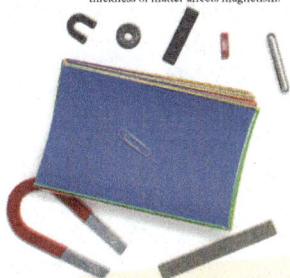

Let's inquire into . . .
Magnetism and Matter
Repeat the Investigation with magnets having different magnetism.
In this Inquiry, you will predict how the strength of magnetic force affects the movement of an object through matter.

257

Materials

- See *Activities* page 209.

> Printer paper or construction paper may be used for this Investigation. To reduce the number of materials needed, you may choose to have the students work in their science groups.

- paper lunch bag
- bar magnet
- paper clips

Engage

- Display the paper bag used in the Lesson 84 introduction. Place the bar magnet in the paper bag with the students watching. Scatter several paper clips on a table. Direct the students' attention to the paper clips as you move the bag toward them.

  What happened to the paper clips? Some of them moved toward the bag.

  Did the magnet need to touch the paper clips to cause them to move? no

  What is between the magnet and the paper clips? Possible answers: paper, air space

  Why does a magnet not need to touch an object in order to move the object? Because of the magnet's magnetic field, the magnet's magnetism pushes and pulls at a distance.

Instruct

Preparation for Reading

- Direct the students to remove pages 209–10 from their *Activities* books. Direct them to read page 257 and *Activities* pages 209–10 silently before beginning.

- Remind the students that it is helpful to look over all the pages before starting an Investigation. Point out that it will help them to see what they are going to do and when they will be writing information down.

Teach for Understanding

- Direct attention to the **Inquiry Skills**.

- Remind the students that scientists use science inquiry skills to gather and use facts.

  Which inquiry skills will you be using in this investigation? Predict, Observe, and Infer

- Review what it means to *predict*, *observe*, and *infer*.

Apply

- Choose a student to read the **Problem** aloud.

- Direct the students to complete the **Hypothesis** to answer the problem.

- Direct attention to the **Materials** list and make the materials available to the students.

- Instruct the students to follow and check off **Procedure** Steps 1–3 as each is completed. Point out that Step 3 is directing them to repeat Steps 1–2 until the magnet no longer moves the paper clip.

Objectives

- Predict the strength of magnetic force.
- Observe and record data.
- Draw conclusions about the strength of magnetic force and the amount of matter that the magnetic force passes through.

Lesson **85** begins here.

- Direct attention to the **Observations** data chart. Remind the students of the importance of accurately recording their observations.
- Direct the students to draw **Conclusions** using their observations. The conclusions may be done individually or with their science group.
- Guide a discussion of the students' conclusions.

Activities

Investigation, pages 209–10

Students will predict how the thickness of matter affects magnetism.

Assess

Rubric

Use the prepared rubric or design a rubric to include your chosen criteria.

Observations

Number of Sheets of Paper	Did the paper clip move when the magnet moved? [Circle the observation.]	
1	yes	no
10	yes	no
20	yes	no
30	yes	no
40	yes	no
50	yes	no
60	yes	no
70	yes	no
80	yes	no
90	yes	no
100	yes	no

Conclusions

1. Did your observations support your hypothesis? ○ yes ○ no
2. The last time the paper clip moved, there were _____ sheets of paper between the magnet and the paper clip.
3. Was the magnet touching the paper clip? ○ yes ○ no
4. What moved the paper clip? _magnetism or magnetic force_
5. What happened to the magnetic force as you added more paper between the magnet and the paper clip? _The magnetic force lessened, or weakened, as more paper was added._
6. What caused the magnetic force to weaken? _The more space, or distance, between the magnet and the paper clip, the weaker the magnetic force._

210 Lesson 85 • Page 257
Investigation

⚠ Helps

Choosing Magnets (Lesson 86)

Avoid using thin, flexible magnetic tape for this activity. Magnetic tape does not have distinct poles. Instead it contains alternating sections or layers of both poles.

Try to include some magnets that have strong magnetic fields.

Be aware of any student who should avoid being around very strong magnets for health reasons.

Magnetism and Matter

Name _____

Problem

Does the strength of a magnet affect its ability to move an object through matter?

Hypothesis

Write a hypothesis that will answer the problem.

Possible hypothesis: A magnet with stronger magnetism will move an object through more matter than a magnet with weaker magnetism will.

Inquiry Skills

• Predict
• Observe
• Infer

Materials

☐ 5 magnets, each a different strength
☐ masking tape
☐ paper clip
☐ 100 sheets of paper

Procedure

☐ 1. Number the magnets 1 to 5 using masking tape.

☐ 2. Place a *paper clip* on top of 1 sheet of paper. Hold the paper *flat and still*.

☐ 3. On the bottom of the paper, move magnet Number *1* to move the *paper clip*.

☐ 4. Repeat the procedure, adding ___ sheets of paper each time between the paper clip and the magnet. Record your observations.

☐ 5. When the magnet stops moving the paper clip, record the *number of sheets of paper* that are between *the magnet and the paper clip*. This number is the magnetism, or magnetic force, of the magnet.

☐ 6. Repeat steps 2–5 for each *magnet*, adding ___ sheets of paper each time.

☐ 7. Record *your observations*.

☐ 8. Make a bar graph that shows the magnetism of each magnet. Use a different color to shade each bar on the graph.

Lesson **86** begins here.

Objectives

• Formulate a hypothesis to predict the magnetic force of several different magnets.
• Develop the steps of the procedure in the Inquiry.
• Observe and record data.
• Organize the data collected onto a graph.
• Draw conclusions about the strength of each magnet's magnetic force using the data collected.

Materials

• See *Activities* page 211. (See Helps, page 328.)

Printer paper or construction paper may be used for this Inquiry. To reduce the number of materials needed, and to encourage collaboration, you may choose to have the students work in their science groups.

Engage

• Discuss observations from the previous day's lesson.
• What force caused the paper clip to move? magnetic force; magnetism
• What happened to the magnetic force as more paper was added between the magnet and the paper clip? The magnetic force became less, or was weakened.

• What did you conclude about magnetic force and the distance between a magnet and a metal object? Magnets act at a distance, or without touching the object they are moving. However, the more distance there is between the magnet and the object, the weaker the magnetic force. Eventually, the magnetism is not strong enough to move the object.

Instruct

Preparation for Reading

• Direct the students to remove pages 211–13 from their *Activities* books. Direct them to read page 257 and *Activities* pages 211–13 silently before beginning.
• Remind the students that it is helpful to look over all the pages before starting an Inquiry lesson. Point out that it will help them to see what they are going to do and when they will be writing information down.

Teach for Understanding

This is an Inquiry lesson. Inquiry lessons allow the students the opportunity to form the hypothesis and design more of the investigation. The *SCIENCE 3 Activities Answer Key* provides possible answers to help the teacher guide the students to success. Encourage the students to collaborate as they work through the Inquiry.

• Direct attention to the **Inquiry Skills**.
• Remind the students that scientists use science inquiry skills to gather and use facts.
  Which inquiry skills will you be using in this lesson? Predict, Observe, and Infer
• Review what it means to *predict*, *observe*, and *infer*.

Apply

• Choose a student to read the **Problem** aloud.
• Guide the students in a discussion as they write the **Hypothesis** together as a class. So each student is testing the same hypothesis, display the students' suggested hypotheses and have the class agree on a single hypothesis to test.
• Direct attention to the **Materials** list and make the materials available to the students.
• Explain that, in this Inquiry lesson, several **Procedure** steps are incomplete and the students are to determine how best to complete each step. Point out that, for help, they can refer back to the Procedure in the Investigation in Lesson 85.
• Point out that Step 6 is directing the students to repeat Steps 2–5 until the magnet stops moving the paper clip.
• Instruct the students to follow and check off Procedure Steps 1–8 as each is completed.

- Direct attention to the **Observations** data chart. Remind the students of the importance of accurately recording their observations.
- Direct the students to graph the magnetism of each magnet using the data recorded on their observations chart. The magnetism on the graph (number of sheets of paper) should be changed to coincide with the number that the students chose for Procedure Steps 4 and 6.

Observations

Record the number of sheets of paper that were between the magnet and the paper clip when the paper clip last moved.

Magnet	Number of Sheets of Paper
1	
2	
3	
4	
5	

Graph the magnetism of each magnet.

Magnetism of Different Magnets

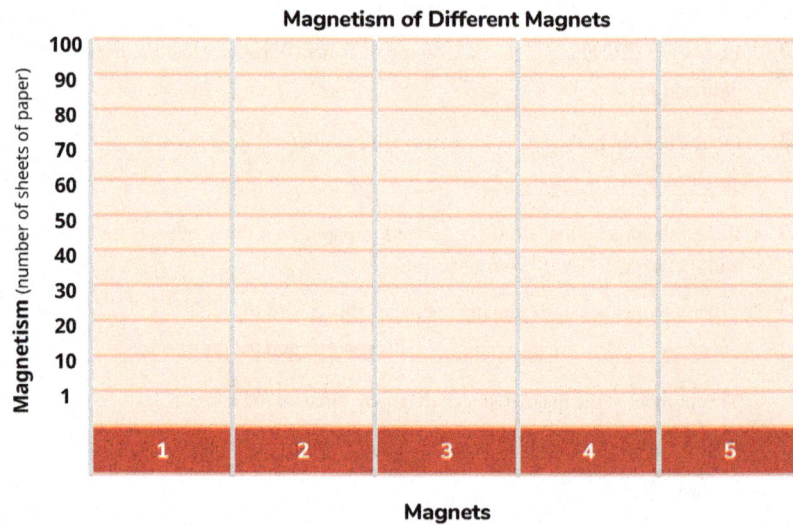

Magnetism (number of sheets of paper)

100 90 80 70 60 50 40 30 20 10 1

Magnets: 1 2 3 4 5

212 Lesson 86 • Page 257
Inquiry

(i) Background

How Scientists Measure Magnetic Fields

The strength of a magnetic field is usually measured in units of gauss or tesla. It takes 10,000 gauss to equal 1 tesla. The earth's magnetic field ranges from approximately 1 gauss at the surface to 25 gauss at the core. In comparison, most MRI magnets range from 0.2 tesla to 7.0 tesla in strength. Scientists use an instrument called a magnetometer to measure magnetic fields.

Magnetism and Matter

Name _____

Conclusions

1. Why did you use the same type of paper and the same paper clip each time? *to control the investigation, or to make it a fair test*

2. Did your observations support your hypothesis?   ○ yes   ○ no

3. What moved the paper clip even though the magnet was not touching it? *magnetism, or magnetic force*

4. What happened to the magnetism as you added more paper between the magnet and the paper clip? *The magnetism weakened as more paper was added.*

5. Which magnet has the strongest magnetism? Magnet number ____

6. Which magnet has the weakest magnetism? Magnet number ____

7. What did you notice about the magnetism with each magnet tested? *Possible answers: Some magnets had more magnetism than other magnets; some magnets moved the paper clip through more paper.*

8. Why is knowing the strength of magnetism helpful? *Possible answers: It helps people to know what magnet they should use to do a specific job; a stronger magnet is needed when there is more matter for the force to pull through; more matter requires more magnetic force.*

• Direct the students to draw **Conclusions** using their observations. The conclusions may be done individually or with their science group.

• Guide a discussion of the students' conclusions.

Activities

Inquiry, pages 211–13

Students will predict how the strength of magnetic force affects the movement of an object.

Assess

Rubric

Use the prepared rubric or design a rubric to include your chosen criteria.

Objectives

- Identify what an electromagnet is.
- Explain how the magnetic force of an electromagnet can be made stronger or weaker.
- Differentiate between a bar magnet and an electromagnet.
- Support the statement that understanding electromagnets has helped people solve real-life problems. **BWS**

Materials

- bar or horseshoe magnet
- 3" or 4" steel nail
- single-stranded insulated wire, 1 m long
- D cell (1.5 volt) battery (See Helps.)
- masking tape
- small paper clips

Vocabulary

- electromagnet
- MRI technologist

Engage

- Display a bar magnet or a horseshoe magnet.
  Can this magnet be turned on or off? no

- Explain that some magnets are magnetized all the time. Other magnets can be turned on and off.

- Point out that in this lesson the students will learn about magnets that can be turned on and off.

Instruct

Preparation for Reading

- Preview and pronounce the vocabulary term *electromagnet*, the name *Hans Christian Oersted*, and the words *temporary* and *maglev*.

- Direct the students to read pages 258–59 silently to find out how electromagnets can be made stronger.

Teach for Understanding

What did Hans Christian Oersted notice about an electric current flowing through a wire? The current made the needle of a nearby compass move.

What did he discover after several more tests? He found that when a current flows through a circuit, it causes a magnetic field. The electric current in the wire made the wire act like a magnet.

What ways did scientists find to make a wire's magnetic field stronger? Possible answers: wrapping the wire many times around a metal object; increasing the amount of electric current

- Point out that Oersted's discoveries led to the development of electromagnets. Explain that what Oersted was doing is what Genesis 1:28 calls exercising dominion or managing the earth by taking part of God's world and making it more useful.

What is an *electromagnet*? a temporary magnet formed when electric current flows through a wire that is wrapped around a piece of metal

---

Electromagnets
Discovering Electromagnets

Over 200 years ago scientists discovered that electricity and magnetism are related. In 1820 Hans Christian Oersted noticed that the needle of a compass moved each time electricity flowed through a wire. He studied this and tested it for several months. He observed that a magnetic field is caused when an electric current flows through a circuit.

Electric current flowing through a wire makes the wire act like a magnet. Oersted's discovery led to other discoveries. Scientists observed that wrapping the wire many times around a metal object makes the magnetic field stronger. An **electromagnet** forms when an electric current flows through a wire that is wrapped around a piece of metal. It is a temporary magnet because electromagnets can be turned on and off. The electromagnet acts like a magnet when the electric current flows through the wire. The magnetic force stops when the current stops.

Electromagnets can be made stronger or weaker. Wrapping the wire more times around the metal object makes the electromagnet stronger. The electromagnet's strength also depends on the amount of electric current flowing through the wire. Increasing the current increases the magnetic field.

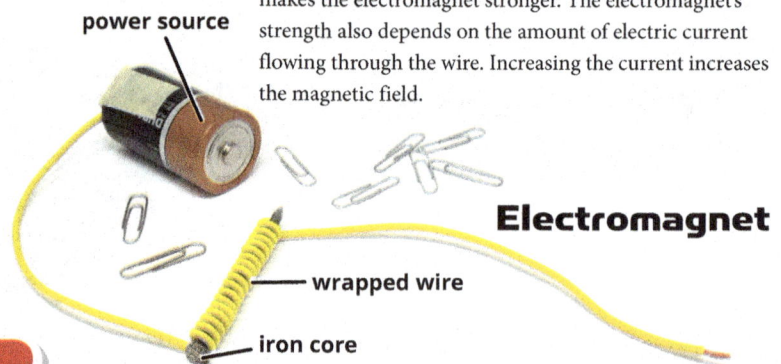

power source

Electromagnet

wrapped wire

iron core

⚠ Helps

Battery Caution

Use alkaline batteries only. Do not use lithium batteries because they deliver larger amounts of current and cause the wires to become hot.

ⓘ Background

Scientist Hans Christian Oersted

Hans Christian Oersted (1777–1851) was a Danish physicist. In 1806 he became a professor of science at Copenhagen University. In April of 1820, as he was demonstrating an experiment to some students, he noticed that the electric current in a wire made the needle of a compass move. Later he tested his findings, and he eventually wrote a paper that presented his discovery of electromagnetism. This discovery made many other discoveries possible.

Solenoids

A solenoid is a type of electromagnet that is usually used to move a piece of metal in a straight line. Solenoids are often used in locks, such as a car's power locks.

Maglev Technology

Maglev trains are used in several countries. Chinese and German trains use electromagnets to achieve levitation, while Japanese trains use super-cooled conducting coils. Maglev elevators are also being developed.

Uses of Electromagnets

Many things use electromagnets. Fire alarms, microwave ovens, toasters, hair dryers, TVs, and computers all use electromagnets. Pushing some types of doorbell buttons completes a circuit and turns on an electromagnet. Releasing the button breaks the circuit, and the magnet turns off. When you turn the current on and off, the electromagnet moves back and forth. As it moves, it hits pieces of metal. This makes the musical sound you hear.

Electromagnets can be extremely strong. They are often used on cranes that lift scrap metal. Very large electromagnets are used on maglev trains. The strong electromagnets lift the train above the rails. Magnetic forces push and pull the train along. There is no friction because the train does not touch the rails as it moves. This allows the train to travel very fast—over 500 kilometers (310 miles) per hour!

✓ **What is an electromagnet?**

Maglev train in Shanghai, China.

259

Activity

Create an Electromagnet

Materials: one D cell (1.5 volt) battery, one 3" or 4" steel nail, 1 m of single-stranded insulated wire, masking tape, several small paper clips

Note: Electrical current will generate heat. An increase in flowing current will generate more heat. If enough heat is generated, a burn injury may result.

Measure 1 m of insulated wire. Strip 1 cm of insulation from both ends of the wire. Measure 10 cm from one end. Using the 10 cm mark as the starting point, wrap the longer length of wire 10 times around the nail, beginning at the nail head. Attach the stripped end of the 10 cm length to the negative end of the battery using a piece of masking tape. With the longer end of the wire, touch the positive end of the battery to "turn on" the electromagnet.

> **WARNING:** DO NOT tape the stripped wire to the positive end of the battery. The wire should only be touching the positive end of the battery for short periods of time.

Touch the tip of the nail to the paper clips and pick up the paper clips attached. Remove the wire from the positive end of the battery to "turn off" the electromagnet, and "release" the paper clips. To pick up more paper clips, increase the strength of the electromagnet. Wrap the wire around the nail an additional 10 times and pick up the paper clips. Compare the number of paper clips picked up. Repeat as desired.

Chapter 11: Electricity and Magnetism

Why do we say that an electromagnet is a temporary magnet? Possible answers: It can be turned on and off; an electromagnet acts like a magnet when electric current flows through the wire and stops acting like a magnet when the current stops.

- Connect the stripped end of the straight piece of insulated wire to the negative end of the battery using masking tape. Touch the other end of the wire to the positive end of the battery. Try to pick up the paper clips with the wire.

- Discuss the wire's lack of ability to pick up the paper clips. Is this an electromagnet? No; to be an electromagnet, the wire needs to be wrapped around a piece of metal.

- Create an electromagnet. Refer to the Activity section and to the picture on page 258 as you build the electromagnet.

- Try to pick up the paper clips using the tip of the nail. Count the number of paper clips the nail picks up.

How does current flow through an electromagnet? The current starts at the power source (battery), flows through the wire, and then goes back to the battery.

How can the circuit be broken? Disconnect one end of the wire from the battery.

How can an electromagnet be made stronger? by wrapping the wire more times around the metal object; by increasing the electric current

Note: Using a larger volt battery will increase the strength of the electromagnet. A 9 volt battery will increase the current and produce a stronger magnetic force than a 1.5 volt battery will.

Note: The larger the 1.5 volt battery, the longer the battery will last. A 1.5 volt D cell battery will last longer than a 1.5 volt AAA, AA, or C cell battery, but it will generate the same amount of energy.

- Wrap more of the wire around the nail. Compare the number of paper clips it can pick up with those picked up before. Repeat as desired.

Teaching for page 259 begins here.

What are the items pictured that use electromagnets? doorbell, scrap-metal magnet, maglev train

- Direct attention to the picture of the maglev train. Ask a volunteer to read the caption aloud.

What is used to move a maglev train? Possible answers: strong electromagnets, magnetic force to push and pull the train

Why is the lack of friction on a maglev train beneficial? It allows the train to travel very fast.

What are some other things that use electromagnets? fire alarms, microwaves, toasters, hair dryers, TVs, computers

✓ **a temporary magnet formed when an electric current flows through a wire that is wrapped around a piece of metal**

Preparation for Reading

- Preview and pronounce the vocabulary term *MRI technologist* and the words *Magnetic Resonance Imaging*.
- Direct the students to read page 260 silently to find out what an MRI technologist does.

Teach for Understanding

What does an MRI use to take pictures of the inside of the human body? It uses a magnetic field and radio waves.

Using an MRI, what, inside the human body, can a doctor see without the use of surgery? tissues and organs, such as bones

- Explain that bones are organs in the body. Remind the students that the bones work together to form the skeletal system (*Science 2*).

What does an MRI technologist learn how to operate? a large tube-like machine

What does the MRI technologist do with the pictures taken during the test? saves them on a computer and sends them to the doctor

Where do most MRI technologists work? in hospitals

- Share the Background information, as needed, at this time.

Activities

Study Guide, page 215

This page reviews Lesson 87. After completion, direct the students to keep it for review for Test 11.

Assess

Quiz 11B

The quiz may be given at any time after completion of this lesson. Material for this quiz comes from the Study Guides in Lessons 84 and 87.

MRI Technologist

An **MRI technologist** operates a large tube-like machine. MRI stands for Magnetic Resonance Imaging. This machine uses a magnetic field and radio waves to take a picture of the inside of the human body. A doctor may send a person to have an MRI to help him know what is wrong with the person. With an MRI, the doctor can see the person's tissues and organs, such as bones, without the use of surgery.

An MRI technologist learns how to operate the machine. She helps the person who is receiving the MRI. She correctly positions the person in the machine. She speaks to the person during the MRI and helps him to be comfortable. She saves the pictures on a computer. She sends the pictures to the doctor to look at. Most MRI technologists work in hospitals.

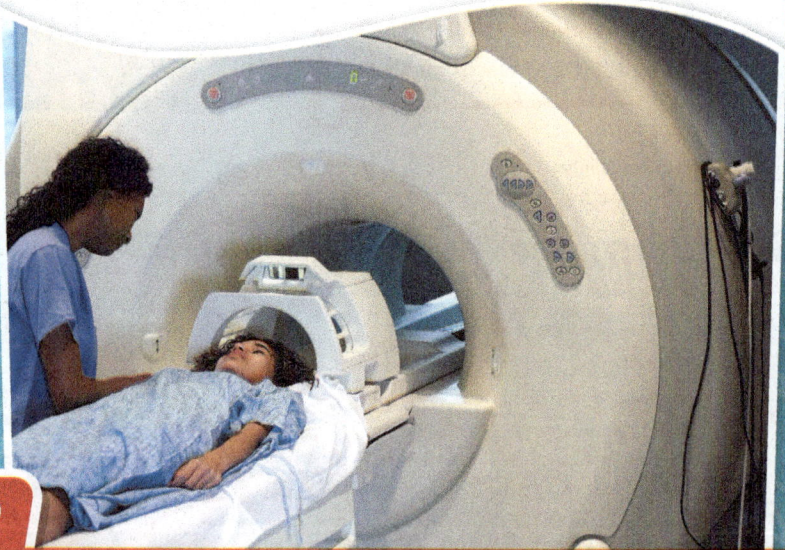

260

ⓘ Background

MRI Technology

An MRI is an examination using magnetic resonance imaging. An MRI scanner uses electromagnets and radio waves to take pictures. The pictures can help the doctor diagnose injuries, bleeding, and other more serious problems. The first MRI done on a human took place on July 3, 1977.

Lesson 88

STEM

Lost and Found

You and your parents are putting together your new bicycle in the garage. A can of screws, nuts, and bolts is knocked over. Many of them roll under the workbench. It is a tight space. No one can reach all of the loose screws, nuts, and bolts.

In this STEM activity, you will design a tool, using a magnet or an electromagnet. The tool will be designed to reach into a tight space. You will test the tool to see whether it will pick up the screws, nuts, and bolts that have rolled out of reach.

261

Objectives

- Design a magnetic or electromagnetic tool, using the engineering design process, that will pick up metal objects from a tight space.
- Create a magnetic or an electromagnetic tool.
- Test and compare models to improve the original design.
- Communicate how the redesign solves the problem.

Materials

- magnets, various types and strengths
- masking tape
- single-stranded insulated wire
- batteries (C or D cell)
- 3" or 4" steel nails
- string
- wire clothes hangers
- dowel rods
- meter sticks
- variety of screws, nuts, and bolts

Teacher Resources

- Visual 2.6: *STEM: The Engineering Design Process*

Engage

Have you ever dropped something and had it roll out of reach?

Do you remember what you used to reach the item?

- Explain that in today's lesson they will design a special tool that will reach into a tight space.

Instruct

Preparation for Reading

- Direct the students to remove *Activities* pages 217–18. Direct the students to read page 261 and *Activities* pages 217–18 silently before beginning.
- Remind the students that it is helpful to look over all the pages before starting a STEM activity. It will help them to see what they are going to do and when they will need to write information.

Teach for Understanding

- Display the *STEM: The Engineering Design Process* visual or direct the students to the copy of *Activities* page 35 in their science notebook.
- Review the design process, as needed.

Apply

- Ask a volunteer to read the *Problem* aloud.
- For this activity, each student will design his own solution to the problem, or the students may work in science groups to collaborate as they design a solution.
- Instruct the students to *Imagine* a solution to the problem and *Plan* a design.
- Remind the students that the design needs to include using a magnet or an electromagnet.

> **CAUTION: Electromagnet**
> Adult supervision will be needed if the students choose to build an electromagnet.

- Explain that electrical current will generate heat. An increase in flowing current will generate more heat. If enough heat is generated, a burn injury may result.
- Explain to the students that they are **not** to tape the stripped wire to the positive end of the battery. The wire should only be *touching* the positive end of the battery for short periods of time.
- Display the materials. Some materials will work better than others. Allow the students to choose materials that they believe will best solve the problem of picking up screws, nuts, and bolts that have rolled out of reach into a tight space.
- Direct the students to list and gather the materials needed for their design.

STEM

Lost and Found

Name _____

Ask

What is the problem?

A can of screws, nuts, and bolts has spilled. The pieces have rolled into a tight space, just out of reach. A magnetic or electromagnetic tool needs to be designed to pick up the screws, nuts, and bolts.

Imagine

Think about possible answers to the problem.

Possible designs should include a tool that is long enough and narrow enough to reach into

the tight space. The tool must also be strong enough to pick up the screws, nuts, and bolts.

_____

Plan

Draw and label your design. List and gather materials.

Accept any reasonable design.

Design	Materials

SCIENCE 3 *Activities*

Lesson 88 • Page 261
STEM

Create

Follow the plan and make a model.

Test and Improve

What changes need to be made to the design?

Accept any reasonable answer.

_____

_____

Improved Design

Share

Tell others how your design solves the problem.

• Direct the students to *Create* a model of their design.
• Instruct the students to *Test* their model.
• After testing their designs, direct the students to compare models and *Improve* their design.
• Invite the students to *Share* their designs with a partner or another science group. Instruct the students to explain how their design solves the problem.

Activities
STEM, pages 217–18
The students will design a tool, using a magnet or an electromagnet, that can reach into a tight space to pick up screws, nuts, and bolts.

Assess

Rubric
Use the prepared rubric or design a rubric to include your chosen criteria.

Objective
• Recall terms and concepts from Chapter 11.

Apply

Review
• Material for Test 11 will come from the Study Guides on *Activities* pages 203–4, 207–8, and 215, as well as Quizzes 11A and 11B. You may review any or all of the material during the lesson.
• You may choose to review Chapter 11 by playing the Keep It Moving review game or a game from the Game Bank (Teacher Resources).

🏆 Review Game

Keep It Moving
Play the game with the entire class. The students represent the conductors along a circuit, and you represent their power source. Each correct answer to a review question keeps the circuit closed and allows the current to move to the next conductor. An incorrect answer breaks the circuit. The next correctly answered question repairs the circuit. Try to see how many conductors the current can flow through before it is broken.

NOTES

NOTES

_____

_____

_____

_____

_____

_____

_____

_____

_____

_____

_____

_____

_____

_____

_____

_____

_____

_____

_____

_____

_____

_____

_____

_____

_____

_____

_____

_____

_____

_____

_____

_____

_____

Objective

• Apply terms and concepts from Chapter 11.

Assess

• Administer Test 11.

Teacher Resources

Explaining the Gospel

One of the greatest desires of Christian teachers is to lead children to faith in the Savior. God has called you to present the gospel to your students so that they may repent and trust Christ, thereby being acceptable to God through Christ.

Relying on the Holy Spirit, you should take advantage of the opportunities that arise for presenting the good news of Jesus Christ. Ask questions to personally apply the Ten Commandments to your students (e.g., What is sin? Have you ever told a lie or taken something that wasn't yours? Are you a sinner?). You may also ask questions to discern the child's sincerity or to reveal any misunderstanding he might have (e.g., What is the gospel? What does it mean to repent? Can you do anything to save yourself?). Read verses from your Bible. You may find the following outline helpful, especially when dealing individually with a child.

1. I have sinned (Romans 3:23).
- Sin is disobeying God's Word (1 John 3:4). I break the Ten Commandments (Exodus 20:2–17) by loving other people or things more than I love God, worshiping other things or people, using God's name lightly, disobeying and dishonoring my parents, lying, stealing, cheating, thinking harmful and sinful thoughts, or wanting something that belongs to somebody else.
- Therefore, I am a sinner (Psalm 51:5; 58:3; Jeremiah 17:9).
- God is holy and must punish me for my sin (Isaiah 6:3; Romans 6:23).
- God hates sin, and there is nothing that I can do to get rid of my sin by myself (Titus 3:5; Romans 3:20, 28). I cannot make myself become a good person.

2. Jesus died for me (Romans 5:8).
- God loves me even though I am a sinner.
- He sent His Son, Jesus Christ, to die on the cross for me. Christ is sinless and did not deserve death. Because of His love for me, Christ took my sin on Himself and was punished in my place (1 Peter 2:24a; 1 Corinthians 15:3; John 1:29).
- God accepted Christ's death as the perfect substitution for the punishment of my sin (2 Corinthians 5:21).
- Three days later, God raised Jesus from the dead. Jesus is alive today and offers salvation to all. This is the gospel of Jesus Christ: He died on the cross for our sins according to the Scriptures, and He rose again the third day according to the Scriptures (1 Corinthians 15:1–4; 2 Peter 3:9; 1 Timothy 2:4).

3. I need to put my trust in Jesus (Romans 10:9–10, 13–14a).
- I must repent (turn away from my sin) and trust only Jesus Christ for salvation (Mark 1:15).
- If I repent and believe in what Jesus has done, I am putting my trust in Jesus.
- Everyone who trusts in Jesus is forgiven of sin (Acts 2:21) and will live forever with God (John 3:16). I am given His righteousness and become a new creation, with Christ living in me (2 Corinthians 5:21; Colossians 1:27).

If a child shows genuine interest and readiness, ask, "Are you ready to put your trust in Jesus and depend on only Christ for salvation?" If he says yes, then ask him to talk to God about this. Perhaps he will pray something like the following:

> God, I know that I've sinned against You and that You hate sin, but that You also love me. I believe that Jesus died to pay for my sin and that He rose from the dead, so I put my trust in Jesus to forgive me and give me a home with You forever. In Jesus' name I pray. Amen.

Show the child how to know from God's Word whether he is forgiven and in God's family (1 John 5:12–13; John 3:18). Encourage the child to follow Jesus by obeying Him each day. Tell the child that whenever he sins, he will be forgiven as soon as he confesses those sins to God (1 John 1:9).

Getting Started

Scheduling

SCIENCE 3 provides material for a one-semester program. Assessment and review days are included, and some lessons have been allotted two days for completion. The instruction of all textual information, as well as the Investigation, Inquiry, Exploration, and STEM activities will adequately fill a half-year science program. Consider the following while planning the schedule.

- The chapters should be taught in order. Vocabulary, concepts, skills, and reading difficulty build throughout the book.
- This 90-lesson course may be taught each day for two quarters or on alternate days for four quarters.
- The lesson plans are designed for 30–40 minute lessons. The text pages may be read as part of the lesson or prior to the lesson at home or in class. Adjust the presentation of the Preparation for Reading sections to precede the students' assigned reading schedule.
- Some Exploration lessons will be introduced in class and then assigned to be completed outside of class.
- Additional activities can be adjusted for use in learning centers.

Science Notebook

It is recommended that each student keep a three-ring notebook for science. The activities pages are three-hole punched to accommodate this suggestion. A notebook will allow the students to organize Instructional Aids, Study Guides, Investigations, Inquiries, Explorations, STEM activities, and other useful information.

A section of the notebook should also serve as a science journal for recording notes, drawings, observations, and thoughts about concepts and activities. A new feature is the Notebook Connection found periodically throughout the *Activities* book. Its purpose is to make a deliberate connection to the student's science notebook and to encourage note taking, creative writing, drawing, summarizing, analyzing and interpreting text and images, and conducting research.

Teaching a Lesson

The teacher is the key to unlocking students' understanding. The teacher edition provides a variety of resources to customize instruction for each student's needs.

The introduction of each lesson provides a short activity or relevant questions that focus on the main topic. Most lessons cover four to six pages of the student edition. The Preparation for Reading section should be presented before the students begin to read the text. The students should then read the material silently on their own. Assign the students to read the text as part of the lesson or prior to the lesson at home or in class. Oral reading of the text should be reserved for short passages only as they relate to the discussion.

The main part of each lesson plan provides questions to help you and the students interact as you identify students' levels of comprehension of the material they have read. This discussion provides an auditory reinforcement of key ideas. Higher-level thinking questions, marked with the gear icon (⚙), help the students apply the information that they have read by relating it to previous knowledge and everyday situations.

Background information and activities located at the bottom of the lesson pages provide additional material for use with the lesson based on the class dynamics and available time.

Differentiated Instruction

Differentiated instruction activities may reinforce or extend a concept taught in the lesson. Remediation may be used to reteach content for students to gain grade-level appropriate competencies. Remediation is appropriate for students who have fallen behind, such as English language learners and struggling students. Enrichment information and activities may be used to expand instruction beyond the scope of the lesson. Enrichment is appropriate for both gifted students and struggling students who are being retaught content in a new way.

Materials

The materials needed to teach SCIENCE 3 can be found in the Materials section at the beginning of each lesson in the teacher edition and at TeacherToolsOnline.com. Materials are also listed by chapter in the Teacher Resources.

The Materials Lists at TeacherToolsOnline.com and in the Teacher Resources specifies by chapter the materials needed to teach the entire program.

Standard school supplies (Bible, pencil, paper, glue stick, scissors, etc.) are not usually listed in the Materials List or in the Materials section of the lesson.

Teacher Resources

The Teacher Resources found in the back of the teacher edition lists the Instructional Aids and Visuals used in teaching the lesson. These resources can also be found at TeacherToolsOnline.com. Instructional Aids are pages designed for use by the students during a lesson. Visuals are designed for display by the teacher during the lesson.

Vocabulary

The Vocabulary section lists the key terms for the lesson. These terms are bold in the student edition and the terms can be found in the Glossary. The terms and their definitions will be reviewed on the Study Guides in each *Activities* chapter. Other important words are in italic. These words can also be found in the Glossary but are not vocabulary terms.

Getting Started

Preparation for Reading

Reading for information is an essential skill for reading success. The Preparation for Reading section prepares the students for reading in the following ways.

- Read and pronounce together vocabulary and other words that may slow the students' reading fluency. Definitions are not discussed at this time so that the students gain an understanding of the meanings from sentence context.

- Emphasize skills used for reading in the content area by directing the students' attention to headings, captions, diagrams, feature boxes, and boldface and italic words as aids to understanding the informational text and as sources of additional information.

- The reading for each lesson is usually divided into two-page segments. One focus statement is provided to direct the students as they read.

Teach for Understanding

In the Teach for Understanding section, you will find questions and comments to help you evaluate the students' understanding of concepts. The questions listed are scaffolded, but they are not exhaustive. Use the questions as a springboard to help you begin the discussion.

- A gear icon (⚙) is used to help you identify the higher-level thinking questions. The answers to higher-level thinking questions are not taken directly from the pages being discussed, thus requiring some analysis by the students. Supply any prompts or background needed to guide the students to answer these questions.

- Develop in the students the expectation of supporting *yes* or *no* responses with an explanation.

Some lessons incorporate brief activities with questions to provide hands-on opportunities to aid students' understanding.

Charts, graphs, photos, illustrations, and diagrams are forms of communication that enhance students' reading and understanding. Guide the students in understanding, interpreting, and using these features.

Graphic organizers are tools to aid the students' understanding. The organizers can help the students see the organization of facts and details. Organizers may be completed during the discussion or as reinforcement following the teaching. They may also be used to help the students take a closer look at the information by making comparisons.

Teaching Investigation, Inquiry, Exploration, and STEM Lessons

It is important to emphasize the value of the Investigation, Inquiry, Exploration, and STEM lessons in *Science 3*. In order for the students to truly know and understand science, they must be able to use it. Knowing when and how to apply science is the basic premise behind the term *scientific literacy*.

The Investigation, Inquiry, Exploration, and STEM lessons allow the students to demonstrate their understanding of science concepts through hands-on activities. Additionally, these activities also allow the students to apply knowledge from other subject areas, such as measurement skills from math, and writing and communication skills from language arts.

Many activities in *Science 3* are open ended. These require the students to incorporate basic science inquiry skills. Because the goal of these activities is to teach a mental process as opposed to specific procedures, the activities may not follow an exact format. An activity may have more than one correct result.

The Explorations, STEM activities, and some of the other activities are project oriented. They sometimes require the students to work outside of class and usually require more creativity than the activities that involve basic experiments. While doing Explorations and STEM activities, the students will use and develop skills such as designing solutions to problems, writing, and sharing orally with a group or partner.

Most of the Investigation, Inquiry, Exploration, and STEM lessons can be completed by students working in groups. Using science groups helps students learn cooperation and management skills as they collaborate. Brainstorming ideas among group members is usually a good problem-solving technique. As they handle materials and interact with one another, the students also demonstrate what they are learning about biblical worldview.

A rubric is provided for each Investigation, Inquiry, Exploration, and STEM lesson for assessment.

Groups for Investigation, Inquiry, and Exploration Activities

Place students in groups of three or four to allow for maximum participation.

A science group should be a mixture of students from all achievement levels. This provides students the opportunity to collaborate with peers on various academic levels.

At first, you may choose to assign tasks to each group member to ensure active participation by all. Later in the year, the students within each group can decide the tasks of each member. Tasks may include handling materials, measuring, recording information, and communicating results.

Groups for STEM Activities

Using groups for STEM activities may require extra input from you. Because the STEM lessons are less structured, students may have a difficult time assigning themselves appropriate tasks.

However, since these lessons often require a high level of creativity, students may be less intimidated in a group setting.

Management Tips

Being prepared is the key to a successful activity. A printable copy of the Materials List is available, by chapter, in the back of the teacher edition and at TeacherToolsOnline.com. Materials are also listed, by lesson, in the teacher edition and in the activities. You will need to determine specific quantities based on the number of science groups or students you have.

Gather materials needed for group Investigations, Inquiries, and Explorations ahead of time. Place the materials needed on a nonbreakable tray or in a nonbreakable container. Distributing and collecting materials in this way will save time and will help with cleanup. Materials for STEM lessons should be placed in a central location to allow students to choose the materials they believe will work best with their design.

Keep paper or plastic on hand to cover work surfaces and buckets or containers in which to dispose of waste material. It is helpful to keep paper towels or rags readily available.

Review and Test Lessons

Included in the *Science 3 Activities* book are Study Guide pages, identified with a colored Review tab. Each chapter's Study Guide pages will help the students review the information in smaller segments. One Study Guide in each chapter includes an application question (Write About It) that requires a longer written response.

A Review lesson is provided for each chapter. Using the suggested game or an alternate game from the Game Bank from the Teacher Resources will review vocabulary and concepts in a fun way.

The Review lesson also gives the teacher an opportunity to check that each student has accurately completed the Study Guides for the chapter. The material for each quiz and chapter assessment comes from the Study Guide pages. A student who understands the material (and does not just know the answers) covered on the review pages will be adequately prepared for the assessments.

The assessment day has been given its own lesson number. There is no material assigned to be taught in this lesson. (*Note:* the *Science 3 Assessments* and *Science 3 Assessments Key* must be purchased separately.)

Getting Started

Science Inquiry Skills

Science is more than just a collection of facts. It is also a demonstration of skills that show an understanding of how science works. Each Investigation, Inquiry, and Exploration lesson in SCIENCE 3 emphasizes two or more basic inquiry skills. The students will use some of the other skills during each activity as well.

Although inquiry skills are emphasized in the Investigation, Inquiry, and Exploration lessons, they are also used in lesson discussions and demonstrations. The basic science inquiry skills are those emphasized in the SCIENCE 3 Activities.

Observe	Use the senses to gather information regarding objects and events.
Measure Using Numbers	Use standard devices and techniques to quantify information.
Infer	Draw conclusions based on previous knowledge or observation.
Classify	Group or order objects based on similarities or differences.
Predict	Forecast an expected result based on prior experience or knowledge.
Communicate	Use written, oral, or graphic means to transmit information to others.

Some of the integrated science inquiry skills are not directly taught in SCIENCE 3, though they are often used. You may choose to incorporate the teaching of these skills as well in preparation for their use in later grades.

Hypothesize	Formulate a statement that can be tested by experimentation.
Identify and control variables	Recognize the changing and unchanging factors in an investigation and adjust one factor to obtain data.
Define operationally	Explain an object or event properly but in terms of each student's observation and experience.
Experiment	Set up and follow a procedure to test a hypothesis.
Collect, record, and interpret data	Gather information about objects and events in an organized and systematic manner.
Make models	Create a physical or mental representation to explain or clarify ideas, objects, or events.

Game Bank

Baseball Challenge

Identify four areas in the classroom to use as bases. Divide the class into two teams and flip a token to determine which team "bats" first. Ask a question to the first student, or "batter." If the batter gets the question right, he may proceed to first base. Players advance to the next base each time a batter answers a question correctly. Correct answers to difficult questions could allow the batter to advance to more bases. A "run" is scored when a batter gets to home base.

If the batter answers incorrectly, someone from the opposing team (the outfield) may answer the question. If the outfielder answers correctly, the batter is out. If the outfielder's answer is incorrect, the batter receives a second chance. Three incorrect answers from the batter equal an out. When a team has received three outs, the teams switch sides.

Basketball

Divide the class into two teams. Each team should choose a spokesperson. Give the teams time to make up several questions about the lesson or chapter. They must know the correct answers to the questions.

The teams will take turns asking each other questions. The team may discuss the answer to a question, but the final answer should come from the team's spokesperson. If the team answers correctly, it receives two points. If desired, the team may also get a chance to make a "basket" by shooting a foam ball into a trashcan or other container.

Concentration

Divide the class into two teams. Display a grid with various point values. There should be two of each value. Cover the point values with consecutively numbered squares. Ask a review question. If a student answers correctly, he may choose two squares. If he finds two of the same point value, he may add those points to his team's score. If he does not find a match, play switches to the other team.

Ducks and Decoys

This game is best used with a multiple choice review. Designate the four corners of the classroom as A, B, C, and D. Count out enough blank index cards so that you have one for each student. Label three-fourths of the index cards "duck" and the remaining cards "decoy." Give one card to each student. The students should not tell others whether they are ducks or decoys. As the review questions are read, ducks should go to the corner that corresponds with the answer they believe is correct. Decoys should intentionally pick a corner that does not have the correct answer. This will encourage students to think of the answer for themselves and not to "follow the flock."

Football

Display a football field with yard lines and end zones marked. Divide the class into two teams. Use a token or marker to represent each team. Place the markers at the 50-yard line. Decide which team goes first. This team receives the first question. If the team answers correctly, its marker advances 10 yards. The opposing team's marker also moves to the new mark (losing ground). If the answer is incorrect, the markers stay in the same place. When two consecutive questions are answered incorrectly, the ball switches to the other team. The second team now receives the questions.

When a team advances its marker to the end zone, a touchdown is scored (6 points). The team has the option of receiving an additional question for an extra point while in the end zone. After a touchdown, place the team markers back at the 50-yard line. The team that did not score the touchdown should now receive the questions.

Four in a Row

Display a grid of five horizontal lines and five vertical lines. Divide the class into two teams: Xs and Os. As team members answer the review questions correctly, they may place their team symbol in a section of the grid. The first team to get four symbols in a row wins.

Jump Start

Group the class into teams. Provide a "jump" chair at the front of the room for each team. The first jumper from each team should sit in his team's chair, keeping both feet flat on the floor and his back against the chair. After the question is read, the seated students who know the answer should jump to their feet and remain standing. The first student to jump up and give the correct answer receives points for his team. Rotate jumpers after each question.

Mix and Match

Group the class into teams. Write several questions on strips of paper. Write the answers to those questions on separate strips of paper. Place the questions in one container and the answers in another container. Mix up the papers. Draw a question from the container and read it aloud. Choose a student to draw an answer. The student should determine whether the answer matches the question he heard. If it does match, the student receives a point for his team. If it does not match, he has the option of stating the correct answer. If he can give the correct answer, he receives a point for his team. Questions and answers should be placed back into their respective containers before the next question is drawn.

Puzzling Questions

Group the class into teams. Write several questions on strips of paper. Write the matching answers on separate strips of paper. Prepare a set of questions and answers for each team. Mix up the papers and give each team a set of questions and corresponding answers. At a given signal, each team should start organizing its papers and matching up the questions with the correct answers. The first team to display all the questions and answers correctly wins.

Materials Lists

The materials under each category are listed by order of use in the chapter. The items listed below are standard science supplies or materials found in a classroom or student's desk.

Standard Science Supplies

balance	metric beakers
calculator	protractor
hand lens	safety goggles
measuring cups with metric markings	spring scale
measuring spoons with metric markings	stopwatch
	thermometer
meter stick	timer

Basic Supplies

buckets	dishpans
cardstock (used for some IA pages)	glue sticks
centimeter ruler	newspaper
colored pencils	notebook paper
construction paper	paper towels
containers	pencils
crayons	scissors
	tape

Visuals and Instructional Aids are found in the SCIENCE 3 *Teacher Edition* and on TeacherToolsOnline.com for the teacher to use electronically, to make a chart, or to print a copy for each student. See the lesson for specific preparation details, such as printing an Instructional Aid on cardstock.

The Additional Activities section lists the materials for the activities located at the bottom of the *Teacher Edition* lesson pages.

These abbreviations are used to identify pages on the Materials List:

IA—Instructional Aid
R—Rubric
V—Visual

Mass and Weight (Dry) Measurement Equivalents

1 ounce		30 grams
16 ounces	1 pound	454 grams
1 kilogram	2.2 pounds	1000 grams

Volume (Liquid) Measurement Equivalents

1 teaspoon		0.33 tablespoon	5 milliliters
1 tablespoon	0.5 ounce	3 teaspoons	15 milliliters
1 cup	8 ounces	16 tablespoons	237 milliliters
1 quart	32 ounces	2 pints	946 milliliters
1 gallon	128 ounces	4 quarts	3.78 liters
1 liter		4.22 cups	1000 milliliters
0.5 liter		2.11 cups	500 milliliters

Chapter 1

Teacher Resources

For each student (or group)

- ❏ IA 1.1: *Bookmarks*, copy one set per student on cardstock
- ❏ IA 1.2: *Solar Mobile Crossbars*, copy one per student on cardstock

For display

- ❏ V 1.1: *Solar System*
- ❏ V 1.2: *Star Pattern*
- ❏ V 1.3: *Ursa Major*
- ❏ V 1.4: *Big Dipper*
- ❏ V 1.5: *Inner Planets*
- ❏ V 1.6: *Planet Order*
- ❏ V 1.7: *Moon Mystery*
- ❏ V 1.8: *Outer Planets*

Teaching Materials

- ❏ video of the solar system to scale
- ❏ large black circle and large white circle, 1 of each (both the same size)
- ❏ 1/2 a large white circle, to fit over the large black circle
- ❏ English pea
- ❏ baseball
- ❏ globe

Explorations

For each student (or group)

- ❏ Solar Mobile planets (*Activities* page 7)
- ❏ *Solar Mobile Crossbars*
- ❏ scissors
- ❏ hole punch
- ❏ 14 pieces of string (30 cm each), per student
- ❏ clear nighttime sky

Additional Activities

- ❏ 100 dried beans (or other small objects)
- ❏ clear container
- ❏ star stickers, per student
- ❏ dark blue construction paper, per student
- ❏ crayons
- ❏ writing paper
- ❏ 1 small bead
- ❏ 1 small marble
- ❏ 2 large marbles
- ❏ 2 softballs
- ❏ 1 soccer ball
- ❏ 1 basketball
- ❏ 3 m string
- ❏ 9 labels
- ❏ meter stick
- ❏ magnetic token, for each team

Assessments

- ❏ Quiz 1A
- ❏ Quiz 1B
- ❏ R: *Solar Mobile*
- ❏ R: *Moon Mystery*
- ❏ Test 1

Chapter 2

Teacher Resources

For display

- ❏ V 2.1: *Weather Map*
- ❏ V 2.2: *Weather Patterns*
- ❏ V 2.3: *Cloud Formations*
- ❏ V 2.4: *Water Cycle*
- ❏ V 2.5: *Tongue Twister*
- ❏ V 2.6: *Severe Weather Safety*
- ❏ V 2.7: *STEM: The Engineering Design Process*
- ❏ V 2.8: *Climate Zones*
- ❏ V 2.9: *Map of Alaska*
- ❏ Video of Climate Change from Institute for Creation Research

Teaching Materials

- ❏ umbrella
- ❏ sunglasses
- ❏ gloves
- ❏ rain gauge
- ❏ thermometer
- ❏ weather vane
- ❏ anemometer
- ❏ blue construction paper, 1 per student
- ❏ liquid glue
- ❏ cotton balls, 6 per student
- ❏ audio clip of severe weather alert
- ❏ video of a derecho
- ❏ map of United States
- ❏ canned food
- ❏ bottled water
- ❏ flashlight
- ❏ batteries
- ❏ weather or battery-powered radio
- ❏ first-aid kit
- ❏ blanket
- ❏ container for emergency supplies
- ❏ other items that are not necessary for emergency kit (candy, toys, soft drinks)
- ❏ world map
- ❏ slips of paper with one country (or city and country) written on each
- ❏ container or jar for slips of paper

Exploration, STEM

For each student (or group)

- ❏ thermometer
- ❏ rain gauge
- ❏ weather vane
- ❏ spray bottle with water
- ❏ 15 cm cardboard square, four per student
- ❏ several rolls of tape
- ❏ wooden craft sticks of various sizes
- ❏ plastic straws
- ❏ aluminum foil
- ❏ wax paper
- ❏ fabric scraps
- ❏ yarn or string
- ❏ roll of paper towels
- ❏ scissors

Additional Activities

- ❏ weather maps from newspapers or printed from websites, one different map per group
- ❏ access to weather map online or weather app

Assessments

- ❏ Quiz 2A
- ❏ Quiz 2B
- ❏ R: *Weather Watcher*
- ❏ R: *Waterproof Roof*
- ❏ Test 2

Chapter 3

Teacher Resources

For each student (or group)

- [ ] IA 3.1: *Dirt Cup*
- [ ] IA 3.2: *Fossilization and Minerals*

For display

- [ ] V 3.1: *Layers of Soil*
- [ ] V 3.2: *Rocks, Rocks, and More Rocks*
- [ ] V 3.3: *Science Inquiry Skills*
- [ ] V 3.4: *Scientific Investigation*
- [ ] V 3.5: *Science Safety Tips*
- [ ] V 3.6: *Ark on Flood Waters*

Teaching Materials

- [ ] 3 soil samples, from three different locations
- [ ] smooth rock
- [ ] rough rock
- [ ] small piece of soap, per student
- [ ] sandpaper, per student
- [ ] paper towel, per student
- [ ] rock samples of limestone, sandstone, obsidian, granite, and marble
- [ ] mineral samples
- [ ] fossil samples, or photos of fossils
- [ ] sponge
- [ ] scissors
- [ ] shallow dish
- [ ] beaker
- [ ] very warm water
- [ ] Epsom salt or table salt
- [ ] spoon

Investigation

For each student (or group)

- [ ] penny
- [ ] steel nail
- [ ] apatite
- [ ] copper
- [ ] gypsum
- [ ] quartz

Additional Activities

- [ ] small spade or shovel
- [ ] handful of soil
- [ ] 1 clear plastic cup, 256 mL (9 oz), per student
- [ ] 2 whole vanilla cookies, per student
- [ ] 227 mL (1 c) of chocolate pudding, per student
- [ ] spoon, per student
- [ ] 1 broken vanilla cookie, per student
- [ ] 1 crushed vanilla cookie, per student
- [ ] clear plastic cup
- [ ] water (multiple lessons)
- [ ] marker
- [ ] freezer
- [ ] clear jar with a lid
- [ ] soil
- [ ] old crayons of various colors for two activities
- [ ] crayon sharpener
- [ ] wax paper for two activities
- [ ] heavy-duty aluminum foil
- [ ] 2 small pieces of wood
- [ ] vise or wood clamps
- [ ] empty soup can
- [ ] saucepan
- [ ] hot plate (or stove)
- [ ] tongs
- [ ] silicone ice cube tray
- [ ] "sedimentary rock," from previous activity
- [ ] piece of fabric
- [ ] iron
- [ ] box of cut-out letters or letter tiles from a game

Assessments

- [ ] Quiz 3A
- [ ] Quiz 3B
- [ ] R: *Hard or Soft?*
- [ ] Test 3

Chapter 4

Teacher Resources

For each student (or group)

❏ IA 4.1: *Look Closer!*
❏ IA 4.2: *Parts of Cells*
❏ IA 4.3: *Layers of the Skin*
❏ IA 4.4: *Layers of the Skin Key*

For display

❏ V 4.1: *Parts of a Microscope*
❏ V 4.2: *Parts of an Animal Cell*
❏ V 4.3: *Parts of a Plant Cell*
❏ V 4.4: *Upper Body Organs*
❏ V 4.5: *Layers of the Skin*
❏ V 4.6: *Fingerprint Patterns*

Teaching Materials

❏ hand lens, one per group
❏ microscope
❏ thinly sliced onion skin, one per group
❏ prepared slides of onion skins
❏ 3 lb. bag of clementines, or other object weighing about 3 lbs.
❏ 1 chicken egg
❏ Kool-Aid packet, for display
❏ prepared Kool-Aid, per student
❏ 3 oz. paper cup, per student
❏ rope for Tug-of-War
❏ blindfold
❏ various items of different textures or temperatures
❏ picture of a knight in armor
❏ soap, for washing hands

Explorations

For each student (or group)

❏ individual flavored gelatin cup
❏ 2 rolled fruit strips
❏ round gumball
❏ small paper plate
❏ plastic knife
❏ washable-ink pad
❏ hand sanitizer
❏ paper towels
❏ magnifying glass

Additional Activities

❏ shoebox or other small container
❏ nail
❏ paper clip
❏ safety pin
❏ thumbtack
❏ hand lens
❏ microscope
❏ various small objects
❏ several yards of sewing elastic, 1" wide
❏ newspaper
❏ 2 index cards
❏ vegetable oil
❏ flour
❏ paper towels
❏ box of adhesive bandages
❏ chart paper

Assessments

❏ Quiz 4A
❏ Quiz 4B
❏ R: *Edible Cell*
❏ R: *Patterns on My Skin*
❏ Test 4

Chapter 5

Teacher Resources

For each student (or group)

- [ ] IA 5.1: *Needs Cards*
- [ ] IA 5.2: *Food Cards*

For display

- [ ] V 5.1: *Sample Plants*
- [ ] V 5.2: *Life Cycle of a Plant*
- [ ] V 5.3: *Plant Parts*
- [ ] V 5.4: *Parts of a Bean Seed*
- [ ] V 5.5: *Basic Photosynthesis*
- [ ] V. 5.6: *Photosynthesis Up Close*
- [ ] V 2.7: *STEM: The Engineering Design Process*

Teaching Materials

- [ ] house plants
- [ ] 2 cut flowers, same kind but different colors
- [ ] world map

Investigation, STEM

For each student (or group)

- [ ] 2 plastic cups, 266 mL (9 oz)
- [ ] marker
- [ ] masking tape
- [ ] potting soil
- [ ] 4 bean seeds
- [ ] water
- [ ] centimeter ruler
- [ ] potted house plant, or a container of potting soil to mimic environment of a potted house plant
- [ ] water bottles
- [ ] plastic bags
- [ ] plastic wrap
- [ ] cotton string
- [ ] nylon rope
- [ ] rubber bands
- [ ] water
- [ ] ice cubes
- [ ] paper towels

Additional Activities

- [ ] food coloring
- [ ] container
- [ ] water
- [ ] white carnation or celery stalk
- [ ] wooden spoon
- [ ] straw basket
- [ ] paper bag
- [ ] cotton dishtowel
- [ ] rope
- [ ] toothpicks
- [ ] metal spoon
- [ ] penny
- [ ] rock
- [ ] seashell
- [ ] other items, as desired

Assessments

- [ ] Quiz 5A
- [ ] Quiz 5B
- [ ] R: *A Place to Grow*
- [ ] R: *Vacation Water Challenge*
- [ ] Test 5

Chapter 6

Teacher Resources

For each student (or group)

- ❏ IA 6.1: *Animal Cards*
- ❏ IA 6.2: *Animal Cards*
- ❏ IA 6.3: *Animal Cards*
- ❏ IA 6.4: *Lizard*

For display

- ❏ V 6.1: *Animal Classification 1*
- ❏ V 6.2: *Life Cycles of Fish*
- ❏ V 6.3: *Stages of Amphibian Metamorphosis*
- ❏ V 6.4: *Life Cycle of Reptiles*
- ❏ V 3.3: *Science Inquiry Skills*
- ❏ V 6.5: *Animal Classification 2*
- ❏ V 6.6: *Animal Classification 3*
- ❏ V 6.7: *Life Cycle of an Insect*
- ❏ V 6.8: *Life Cycle of a Spider*
- ❏ V 6.9: *Life Cycle of Mealworms*

Teaching Materials

- ❏ assortment of pencil erasers
- ❏ pictures of fish from different habitats, for display
- ❏ picture of an animal camouflaged in its habitat, per group (or video of camouflaged animals)
- ❏ white sheet of construction paper
- ❏ dark brown or black sheet of construction paper
- ❏ 2 thermometers
- ❏ tape
- ❏ timer
- ❏ leather belt or other piece of leather
- ❏ piece of scrap paper, per student
- ❏ timer or watch with a second hand
- ❏ piece of cotton clothing or other clothing made from plant or animal fibers
- ❏ piece of polyester clothing
- ❏ video of a ballooning spider
- ❏ camera or cellphone camera (optional)

Investigations

For each student (or group)

- ❏ 2 paper lizards (IA 6.4)
- ❏ scissors
- ❏ tape
- ❏ 2 thermometers
- ❏ mealworms
- ❏ clear container with a lid
- ❏ disposable (latex-free) gloves
- ❏ uncooked oatmeal
- ❏ several baby carrots
- ❏ hand lens
- ❏ strips of construction paper (optional)
- ❏ tape (optional)

Additional Activities

- ❏ aluminum pie pan
- ❏ table salt
- ❏ metal spoon
- ❏ full page pictures of a fish, an amphibian, a reptile, an insect, and a spider

Assessments

- ❏ Quiz 6A
- ❏ Quiz 6B
- ❏ R: *Leaping Lizards*
- ❏ R: *Changes in Dark Places*
- ❏ Test 6

Chapter 7

Teacher Resources

For each student (or group)

❏ IA 6.1: *Animal Cards*
❏ IA 6.2: *Animal Cards*
❏ IA 6.3: *Animal Cards*
❏ IA 7.1: *Penguin*
❏ IA 7.2: *Bird or Mammal?*
❏ IA 7.3: *Bird or Mammal? Key*
❏ IA 7.4: *Unusual Animals*
❏ IA 7.5: *Unusual Animals*
❏ IA 7.6: *Unusual Animals Key*
❏ IA 7.7: *Unusual Animals Key*

For display

❏ V 7.1: *Animal Classification 4*
❏ V 7.2: *Life Cycle of a Bird*
❏ V 7.3: *Fruit Bat*
❏ V 7.4: *Animal Classification 5*
❏ V 7.5: *Life Cycle of a Mammal*
❏ V 7.6: *Compare and Contrast*
❏ audio of a Yao hunter calling a honeyguide (TeacherToolsOnline.com)
❏ video of a male sandgrouse collecting water (TeacherToolsOnline.com)

Teaching Materials

❏ oral thermometer
❏ hanging bird feeder
❏ bird seed
❏ picture of Air Force One (plane)
❏ picture of a blue whale
❏ piece of scrap paper, one per student
❏ nest-building materials such as clay, straw, grass, twigs, yarn, leaves, per student or group of students
❏ tray to make distribution of nest-building materials easier, per student or group
❏ piece of paper with "Limited Food Supply" written on it
❏ piece of paper with "Plenty of Food" written on it
❏ 2 or 3 pictures of several different kinds of birds
❏ 2 or 3 pictures of several different kinds of mammals

❏ 2 or 3 pictures of several mammals that have blubber (sea lions, seals, whales, manatees, dolphins)

Explorations

For each student (or group)

❏ bird feeder
❏ bird seed
❏ 2 containers of ice water
❏ 6 plastic sandwich bags, per each group of 3 students
❏ 2 thermometers, per group
❏ stopwatch
❏ feathers
❏ fur
❏ shortening or butter
❏ plastic spoon
❏ paper towels

Additional Activities

❏ jellybeans
❏ black, white, and orange crayons
❏ spray bottle with water
❏ paper towel
❏ water
❏ index card or piece of cardboard

Assessments

❏ Quiz 7A
❏ Quiz 7B
❏ R: *Feathery Friends*
❏ R: *Blubber, Feathers, and Fur*
❏ Test 7

Chapter 8

Teacher Resources

For each student (or group)

- ❏ IA 8.1: *Food Chain Links*
- ❏ IA 8.2: *Food Chain Links*
- ❏ IA 8.3: *Tag Identity Badges*
- ❏ IA 8.4: *Tag Identity Badges*
- ❏ IA 8.5: *Tag Life Cards*
- ❏ IA 8.6: *Word Chains Cards*
- ❏ IA 8.7: *Word Chains Cards*
- ❏ IA 8.8: *Word Chains Cards*

For display

- ❏ V 8.1: *Fitting Together Organizer*
- ❏ V. 2.7: *STEM: The Engineering Design Process*
- ❏ V 8.2: *Food Chain*
- ❏ V 8.3: *Food Web*

Teaching Materials

- ❏ pictures of plants, animals, and people
- ❏ mouse, picture or live pet mouse
- ❏ paper chain with 4 links, per student

Exploration, STEM

For each student (or group)

- ❏ one 2 liter soda bottle, per student or group
- ❏ *Tag Identity Badges* (IA 8.3–8.4)
- ❏ *Tag Life Cards* (IA 8.5)
- ❏ safety pins, one per student
- ❏ posterboard
- ❏ marker
- ❏ large plastic hoops
- ❏ stopwatch
- ❏ whistle

Additional Activities

- ❏ blank drawing paper, per student
- ❏ 5 pieces of paper
- ❏ marker
- ❏ hole punch
- ❏ yarn
- ❏ 15 envelopes

Assessments

- ❏ Quiz 8A
- ❏ Quiz 8B
- ❏ R: *Habitat Help!*
- ❏ R: *Ecosystem Tag*
- ❏ Test 8

Chapter 9

Teacher Resources

For each student (or group)

❏ IA 9.1: *Instrument Cards*
❏ IA 9.2: *Instrument Cards*

For display

❏ V 9.1: *Changing States*
❏ V 9.2: *Speed of Sound*
❏ V 9.3: *Parts of the Ear*
❏ *Musical Bottles* video (TeacherToolsOnline)

Teaching Materials

❏ 5 clear containers of various shapes and sizes, each large enough to hold 250 mL
❏ beaker that holds at least 250 mL
❏ water (multiple lessons)
❏ decorative pillow
❏ toy truck
❏ whole apple
❏ knife
❏ plate or cutting board
❏ 9" latex balloon, per student
❏ ice cube, per student
❏ paper towel, per student
❏ unpopped popcorn kernels
❏ small cup or bowl, per student
❏ hot air popcorn popper
❏ melted butter, optional
❏ salt, optional
❏ additional popcorn toppings, optional
❏ empty drinking glass or jar
❏ ice
❏ self-sealing sandwich bag, per student
❏ 3 or more ingredients to make a snack mix, per student
❏ serving spoon for each ingredient
❏ bell
❏ 2 wooden pencils
❏ small rock or pebble
❏ large shallow container, partially filled with water
❏ sound clips of alarm clock, siren, flute, and lawnmower
❏ plastic containers of various shapes and sizes, one per student

❏ aluminum foil
❏ balloons, not inflated
❏ rubber bands
❏ squares of cardboard
❏ whistle

Exploration, Investigation, Inquiry

For each student (or group)

❏ 2 solid objects of different shapes and sizes
❏ balance
❏ clear container
❏ beaker or metric measuring cup (multiple lessons)
❏ water (multiple lessons)
❏ balloon, not fully inflated
❏ 3 narrow-necked plastic 2-liter bottles (multiple lessons)
❏ funnel (multiple lessons)
❏ 1 narrow-necked plastic 500 mL bottle
❏ colored pencils, blue and red

Additional Activities

❏ 2 clear plastic cups
❏ permanent marker
❏ beaker or metric measuring cup (multiple lessons
❏ rubbing alcohol
❏ water (multiple lessons)
❏ white glue
❏ borax
❏ wooden craft stick, per student
❏ plastic cups, 2 per student
❏ metric or standard measuring spoon

Assessments

❏ Quiz 9A
❏ Quiz 9B
❏ R: *Which Kind of Matter?*
❏ R: *Musical Bottles* (Investigation)
❏ R: *Musical Bottles* (Inquiry)
❏ Test 9

Chapter 10

Teacher Resources

For each student (or group)

❏ IA 10.1: *Doing Work*
❏ IA 10.2: *Doing Work Key*
❏ IA 10.3: *Pattern of Motion Cards*

For display

❏ V 10.1: *Cause and Effect*
❏ V 10.2: *Kinds of Forces*
❏ V 10.3: *Balanced or Unbalanced?*
❏ V 2.7: *STEM: The Engineering Design Process*
❏ V 10.4: *Equations*

Teaching Materials

❏ dart board game with balls using hook-and-loop fasteners
❏ jump rope
❏ soccer ball
❏ scrap of paper, per student
❏ scale
❏ object to be weighed
❏ strong magnet
❏ iron filings, nails, or paper clips
❏ foam ball
❏ chair
❏ book
❏ table
❏ rope for tug of war
❏ toy truck
❏ 2 small balls with different masses
❏ large plastic or metal bowl
❏ stapler
❏ globe

Investigation, STEM

For each student (or group)

❏ Ping-Pong ball™
❏ ruler
❏ marble, per student
❏ variety of paper tubes
❏ ball of string
❏ ball of yarn
❏ different-sized paper clips
❏ sheets of cardboard or posterboard
❏ sheets of paper
❏ tape
❏ glue
❏ different-sized dowels
❏ different-sized boxes

Additional Activities

❏ 2 books
❏ string, the length of the circumference of the books plus at least 5 cm
❏ rubber band
❏ 6 pencils
❏ ruler
❏ drawing paper
❏ 20 feet of string
❏ 18 strips of paper
❏ masking tape

Assessments

❏ Quiz 10A
❏ Quiz 10B
❏ R: *Balanced and Unbalanced Forces*
❏ R: *A-Maze-ing Marble Game*
❏ Test 10

Chapter 11

Teacher Resources

For each student (or group)

❏ IA 11.1: *Charged Objects*
❏ IA 11.2: *Indoor Safety Counts!*
❏ IA 11.3: *Outdoor Safety Counts!*

For display

❏ V 2.7: *STEM: The Engineering Design Process*

Teaching Materials

❏ picture of socks, or other articles of clothing, attached to clothing by means of static electricity, for display
❏ resealable sandwich bag, per student for *Charged Objects* instructional aid cards
❏ 9" balloons, 1 for teacher and 1 per student
❏ paper confetti, small amount per student
❏ salt, small amount per student
❏ pepper, small amount per student
❏ toaster
❏ bread
❏ insulated wire with some insulation removed
❏ pliers or wire cutters with insulated grips
❏ 2 bar magnets, per group of students
❏ paper lunch bag with bar magnet inside (multiple lessons)
❏ paper clips (multiple lessons)
❏ iron filings
❏ sheet of white paper
❏ plastic bag or plastic wrap
❏ document camera (optional)
❏ variety of magnets (bar, horseshoe, disc) (multiple lessons)
❏ 3" or 4" steel nail
❏ single strand insulated wire, 1 m long
❏ D cell (1.5 volt) battery
❏ masking tape
❏ small paper clips

Exploration, Investigation, Inquiry, STEM

For each student (or group)

❏ 3 pieces of insulated wire, each 20 cm long
❏ miniature lamp and socket with 2 terminals (1.2 volt bulb)

❏ batteries (1.5 volt C or D cell) (multiple lessons)
❏ masking tape (multiple lessons)
❏ aluminum foil
❏ button
❏ craft stick
❏ metal key
❏ penny
❏ horseshoe or bar magnet
❏ paper clip (multiple lessons)
❏ 100 sheets of paper (multiple lessons)
❏ 5 magnets, various types and strengths (multiple lessons)
❏ single strand insulated wire
❏ 3" or 4" steel nails
❏ string
❏ wire clothes hangers
❏ dowel rods
❏ meter sticks
❏ variety of screws, bolts, and nuts

Additional Activities

❏ several plastic or nylon combs
❏ wool (optional)
❏ several table-tennis balls
❏ bendable drinking straw
❏ lump of clay
❏ 1" x 6" strip of tissue paper or lightweight aluminum foil
❏ tape
❏ inflated balloon
❏ piece of wool or fur
❏ 11 craft sticks
❏ 2 toy cars
❏ 2 disc magnets
❏ electrical tape

Assessments

❏ Quiz 11A
❏ Quiz 11B
❏ R: *Conductors Needed*
❏ R: *Magnetism and Matter (Investigation)*
❏ R: *Magnetism and Matter (Inquiry)*
❏ R: *Lost and Found*
❏ Test 11

Instructional Aids

Bookmarks

Copy on cardstock and cut out. Distribute one pair to each student.

Solar Mobile Crossbars

Name _____

Copy on cardstock. Distribute a cardstock copy to each student.

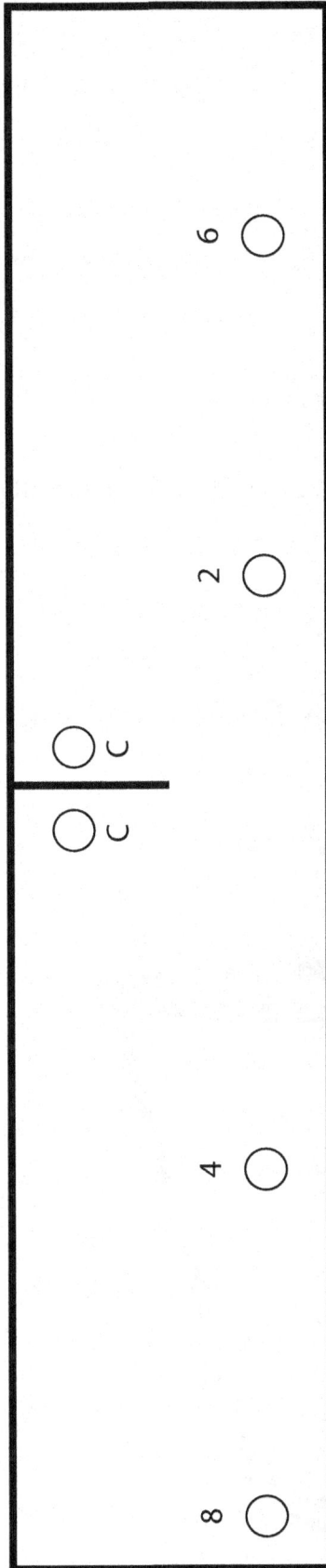

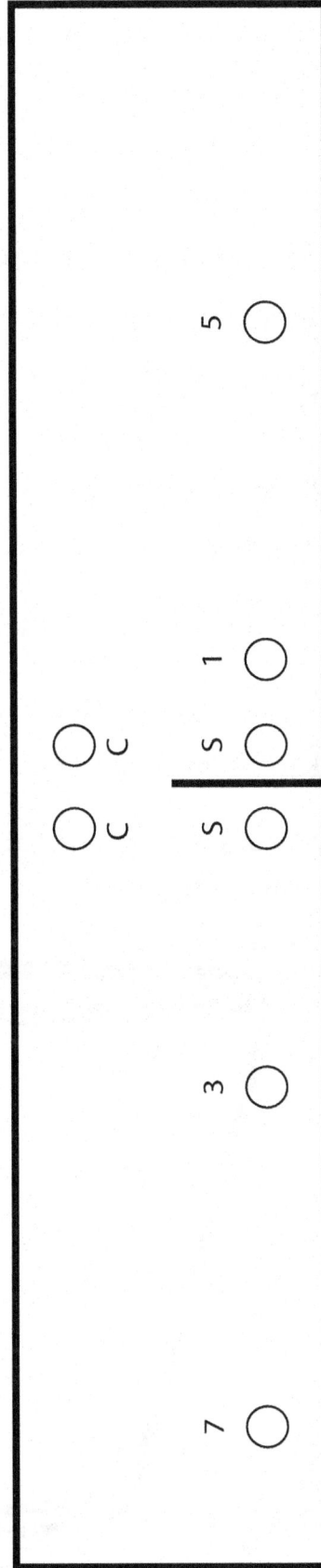

6

2

5

1

S

S

4

3

8

7

Dirt Cup

Purpose

Model the layers of soil.

Procedure

☐ 1. Place 2 whole cookies in the bottom of the empty cup.

☐ 2. Spoon half the pudding on top of the cookie.

☐ 3. Add the broken cookie to the pudding in the cup and stir the pudding and broken cookie.

☐ 4. Stir the crushed cookie into the leftover pudding.

☐ 5. Spoon the pudding and crushed cookie mixture on top of the broken-cookie layer.

Materials

☐ 1 clear plastic cup, 256 mL (9 oz)
☐ 2 whole vanilla cookies
☐ 227 mL (1 cup) of chocolate pudding
☐ spoon
☐ 1 broken vanilla cookie
☐ 1 crushed vanilla cookie

Draw the different layers of your dirt cup. Label each layer *topsoil*, *subsoil*, or *bedrock*.

Fossilization and Minerals

Name _____

1. Is the sponge hard or soft?   hard   soft

2. What sound did the sponge make when it was broken in half?

   _____

3. What liquid filled the spaces in the sponge? _____

4. When the salt water evaporated, what was left behind in the spaces?

   _____

5. Is salt a mineral?   yes   no

6. What are rocks made of? _____

7. How many days did it take for the sponge to dry up and turn to a fossil?

   _____ days

8. How does the fossil model help to support the biblical view for how

   most fossils formed? _____

   _____

   _____

   _____

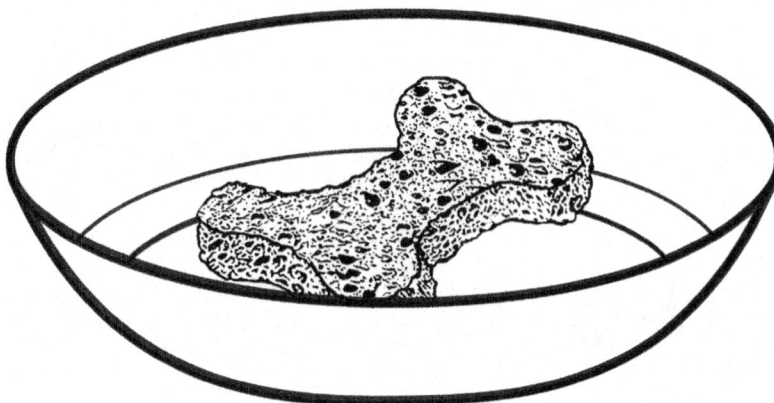

Look Closer!

Use a hand lens to look at the sample. Draw what you observe in the box with the picture of the hand lens. Use a microscope to look at the sample. Draw what you observe in the box with the picture of the microscope.

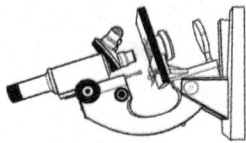

Name _____

Parts of Cells

Name _____

cell membrane   cell wall   cytoplasm   nucleus

Complete the cell diagrams, using the terms from the word bank.
Three terms will be used more than once.

Animal Cell

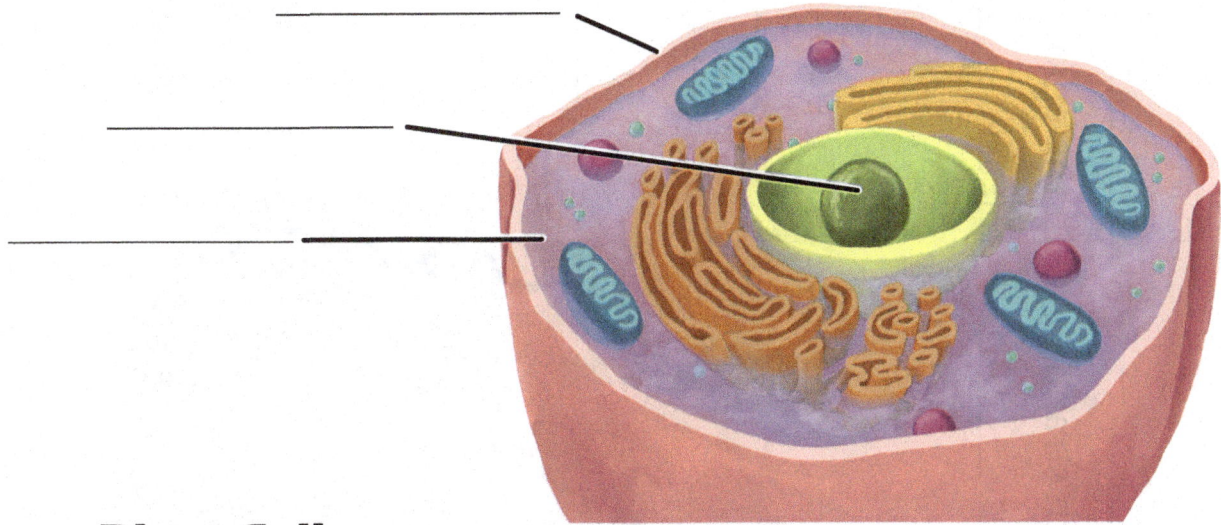

Plant Cell

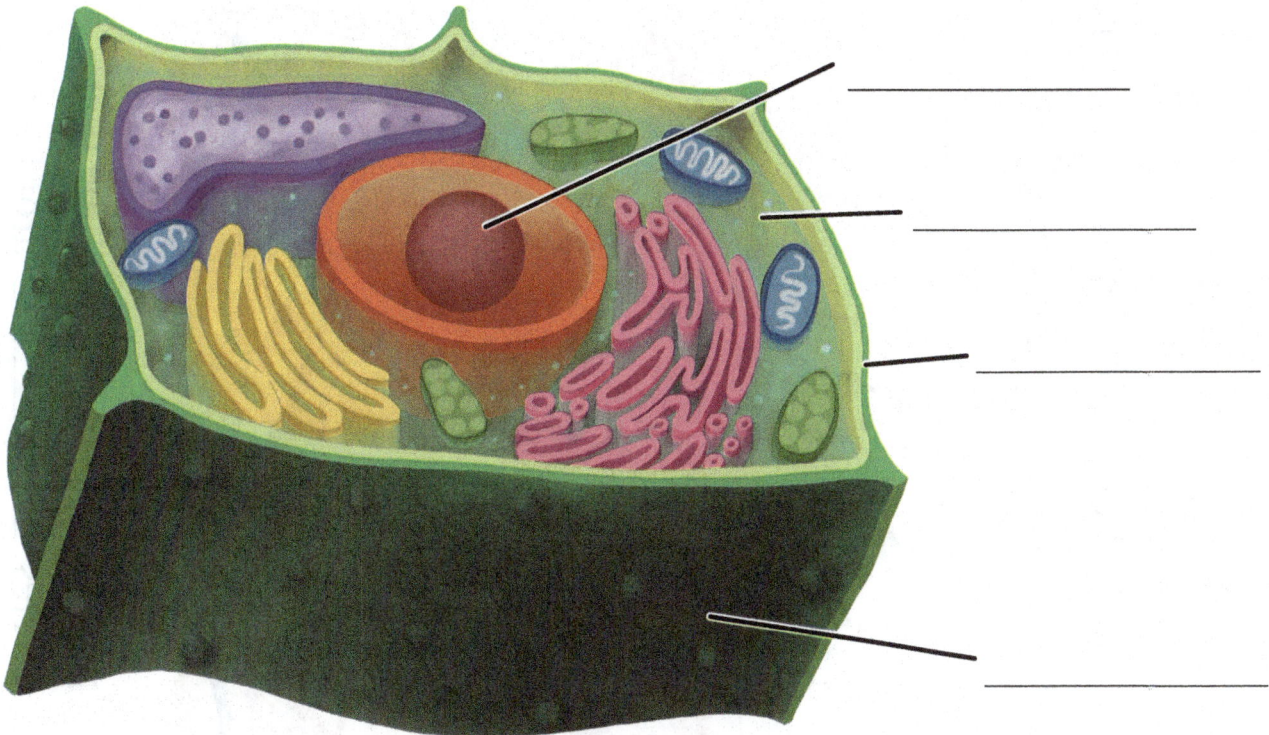

Layers of the Skin

Name ─────────

─── the tiny openings

─── in the epidermis where the body

─── releases

─── where each strand of

─── hair

─── releases oil, which helps

─── your body

─── layer

─── layer

─── special

─── part in the

─── that makes sweat and moves

─── it to the skin's surface

─── tiny pathway that

─── between the

─── brain and other parts of the body

─── carries

─── thin tube that

─── around

─── carries

─── your body

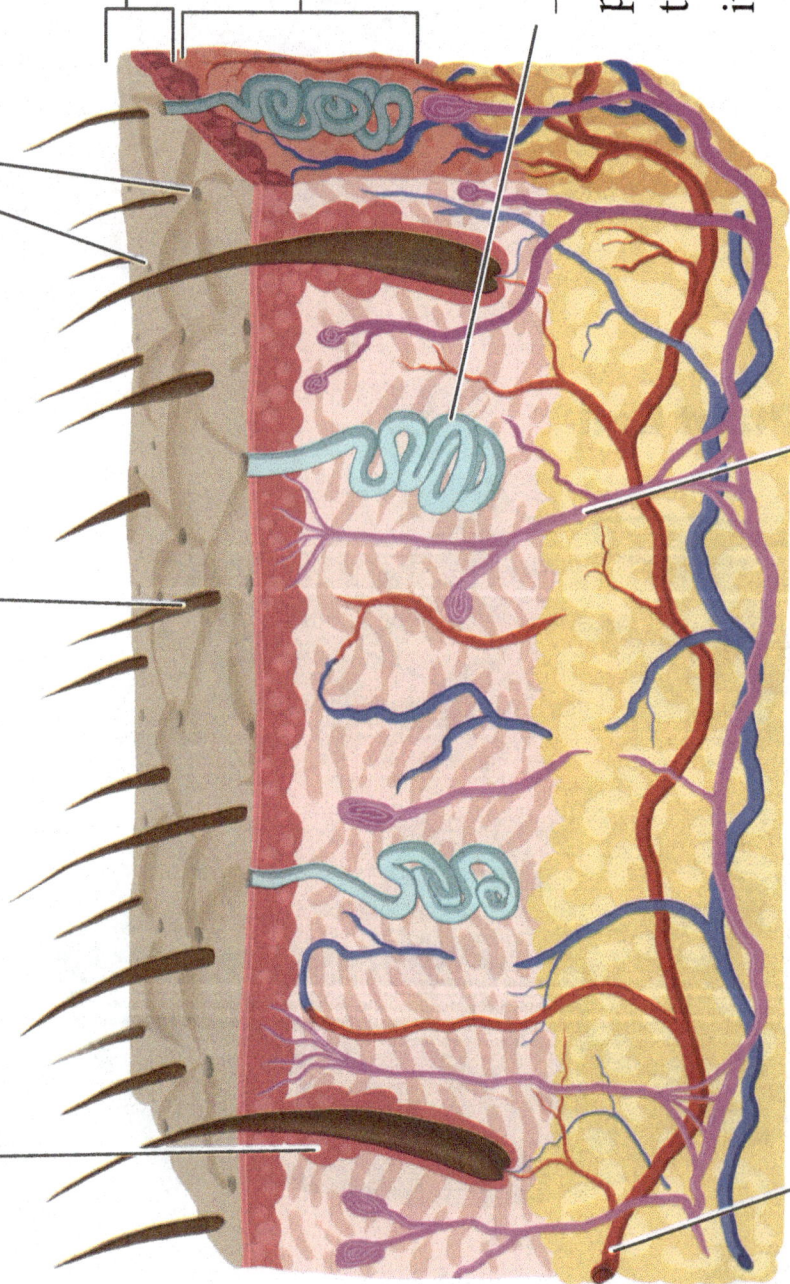

Layers of the Skin Key

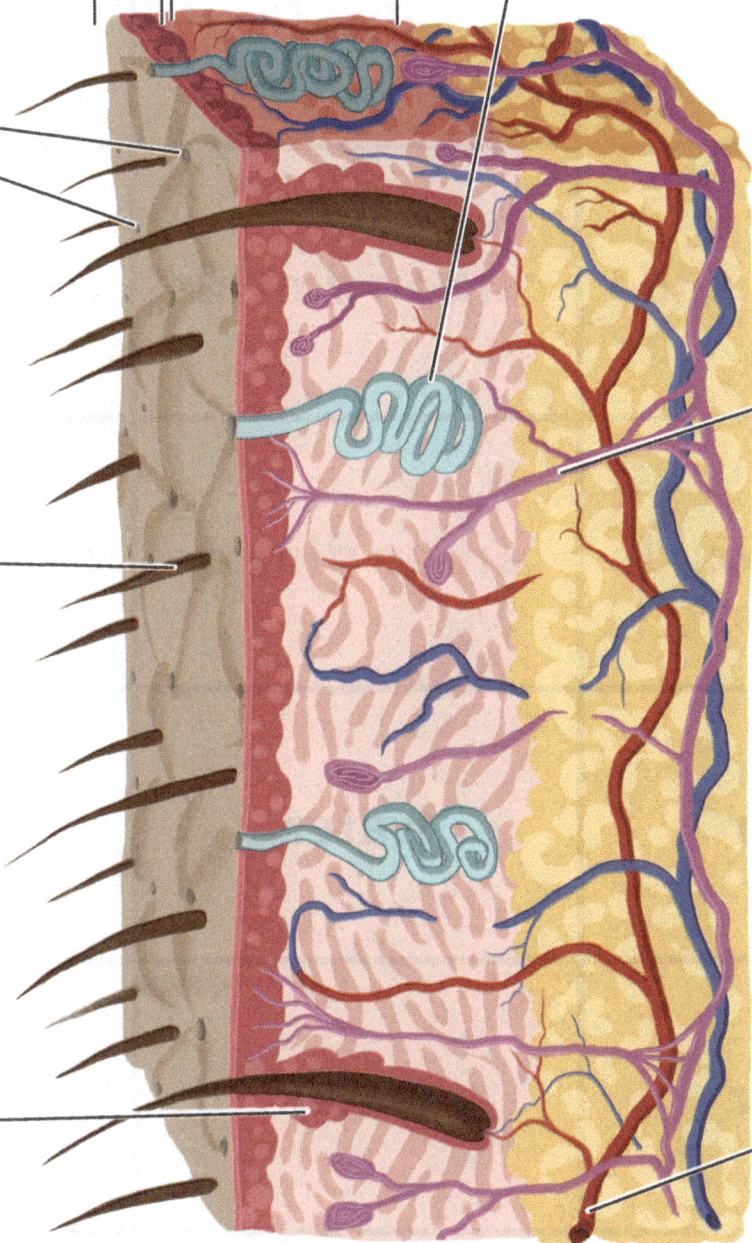

oil gland — releases oil, which helps *protect* your body

hair follicle — where each strand of hair *starts*

pores — the tiny openings in the epidermis where the body releases *sweat*

epidermis

top — layer

dermis

second — layer

sweat gland — special part in the *dermis* that makes sweat and moves it to the skin's surface

nerve — tiny pathway that carries *messages* between the brain and other parts of the body

blood vessel — thin tube that carries *blood* around your body

Needs Cards

Name _____

Print on cardstock and cut apart.

sunlight	sunlight	sunlight
sunlight	sunlight	sunlight
carbon dioxide	carbon dioxide	carbon dioxide
carbon dioxide	carbon dioxide	carbon dioxide
water	water	water
water	water	water

Food Cards

Print on cardstock and cut apart.

food	food	food
food	food	food
food	food	food
food	food	food
food	food	food
food	food	food

Animal Cards

alligator

Gila monster

banded California snake

wolf spider

black widow spider

tarantula

Animal Cards

trout

angelfish

chicken

ostrich

cow

beaver

Animal Cards

frog

salamander

toad

ant

bee

butterfly

Lizard

Copy and cut out two lizards.

Penguin

Name _____

Bird or Mammal?

Read each verse. Write what kind of animal is mentioned in the verse.

1. Genesis 8:9 _____

2. Genesis 24:11 _____

3. Judges 7:5 _____

4. Judges 15:4 _____

5. Job 39:26 _____

6. Isaiah 40:31 _____

7. Jeremiah 13:23 _____

8. Luke 12:6 _____

Write each animal from above in the correct group.

Birds	Mammals

Bird or Mammal? Key

Name _____

Read each verse. Write what kind of animal is mentioned in the verse.

1. Genesis 8:9   *dove* _____

2. Genesis 24:11   *camel* _____

3. Judges 7:5   *dog* _____

4. Judges 15:4   *fox* _____

5. Job 39:26   *hawk* _____

6. Isaiah 40:31   *eagle* _____

7. Jeremiah 13:23   *leopard* _____

8. Luke 12:6   *sparrow* _____

Write each animal from above in the correct group.

Birds	Mammals
dove	*camel*
eagle	*dog*
hawk	*fox*
sparrow	*leopard*

Unusual Animals

Name _____

Some animals have similar characteristics. Scientists group animals with similar characteristics together. Most animals easily fit into one of the groups: fish, amphibians, reptiles, birds, or mammals. But some animals do not fit neatly into any one group!

Read the paragraph. Write which group you think the animal belongs in. List one reason why you chose that group.

When this animal was first discovered in the late 1700s, scientists thought it was a hoax, or a trick. A platypus has a furry body. Its rubbery bill and webbed feet are like a duck's bill and feet. Its tail is like a beaver's. A platypus lays eggs on land and breathes air through lungs. The mother platypus feeds her babies with milk.

Group: _____

Reason: _____

Armadillo is a Spanish word meaning "little armored one." The Brazilian three-banded armadillo has bony scales that cover most of its body. The scales are connected by bands of skin. Its belly and the insides of its legs have no armor. These parts are covered with long hair. The armadillo can roll up into a tight ball for protection. It finds food by smelling with its nose close to the ground. When it finds prey, the armadillo digs a hole. Then it uses its long, sticky tongue to lap up ants and termites. The armadillo has one live baby at a time.

Group: _____

Reason: _____

Unusual Animals

Name _____

Mudpuppies live in lakes, ponds, and rivers. They spend all their lives in water. They can grow to be as long as 48 cm (19 in.) long. They can eat small fish, snails, and insects. Mudpuppies have lungs, but they do not use them to breathe. They breathe through feathery red gills on the outside of their body. They have thin, smooth, moist skin and four toes on each of their feet. Mudpuppies lay their eggs in water.

Group: _____

Reason: _____

A cassowary is a large animal! This animal can weigh about 60 kg (130 lb) and be over 2 m (6½ ft) tall. It has two strong legs, and its feet have three toes and sharp claws. A cassowary protects itself by kicking. It has black feathers and lays pale green eggs. A cassowary cannot fly but can swim. It eats fruit, insects, and small animals that are found near the ground.

Group: _____

Reason: _____

Unusual Animals Key

Name _____

Some animals have similar characteristics. Scientists group animals with similar characteristics together. Most animals easily fit into one of the groups: fish, amphibians, reptiles, birds, or mammals. But some animals do not fit neatly into any one group!

Read the paragraph. Write which group you think the animal belongs in. List one reason why you chose that group.

When this animal was first discovered in the late 1700s, scientists thought it was a hoax, or a trick. A platypus has a furry body. Its rubbery bill and webbed feet are like a duck's bill and feet. Its tail is like a beaver's. A platypus lays eggs on land and breathes air through lungs. The mother platypus feeds her babies with milk.

Group: _*mammals*_____

Reason: _*Possible answers: has fur, breathes air through lungs, feeds babies with milk*_____

Armadillo is a Spanish word meaning "little armored one." The Brazilian three-banded armadillo has bony scales that cover most of its body. The scales are connected by bands of skin. Its belly and the insides of its legs have no armor. These parts are covered with long hair. The armadillo can roll up into a tight ball for protection. It finds food by smelling with its nose close to the ground. When it finds prey, the armadillo digs a hole. Then it uses its long, sticky tongue to lap up ants and termites. The armadillo has one live baby at a time.

Group: _*mammals*_____

Reason: _*Possible answers: has hair, has live babies*_____

Unusual Animals Key

Name _____

Mudpuppies live in lakes, ponds, and rivers. They spend all their lives in water. They can grow to be as long as 48 cm (19 in.) long. They can eat small fish, snails, and insects. Mudpuppies have lungs, but they do not use them to breathe. They breathe through feathery red gills on the outside of their body. They have thin, smooth, moist skin and four toes on each of their feet. Mudpuppies lay their eggs in water.

Group: _amphibians_

Reason: _Possible answers: have thin, smooth, moist skin; breathe with gills_

A cassowary is a large animal! This animal can weigh about 60 kg (130 lb) and be over 2 m (6½ ft) tall. It has two strong legs, and its feet have three toes and sharp claws. A cassowary protects itself by kicking. It has black feathers and lays pale green eggs. A cassowary cannot fly but can swim. It eats fruit, insects, and small animals that are found near the ground.

Group: _birds_

Reason: _Possible answers: has feathers, lays eggs_

Food Chain Links

Name _____

Print on cardstock and cut apart.

grass  gets energy from the sun	**nuts**  gets energy from the sun
rabbit  eats grass	**snake**  eats mice, rabbits, squirrels, and small birds
owl  eats mice, rabbits, snakes, squirrels, raccoons, skunks, and small birds	**mouse**  eats seeds, grasses, and leaves

Food Chain Links

Print on cardstock and cut apart.

robin

eats seeds, worms, and insects

squirrel

eats nuts, seeds, and insects

grasshopper

eats grass, clover, and other plants

the sun

skunk

eats insects, grasses, nuts, fruit, mice, and young squirrels and rabbits

raccoon

eats nuts, insects, snakes, bird eggs, young birds, squirrels, and mice

Tag Identity Badges

Name _____

Tag Identity Badges

Name _____

Tag Life Cards

Name _____

1 life	1 life	1 life	1 life
1 life	1 life	1 life	1 life
1 life	1 life	1 life	1 life

Word Chains Cards

Copy on cardstock, cut apart, and sort into terms.

E	C	O	S	Y	S	T
E	M	E	N	V	I	R
O	N	M	E	N	T	P
O	P	U	L	A	T	I
O	N	H	A	B	I	T
A	T	C	O	M	M	U

Science 3 • Teacher Resources

Word Chains Cards

Name _____

Copy on cardstock, cut apart, and sort into terms.

N	I	T	Y	P	R	O
D	U	C	E	R	C	O
N	S	U	M	E	R	D
E	C	O	M	P	O	S
E	R	H	E	R	B	I
V	O	R	E	O	M	N

Word Chains Cards

Name _____

Copy on cardstock, cut apart, and sort into terms.

I	V	O	R	E	C	A
R	N	I	V	O	R	E
F	O	O	D	C	H	A
I	N	P	R	E	D	A
T	O	R	P	R	E	Y
F	O	O	D	W	E	B

Instrument Cards

Name _____

Print on cardstock and cut apart.

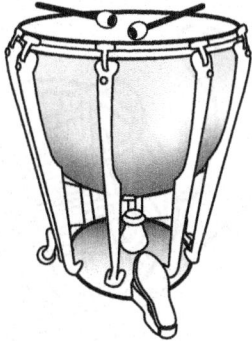

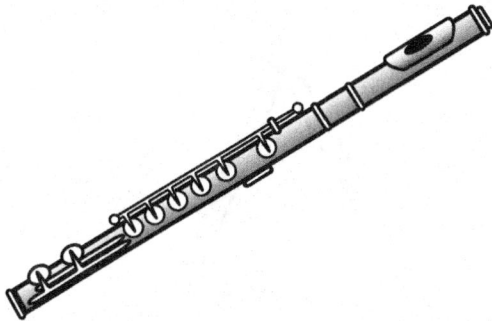

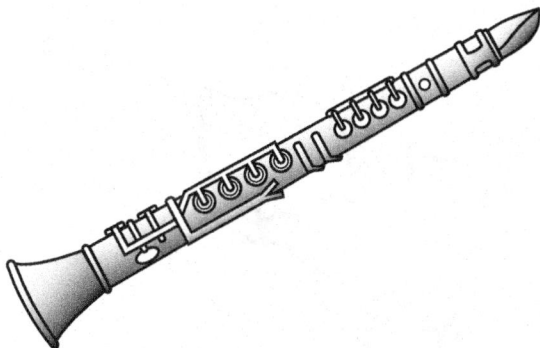

Instrument Cards

Name _____

Print on cardstock and cut apart.

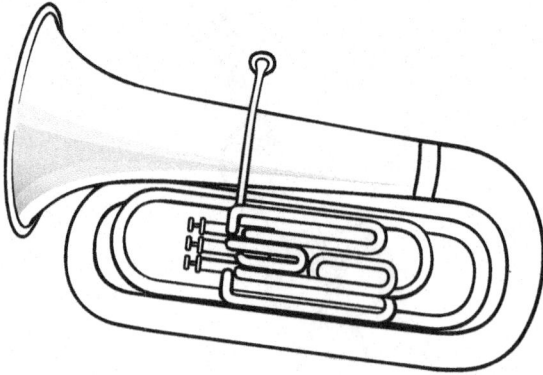

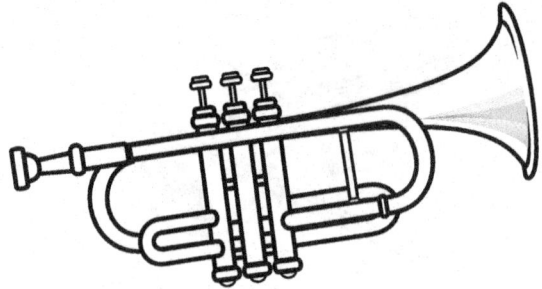

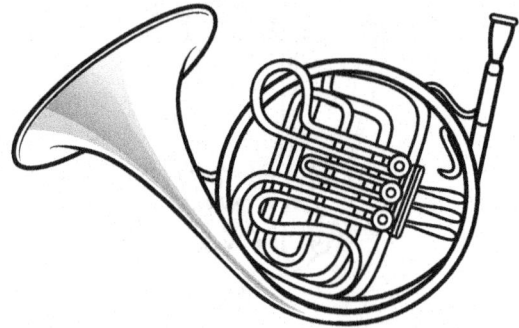

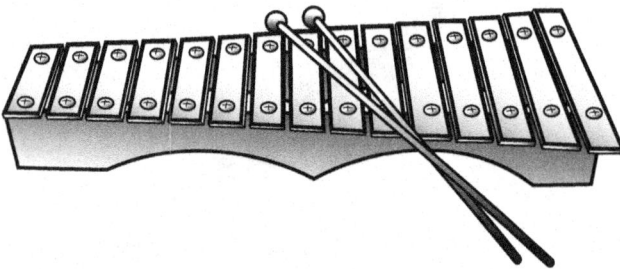

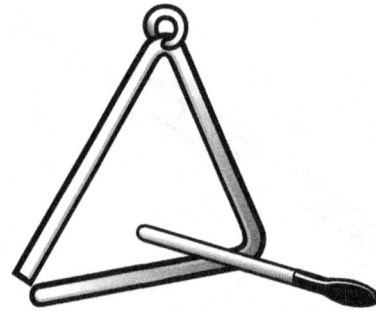

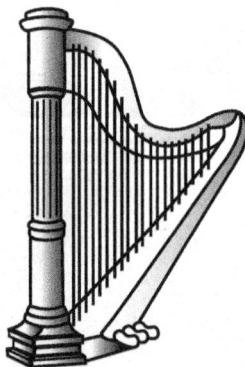

Doing Work

Name _____

Scientists say that work is done when a force moves something. When you are helping others and working for God, you might think about building a church or telling people about Jesus. But many other kinds of work can also bring God glory.

Read the verses. Write what is doing the work in the verses.

animal   machine   people   water   wind

Bible Verse	Work Done	What Is Doing Work
Exodus 10:19	carried away the locusts	_____
Exodus 14:28	covered the chariots, horsemen, and all of Pharaoh's army	_____
Ruth 2:2–3	gleaned grain from the fields	_____
2 Kings 5:9	pulled Naaman's chariot	_____
Jonah 1:17	swallowed Jonah	_____
Matthew 4:18	casting a net into the sea	_____
Luke 19:2–4	ran and climbed a sycamore tree to see Jesus	_____
Luke 19:35	carried Jesus	_____

Complete the section.

1. Name one kind of work you will do today. _____

2. Why should you be willing to do this work? Use the Bible's teaching in your answer.

_____

Doing Work Key

Name _____

Scientists say that work is done when a force moves something. When you are helping others and working for God, you might think about building a church or telling people about Jesus. But many other kinds of work can also bring God glory.

Read the verses. Write what is doing the work in the verses.

| animal | machine | people | water | wind |

Bible Verse	Work Done	What Is Doing Work
Exodus 10:19	carried away the locusts	*wind*
Exodus 14:28	covered the chariots, horsemen, and all of Pharaoh's army	*water*
Ruth 2:2–3	gleaned grain from the fields	*people*
2 Kings 5:9	pulled Naaman's chariot	*animal*
Jonah 1:17	swallowed Jonah	*animal*
Matthew 4:18	casting a net into the sea	*people*
Luke 19:2–4	ran and climbed a sycamore tree to see Jesus	*people*
Luke 19:35	carried Jesus	*animal*

Complete the section.

1. Name one kind of work you will do today. _____

2. Why should you be willing to do this work? Use the Bible's teaching in your answer.
   All my work should bring glory to God.

Pattern of Motion Cards

Name _____

Copy the page enough times so that each student will have two cards.

zigzag	**zigzag**	**zigzag**
back-and-forth	**back-and-forth**	**back-and-forth**
up-and-down	**up-and-down**	**up-and-down**
S **pattern**	*S* **pattern**	*S* **pattern**
straight line	**straight line**	**straight line**
circle	**circle**	**circle**
zigzag	**zigzag**	**zigzag**
back-and-forth	**back-and-forth**	**back-and-forth**
up-and-down	**up-and-down**	**up-and-down**
S **pattern**	*S* **pattern**	*S* **pattern**
straight line	**straight line**	**straight line**
circle	**circle**	**circle**

Charged Objects

Make one copy for each student. Cut apart into cards.

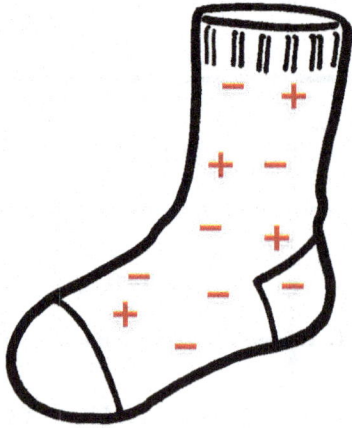

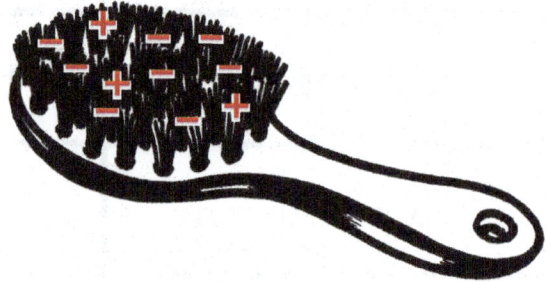

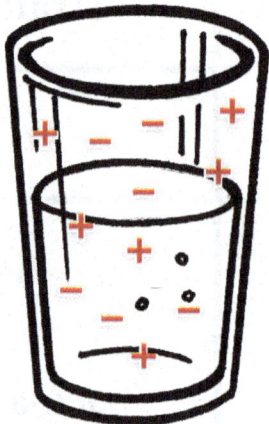

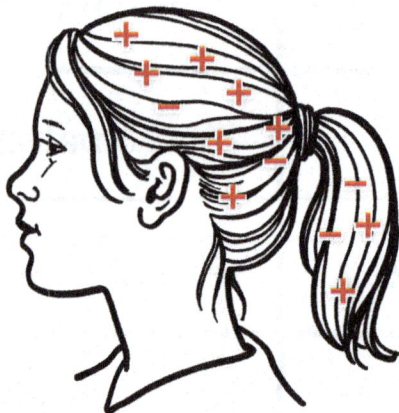

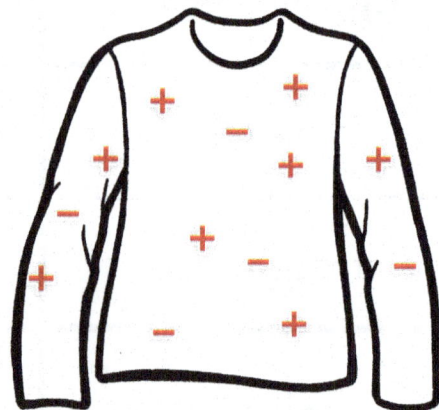

Science 3 • Teacher Resources

Indoor Safety Counts!

Name _____

1 Don't plug too many items into one outlet or extension cord.

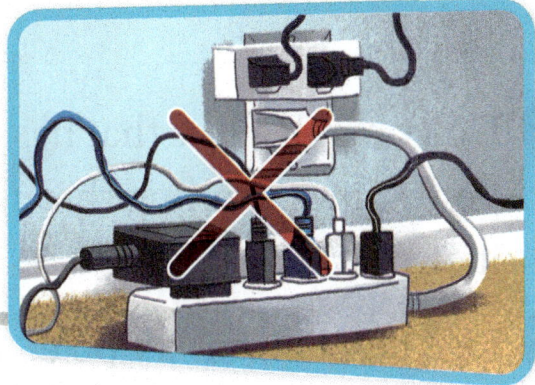

2 Put electrical cords where people will not trip over them. Don't place anything on an electrical cord. Never run a cord under rugs or furniture.

3 Don't use items that have frayed or damaged cords.

4 Keep electrical cords away from heat and water.

5 Don't touch light switches or electrical appliances if you are touching water or if you are standing on a wet or damp surface.

Instructional Aid 11.2
For use with Lesson 82

Outdoor Safety Counts!

1 Don't climb a tree that has power lines running through it or near it.

2 Don't climb a utility pole or tower.

3 Don't fly kites near power lines.

4 Don't swim during an electrical storm.

5 Stay away from power lines that are lying near to or on the ground.

Visuals

Solar System

Star Pattern

Ursa Major

Big Dipper

Inner Planets

Mercury

This planet is _____ to the sun and the

_____ planet.

Venus

A thick layer of _____ makes this planet the

_____ planet in our solar system.

Earth

This is the only planet with an _____ and

_____ water that make _____

possible.

Mars

This planet is known as the _____ planet,

and its surface is covered with red _____.

Planet Order

My Very Educated Mother Just Served Us Nachos.

Moon Mystery

Observation Chart		
Week	**Date**	**Moon Phase**
1		
2		
3		
4		

Outer Planets

Jupiter

This is the _____ planet and has

a giant _____ spot.

Saturn

This planet is best known for its _____.

Uranus

This planet spins on its _____.

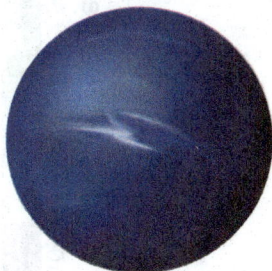

Neptune

This planet is known as a _____ planet

and is _____ from the sun.

Weather Map

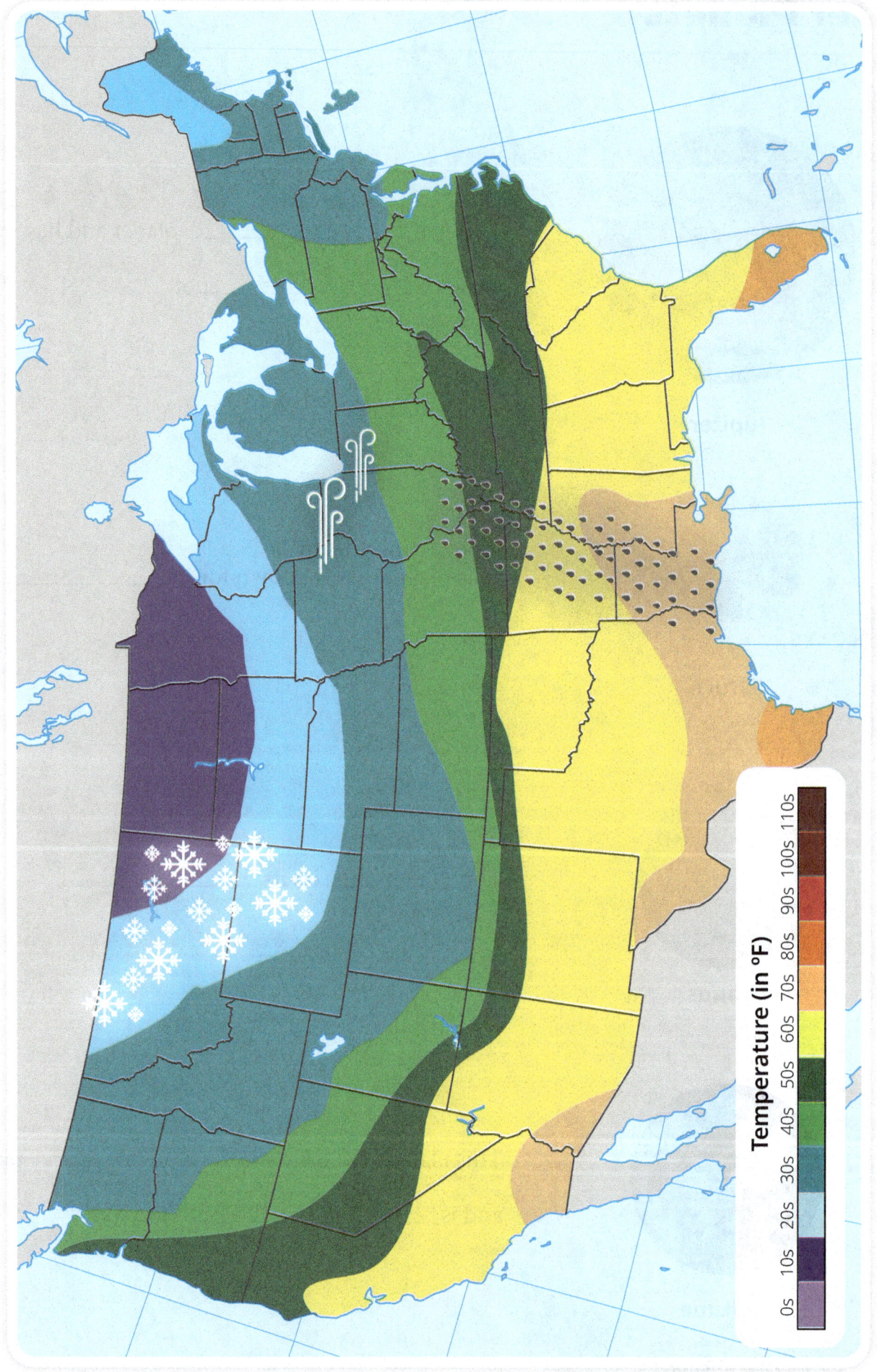

Temperature (in °F)

| 0s | 10s | 20s | 30s | 40s | 50s | 60s | 70s | 80s | 90s | 100s | 110s |

Weather Patterns

SUN	MON	TUES	WED	THUR	FRI
Mostly Cloudy	Snow Showers	Mostly Cloudy	Some Clouds	Sunny	Some Clouds
H 39° L 18°	H 31° L 20°	H 34° L 16°	H 36° L 18°	H _____ L 20°	H _____ L 22°

MON	TUES	WED	THUR	FRI	SAT
Snow Showers	Windy	Windy	Some Clouds	_____ _____	_____ _____
H 31° L 18°	H 27° L 14°	H 25° L 14°	H 30° L 17°	H 29° L 14°	H 26° L 16°

Cloud Formations

stratus clouds

cirrus clouds

cumulus clouds

fog

Water Cycle

Whether the weather be fine,

Or whether the weather be not,

Whether the weather be cold,

Or whether the weather be hot,

We'll weather the weather,

Whatever the weather,

Whether we like it or not!

Severe Weather Safety

If there is a thunderstorm or tornado warning, you should:

Go Indoors:

Take shelter in the nearest house or building. Stay indoors until the severe weather has passed.

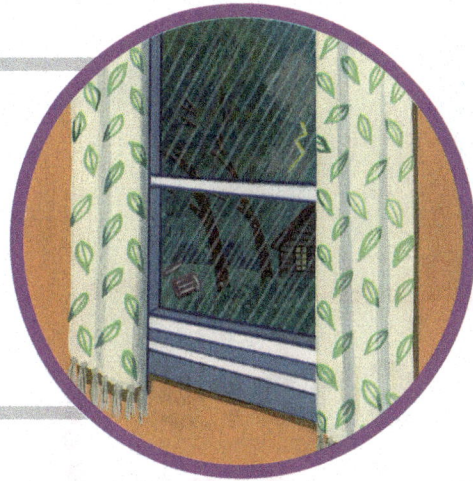

Stay Away:

From doors and windows!

Go Downstairs:

Go to the lowest level of the building. If the house or building has a basement, go to the basement until the severe weather has passed.

Take Cover:

Wear a bicycle or sports helmet to protect your head. If possible, sit under a sturdy desk or table.

Engineering Design Process

STEM

Ask

Imagine

Share

Plan

Test and Improve

Create

Visual 2.7
For use with Lesson 11

Climate Zones

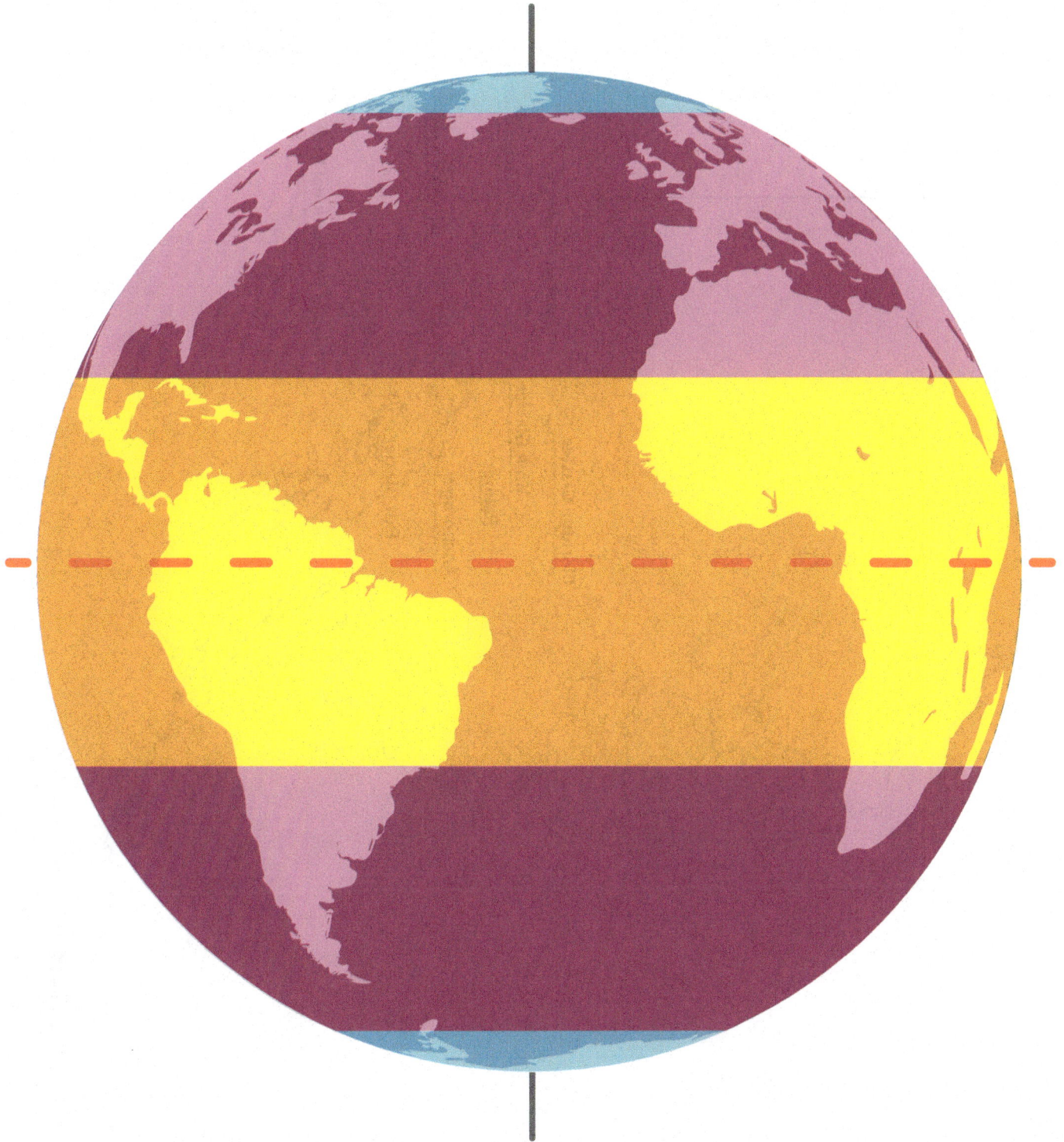

Map of Alaska

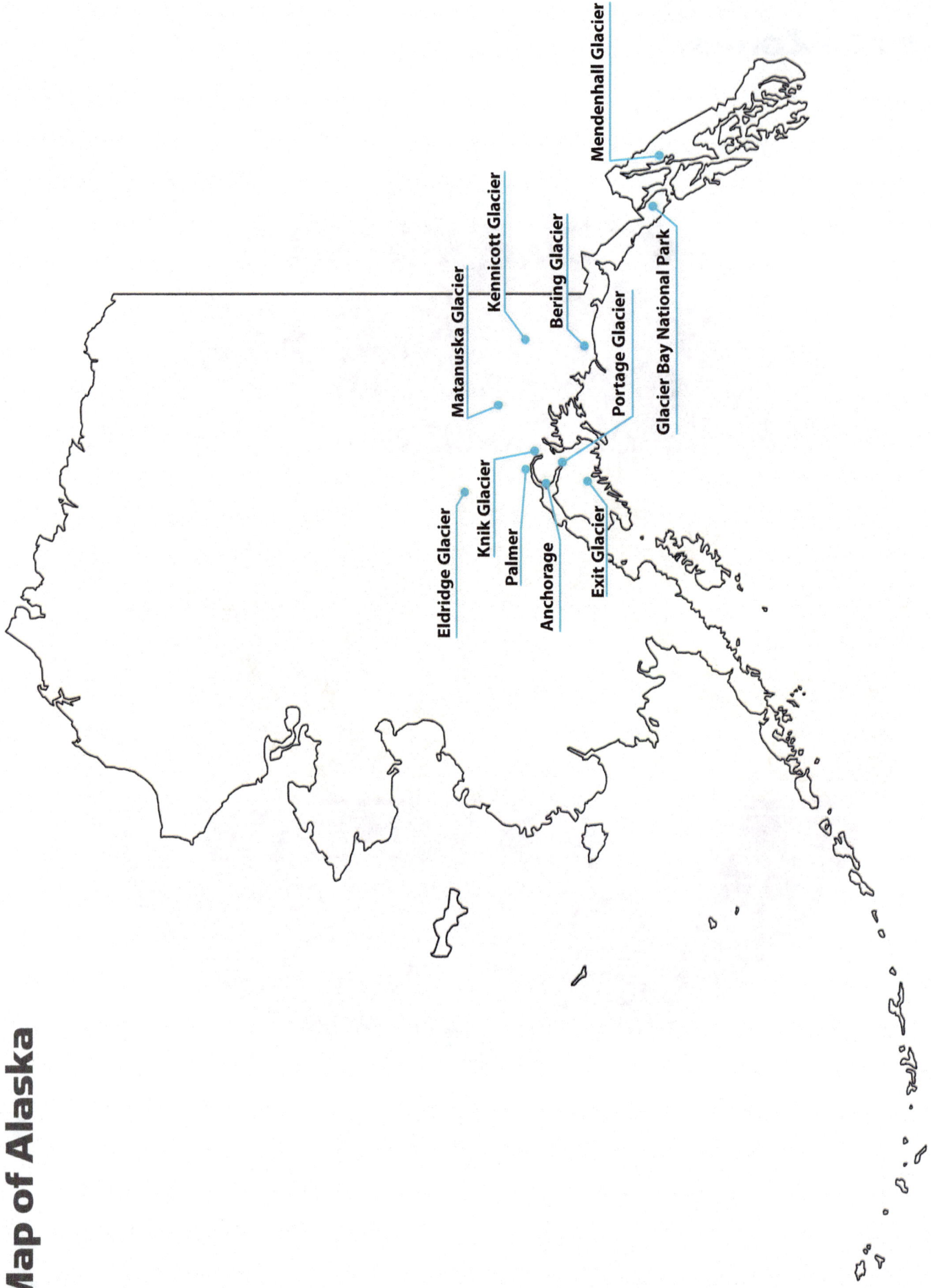

Matanuska Glacier

Kennicott Glacier

Bering Glacier

Mendenhall Glacier

Portage Glacier

Glacier Bay National Park

Eldridge Glacier

Knik Glacier

Palmer

Anchorage

Exit Glacier

Layers of Soil

Rocks, Rocks, and More Rocks

Sedimentary

forms when layers of sediment are pressed together and harden

Igneous

forms when melted rock cools

Rocks

hard pieces of the earth's surface

Metamorphic

forms when sedimentary and igneous rocks are changed by great heat and pressure

Science Inquiry Skills

Observe
to use your senses to find out something

Classify
to group things that are alike

Communicate
to share what you know; to give information about something

Measure
to find the size, number, or amount of something

Infer
to use what you know to tell why things happen

Predict
to use what you know to say what may happen

SCIENTIFIC INVESTIGATION

An **investigation** is a scientific test to solve a problem.

Inquiry Skills

Each investigation will direct attention to two or three **inquiry skills**.

Hypothesis

A **hypothesis** is a possible answer to the problem. The investigation tests the hypothesis to find whether it is correct. The hypothesis is written as a statement.

Procedure

The **procedure** is the steps of an investigation. The procedure must be followed carefully to have correct results.

Conclusions

The **conclusions** are the answers to the test. They tell what was learned from the investigation and whether the hypothesis was correct.

Problem

An investigation begins with a problem. The **problem** comes from an observation you have made. The problem is a question you want answered.

Materials

The **materials** are the supplies needed to do the investigation.

Observations

Scientists observe what happens, using the five senses. Scientists record data from their **observations** in charts, lists, or graphs.

Science Safety Tips

Follow your teacher's directions. **1**

Use tools the way they were made to be used. **2**

Handle supplies with care. **3**

Wear safety goggles when needed. **4**

Tell your teacher about accidents. **5**

Wait your turn. **6**

Clean up. **7**

Ark on Flood Waters

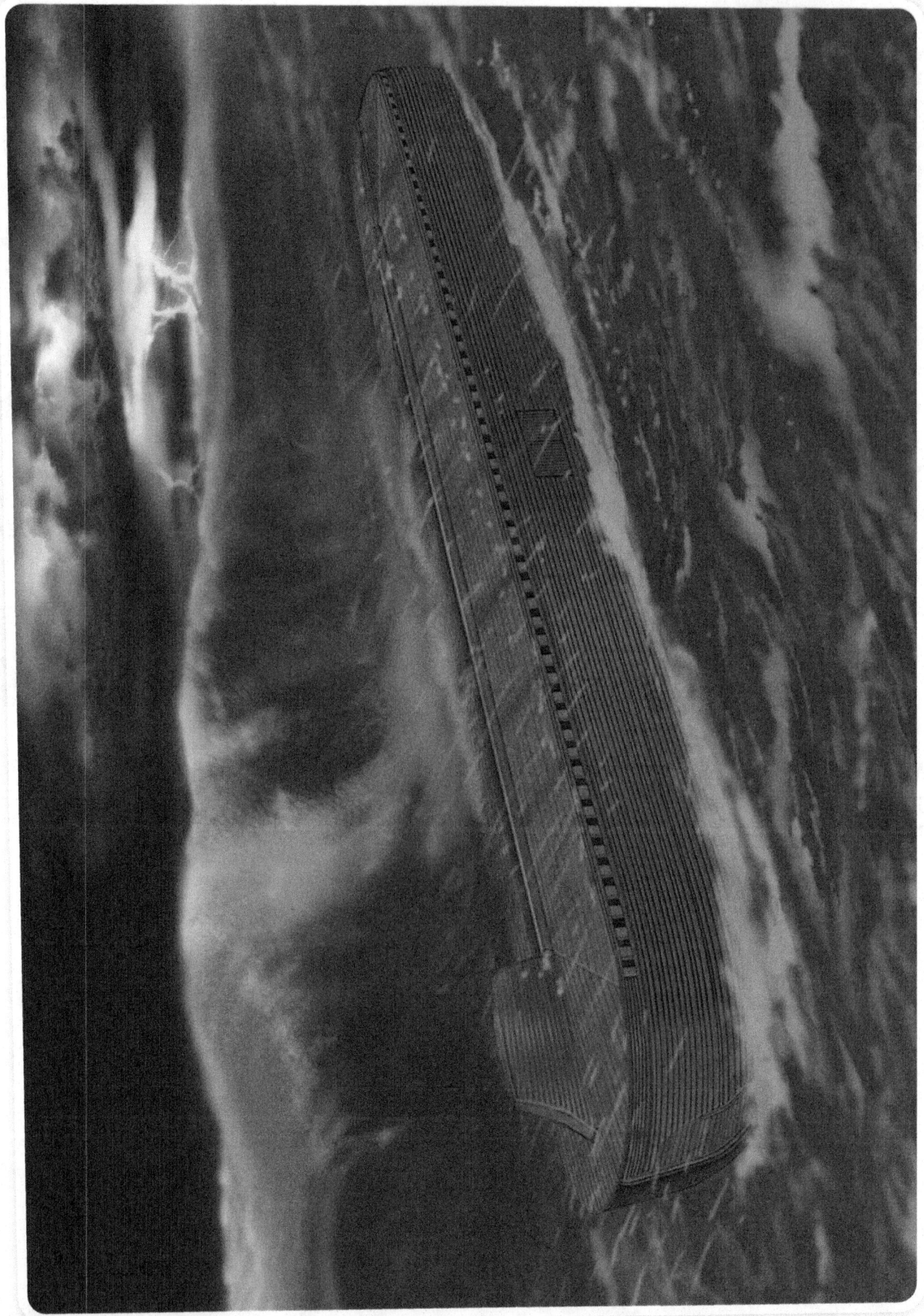

Parts of a Microscope

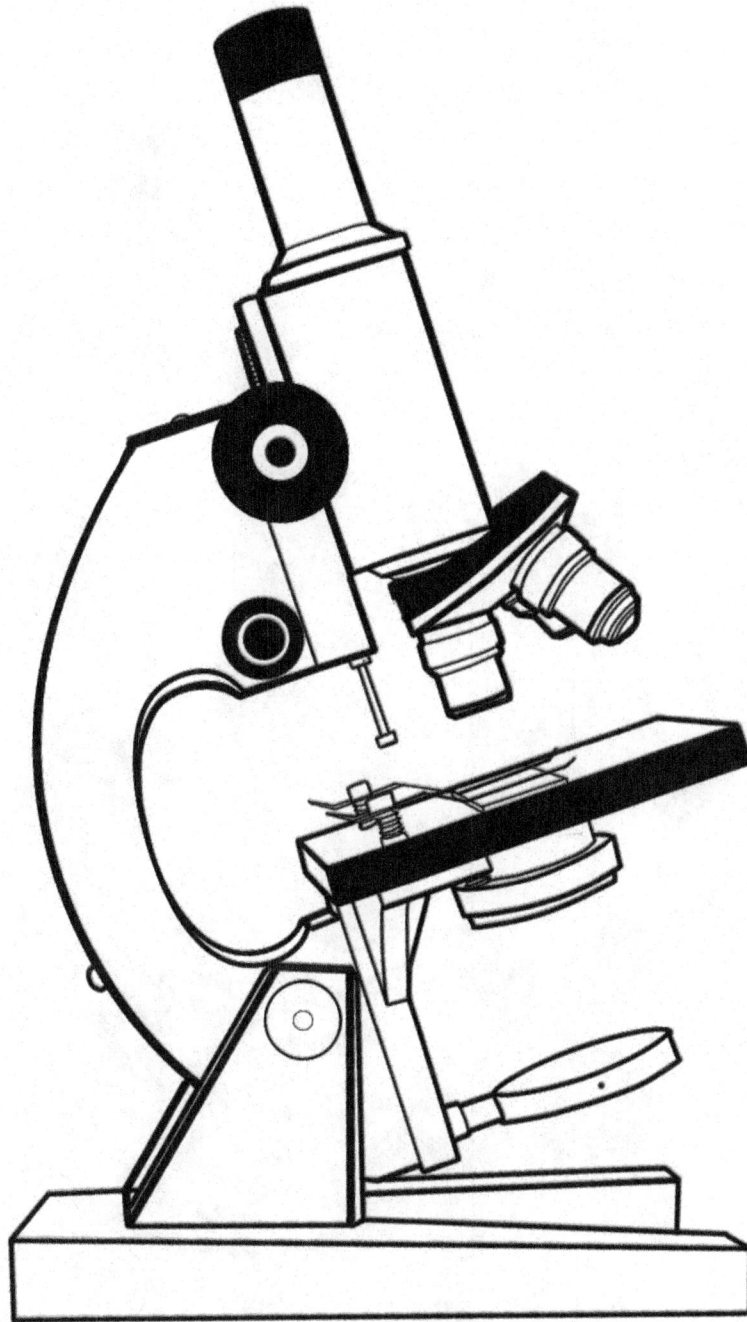

Parts of an Animal Cell

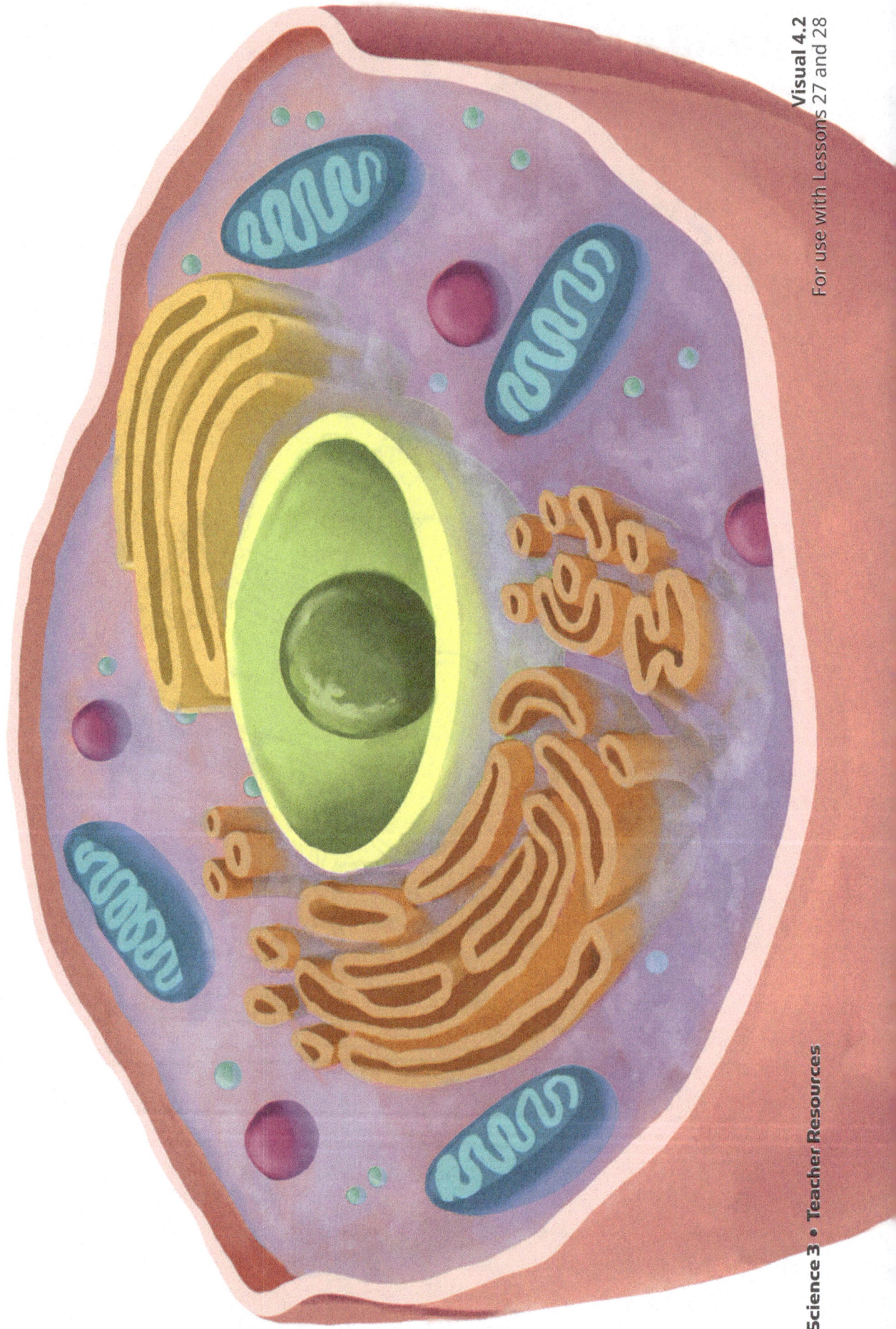

Parts of a Plant Cell

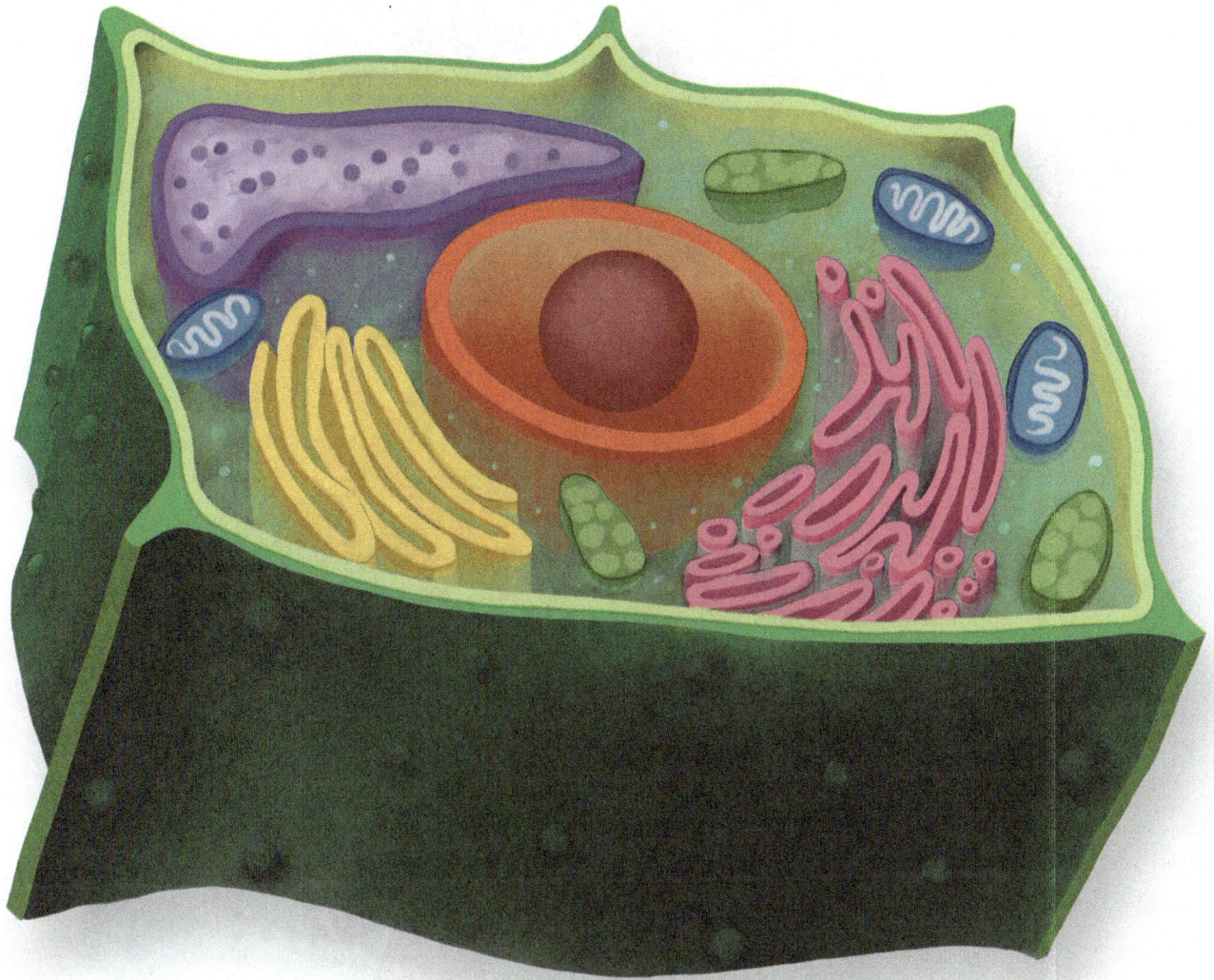

Upper Body Organs

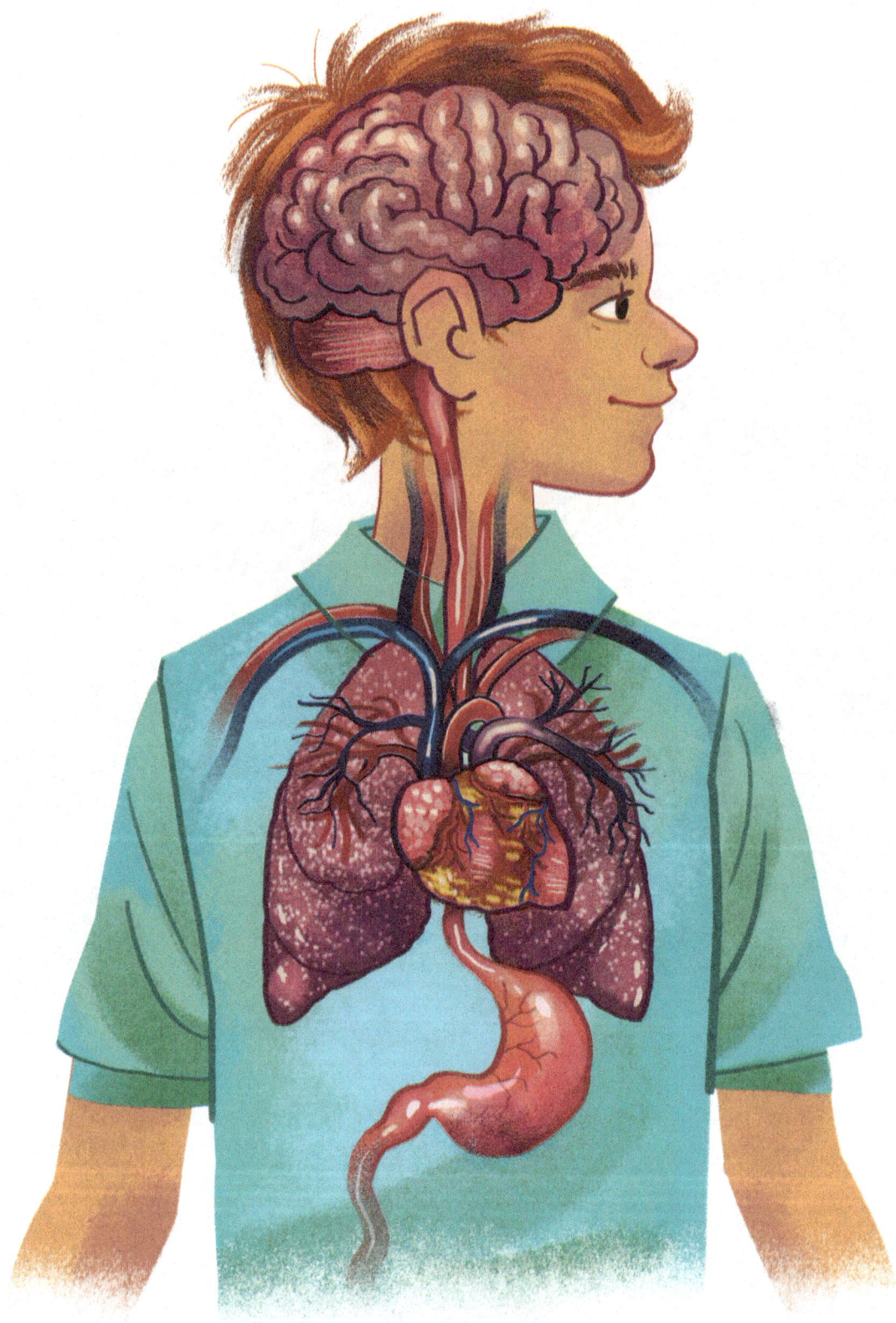

Layers of the Skin

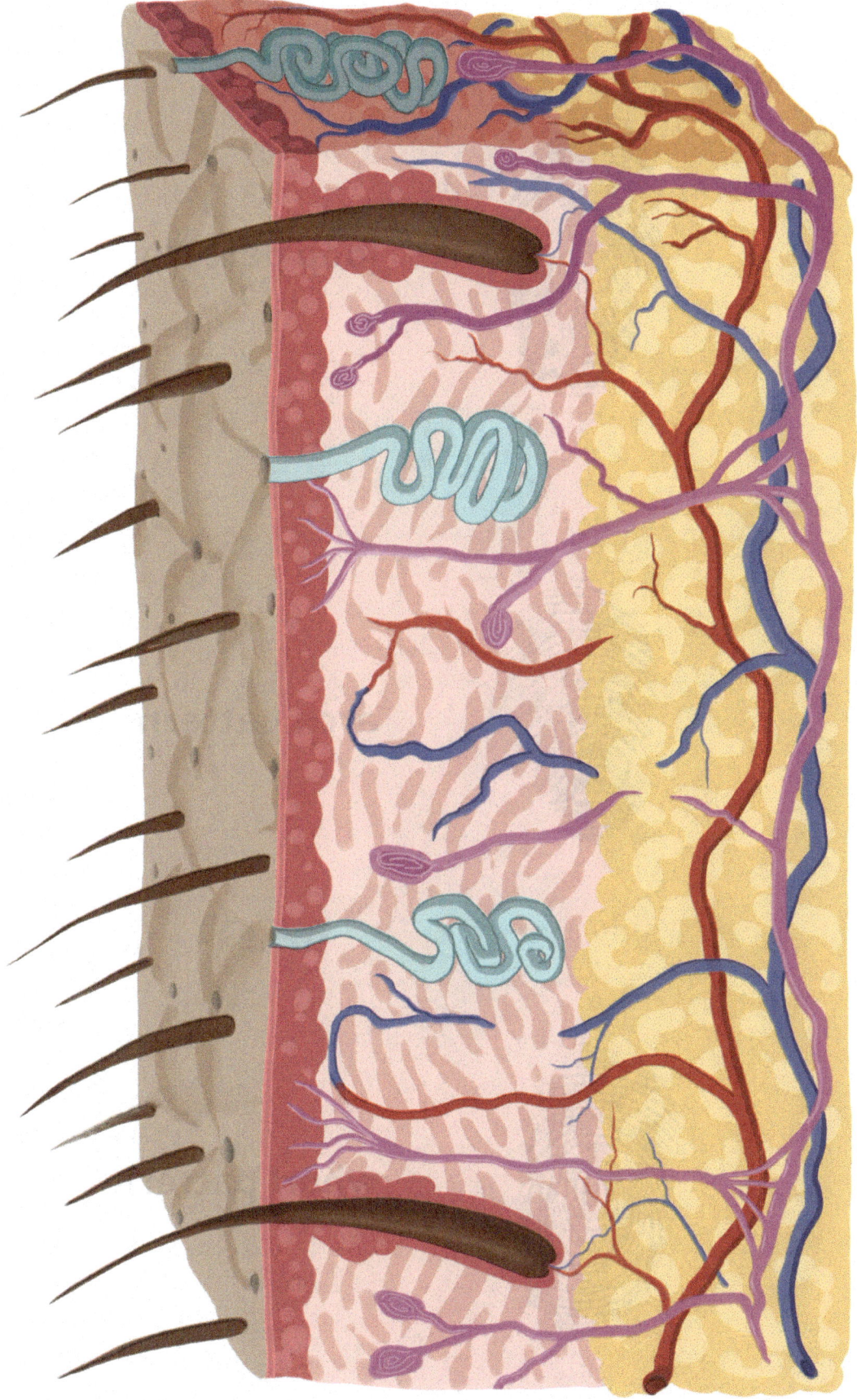

Fingerprint Patterns

 loop pattern

 arch pattern

 whorl pattern

Sample Plants

Life Cycle of a Plant

Plant Parts

Parts of a Bean Seed

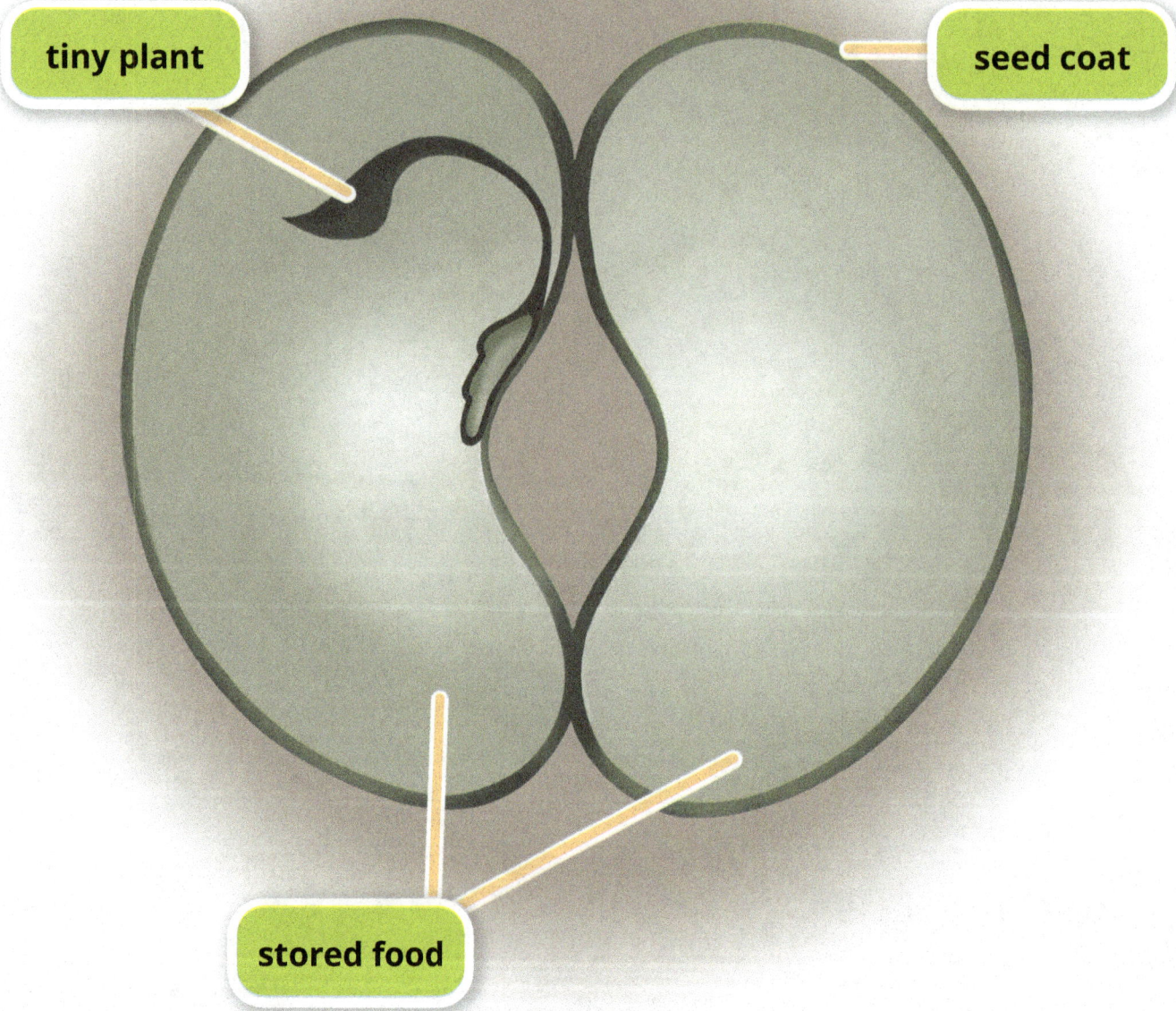

tiny plant

seed coat

stored food

Basic Photosynthesis

Photosynthesis Up Close

Animal Classification 1

Animals

_____-Blooded

body temperature _____

depending on its surroundings

_____-Blooded

body temperature stays _____

_____ all the time

Vertebrates   **Invertebrates**   **Vertebrates**

_____

- live in _____
- breathe with _____
- fins and _____
- slime _____
  their bodies
- reproduce by laying _____
  or by giving birth to

  _____ offspring

_____

- live in water and on _____
- thin, smooth, _____ skin
- young breathe with _____
- adults breathe with _____
- reproduce by laying _____
- go through _____

Some Survival and Growth Features

- _____ to blend in
- body colors to help catch

  _____
- body produces _____
- spines with _____

Some Survival and Growth Features

- live in _____ habitats
  to keep their skin from drying out
- eyelids and _____ as
  adults
- poisonous and _____
  colored skin

Life Cycles of Fish

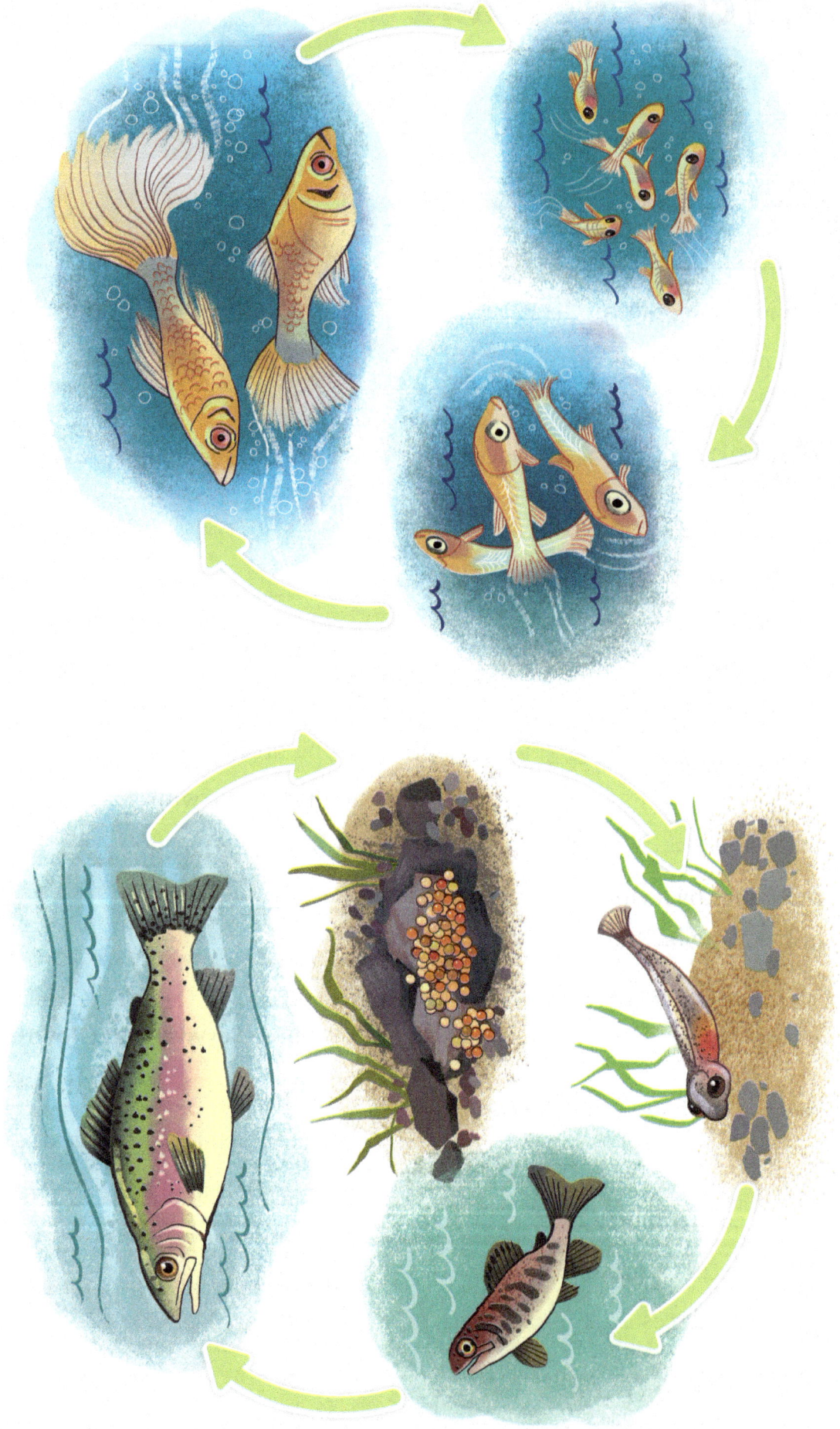

Stages of Amphibian Metamorphosis

Life Cycle of Reptiles

Animal Classification 2

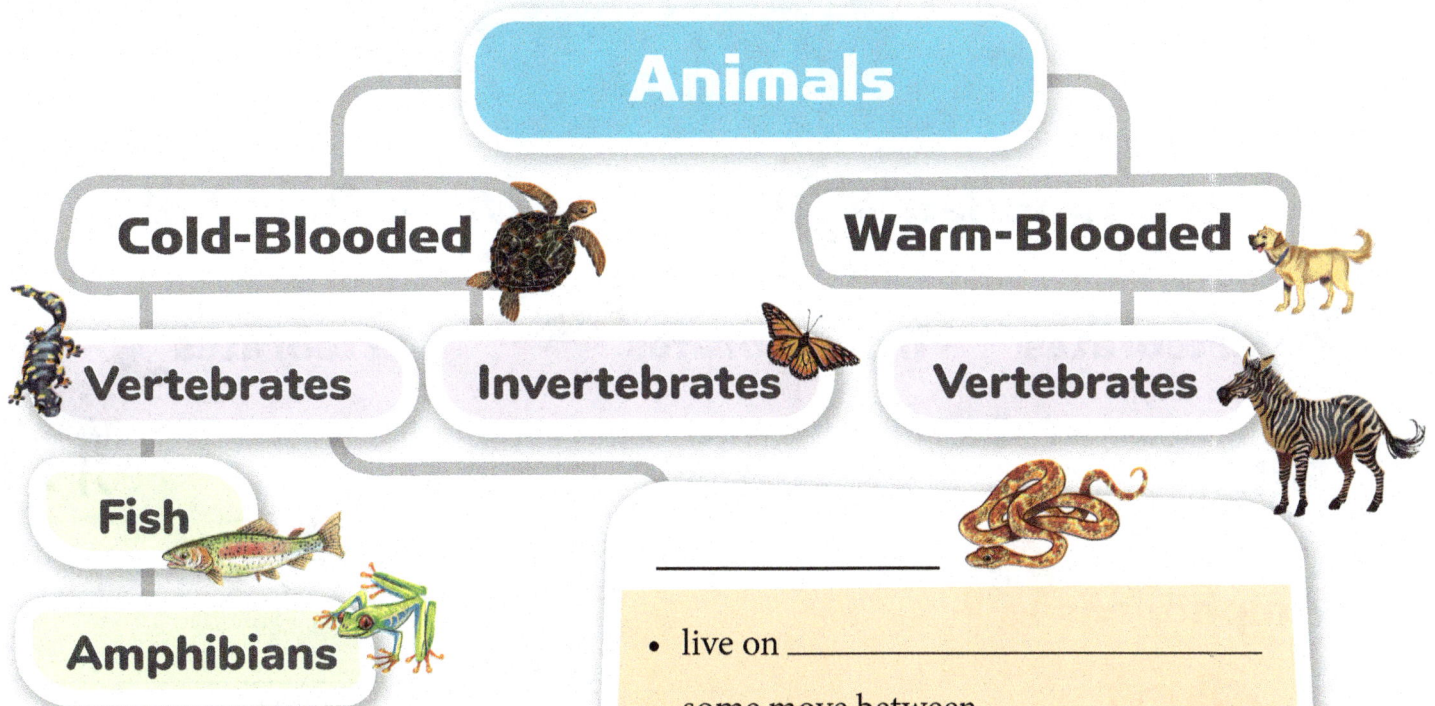

Animals

Cold-Blooded

Warm-Blooded

Vertebrates

Invertebrates

Vertebrates

Fish

Amphibians

_____

- live on _____
- some move between _____
  and _____
- tough, dry, scaly _____
- breathe oxygen with _____
- reproduce by _____

Some Survival and Growth Features

- skin protects reptile from
  _____

- skin protects reptile as it
  _____

- skin prevents the reptile's body from
  _____ out

- skin keeps _____ in

- _____ to help it blend in

- does not move much in the
  _____

- movable upper _____

- forked _____

- some can lose and regrow
  _____

Animal Classification 3

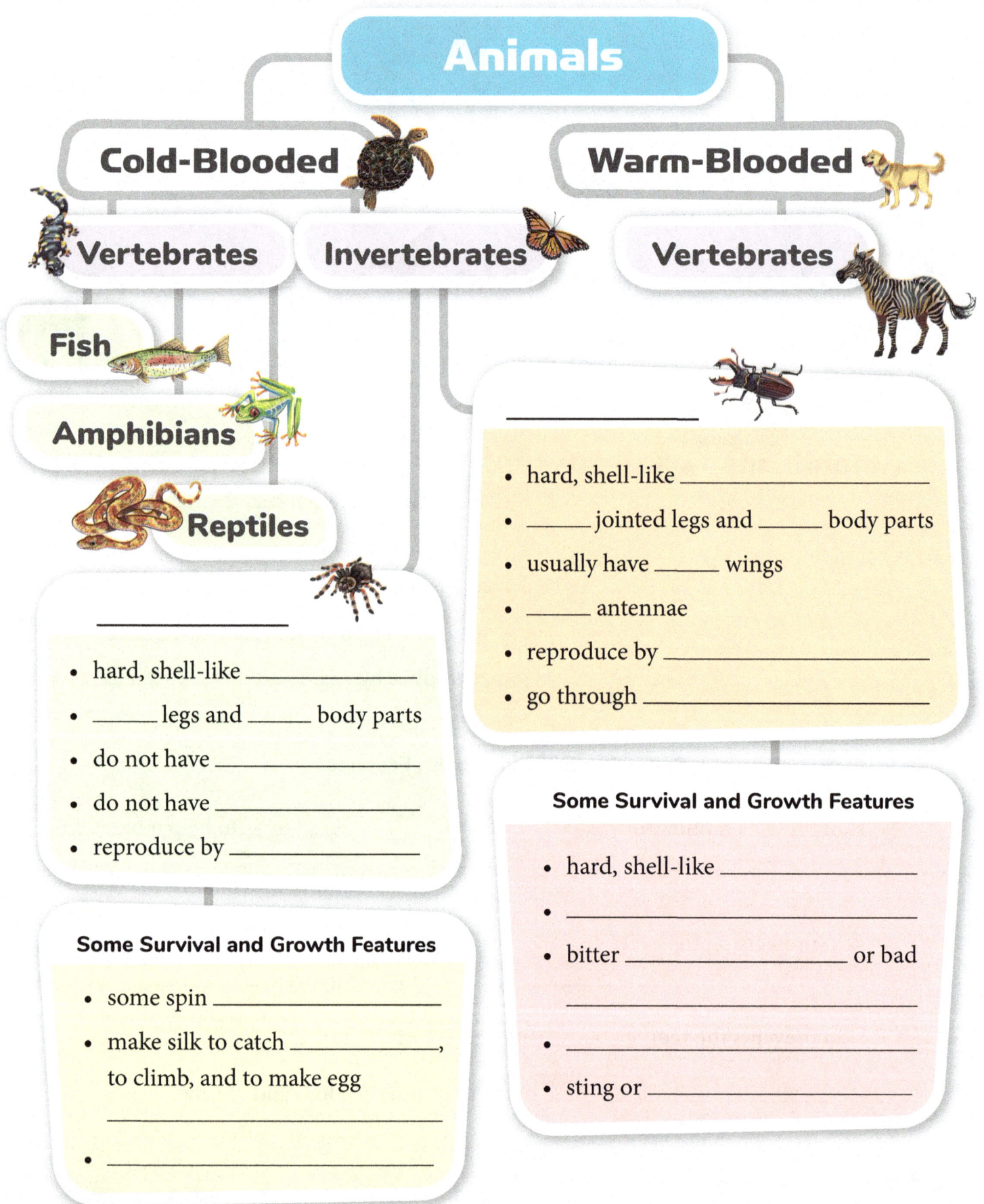

Animals

Cold-Blooded

Warm-Blooded

Vertebrates   **Invertebrates**   **Vertebrates**

Fish

Amphibians

Reptiles

_____

- hard, shell-like _____
- _____ legs and _____ body parts
- do not have _____
- do not have _____
- reproduce by _____

Some Survival and Growth Features

- some spin _____
- make silk to catch _____, to climb, and to make egg _____
- _____

_____

- hard, shell-like _____
- _____ jointed legs and _____ body parts
- usually have _____ wings
- _____ antennae
- reproduce by _____
- go through _____

Some Survival and Growth Features

- hard, shell-like _____
- _____
- bitter _____ or bad _____
- _____
- sting or _____

Life Cycle of an Insect

Life Cycle of a Spider

Life Cycle of Mealworms

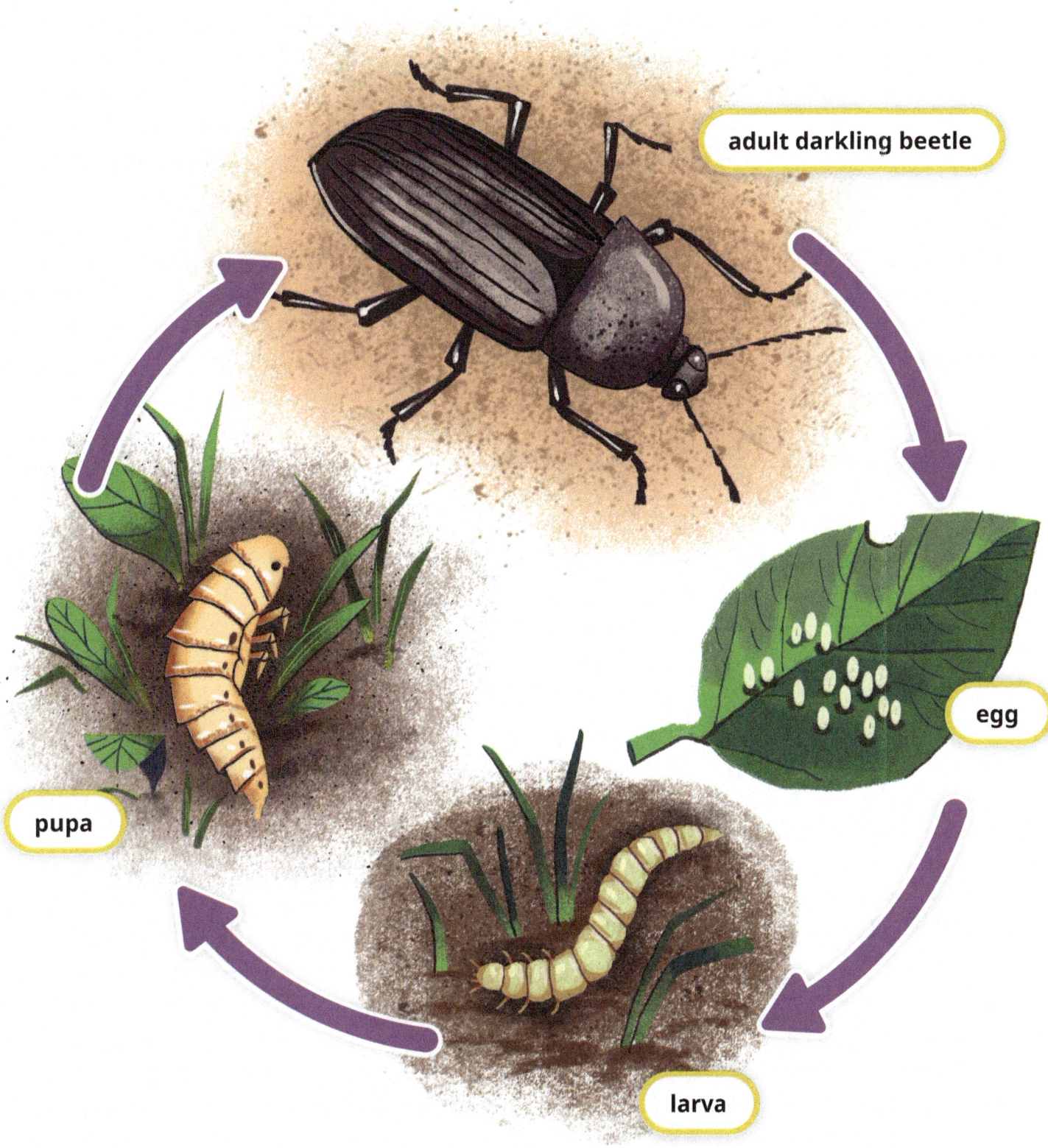

adult darkling beetle

egg

larva

pupa

Animal Classification 4

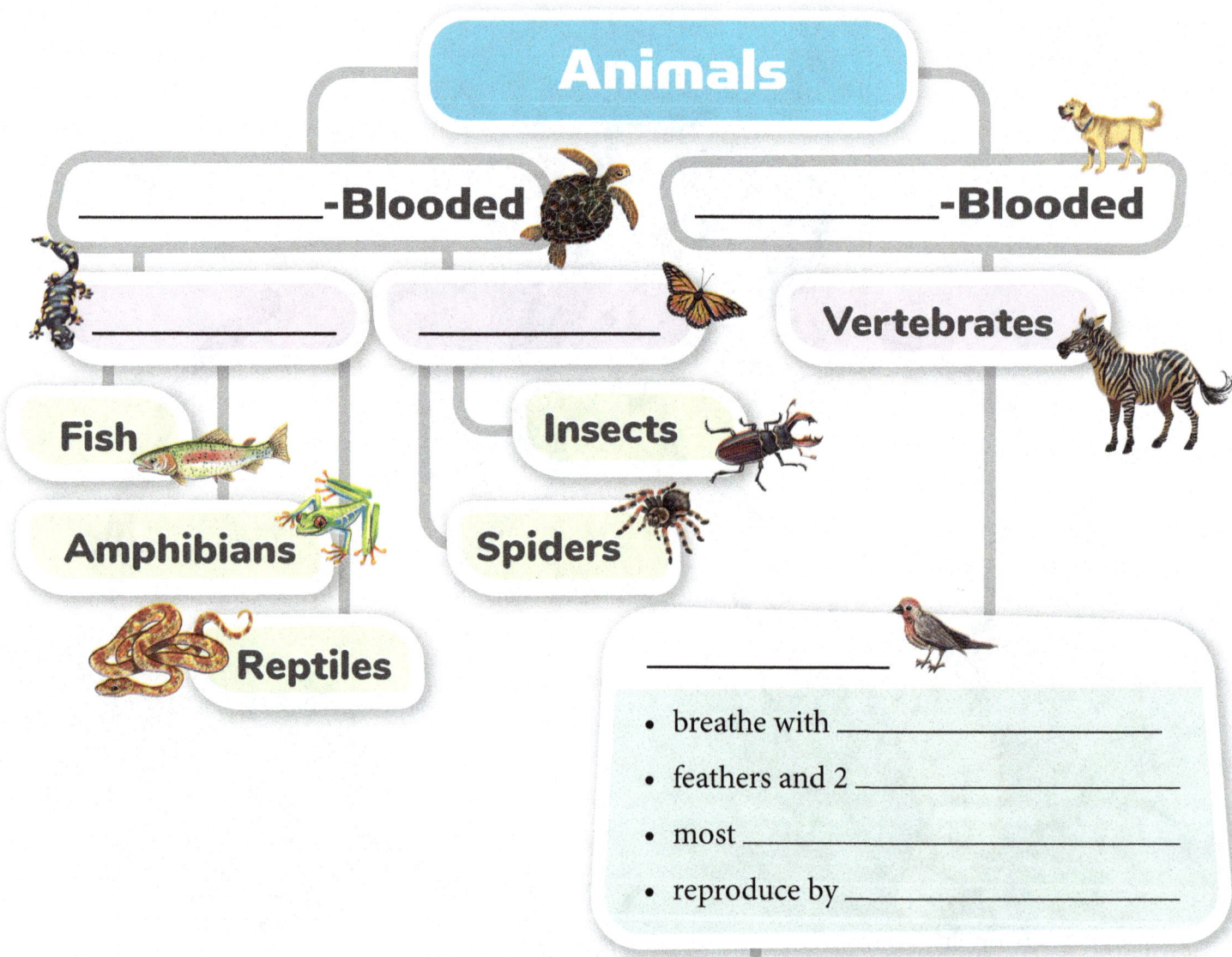

Animals

_____-Blooded

_____-Blooded

_____

_____

Vertebrates

Fish

Amphibians

Reptiles

Insects

Spiders

_____

- breathe with _____
- feathers and 2 _____
- most _____
- reproduce by _____

Some Survival and Growth Features

- fluffing _____ for warmth
- _____ together for warmth
- to find food, some follow animals that _____
- feet for walking, _____, or _____

- claws for hanging on tree trunks or killing _____
- beaks for eating _____, building _____, killing prey, or preening
- camouflaged _____ and _____

Life Cycle of a Bird

chick

egg

adult

Fruit Bat

Animal Classification 5

Complete the organizer.

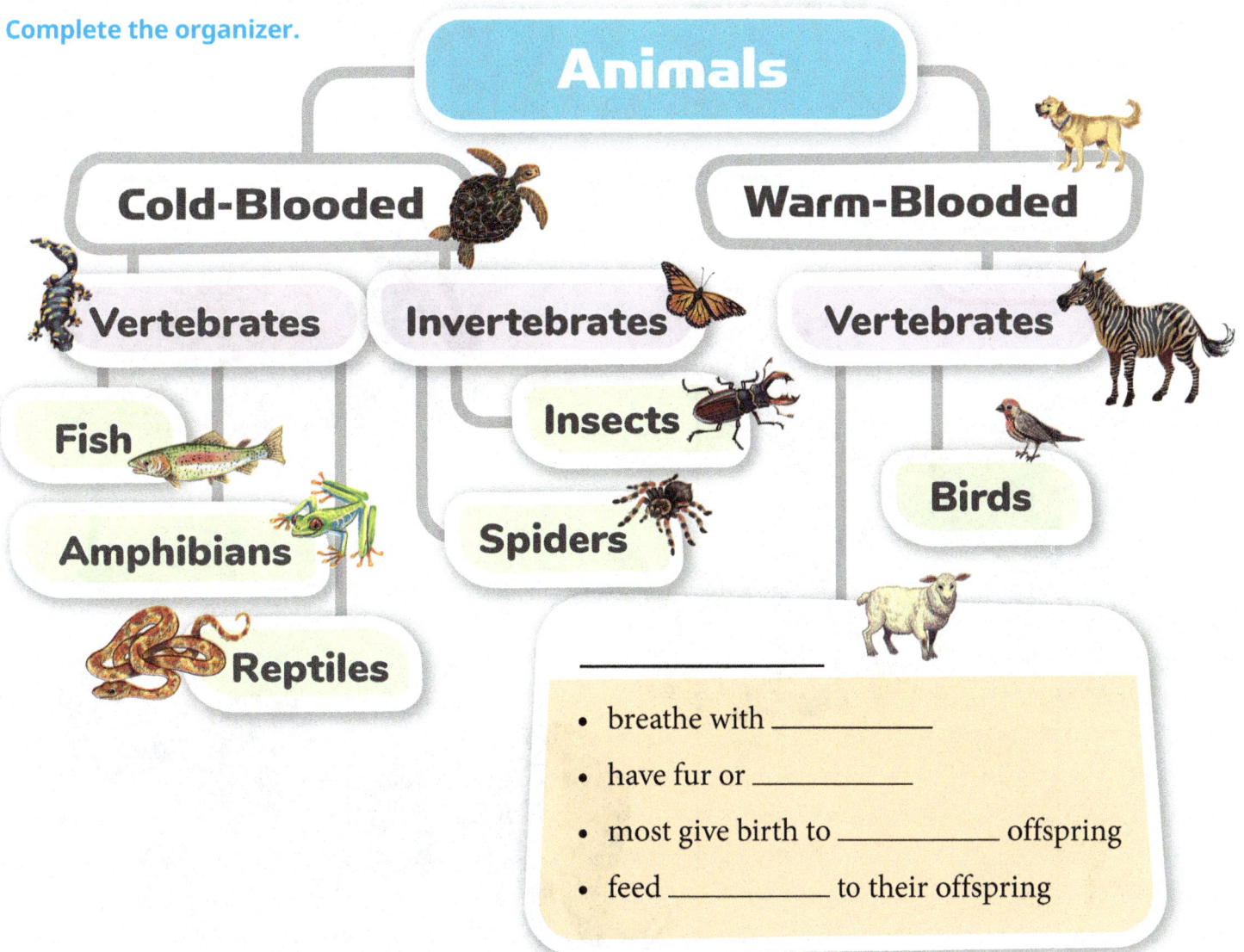

Animals

Cold-Blooded

Warm-Blooded

Vertebrates

Invertebrates

Vertebrates

Fish

Insects

Birds

Amphibians

Spiders

Reptiles

_____

- breathe with _____
- have fur or _____
- most give birth to _____ offspring
- feed _____ to their offspring

Some Survival and Growth Features

- in cold climates have thick _____
- in cold climates have a thick layer of _____
- in cold climates with little food are able to go into a deep _____
- sweat _____ to survive with very little water

- escape predators by _____ very fast
- escape predators by _____ very still
- use _____ to hide
- have spines or _____ for protection
- use _____ for protection

Life Cycle of a Mammal

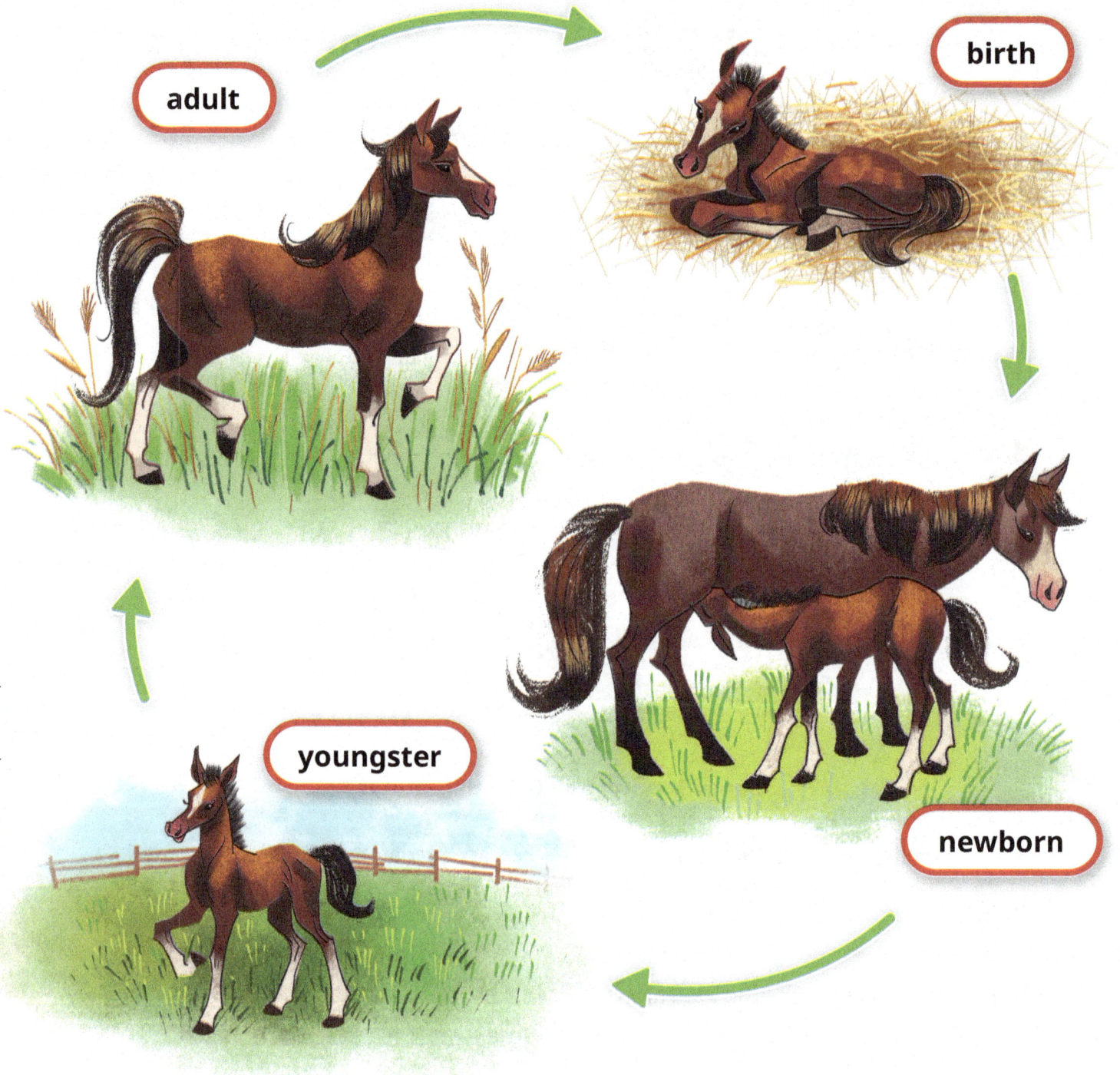

adult

birth

newborn

youngster

Compare and Contrast

	Inherited Traits	Instincts	Learned Behaviors
Definitions	a characteristic that is passed down from _____ to _____	a God-given _____ that an animal is _____	something that an animal _____ to do
What animals have this?	_____ and _____ animals	_____ and _____ animals	_____ and _____ animals
Are animals born with this?	_____	_____	_____
Where does it come from?	Animals are _____ by _____ with inherited traits.	Animals are _____ by _____ with instincts.	Animals _____ behaviors from experience, _____, or other animals.
Example	_____ _____ _____	_____ _____ _____ _____	_____ _____ _____ _____ _____ _____

Fitting Together Organizer

Ecosystem

All the _____
and _____
things and the
_____
in which they
_____.

Habitat

A _____ where
a _____ thing
_____.

Community

All the _____ things
in _____ area.

Population

All the _____ things
of _____ kind that
live in _____ area.

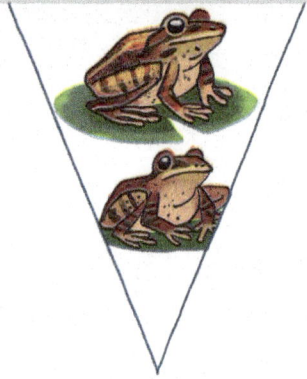

Food Chain

= movement of energy

Food Web

= movement of energy

wood frog

garter snake

red fox

grasshopper

berry bush

Changing States

warmer

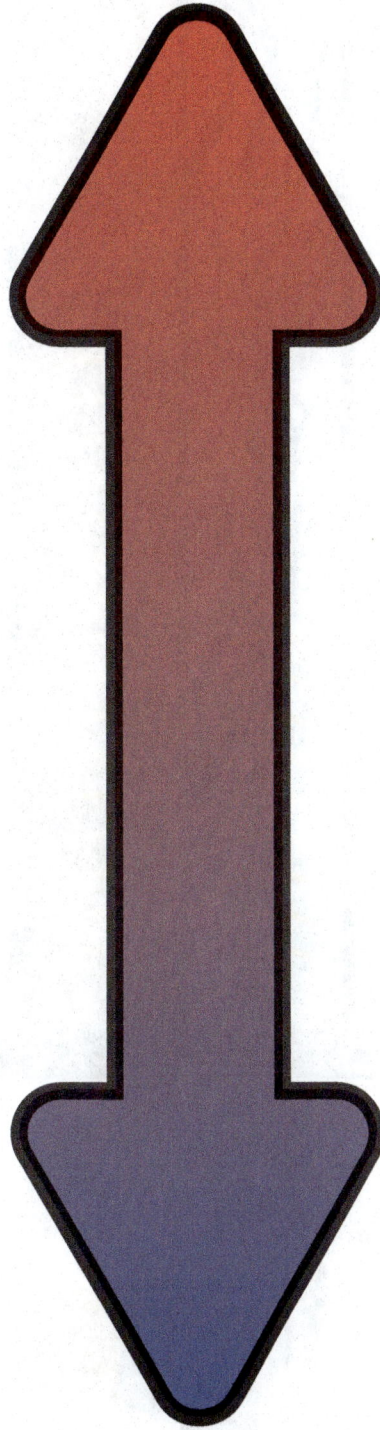

cooler

Speed of Sound

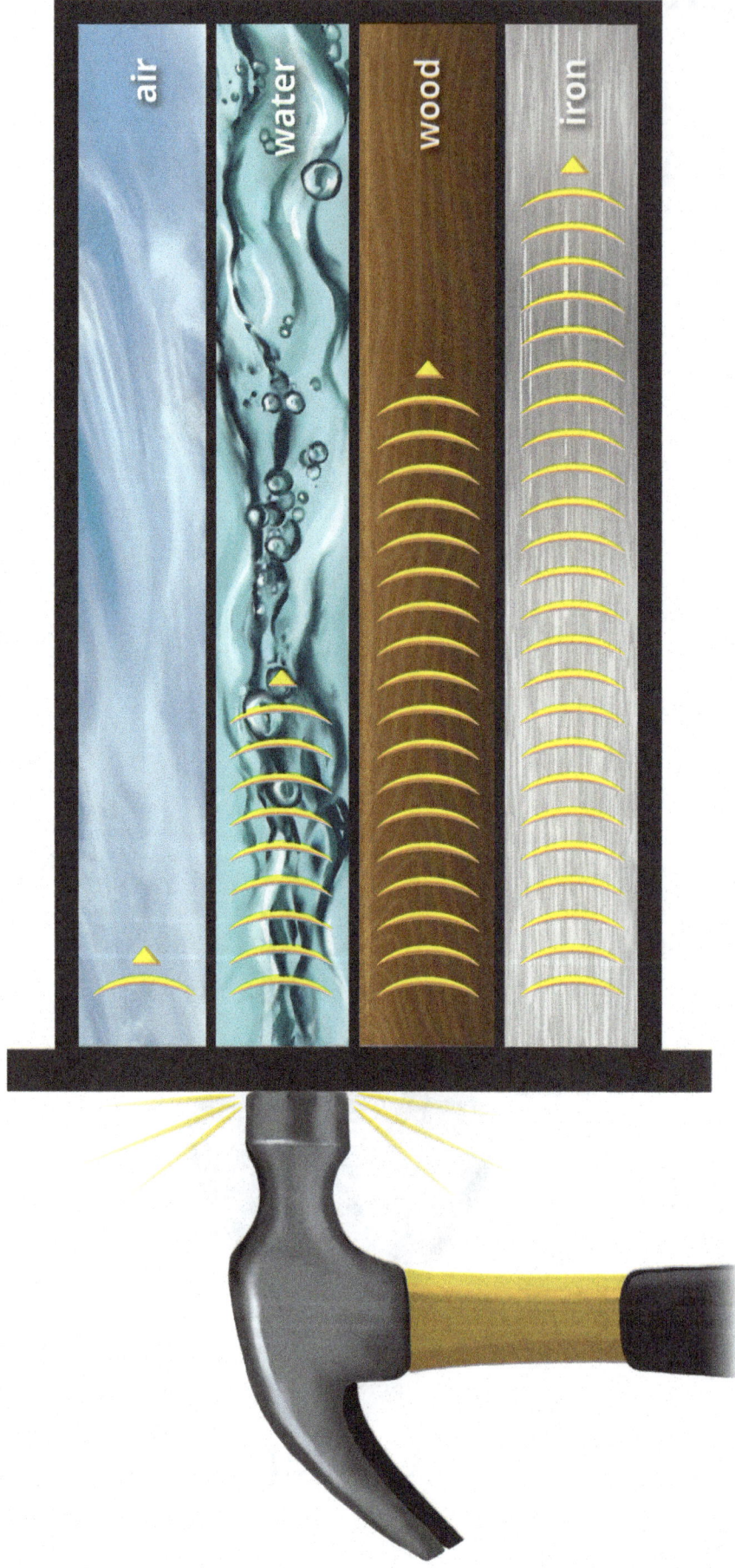

Parts of the Ear

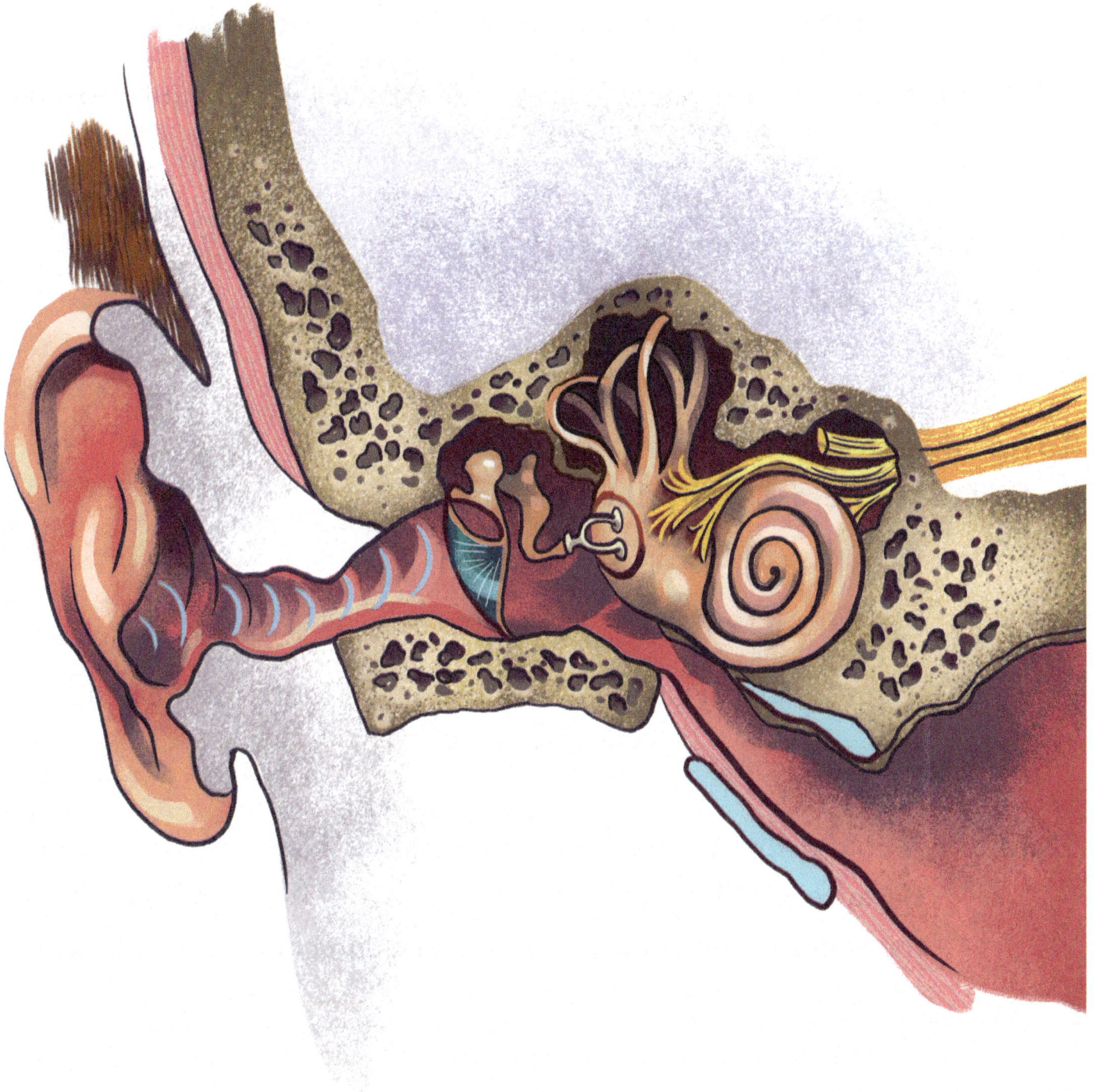

Cause and Effect

Cause and effect shows both what happened and why it happened.

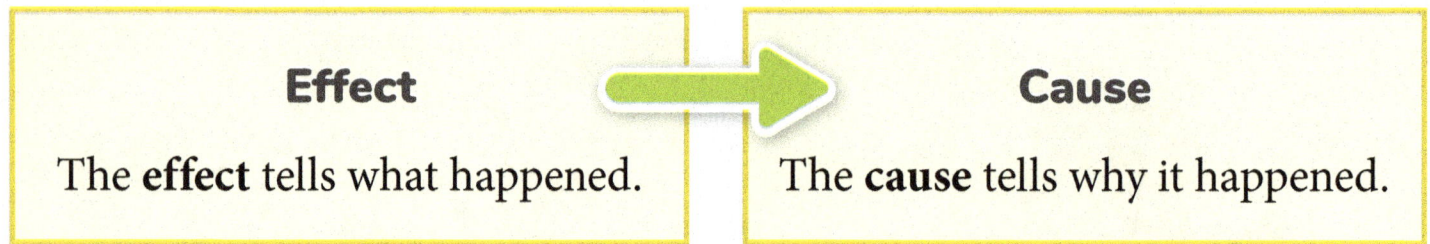

Effect	→	Cause
The **effect** tells what happened.		The **cause** tells why it happened.

Example: The water boiled in the teakettle.
Because the water was heated,
the water became a gas.

To find the **effect**, ask "**What happened?**"
To find the **cause**, ask "**Why did it happen?**"

Effect	→	Cause
The water became a gas.		The water was heated.

Example: A cold drink was poured into a glass on a hot day.
Because a cold drink cooled the glass and the air around it,
the water vapor in the air condensed and changed to a liquid.

Effect	→	Cause

Kinds of Forces

_____
- the force that slows or stops motion

Contact

a force that _____ only when

one object _____ another

Kinds of Force

Noncontact

a force that _____ between

objects that are _____

_____
- the force that _____

all things toward the

_____ of the earth

_____
- the _____ or

_____ force caused

by a _____

Balanced or Unbalanced?

Equations

< less than   > greater than   = equal

1. _____

  ○ balanced forces showing no motion
  ○ unbalanced forces showing motion

2. _____

  ○ balanced forces showing no motion
  ○ unbalanced forces showing motion

3. _____

  ○ balanced forces showing no motion
  ○ unbalanced forces showing motion

4. _____

  ○ balanced forces showing no motion
  ○ unbalanced forces showing motion

Index

decomposer, 177

derecho, 37

dermis, 98, 102, 104

direction. *See under* force

distance. *See under* force

dominion, 158

drought, 35, 175

dwarf planet, 21

E

ear, 218, 219–220

Earth, 6, 7, 13, 14

earthquake, 39

echo, 213

ecosystem, 168–191
  balance, 189–190, 191
  changes, 186–188
  definition, 171
  food chain, 182–183
  groups of animals, 181
  living together, 172–173
  plants and animals, 176–177
  resources, 174
  types of consumers, 178–180

effect. *See* cause and effect

electric charge, 245

electricity, 243–252
  charges, 244–245, 246–247, 248
  circuits, 250–251, 252, 259
  current, 248–251, 258
  static, 246–247

electromagnet, 258–259, 261

energy, 3, 7, 49, 111, 113, 176, 177, 178,
  182–183

engineering design process, 40

environment, 75, 171, 186

epidermis, 98–99, 104

equator, 20, 40, 41, 45

Eris, 21

erosion, 57–59

evaporation, 30–31, 204

evolution, 78, 128

exert, 224, 228, 230, 233

Exploration
  Blubber, Feathers, and Fur, 167
  Conductors Needed, 252
  Ecosystem Tag, 191
  Edible Cell, 94
  Feathery Friends, 152
  Moon Mystery, 17
  Patterns on My Skin, 101
  Solar Mobile, 11
  Weather Watcher, 34
  Which Kind of Matter?, 201

extinct, 74, 75, 79

F

Fall, 136, 179
  See also under worldview

Fantastic Facts
  carnivorous plant, 179
  chlorophyll, 113
  cow magnet, 256
  current electricity, 251
  day on planets, 13
  derecho, 37
  dermal ridges, 100
  echoes, 214

Teacher Edition Index

Photo Credits

Key (t) top; (c) center; (b) bottom; (l) left; (r) right; (bg) background; (i) inset

Cover

front (car) TurboSquid.com; front (hummingbird) ssucsy/iStock/Getty Images Plus/Getty Images; front (lizard) PetlinDmitry/iStock/Getty Images Plus/Getty Images; back (aloe vera) Elenathewise/iStock/Getty Images Plus/Getty Images

Front Matter

iiit Tatiana Epifanova/iStock/Getty Images Plus/Getty Images; iiic Capitano Productions Eye/Shutterstock.com; iiib Elena11/Shuttterstock.com; iv (hummingbird) Kenneth Canning/iStock/Getty Images Plus/Getty Images vt boule13/iStock/Getty Images Plus/Getty Images; vbl Olga Shestakova/iStock/Getty Images Plus/Getty Images; vbr Wittayayut/iStock/Getty Images Plus/Getty Images

Chapter 1

vi–1 Sino Images/500px Asia/Getty Images; 2–3 idizimage/iStock/Getty Images Plus/Getty Images; 6l Mr Aesthetics/Shutterstock.com; 6r youngID/DigitalVision Vectors/Getty Images; 7t Rainer Fuhrmann/EyeEm/Getty Images; 7b Robert George Young/Photodisc/Getty Images; 8–9 (child looking through telescope) lucentius/E+/Getty Images; 9t Al Fenn/The LIFE Picture Collection/Getty Images; 9b Stocktrek Images/Getty Images; 10t Capitano Productions Eye/Shutterstock.com; 10b "John W. Young on the Moon"/Nasa/Public Domain; 5, 12, 14, 16, 18, 20 (Mercury) Elena11/Shuttterstock.com; 5, 12, 13, 14, 16, 18, 20 (Venus), 5, 13, 15, 17, 19, 20, 21 (Neptune) NASA/JPL; 5, 12, 14, 15, 16, 18, 20 (Earth) ASPARINGGA/Shutterstock.com; 14, 15 (moon and moon phases) Indiloo/Shutterstock.com; 5, 12, 14, 16, 18, 20 (Mars) Pike-28/Shutterstock.com; 5, 13, 15, 17, 18, 19, 21 (Jupiter); Pe3k/Shutterstock.com; 5, 13, 15, 17, 19, 21 (Saturn) NASA, ESA, A. Simon (GSFC) and the OPAL Team; 5, 13, 15, 17, 19, 20, 21 (Uranus) NASA images/Shutterstock.com; 5, 12, 14, 15, 16, 18, 20 (Sun) Nerthuz/Shutterstock.com; 18 (Great Red Spot) NASA/JPL–Caltech/SwRI/MSSS/Gerald Eichstädt/Seán Doran © CC NC SA; 19 (Saturn's Rings) NASA/JPL-Caltech/Space Science Institute; 21t NASA/JPL–Caltech; 21c Dotted Yeti/Shutterstock.com

Chapter 2

22–23 DeepDesertPhoto/RooM/Getty Images; 24t cworthy/E+/Getty Images; 24c shayes17/iStock/Getty Images Plus/Getty Images; 24b MediaProduction/iStock/Getty Images Plus/Getty Images; 25t NASA; 25b dmac/Alamy Stock Photo; 26t chrupka/Shutterstock.com; 26b filo/DigitalVision Vectors/Getty Images; 27t BMJ/Shutterstock.com; 27bl onurdongel/iStock/Getty Images Plus/Getty Images; 27br Viktor Wallon-Hars/iStock/Getty Images Plus/Getty Images; 28bg korinnna/iStock/Getty Images Plus/Getty Images; 28bl krithnarong Raknagn/Shutterstock.com; 29t Eastcott Momatiuk/The Image Bank/Getty Images; 29tc Baac3nes/Moment/Getty Images; 29bc Visun Khankasem/Shutterstock.com; 29b PhotoAlto/Odilon Dimier/PhotoAlto Agency RF Collections/Getty Images; 32t Nick Brundle Photography/Moment Open/Getty Images; 32c amriphoto/E+/Getty Images; 32bl DavidPrahl/iStock/Getty Images Plus/Getty Images; 32br Gregory_DUBUS/iStock/Getty Images Plus/Getty Images; 33t alfimimnill/iStock/Getty Images Plus/Getty Images; 33b SandraKavas/iStock/Getty Images Plus/Getty Images; 35t Westend61/Getty Images; 35b Martin Lisius/The Image Bank/Getty Images; 36t David Chua/500px Plus/Getty Images; 36b Jason Persoff Stormdoctor/Image Source/Getty Images; 37t Ryan McGinnis/Moment/Getty Images; 37c Stocktrek Images/Getty Images; 37b GeoStock/The Image Bank/Getty Images; 38t Africa Studio/Shutterstock.com; 38b Emily Heinz; 39 Kosamtu/E+/Getty Images; 40 Sneksy/iStock/Getty Images Plus/Getty Images; 41 NPeter/Shutterstock.com; 42–43 moodboard/Brand X Pictures/Getty Images; 44 JamesBrey/E+/Getty Images; 45t agustavop/iStock/Getty Images Plus/Getty Images; 45b Maudib/iStock/Getty Images Plus/Getty Images; 47 carterdayne/iStock/Getty Images Plus/Getty Images; 48l T.J. Hileman, courtesy of Glacier National Park Archives; 48r Lisa McKeon, USGS Northern Rocky Mountain Science Center; 49 Philippe Bourseiller/The Image Bank/Getty Images; 50t Seth Joel/Photographer's Choice RF/Getty Images; 50b Larry Mayer/Stockbyte/Getty Images; 51t Jose Luis Pelaez Inc/DigitalVision/Getty Images; 51b matkovci/Shutterstock.com

Chapter 3

52–53 Checubus/Shutterstock.com; 54–55 (soil cross section) ifong/Shutterstock.com; 54 (earthworm) kzww/Shutterstock.com; 54bl borchee/E+/Getty Images; 54br Atomazul/Shutterstock.com; 55bl SharonDay/iStock/Getty Images Plus/Getty Images; 55br chinaface/E+/Getty Images; 57 Evgeniya Moroz/Shutterstock.com; 58t Mac99/E+/Getty Images; 58b Eloi_Omella/E+/Getty Images; 58b Eloi_Omella/E+/Getty Images; 59t David Forster/Alamy Stock Photo; 59bl Rvo233/iStock/Getty Images Plus/Getty Images; 59br, 62b, 69l, 69cl, 69cr, 69r, 71, 81b Emily Heinz; 60t JamesBrey/E+/Getty Images; 60c Dennis Cox/Alamy Stock Photo; 60bl oksanatukane/Shutterstock.com; 60br terra24/iStock/Getty Images Plus/Getty Images; 61t claylib/E+/Getty Images; 61b raphoto/iStock/Getty Images Plus/Getty Images; 62t A_Pobedimskiy/iStock/Getty Images Plus/Getty Images; 62tc Adobe Stock/siimsepp; 62bc Gary Ombler/Dorling Kindersley/Science Source; 63 (Erta Ale) Ivoha/Alamy Stock Photo; 63 (Volcán de Fuego) MarcoRof/iStock/Getty Images Plus/Getty Images; 63 (granite) agefotostock/Alamy Stock Photo; 63 (obsidian) mahirart/Shutterstock.com; 63 (Mt. Rushmore) Andrew Zarivny/Shutterstock.com; 64t jaym-z/iStock/Getty Images Plus/Getty Images; 64c, 64b Tyler Boyes/Shutterstock.com; 65 (gold coins) Myotis/Shutterstock.com; 66 (talc) Nyura/Shutterstock.com; 66 (gypsum) Susan E. Degginger/Alamy Stock Photo; 66 (calcite) Jasius/Moment/Getty Images; 66, 68 (fluorite) Delennyk/Shutterstock.com; 66 (apatite) Karol Kozlowski/Shutterstock.com; 66 (hand pointing) Denys Prykhodov/Shutterstock.com; 66 (penny) United States Mint Image; 66 (mirror) jocic/Shutterstock.com; 67 (Friedrich Mohs) "Friedrich Mohs" by Josef Kriehuber/Wikimedia Commons/Public Domain; 67 (feldspar) Aleksandr Pobedimskiy/Shutterstock.com; 67, 70 (quartz) Mara Fribus/Shutterstock.com; 67 (topaz) Alan Curtis/Alamy Stock Photo; 67 (corundum)

SCIENCE 3 Activities Photo Credits